NEW ASPECTS OF
POSITIVE-STRAND RNA VIRUSES

NEW ASPECTS OF POSITIVE-STRAND RNA VIRUSES

EDITED BY

MARGO A. BRINTON

Biology Department, Georgia State University, Atlanta, Georgia

AND

FRANZ X. HEINZ

Institute of Virology, University of Vienna, Vienna, Austria

American Society for Microbiology
Washington, D.C.

Copyright © 1990   American Society for Microbiology
1325 Massachusetts Ave., N.W.
Washington, DC 20005

New aspects of positive-strand RNA viruses / [edited by] Margo A. Brinton and Franz X. Heinz.
    p. cm.
  Based on a symposium held in Vienna, Austria in June of 1989.
  ISBN 1-55581-022-5
  1. RNA viruses—Congresses. 2. Virus diseases—Congresses.
I. Brinton, Margo A. II. Heinz, Franz X.
  [DNLM: 1. RNA Viruses—congresses. QW 168 N532 1989]
QR395.N48 1990
576′.64—dc20
DNLM/DLC
for Library of Congress   90-435
                                                    CIP

ISBN 1-55581-022-5

CONTENTS

IX. Strategies for Control of Virus Disease

CONTRIBUTORS

Ravindra Acharya, Laboratory of Molecular Biophysics, South Parks Road, Oxford OX1 3QU, United Kingdom **(31)***

Vadim I. Agol, Institute of Poliomyelitis and Viral Encephalitides, USSR Academy of Medical Sciences, Moscow Region 142782, and Moscow State University, Moscow 119898, USSR **(46)**

Alexei A. Agranovsky, A. N. Belozersky Laboratory at Moscow State University, Moscow 119899, USSR **(5)**

Barbara Alstein, Department of Microbiology and Immunology, Hahnemann University School of Medicine, Broad and Vine, Philadelphia, Pennsylvania 19102 **(40)**

Koen Andries, Janssen Research Foundation, Turnhoutseweg 30, B-2340 Beerse, Belgium **(54)**

David J. Barton, Department of Microbiology, Medical College of Ohio, Toledo, Ohio 43699 **(12)**

David Baulcombe, The Sainsbury Laboratory, Colney Lane, Norwich NR4 7UH, United Kingdom **(51)**

Ewald Beck, Zentrum für Molekulare Biologie Heidelberg, D-6900 Heidelberg, Federal Republic of Germany **(27)**

Danièle Benichou, Laboratory of Molecular Virology, Centre National de la Recherche Scientifique, URA 545, Pasteur Institute, 25, rue du Dr. Roux, 75724 Paris Cedex 15, France **(47)**

Ingrid E. Bergmann, Centro de Virologia Animal, Serrano 661 (1414), Buenos Aires, Argentina **(27)**

Dieter Blaas, Institut für Biochemie der Universität Wien, Währingerstrasse 17, A 1090 Vienna, Austria **(26)**

T. Bodwell, Department of Neurology, University of Chicago School of Medicine, Chicago, Illinois 60637 **(19)**

Peter J. Bredenbeek, Institute of Virology, Veterinary Faculty, State University of Utrecht, Yalelaan 1, 3584 CL Utrecht, The Netherlands **(2)**

M. A. Brinton, Department of Biology, Georgia State University, Atlanta, Georgia 30303 **(6)**

Fred Brown, Wellcome Biotech, Langley Court, Beckenham, Kent BR3 3BS, United Kingdom **(31)**

Jack F. Bukowski, Division of Infectious Diseases, Department of Medicine, University of Massachusetts Medical Center, Worcester, Massachusetts 01655 **(44)**

M. A. Calenoff, Department of Neurology, Northwestern University Medical School, Chicago, Illinois 60611 **(48)**

Adina Candrea, Laboratory of Medical Virology, Pasteur Institute, 25, rue du Dr. Roux, 75724 Paris Cedex 15, France **(47)**

Annie Canu, Department of Virus Research, Max Planck Institute for Biochemistry, D-8033 Martinsried, Federal Republic of Germany **(50)**

James C. Carrington, Department of Biology, Texas A&M University, College Station, Texas 77843 **(25)**

Mei Chao, Fox Chase Cancer Center, Philadelphia, Pennsylvania 19111 **(3)**

Zhongguo Chen, Department of Biological Sciences, Purdue University, West Lafayette, Indiana 47906 **(32)**

Ewan D. Chirnside, Institute of Virology, Veterinary Faculty, State University of Utrecht, Yalelaan 1, 3584 CL Utrecht, The Netherlands **(2)**

Marie Chow, Department of Biology, Massachusetts Institute of Technology, 77 Massachusetts Avenue, Cambridge, Massachusetts 02139 **(23)**

Philip S. Collis, Department of Immunology and Medical Microbiology, University of Florida, College of Medicine, Gainesville, Florida 32610-0266 **(9)**

Richard J. Colonno, Department of Virus and Cell Biology, Merck Sharp and Dohme Research Laboratories, West Point, Pennsylvania 19486 **(37)**

Susan R. Compton, Department of Molecular, Cellular and Developmental Biology, University of Colorado, Boulder, Colorado 80309 **(35, 36)**

R. G. H. Cotton, Olive Miller Protein Laboratory, Murdoch Institute, Royal Children's Hospital, Parkville, Melbourne, Victoria 3052, Australia **(7)**

Thérèse Couderc, Laboratory of Medical Virology, Pasteur Institute, 25, rue du Dr. Roux, 75724 Paris Cedex 15, France **(47)**

Radù Crainic, Laboratory of Medical Virology, Pasteur Institute, 25, rue du Dr. Roux, 75724 Paris Cedex 15, France **(47)**

Richard L. Crowell, Department of Microbiology and Immunology, Hahnemann University School of Medicine, Broad and Vine, Philadelphia, Pennsylvania 19102 **(40)**

Lisa Curry, Department of Applied Biological Sciences, Massachusetts Institute of Technology, 77 Massachusetts Avenue, Cambridge, Massachusetts 02139 **(23)**

N. L. Davis, Department of Microbiology and Immunology, University of North Carolina, Chapel Hill,

---

* Parentheses indicate corresponding chapter(s) in this volume.

North Carolina 27599-7290 **(49)**

Lina De, Division of Viral Diseases, Center for Infectious Diseases, Centers for Disease Control, Atlanta, Georgia 30333 **(52)**

José de Chastonay, Institute for Medical Microbiology, University of Bern, Friedbühlstrasse 51, CH-3010 Bern, Switzerland **(15)**

Johan A. den Boon, Institute of Virology, Veterinary Faculty, State University of Utrecht, Yalelaan 1, 3584 CL Utrecht, The Netherlands **(2)**

Martine Devic, The Sainsbury Laboratory, Colney Lane, Norwich NR4 7UH, United Kingdom **(51)**

Antonius A. F. de Vries, Institute of Virology, Veterinary Faculty, State University of Utrecht, Yalelaan 1, 3584 CL Utrecht, The Netherlands **(2)**

Bart Dewindt, Janssen Research Foundation, Turnhoutseweg 30, B-2340 Beerse, Belgium **(54)**

Corrille M. DeWitt, Department of Virus and Cell Biology, Merck Sharp and Dohme Research Laboratories, West Point, Pennsylvania 19486 **(37)**

Guy D. Diana, Department of Medicinal Chemistry, Sterling Research Group, Rensselaer, New York 12144 **(53)**

Valerian V. Dolja, A. N. Belozersky Laboratory at Moscow State University, Moscow 119899, USSR **(5)**

William G. Dougherty, Department of Microbiology, Oregon State University, Corvallis, Oregon 97331-3804 **(25)**

Frank J. Dutko, Department of Virology and Oncopharmacology, Sterling Research Group, Rensselaer, New York 12144 **(53)**

R. Eggen, Department of Molecular Biology, Agricultural University, Dreijenlaan 3, 6703 HA Wageningen, The Netherlands **(17)**

Francis A. Ennis, Division of Infectious Diseases, Department of Medicine, University of Massachusetts Medical Center, Worcester, Massachusetts 01655 **(44)**

K. S. Faaberg, Department of Neurology, Northwestern University Medical School, Chicago, Illinois 60611 **(48)**

B. Falgout, Molecular Viral Biology Section, Laboratory of Infectious Diseases, National Institute of Allergy and Infectious Diseases, Bethesda, Maryland 20892 **(29)**

Matthias M. Falk, Zentrum für Molekulare Biologie Heidelberg, D-6900 Heidelberg, Federal Republic of Germany **(27)**

Y.-X. Feng, Laboratory of Molecular Genetics, National Institute of Child Health and Human Development, Bethesda, Maryland 20892 **(28)**

Frederike Fessl, Ernst Boehringer Institut für Arzneimittelforschung, Dr. Boehringergasse 5–11, A 1120 Vienna, Austria **(26)**

D. J. Filman, Department of Molecular Biology, Research Institute of Scripps Clinic, 10666 North Torrey Pines Road, La Jolla, California 92037 **(30)**

Andrew Fisher, Department of Biological Sciences, Purdue University, West Lafayette, Indiana 47906 **(32)**

J. Bert Flanegan, Department of Immunology and Medical Microbiology, University of Florida, College of Medicine, Gainesville, Florida 32610-0266 **(9)**

O. Flore, University of Cagliari, Cagliari, Italy **(30)**

D. C. Flynn, Department of Microbiology, University of Virginia, School of Medicine, Charlottesville, Virginia 22908 **(49)**

Graham Fox, Wellcome Biotech, Langley Court, Beckenham, Kent BR3 3BS, United Kingdom **(31)**

Marion Freistadt, Department of Microbiology, Columbia University College of Physicians and Surgeons, 701 West 168th Street, New York, New York 10032 **(41)**

C. E. Fricks, County USC Medical Center, Los Angeles, California 90033 **(30)**

Elizabeth Fry, Laboratory of Molecular Biophysics, South Parks Road, Oxford OX1 3QU, United Kingdom **(31)**

J. Fu, Department of Neurology, University of Chicago School of Medicine, Chicago, Illinois 60637 **(19)**

Henrik Garoff, Department of Molecular Biology, Center for Biotechnology, The Karolinska Institute, Huddinge University Hospital, S-141 86 Huddinge, Sweden **(24)**

Cristina Giachetti, Department of Microbiology and Molecular Genetics, College of Medicine, University of California, Irvine, California 92717 **(13)**

Marc Girard, Laboratory of Molecular Virology, Centre National de la Recherche Scientifique, URA 545, Pasteur Institute, 25, rue du Dr. Roux, 75724 Paris Cedex 15, France **(47)**

E. K. Godeny, Department of Biology, Georgia State University, Atlanta, Georgia 30303 **(6)**

Rob Goldbach, Department of Virology, Agricultural University, Binnenhaven 11, 6709 PD Wageningen, The Netherlands **(1, 17)**

R. J. Gorelick, BRI, Basic Research Program, National Cancer Institute-Frederick Cancer Research Facility, Frederick, Maryland 21701 **(28)**

Arash Grakoui, Department of Molecular Microbiology, Box 8230, Washington University School of

Medicine, 660 South Euclid Avenue, St. Louis, Missouri 63110-1093 (16)

Jeffrey M. Greve, Molecular Therapeutics Inc., Miles Research Center, 400 Morgan Lane, West Haven, Connecticut 06516 (38)

Pablo R. Grigera, Centro de Virologia Animal, Serrano 661 (1414), Buenos Aires, Argentina (27)

F. Guirakhoo, Institute of Virology, University of Vienna, Kinderspitalgasse 15, A-1095 Vienna, Austria (43)

Kimiko Hagino-Yamagishi, Department of Microbiology, The Tokyo Metropolitan Institute of Medical Science, Honkomagome, Bunkyo-ku, Tokyo 113, Japan (14)

Timothy C. Hall, Biology Department, Texas A&M University, College Station, Texas 77843-3258 (8)

Simon J. Hambidge, Department of Biochemistry, Biophysics and Genetics and Department of Microbiology and Immunology, University of Colorado Health Sciences Center, Denver, Colorado 80262 (22)

D. L. Hatfield, Laboratory of Experimental Carcinogenesis, National Cancer Institute, Bethesda, Maryland 20892 (28)

F. X. Heinz, Institute of Virology, University of Vienna, Kinderspitalgasse 15, A-1095 Vienna, Austria (43)

Christopher U. T. Hellen, Department of Microbiology, State University of New York at Stony Brook, Stony Brook, New York 11794 (20)

Peter Hans Hofschneider, Department of Virus Research, Max Planck Institute for Biochemistry, D-8033 Martinsried, Federal Republic of Germany (50)

J. M. Hogle, Department of Molecular Biology, Research Institute of Scripps Clinic, 10666 North Torrey Pines Road, La Jolla, California 92037 (30)

Kathryn V. Holmes, Department of Pathology, Uniformed Services University of the Health Sciences, Bethesda, Maryland 20814-4799 (11, 36)

Christina Hölscher, Department of Microbiology, State University of New York at Stony Brook, Stony Brook, New York 11794 (20)

H. Holzmann, Institute of Virology, University of Vienna, Kinderspitalgasse 15, A-1095 Vienna, Austria (43)

Florian Horaud, Laboratory of Medical Virology, Pasteur Institute, 25, rue du Dr. Roux, 75724 Paris Cedex 15, France (47)

Marian C. Horzinek, Institute of Virology, Veterinary Faculty, State University of Utrecht, Yalelaan 1, 3584 CL Utrecht, The Netherlands (2)

Michael T. Howell, Department of Biochemistry, University of Cambridge, Cambridge CB2 1QW, United Kingdom (21)

Sen-Yung Hsieh, Fox Chase Cancer Center, Philadelphia, Pennsylvania 19111 (3)

Kuo-Hom Lee Hsu, Wyeth Ayrest Research, P.O. Box 8299, Philadelphia, Pennsylvania 19101 (40)

Henry V. Huang, Department of Molecular Microbiology, Box 8230, Washington University School of Medicine, 660 South Euclid Avenue, St. Louis, Missouri 63110-1093 (16)

Clayton C. Huntley, Biology Department, Texas A&M University, College Station, Texas 77843-3258 (8)

J. P. Icenogle, Centers for Disease Control, Atlanta, Georgia 30333 (30)

Bruce L. Innis, Department of Virology, Armed Forces Research Institute of Medical Sciences, Bangkok, Thailand (44)

Richard J. Jackson, Department of Biochemistry, University of Cambridge, Cambridge CB2 1QW, United Kingdom (21)

Martine Jaegle, Department of Molecular Biology, Institute for Plant Science Research, Maris Lane, Trumpington, Cambridge CB2 2LQ, United Kingdom (51)

Sung-Key Jang, Department of Microbiology, State University of New York at Stony Brook, Stony Brook, New York 11794 (20)

Jurand Janus, Division of Infectious Diseases, Department of Medicine, University of Massachusetts Medical Center, Worcester, Massachusetts 01655 (44)

John E. Johnson, Department of Biological Sciences, Purdue University, West Lafayette, Indiana 47906 (32)

Robert E. Johnston, Department of Microbiology and Immunology, University of North Carolina, Chapel Hill, North Carolina 27599 (25, 49)

Paul Kaesberg, Institute for Molecular Virology and Department of Biochemistry, University of Wisconsin-Madison, Madison, Wisconsin 53706

Ann Kaminski, Department of Biochemistry, University of Cambridge, Cambridge CB2 1QW, United Kingdom (21)

Reinhard Kandolf, Department of Virus Research, Max Planck Institute for Biochemistry, D-8033 Martinsried, Federal Republic of Germany (50)

Gerardo Kaplan, Department of Microbiology, Columbia University College of Physicians and Surgeons, 701 West 168th Street, New York, New York 10032 (41)

Alexander V. Karasev, A. N. Belozersky Laboratory at Moscow State University, Moscow 119899,

USSR (5)

Olen M. Kew, Division of Viral Diseases, Center for Infectious Diseases, Centers for Disease Control, Atlanta, Georgia 30333 (52)

Karla Kirkegaard, Department of Molecular, Cellular and Developmental Biology, University of Colorado, Boulder, Colorado 80309 (35)

Philip Kirschner, Department of Virus Research, Max Planck Institute for Biochemistry, D-8033 Martinsried, Federal Republic of Germany (50)

Karin Klingel, Department of Virus Research, Max Planck Institute for Biochemistry, D-8033 Martinsried, Federal Republic of Germany (50)

D. A. Knorr, Department of Plant Pathology, University of California, Berkeley, California 94720 (18)

W. Kong, Department of Neurology, University of Chicago School of Medicine, Chicago, Illinois 60637 (19)

Udo Kontny, Division of Infectious Diseases, Department of Medicine, University of Massachusetts Medical Center, Worcester, Massachusetts 01655 (44)

Hans-Georg Kräusslich, Deutsches Krebsforschungszentrum, Institut für Virusforschung, Im Neunheimer Feld 280, D-69 Heidelberg 1, Federal Republic of Germany (20)

Ernst Kuechler, Institut für Biochemie der Universität Wien, Währingerstrasse 17, A 1090 Vienna, Austria (26)

Shusuke Kuge, Gene Regulation Laboratory, Imperial Cancer Research Fund, Lincoln's Inn Fields, London WC2A 3PX, United Kingdom (14)

Richard J. Kuhn, Division of Biology, California Institute of Technology, Pasadena, California 91125 (10)

C. Kunz, Institute of Virology, University of Vienna, Kinderspitalgasse 15, A-1095 Vienna, Austria (43)

Mark Kuo, Fox Chase Cancer Center, Philadelphia, Pennsylvania 19111 (3)

Ichiro Kurane, Division of Infectious Diseases, Department of Medicine, University of Massachusetts Medical Center, Worcester, Massachusetts 01655 (44)

Robert L. LaFemina, Department of Virus and Cell Biology, Merck Sharp and Dohme Research Laboratories, West Point, Pennsylvania 19486 (37)

C.-J. Lai, Molecular Viral Biology Section, Laboratory of Infectious Diseases, National Institute of Allergy and Infectious Diseases, Bethesda, Maryland 20892 (29)

Julian L. Leibowitz, Department of Pathology and Laboratory Medicine, University of Texas Medical School, P.O. Box 20708, Houston, Texas 77225 (11)

J. G. Levin, Laboratory of Molecular Genetics, National Institute of Child Health and Human Development, Bethesda, Maryland 20892 (28)

Robin Levis, National Cancer Institute, Building 41, Room D909, Bethesda, Maryland 20892 (16)

Paul J. Lewi, Janssen Research Foundation, Turnhoutseweg 30, B-2340 Beerse, Belgium (54)

Peter Liljeström, Department of Molecular Biology, Center for Biotechnology, The Karolinska Institute, Huddinge University Hospital, S-141 86 Huddinge, Sweden (24)

S.-C. Lin, Department of Microbiology and Immunology, University of North Carolina, Chapel Hill, North Carolina 27599-7290 (49)

H. L. Lipton, Department of Neurology, University of Colorado School of Medicine, 4200 East 9th Avenue, Denver, Colorado 80262 (48)

Mario Lobigs, Department of Molecular Biology, Center for Biotechnology, The Karolinska Institute, Huddinge University Hospital, S-141 86 Huddinge, Sweden (24)

Derek Logan, Laboratory of Molecular Biophysics, South Parks Road, Oxford OX1 3QU, United Kingdom (31)

Dennis G. Macejak, Department of Biochemistry, Biophysics and Genetics and Department of Microbiology and Immunology, University of Colorado Health Sciences Center, Denver, Colorado 80262 (22)

C. Mandl, Institute of Virology, University of Vienna, Kinderspitalgasse 15, A-1095 Vienna, Austria (43)

Daniel Marc, Laboratory of Molecular Virology, Centre National de la Recherche Scientifique, URA 545, Pasteur Institute, 25, rue du Dr. Roux, 75724 Paris Cedex 15, France (47)

Steven D. Marlin, Department of Immunology, Boehringer Ingelheim Pharmaceuticals, Inc., 90 East Ridge, P.O. Box 368, Ridgefield, Connecticut 06877 (39)

Loren E. Marsh, Biology Department, Texas A&M University, College Station, Texas 77843-3258 (8)

Annette Martin, Laboratory of Molecular Virology, Centre National de la Recherche Scientifique, URA 545, Pasteur Institute, 25, rue du Dr. Roux, 75724 Paris Cedex 15, France (47)

Ingrid Maurer-Fogy, Ernst Boehringer Institut für Arzneimittelforschung, Dr. Boehringergasse 5–11, A 1120 Vienna, Austria (26)

Alan McClelland, Molecular Therapeutics Inc., Miles Research Center, 400 Morgan Lane, West Haven, Connecticut 06516 (38)

Mark A. McKinlay, Department of Virology and Oncopharmacology, Sterling Research Group,

Rensselaer, New York 12144 (53)

Anthony Meager, Division of Immunobiology, National Institute for Biological Standards and Control, Hertfordshire EN6 3QG, United Kingdom (44)

Cathy L. Mendelsohn, Department of Microbiology, Columbia University College of Physicians and Surgeons, 701 West 168th Street, New York, New York 10032 (41)

Vincent J. Merluzzi, Department of Immunology, Boehringer Ingelheim Pharmaceuticals, Inc., 90 East Ridge, P.O. Box 368, Ridgefield, Connecticut 06877 (26, 39)

Jürgen Mertsching, Department of Virus Research, Max Planck Institute for Biochemistry, D-8033 Martinsried, Federal Republic of Germany (50)

Kalervo Metsikkö, Department of Molecular Biology, Center for Biotechnology, The Karolinska Institute, Huddinge University Hospital, S-141 86 Huddinge, Sweden (24)

W. J. Meyer, Department of Microbiology and Immunology, University of North Carolina, Chapel Hill, North Carolina 27599-7290 (49)

Gregor Meyers, Federal Research Centre for Virus Diseases of Animals, D-7400 Tübingen, Federal Republic of Germany (4)

Henri Moereels, Janssen Research Foundation, Turnhoutseweg 30, B-2340 Beerse, Belgium (54)

B. Joan Morasco, Department of Immunology and Medical Microbiology, University of Florida, College of Medicine, Gainesville, Florida 32610-0266 (9)

T. J. Morris, Department of Plant Pathology, University of California, Berkeley, California 94720 (18)

Mary Morrison, Department of Microbiology, Columbia University College of Physicians and Surgeons, 701 West 168th Street, New York, New York 10032 (41)

Nicola Moscufo, Department of Biology, Massachusetts Institute of Technology, 77 Massachusetts Avenue, Cambridge, Massachusetts 02139 (23)

Eric G. Moss, Department of Microbiology, Columbia University College of Physicians and Surgeons, 701 West 168th Street, New York, New York 10032 (41)

Lyle Najita, Department of Biochemistry, Biophysics and Genetics and Department of Microbiology and Immunology, University of Colorado Health Sciences Center, Denver, Colorado 80262 (22)

Hubert G. M. Niesters, Division of Biology, California Institute of Technology, Pasadena, California 91125 (10)

Suchitra Nimmannitya, Children's Hospital, Bangkok, Thailand (44)

Ananda Nisalak, Department of Virology, Armed Forces Research Institute of Medical Sciences, Bangkok, Thailand (44)

Hans Nitschko, Department of Molecular Microbiology, Washington University School of Medicine, St. Louis, Missouri 63110-1093 (34)

Akio Nomoto, Department of Microbiology, The Tokyo Metropolitan Institute of Medical Science, Honkomagome, Bunkyo-ku, Tokyo 113, Japan (14)

Ans F. H. Noten, Institute of Virology, Veterinary Faculty, State University of Utrecht, Yalelaan 1, 3584 CL Utrecht, The Netherlands (2)

Baldev K. Nottay, Division of Viral Diseases, Center for Infectious Diseases, Centers for Disease Control, Atlanta, Georgia 30333 (52)

Jürg P. F. Nüesch, Institute for Medical Microbiology, University of Bern, Friedbühlstrasse 51, CH-3010 Bern, Switzerland (15)

M. Steven Oberste, Department of Immunology and Medical Microbiology, University of Florida, College of Medicine, Gainesville, Florida 32610-0266 (9)

Emilia L. Oleszak, Department of Pathology and Laboratory Medicine, University of Texas Medical School, P.O. Box 20708, Houston, Texas 77225 (11)

Severo Paglini, Institute of Virology, University of Cordoba, Corboda, Argentina (40)

Peter Pallai, Boehringer Ingelheim Pharmaceuticals, Inc., 90 East Ridge, P.O. Box 368, Ridgefield, Connecticut 06877 (26)

Mark A. Pallansch, Division of Viral Diseases, Center for Infectious Diseases, Centers for Disease Control, Atlanta, Georgia 30333 (52)

T. Dawn Parks, Department of Microbiology, Oregon State University, Corvallis, Oregon 97331-3804 (25)

D. F. Pence, Department of Microbiology and Immunology, University of North Carolina, Chapel Hill, North Carolina 27599-7290 (49)

Daniel C. Pevear, Department of Virology and Oncopharmacology, Sterling Research Group, Rensselaer, New York 12144 (53)

Michael Pfleiderer, Institute of Virology, Versbacher Strasse 7, 8700 Würzburg, Federal Republic of Germany (42)

Gregory P. Pogue, Biology Department, Texas A&M University, College Station, Texas 77843-3258 (8)

J. M. Polo, Department of Microbiology and Immunology, University of North Carolina, Chapel Hill, North Carolina 27599-7290 (49)

Vincent R. Racaniello, Department of Microbiology, Columbia University College of Physicians and Surgeons, 701 West 168th Street, New York, New York 10032 (41)

Ramaswamy Raju, Department of Molecular Microbiology, Box 8230, Washington University School of Medicine, 660 South Euclid Avenue, St. Louis, Missouri 63110-1093 **(16)**

A. L. N. Rao, Biology Department, Texas A&M University, College Station, Texas 77843-3258 **(8)**

A. Rein, BRI, Basic Research Program, National Cancer Institute-Frederick Cancer Research Facility, Frederick, Maryland 21701 **(28)**

Ruibao Ren, Department of Microbiology, Columbia University College of Physicians and Surgeons, 701 West 168th Street, New York, New York 10032 **(41)**

Quentin Reuer, Department of Microbiology, State University of New York at Stony Brook, Stony Brook, New York 11794 **(20)**

Carol Reynolds, Department of Applied Biological Sciences, Massachusetts Institute of Technology, 77 Massachusetts Avenue, Cambridge, Massachusetts 02139 **(23)**

Charles M. Rice, Department of Molecular Microbiology, Box 8230, Washington University School of Medicine, 660 South Euclid Avenue, St. Louis, Missouri 63110-1093 **(16)**

Rebeca Rico-Hesse, Department of Epidemiology and Public Health, Yale University School of Medicine, P.O. Box 3333, New Haven, Connecticut 06510 **(52)**

R. P. Roos, Department of Neurology, University of Chicago School of Medicine, Chicago, Illinois 60637 **(19)**

L. Rosenstein, Department of Neurology, University of Chicago School of Medicine, Chicago, Illinois 60637 **(19)**

Michael G. Rossmann, Department of Biological Sciences, Lilly Hall of Life Sciences, Purdue University, West Lafayette, Indiana 47907 **(53)**

Robert Rothlein, Department of Immunology, Boehringer Ingelheim Pharmaceuticals, Inc., 90 East Ridge, P.O. Box 368, Ridgefield, Connecticut 06877 **(39)**

Alan L. Rothman, Division of Infectious Diseases, Department of Medicine, University of Massachusetts Medical Center, Worcester, Massachusetts 01655 **(44)**

M. Routbort, Department of Neurology, University of Chicago School of Medicine, Chicago, Illinois 60637 **(19)**

Edward Routledge, Institute of Virology, Versbacher Strasse 7, 8700 Würzburg, Federal Republic of Germany **(42)**

David Rowlands, Wellcome Biotech, Langley Court, Beckenham, Kent BR3 3BS, United Kingdom **(31)**

Tillmann Rümenapf, Federal Research Centre for Virus Diseases of Animals, D-7400 Tübingen, Federal Republic of Germany **(4)**

D. L. Russell, Department of Biology, Washington and Lee University, Lexington, Virginia 24450 **(49)**

Peter Sarnow, Department of Biochemistry, Biophysics and Genetics and Department of Microbiology and Immunology, University of Colorado Health Sciences Center, Denver, Colorado 80262 **(22)**

Dorothea L. Sawicki, Department of Microbiology, Medical College of Ohio, Toledo, Ohio 43699 **(12)**

Stanley G. Sawicki, Department of Microbiology, Medical College of Ohio, Toledo, Ohio 43699 **(12)**

T. S. Schaefer, Department of Agricultural Chemistry, Oregon State University, Corvallis, Oregon 97331 **(28)**

Sondra Schlesinger, Department of Molecular Microbiology, Box 8230, Washington University School of Medicine, 660 South Euclid Avenue, St. Louis, Missouri 63110-1093 **(16, 34)**

Alan L. Schmaljohn, Virology Division, U.S. Army Medical Research Institute for Infectious Diseases, Fort Detrick, Frederick, Maryland 21701 **(45)**

Tim Schmidt, Department of Biological Sciences, Purdue University, West Lafayette, Indiana 47906 **(32)**

Heike Schönke, Department of Virus Research, Max Planck Institute for Biochemistry, D-8033 Martinsried, Federal Republic of Germany **(50)**

Bert L. Semler, Department of Microbiology and Molecular Genetics, College of Medicine, University of California, Irvine, California 92717 **(13, 19)**

Lamia Sharmeen, Fox Chase Cancer Center, Philadelphia, Pennsylvania 19111 **(3)**

Stuart G. Siddell, Institute of Virology, Versbacher Strasse 7, 8700 Würzburg, Federal Republic of Germany **(42)**

Günter Siegl, Institute for Medical Microbiology, University of Bern, Friedbuhlstrasse 51, CH-3010 Bern, Switzerland **(15)**

John Simons, Department of Applied Biological Sciences, Massachusetts Institute of Technology, 77 Massachusetts Avenue, Cambridge, Massachusetts 02139 **(23)**

Tim Skern, Institut für Biochemie der Universität Wien, Währingerstrasse 17, A 1090 Vienna, Austria **(26)**

David E. Slade, Department of Microbiology, North Carolina State University, Raleigh, North Carolina 27695 **(25)**

Holly A. Smith, Department of Microbiology, Oregon State University, Corvallis, Oregon 97331-3804 **(25)**

J. F. Smith, Division of Virology, U.S. Army Medical Research Institute of Infectious Diseases, Frederick, Maryland 21701 **(49)**

Eric J. Snijder, Institute of Virology, Veterinary Faculty, State University of Utrecht, Yalelaan 1, 3584 CL Utrecht, The Netherlands **(2)**

Jerry Snoeks, Janssen Research Foundation, Turnhoutseweg 30, B-2340 Beerse, Belgium **(54)**

Wolfgang Sommergruber, Ernst Boehringer Institut für Arzneimittelforschung, Dr. Boehringergasse 5–11, A 1120 Vienna, Austria.

Willy J. M. Spaan, Institute of Virology, Veterinary Faculty, State University of Utrecht, Yalelaan 1, 3584 CL Utrecht, The Netherlands **(2)**

D. W. Speicher, Wistar Institute, 3601 Spruce Street, Philadelphia, Pennsylvania 19104 **(6)**

Roland Stauber, Institute of Virology, Versbacher Strasse 7, 8700 Würzburg, Federal Republic of Germany **(42)**

Cynthia Stauffacher, Department of Biological Sciences, Purdue University, West Lafayette, Indiana 47906 **(32)**

David S. Stec, Department of Microbiology and Immunology, University of Maryland School of Medicine, Baltimore, Maryland 21201 **(45)**

S. Stein, Department of Neurology, University of Chicago School of Medicine, Chicago, Illinois 60637 **(19)**

Ellen G. Strauss, Division of Biology, California Institute of Technology, Pasadena, California 91125 **(10, 45)**

James H. Strauss, Division of Biology, California Institute of Technology, Pasadena, California 91125 **(10, 45)**

David Stuart, Laboratory of Molecular Biophysics, South Parks Road, Oxford OX1 3QU, United Kingdom **(31)**

R. Syed, Department of Molecular Biology, Research Institute of Scripps Clinic, 10666 North Torrey Pines Road, La Jolla, California 92037 **(30)**

John Taylor, Fox Chase Cancer Center, Philadelphia, Pennsylvania 19111 **(3)**

Heinz-Jürgen Thiel, Federal Research Centre for Virus Diseases of Animals, D-7400 Tübingen, Federal Republic of Germany **(4)**

Gregory J. Tobin, Department of Immunology and Medical Microbiology, University of Florida, College of Medicine, Gainesville, Florida 32610-0266 **(9)**

Joanne E. Tomassini, Department of Virus and Cell Biology, Merck Sharp and Dohme Research Laboratories, West Point, Pennsylvania 19486 **(37)**

W. Tuma, Institute of Virology, University of Vienna, Kinderspitalgasse 15, A-1095 Vienna, Austria **(43)**

Sylvie van der Werf, Laboratory of Molecular Virology, Centre National de la Recherche Scientifique, URA 545, Pasteur Institute, 25, rue du Dr. Roux, 75724 Paris Cedex 15, France **(47)**

A. van Kammen, Department of Molecular Biology, Agricultural University, Dreijenlaan 3, 6703 HA Wageningen, The Netherlands **(17)**

J. Verver, Department of Molecular Biology, Agricultural University, Dreijenlaan 3, 6703 HA Wageningen, The Netherlands **(17)**

Peter Volkmann, Ernst Boehringer Institut für Arzneimittelforschung, Dr. Boehringergasse 5–11, A 1120 Vienna, Austria **(26)**

Johanna Wahlberg, Department of Molecular Biology, Center for Biotechnology, The Karolinska Institute, Huddinge University Hospital, S-141 86 Huddinge, Sweden **(24)**

Barbara Weiss, Department of Molecular Microbiology, Washington University School of Medicine, St. Louis, Missouri 63110-1093 **(34)**

Susan R. Weiss, Department of Microbiology, University of Pennsylvania Medical School, Philadelphia, Pennsylvania 19104-6076 **(11)**

J. Wellink, Department of Molecular Biology, Agricultural University, Dreijenlaan 3, 6703 HA Wageningen, The Netherlands **(17)**

Gerd Wengler, Institut für Virologie, Justus-Liebig-Universität Giessen, D-6300 Giessen, Federal Republic of Germany **(33)**

Richard K. Williams, Department of Pathology, Uniformed Services University of the Health Sciences, Bethesda, Maryland 20814 **(36)**

L. Willis, Department of Biochemistry, North Carolina State University, Raleigh, North Carolina 27695 **(49)**

Eckard Wimmer, Department of Microbiology, State University of New York at Stony Brook, Stony Brook, New York 11794 **(20)**

Maureen G. Woods, Department of Virology and Oncopharmacology, Sterling Research Group, Rensselaer, New York 12144 **(53)**

Peter J. Wright, Department of Microbiology, Monash University, Clayton, Victoria 3168, Australia **(7)**

Chen-Fu Yang, Division of Viral Diseases, Center for Infectious Diseases, Centers for Disease Control, Atlanta, Georgia 30333 **(52)**

Dorothy C. Young, Department of Immunology and Medical Microbiology, University of Florida, College of Medicine, Gainesville, Florida 32610-0266 **(9)**

Roland Zell, Department of Virus Research, Max Planck Institute for Biochemistry, D-8033 Martinsried, Federal Republic of Germany **(50)**

Andree Zibert, Zentrum für Molekulare Biologie Heidelberg, D-6900 Heidelberg, Federal Republic of Germany **(27)**

Phillip W. Zoltick, Department of Microbiology, University of Pennsylvania Medical School, Philadelphia, Pennsylvania 19104-6076 **(11)**

Manfred Zorn, Institut für Biochemie der Universität Wien, Währingerstrasse 17, A 1090 Vienna, Austria **(26)**

PREFACE

The First International Symposium on Positive Strand RNA Viruses was organized under the UCLA Symposia Series and held in Keystone, Colorado, in April, 1986. The proceedings of this symposium (M. A. Brinton and R. R. Rueckert, ed., *Positive Strand RNA Viruses*, Alan R. Liss, Inc., New York, 1987) documented the rapid advances that had been made in the understanding of positive-strand RNA virus biology through the application of molecular biological technology. Similarities between divergent positive-strand RNA virus systems were emphasized through the organization of sessions by topics rather than by virus families. The success of the first meeting has led to establishment of a Positive Strand RNA Virus Symposium Series held triannually and alternating between the United States and Europe. The present book contains the proceedings of the second symposium, held in Vienna, Austria, in June of 1989. This meeting was attended by more than 300 scientists from all over the world, including Eastern Europe and the USSR. The subsequent dramatic political changes which have resulted in the opening of the borders between Western and Eastern Europe should increase participation by Eastern European and USSR scientists in the next Positive Strand RNA Virus Symposium.

The proceedings of the second symposium indicate that the first meeting was successful in encouraging new scientific interactions and technical approaches to the study of positive-strand RNA viruses. Advances presented in the areas of viral evolution, viral replication, and viral structure were especially striking. Bridges between new basic research data and the application of this information to combating a wide variety of serious viral diseases were exemplified by the presentations on new picornavirus antiviral drugs.

Chapters from most of the 47 plenary speakers and from 15 speakers selected from submitted abstracts are contained in this book. The minireview format of the contributions precludes comprehensive literature citation. We gratefully acknowledge the cooperation shown by those submitting manuscripts and regret that the abstracts from the 150 posters presented could not be included. We thank Tim Woltering for secretarial assistance in the preparation of this book.

Franz X. Heinz

Margo A. Brinton

ACKNOWLEDGMENTS

The Second International Symposium on Positive Strand RNA Viruses would not have been possible without substantial financial support from private companies and government institutions. We gratefully acknowledge support from:

Aigner

Abbott Laboratories

Beckman

Behring

Biomedica

Boehringer Ingelheim

Boehringer Mannheim

Chemomedica

E. I. Du Pont de Nemours & Company

Eisenhut and Jarolin

Haack

Immuno Ag

Laevosan

Mayrhoter

Merck Corporation

National Heart, Lung and Blood Institute

National Institute of Allergy and Infectious Diseases
(U.S. Public Health Service grant AI29291)

PAA-Labor

The Perkin-Elmer Corporation

Pharmacia

Schoeller-Pharma

Searle

Sterling Winthrop Corporation

Szabo

U.S. Army Medical Research and Development Command
(grant no. DAMD17-89-Z-9027)

Positive-Stranded RNA Viruses: Early History and the Role of Model Viruses

Paul Kaesberg

It is a great honor and pleasure for me to open this meeting and to be in this beautiful, historic city. I was last in Vienna for scientific purposes more than 30 years ago, in 1958, as a symposium speaker at the International Biochemical Congress. The city has changed physically, but spiritually it remains invariant. On behalf of speakers and listeners, I express our gratitude to the organizers of this meeting and especially to our esteemed host, Professor Christian Kunz, to Dr. Franz Heinz and his committee, and to Drs. Margo Brinton and Roland Rueckert.

Why are we here? Why is it scientifically valid for us to be together to discuss positive-stranded RNA viruses? Clearly, they have many similarities whose facets we can extrapolate to illuminate our own viruses. But it goes deeper than that. It is now evident that diverse groups of positive-stranded RNA viruses of eucaryotes have evolutionary pathways in common, and perhaps even a common origin. When we compare biological entities that have similarities but no obvious recent common history, say, RNA viruses and ribosomes or eucaryotic RNA viruses and RNA bacteriophages, their similarities arise primarily from the chemical and physical nature of their constituents and from the geometrical constraints of their assemblages. But biological entities, related by evolutionary pathway, have additional common features, e.g., similar RNA replication mechanisms, similar proteolytic pathways for virion maturation, and similar regulation of genome expression—features not exclusively dictated by chemistry, physics, and geometry.

Not all positive-stranded RNA viruses have discernible evolutionary pathways. Consider the RNA phages. RNA phages were the very cornerstone of the foundations of the science of molecular biology. They provided the first evidence for the mechanism of translation and its regulation. They provided proofs for the reality of the genetic code. They were the principal test objects for the development of nucleic acid sequencing. They provided the first infectious nucleic acid synthesized in vitro. Of all viruses, they rank first in impact upon molecular biology. But to date, their evolutionary relationship to other positive-stranded RNA viruses rests primarily on their use of RNA-dependent RNA replication, trademarked by the ubiquitous Gly-Asp-Asp sequence of their encoded replicase subunit. Nowadays, with fundamental processes of molecular biology well in hand, one is inclined to pass RNA phages by as exemplars of positive-stranded RNA viruses. They are not the models to which one turns to predict unknown properties of other positive-stranded viruses.

Similarly, consider the retroviruses. They surely have derived from entities having central roles in biology, although the precise nature of these entities is still obscure. Will retroviruses illuminate the cellular constituents from which they sprang, or will they become the most intensely studied of all viruses and thus become the model with which all other viruses are compared and contrasted? For the moment, the evolutionary relationship of retroviruses to other positive-stranded viruses rests mostly on the same Gly-Asp-Asp sequence (disguised as Met-Asp-Asp) of the RNA phages, and so the lessons they teach are not specifically related to the positive-stranded viruses of this meeting.

It may be that, at present, the alphaviruses, which have a constituency both in insects and in higher animals and have demonstrable relatives

Paul Kaesberg, Institute for Molecular Virology and Department of Biochemistry, University of Wisconsin-Madison, Madison, Wisconsin 53706.

among the plant viruses, have the greatest potential to represent the essence of positive-stranded viruses.

Therefore, I see as a theme for this meeting the recognition of evolutionary relationships among positive-stranded RNA viruses and the exploitation of these relationships for the prediction of virus properties. Thus, we have an opening session that describes evolutionary relationships, and we have succeeding sessions that focus on specific viruses. Some viruses will be discussed because they are typical, and others because they are unique in their importance to our health, our economy, or our well-being.

With tobacco mosaic virus (TMV) as the model virus, Wendall Stanley (1935) had already shown in the 1930s that viruses are discrete physical entities capable of isolation in pure form and capable of precise chemical characterization, certainly the premier achievement in virology in that era. In the 1950s, the only viruses susceptible to large-scale isolation and purification were the plant viruses. They were not only the exemplar viruses of that day, but also the only uniform particles available in their size range as test objects for development of electron microscopy and such physical methods as sedimentation and diffusion and X-ray and light scattering. In that period, Heinz Fraenkel-Conrat and Robley Williams (1955) showed that TMV could be taken apart and reassembled and that the genetic material was assuredly RNA, indeed infectious RNA, certainly the premier achievement in virology in this decade.

Studies on viral icosahedra had their beginnings in 1956 in three papers. In *Nature*, Watson and Crick (1956) proposed that viruses are composed of a definite number of protein subunits in a regular arrangement, possibly with the symmetry of an icosahedron. In the same issue, D. L. D. Caspar (1956) showed that crystals of tomato bushy stunt virus gave X-ray diffraction patterns having icosahedral symmetry. Also at that time, I reported in *Science* that a number of small plant viruses, when examined in the electron microscope, look remarkably like small icosahedra (Kaesberg, 1956). Caspar's diffraction results might have been an isolated case, and my micrographs left some doubt, because the viruses used were exceptionally small, weighing only some 5 million to 9 million daltons. But then, in 1958, Robley Williams and Kenneth Smith, with tipula iridescent virus, some 300 times as massive, demonstrated clearly that viruses could be icosahedral in external appearance (Williams and Smith, 1958).

I talked about icosahedral viruses at the Vienna meeting and also about our X-ray scattering evidence (on turnip yellow mosaic virus top and bottom components) that the protein portion of viruses was on the outside and enclosed and interdigitated an interior composed of the viral genome. Shortly thereafter, at the first International Meeting of Electron Microscopists in Berlin, Sidney Brenner and Robert Horne reported their invention of the method of negative staining, with which they provided the first visual evidence of the subunit structure of virion capsids.

In 1962, Caspar and Klug presented their classic, theoretical paper on the quasi-equivalence structure of viruses (Caspar and Klug, 1962). Although we now know that viruses are only quasi-quasi-equivalent, that theory remains as the cornerstone of modern virus crystallography and electron microscopy.

Positive-stranded RNA viruses emerged from the realm of plant virology in the early 1960s, with Loeb and Zinder's discovery of the RNA phages (Loeb and Zinder, 1961). I have already alluded to their tremendous impact upon molecular biology. And let nothing I have said detract from their impact on virology. For more than a decade, phage MS2 was the only virus whose genome nucleotide sequence was known. After more than two decades, Qβ replicase remains the only purified (viral) RNA replicase.

However, it was in translation, one of the areas of greatest triumph for research on RNA phages, that doubts were raised regarding their suitability as models for RNA viruses indigenous to eucaryotes. Aach and his associates (1964) had shown that in *Escherichia coli* extracts even the best-known eucaryotic messenger, that of TMV, failed to induce synthesis of a protein it certainly encoded, that of TMV coat protein. Thus, new models were sought, first among the plant viruses and subsequently among the animal viruses. These models led to new insights into translation of eucaryotic RNAs and its regulation. With alfalfa mosaic virus it led to the discovery of divided genomes (Van Vloten-Doting et al., 1970), with bromoviruses it led to the discovery of silent cistrons and subgenomic messengers (Shih and Kaesberg, 1973), and with picornaviruses it led to the discovery of posttranslational, proteolytic cleavage (Summers and Maizel, 1968). Collectively, these features of viral RNAs and their translation allow the expression of multiple proteins encoded in the genomes of eucaryotic viruses without the necessity of translating internal cistrons.

The failure to directly translate internal cistrons is popularly known as the eucaryotic rule: in eucaryotes, internal cistrons are not translated. For a long time, this rule seemed strange. What could be the selective advantage of prohibiting a particular process? Now virologists understand that subgenomic messengers, posttranslational processing, and readthrough are features that provide fine tuning for the temporal and quantitative control of translational expression. And it is not the prohibition of translation of internal cistrons that matters. The important thing is that there exist these regulating features, features that are the very soul of the life cycle of viruses.

Most of my remarks refer to events that predate 1975, and for most of you in the audience this is history. Only for the youngest of you is post-1975 history. Thus, I do not want to here recount or evaluate the tremendous advances in positive-strand virology that have taken place since that date. Let me say, though, that the period centering around 1975 represented a turning point, not so much as a period of successes but as a period of incubation. The failure of intact TMV RNA to induce synthesis of TMV coat protein in *E. coli* extracts turned many investigators to technically more difficult, homologous protein-synthesizing systems, a procedure subsequently found to be unnecessary. Indeed, it has been found that some RNA viruses that are normally restricted to the animal kingdom can be synthesized completely faithfully in plants (Selling et al., 1990). RNA-sequencing developments were considerable, but only Sanger and Fiers successfully overcame the immense efforts required to get meaningful results. Many new systems were tried, but RNA-dependent RNA synthesis remained an enigma. But then, what had seemed like wheel spinning soon bore fruit with the emergence of RNA dideoxy methods, direct cDNA sequencing, recombinant DNA chemistry applied to RNA, crystallography of RNA viruses, and infectious RNA transcribed from cDNA. And all of these advances are the stuff of which this meeting is made.

REFERENCES

Aach, H., G. Funatsu, M. Nirenberg, and H. Fraenkel-Conrat. 1964. Further attempts to characterize products of TMV-RNA-directed protein synthesis. *Biochemistry* **3**:1362–1366.

Caspar, D. L. D. 1956. Structure of bushy stunt virus. *Nature* (London) **177**:476–477.

Caspar, D. L. D., and A. Klug. 1962. Physical principles in the construction of regular viruses. *Cold Spring Harbor Symp. Quant. Biol.* **27**:1–24.

Fraenkel-Conrat, H., and R. C. Williams. 1955. Reconstitution of active tobacco mosaic virus from its inactive protein and nucleic acid components. *Proc. Natl. Acad. Sci. USA* **41**:690–698.

Kaesberg P. 1956. Structure of small "spherical" viruses. *Science* **124**:626–628.

Loeb, T., and N. D. Zinder. 1961. A bacteriophage containing RNA. *Proc. Natl. Acad. Sci. USA* **47**:282–289.

Selling, B., R. Allison, and P. Kaesberg. 1990. Genomic RNA of an insect virus directs synthesis of infectious virions in plants. *Proc. Natl. Acad. Sci. USA* **87**:434–438.

Shih, D. S., and P. Kaesberg. 1973. Translation of brome mosaic viral cell-free system derived from wheat embryo. *Proc. Natl. Acad. Sci. USA* **70**:1799–1803.

Stanley, W. M. 1935. Isolation of a crystalline protein possessing the properties of tobacco mosaic virus. *Science* **81**:644–645.

Summers, D. F., and J. V. Maizel. 1968. Evidence for large precursor proteins in poliovirus synthesis. *Proc. Natl. Acad. Sci. USA* **59**:966–971.

Van Vloten-Doting, L., A. Dingjan-Versteegh, and E. Jaspars. 1970. Three nucleoprotein components of alfalfa mosaic virus necessary for infectivity. *Virology* **40**:419–430.

Watson, J. D., and F. Crick. 1956. Structure of small viruses. *Nature* (London) **177**:473–475.

Williams, R., and K. Smith. 1958. The polyhedral form of the tipula iridescent virus. *Biochim. Biophys. Acta* **28**:464–469.

I. VIRAL EVOLUTION

Chapter 1

Genome Similarities between Positive-Strand RNA Viruses from Plants and Animals

Rob Goldbach

More than 75% of the plant viruses described so far, i.e., at least 500 distinct viruses, have a single-stranded RNA genome of positive polarity. These viruses have been classified into 28 taxonomic groups (Table 1) on the basis of particle morphology, serology, genome strategy, and biological properties such as vector specificity and host range. Despite the fact that the plant RNA viruses may greatly differ in these aspects, computer-assisted sequence comparisons of the proteins they encode have demonstrated that most if not all of them are genetically interrelated and, moreover, that they have remote relatives among the RNA viruses of animals. Hence, the plant como-, nepo-, and potyviruses are genetically related to the picornaviruses, while an even larger number of plant virus groups, such as the tobamo- bromo-, cucumo, tobra-, and furoviruses, are related to the alphaviruses. Evidence is presented that two major evolutionary mechanisms underlie the genetic relationships reported so far, i.e., divergency from common ancestors and interviral recombination.

GENETIC RELATIONSHIPS BETWEEN POSITIVE-STRAND RNA VIRUSES OF PLANTS AND ANIMALS

Although it was known for many years that some RNA virus groups bridge the gap between plant and animal kingdoms (e.g., rhabdoviruses and reoviruses [Matthews, 1981]), the finding that CPMV (for virus abbreviations, see footnote *a*, Table 1), a plant RNA virus, is genetically linked to the animal picornaviruses (Franssen et al., 1984) and that three other plant viruses, TMV, AlMV, and BMV, are related to the alphaviruses (Haseloff et al., 1984) was rather unexpected. With the elucidation of more plant viral genome sequences, the idea that most plant RNA viruses are in fact wide variations on only a few themes has become more and more valid. From 20 of the 28 groups of plant positive-strand RNA viruses, one or more members have been analyzed with respect to genome sequence (Table 1). As far as is known, most if not all of these viruses may be classified into two super-groups (Goldbach, 1986, 1987), each containing a number of classical taxonomic groups, namely, the picornalike plant viruses (como-, nepo-, and potyviruses; Fig. 1), which are related to the picornaviruses, and the Sindbis-like plant viruses (containing, among others, the tobamo-, bromo-, tobra-, and furoviruses; Fig. 2), which are related to the alphaviruses.

The plant viruses forming the supergroup of picornavirus-related plant viruses share the following properties with the true picornaviruses of animals: (i) their genomic RNAs are supplied with a genome-linked protein (VPg) at the 5' end and have a 3' poly(A) tail; (ii) their RNAs are expressed via the production of polyproteins from which smaller, functional proteins are generated by proteolytic processing; (iii) they encode a number of nonstructural proteins exhibiting amino acid sequence homology (shaded areas in Fig. 1); and (iv) these conserved proteins are encoded by similarly arranged gene sets. All of these similarities indicate a genetic relationship between these plant viruses and the picornaviruses. It should also be mentioned here that, in some aspects, the picornalike plant viruses are clearly dissimilar to the picornavi-

Rob Goldbach, Department of Virology, Agricultural University, Binnenhaven 11, 6709 PD Wageningen, The Netherlands.

TABLE 1
The Groups of Plant Positive-Strand RNA Viruses[a]

Group	Member(s) sequenced
Monopartite genome	
Tymoviruses	TYMV
Sobemoviruses	SBMV
Tombusviruses	TBSV, CNV
Tobacco necrosis virus	Tobacco necrosis virus
Luteoviruses	BYDV, BWYV, PLRV
Tobamoviruses	TMV, ToMV
Potexviruses	PVX, WClMV, PMV, NMV
Potyviruses	PVY, TEV, TVMV, PPV
Closteroviruses	
Maize chlorotic dwarf virus	
Carmoviruses	CarMV, TCV
Marafiviruses	
Parsnip yellow fleck virus	
Carlaviruses	LSV, PVS
Capilloviruses	
Tenuiviruses	
Bipartite genome	
Comoviruses	CPMV, RCMV
Nepoviruses	TBRV
Tobraviruses	TRV, PEBV
Dianthoviruses	RCNMV
Pea enation mosaic virus	
Furoviruses	BNYVV
Fabaviruses	
Tripartite genome	
Alfalfa mosaic virus	AlMV
Ilarviruses	
Bromoviruses	BMV, CCMV
Cucumoviruses	CMV
Hordeiviruses	BSMV

[a] Abbreviations: TYMV, turnip yellow mosaic virus; SBMV, Southern bean mosaic virus; TBSV, tomato bushy stunt virus; CNV, cucumber necrosis virus; BYDV, barley yellow dwarf virus; BWYV, beet Western yellowing virus; PLRV, potato leafroll virus; TMV, tobacco mosaic virus; ToMV, tomato mosaic virus; PVX, potato virus X; WClMV, white clover mosaic virus; PMV, papaya mosaic virus; NMV, narcissus mosaic virus; PVY, potato virus Y; TEV, tobacco etch virus; TVMV, tobacco vein mottling virus; PPV, plum pox virus; CarMV, carnation mottle virus; TCV, turnip crinkle virus; LSV, lily symptomless virus; PVS, potato virus S; CPMV, cowpea mosaic virus; RCMV, red clover mottle virus; TBRV, tomato black ring virus; TRV, tobacco rattle virus; PEBV, pea early browning virus; RCNMV, red clover necrotic mosaic virus; BNYVV, beet necrotic yellow vein virus; AlMV, alfalfa mosaic virus; BMV, brome mosaic virus; CCMV, cowpea chlorotic mottle virus; CMV, cucumber mosaic virus; BSMV, barley stripe mosaic virus.

ruses. Thus, como- and nepoviruses have a bipartite genome, whereas the genome of picornaviruses is monopartite. Furthermore, the potyviruses have a rod-shaped protein particle, built up from a single type of coat protein, whereas the other plant viruses within the supergroup have a capsid architecture very similar to that of the picornaviruses. Also, the position of the coat protein gene in the genetic map of potyviruses differs from that of the other picornalike plant viruses. The relevance of such differences for our understanding of RNA virus evolution will be discussed below.

The second major cluster of plant RNA viruses includes the Sindbis-like plant viruses (summarized in Fig. 2), which are all (to a lesser or greater extent) related to the *Alphaviridae*, of which Sindbis virus is the best-studied example

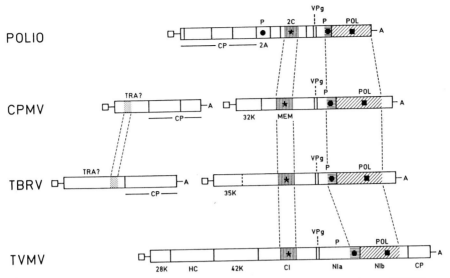

FIGURE 1. Comparison of the genomes of picornaviruses (poliovirus), comoviruses (CPMV), nepoviruses (TBRV), and potyviruses (TVMV). Coding regions are indicated as open bars, VPg as open squares, and polyadenylate sequences as A. Regions of amino acid sequence homology in the gene products are indicated by similar shading. Other notations: CP, capsid protein(s); TRA, putative transport function; P, protease; MEM, membrane binding; POL, core RNA-dependent RNA polymerase; HC, helper component; ★, nucleotide-binding domain; ●, cysteine protease domain; ■, conserved polymerase domain.

(Strauss and Strauss, 1986). The various members of this supergroup have distinct morphologies and show a more extensive variation with respect to genome structure and expression than do the members of the picornalike plant viruses. Viruses belonging to this group include the tobamoviruses (e.g., TMV), tobraviruses (TRV), bromoviruses, cucumoviruses, AlMV, hordeiviruses (BSMV), and furoviruses (BNYVV) and a number of plant viruses with a smaller, monopartite genome (Fig. 2). All of the viruses belonging to this supergroup produce their capsid proteins from a subgenomic mRNA, but in other aspects the translation strategy can be very diverse. Unlike the ease with the first supergroup, the 3'-terminal structure of the genome is not conserved, whereas the 5'-terminal structure is usually a cap but sometimes a VPg (sobemo- and luteoviruses). Despite these differences, all members (including the alphaviruses) are genetically related, since they specify three proteins (or protein domains, depending on the translation strategy) coded for by genes which are similarly arranged (Fig. 2). With more RNA sequences becoming available, the supergroup of Sindbis-related plant viruses becomes more and more heterogenous. In particular, the position of the viruses with the smaller RNA genomes, i.e., the carmo-, tombus-, and luteoviruses, is debatable, since they possess only one of the three conserved genes shared among the other viruses of the supergroup.

FUNCTIONS OF THE CONSERVED PROTEINS

For both supergroups of picornalike and Sindbis-like viruses, the conserved proteins have all been shown or suggested to be involved in the viral RNA replication process. This is most evident for the proteins shared among the picornalike viruses, of which CPMV and poliovirus have been studied most extensively (Goldbach and Van Kammen, 1985; Takegami et al., 1983). The genetic organization of the larger B RNA of CPMV strikingly resembles that of the P2 and P3 regions of the poliovirus genome (Fig. 3). The proteins encoded in these regions show the greatest homology, the comoviral 58-kilodalton membrane-bound protein, the genomelike protein VPg, the 24-kilodalton protease, and the core polypeptide of the viral polymerase clearly corresponding to picornaviral proteins 2C, VPg, 3C, and 3D, all proteins proposed to be involved in a membrane-bound replicase complex.

The most prominent sequence motifs in the

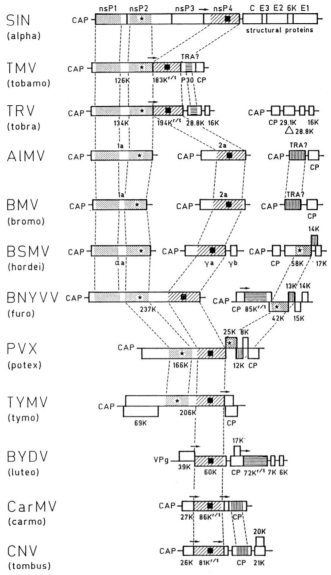

FIGURE 2. Comparison of the genomes of Sindbis virus (SIN) (*Alphaviridae*) and the Sindbis-like plant viruses. The supergroup of Sindbis-like viruses includes the tobamoviruses (TMV), tobraviruses (TRV), AlMV group, bromoviruses (BMV), hordeiviruses (BSMV), furoviruses (BNYVV), potexviruses (PVX), tymoviruses (TYMV), luteoviruses (BYDV), carmoviruses (CarMV), tombusviruses (CNV), and sobemoviruses (SBMV; shown in Fig. 4). Coding regions in the genomes are indicated as open bars; regions of amino acid sequence homology in the gene products are indicated by similar shading. Closely adjoining or slightly overlapping cistrons in the TMV and CarMV genomes are drawn contiguously. →, Leaky termination codon; r/t, readthrough; for other notations, see legend to Fig. 1.

domains conserved in these proteins are indicated in Fig. 1. All (putative) RNA-dependent RNA polymerases of the group contain a remarkable domain consisting of two conserved blocks (Kamer and Argos, 1984; Goldbach, 1986). The consensus sequence of the first block is S/GTXXXTXXXNT/S, where X may be any amino acid residue. The second block, 21 to 37 residues downstream of this first block, is even more prominent, consisting of a GDD sequence

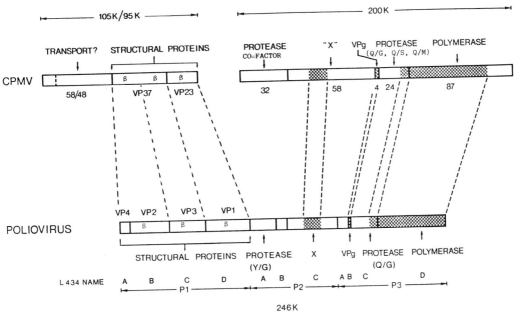

FIGURE 3. Comparison of the polyproteins encoded by CPMV (top) and poliovirus (bottom). The (proposed) functions of the cleavage products derived from these polyproteins are indicated, as are their sizes in kilodaltons (K). Regions of significant (>20%) sequence homology are cross-hatched. β indicates the presence of a β barrel.

flanked by hydrophobic residues. This domain appears to be characteristic of all viral polymerases and is also found in the (putative) core polymerase polypeptides specified by the Sindbis-like viruses (Fig. 2).

A second characteristic motif, centered about the sequence GKS/T, is found in another conserved domain, and again this motif is shared not only by the picornalike viruses but also by the Sindbis-like viruses (Fig. 1 and 2). This motif probably represents an ATP-binding domain, since it is also found in ATP-binding proteins (Gorbalenya et al., 1985). Further sequence comparisons, moreover, revealed that the proposed ATP-binding proteins of the Sindbis-like viruses have distant homology to a family of *Escherichia coli* helicases, i.e., UvrD, Rep, RecB, and RecD (Gorbalenya et al., 1988; Hodgman, 1988a, 1988b). Strikingly, some viruses, e.g., furo- and potexviruses, encode two of such proteins (Fig. 2). Gorbalenya et al. (1988) have postulated that the Sindbis-like viruses indeed may specify proteins with helicase activity, a function that may be involved in unwinding of replicative-form molecules during replication. Recently, Lain et al. (submitted) have shown that the cylindric inclusion proteins (Fig. 1) of potyviruses and flaviviral NS3 proteins likewise show homology to a second family of proteins with helicase activity (e.g., mouse eIF-4A and human nuclear protein p68). Hence, although no homology has been found yet between the nucleotide-binding proteins of picorna-, como-, and nepoviruses and any known helicase protein family, the production of a protein with helicase activity and the involvement of such activity in genome replication may turn out to be a general property of eucaryotic RNA viruses.

In addition to a core polymerase and a putative nucleotide-binding protein, the picornalike viruses share two other conserved proteins, VPg, supposed to be involved in priming of RNA synthesis, and a 3C-type viral protease (Fig. 1).

EVOLUTIONARY MECHANISMS

From the comparisons made above and summarized in Fig. 1 and 2, the conclusion seems justified that the positive-strand RNA viruses of plants whose genome sequences have been determined so far can be clustered into two supergroups, one smaller group being related to the picornaviruses and another, larger and more heterogeneous group being related to the alphaviruses. The question arises as to how present-day RNA viruses, which infect organisms from different eucaryotic kingdoms that diverged 1.2 $\times$ 10^9 years ago, can encode proteins with homologous sequences.

In theory, four different evolutionary pathways may be considered to account for the interviral relationships that have been detected so far: (i) convergent evolution, (ii) transduction of host genes, (iii) common ancestry, and (iv) interviral recombination. The first two mechanisms seem difficult to imagine, since they do not explain the colinearity of the genetic maps of the various viruses. Hence, the third and fourth mechanisms, i.e., divergency from common ancestors and interviral recombination, are the only possibilities that remain, and evidence is presented below that evolution of RNA viruses is probably based on both.

Divergency as Evolutionary Mechanism

Why consider divergent evolution, i.e., the evolvement of related plant and animal viruses from common ancestors? This option can be best illustrated by comparing all similarities between comovirus CPMV and poliovirus, one of the best-studied combinations of a plant and animal virus (Fig. 3). CPMV and poliovirus, apart from the properties characteristic for all picornalike viruses mentioned above, also share their capsid architecture. Although CPMV has two capsid proteins, these proteins fold into three β-barrel domains which mimic in many details the three independent β-barrel domains of the major proteins in the picornavirus capsid (Stauffacher et al., 1987). Moreover, the arrangement of these domains reflects for both viruses the alignment of the gene order. In other words, by placing M RNA of CPMV left of B RNA, the general gene order in the (divided) genome of this virus is colinear with that of poliovirus, which is certainly true if we consider a protease cofactor (the 32-kilodalton protein of CPMV) evolutionarily very close to a protease itself (the poliovirus 2A protease). Hence, CPMV may be regarded as a split picornavirus, with the P1 region (M RNA) physically separated from the P2 and P3 regions (B RNA). Obviously, in view of all similarities, common ancestry seems to be the only plausible evolutionary mechanism to underlie the genetic relationship between como- and picornaviruses. In this context, it is worthwhile mentioning that recently a number of plant viruses (e.g., dandelion yellow mosaic, anthriscus yellows, and parsnip yellow fleck) (Murant et al., 1987) have been identified as putative picornaviruses. Hence, this virus group may bridge the gap between plant and animal kingdom, thus making

common ancestry an even more credible explanation for all homologies found between the picornalike viruses.

As discussed in detail elsewhere (Goldbach, 1986), the high mutation rates generally observed for RNA genomes make it very improbable that common ancestral viruses predated the evolutionary separation of plant and animal cells approximately 1.2 $\times$ 10^9 years ago. It seems, therefore, more likely that when divergency has been a major mechanism during RNA virus evolution, common ancestors of plant and animal viruses existed in much more recent times. If this is true, then there are a number of strong arguments for supposing that insects may have played a major role in RNA virus evolution, either as sources of ancestral viruses or as carriers that introduced animal viruses into plants or vice versa. (i) The host ranges of plant and animal viruses overlap in insects. A large number of plant viruses are transmitted by insects, and a number of them even multiply in their respective insect vectors (Matthews, 1981). On the other hand, some insect viruses use plants as reservoirs for horizontal transmission, e.g., *Rhopalosiphum padi* virus in barley (Gildow and D'Arcy, 1988). (ii) There are many RNA viruses in insects, with a great variety in morphology and genome properties (Moore et al., 1985). Hence, insects are indeed potential sources of ancestral viruses. (iii) Some plus-strand RNA viruses of insects have divided genomes (the *Nodaviridae*). Since plus-strand RNA viruses of plants often have divided genomes whereas the *Nodaviridae* are, in this aspect, unique among the animal viruses, this insect virus group may represent an important link in the evolution of plant and animal RNA viruses.

Recombination as Evolutionary Mechanism

Divergency from common ancestors as the sole evolutionary mechanism, however, cannot explain all similarities and, more important, the differences between related plant and animal RNA viruses. For instance, the unique gene order in the potyviral genome, with a coat protein gene downstream of the conserved gene set typical for all picornalike viruses, and, moreover, a number of genes upstream of this set for which no counterparts are found for the como- and nepoviruses make this virus group very distinct from the others. Also, other viruses appear to have, in addition to their conserved genes, one or more extra genes not found in the genomes of related viruses (e.g., the transport gene TRA in plant viral genomes for which no

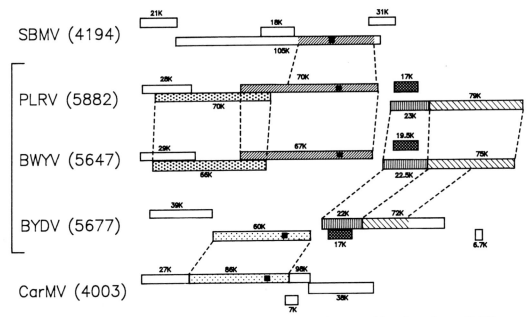

FIGURE 4. Comparison of the open reading frames in the genomic RNAs of three luteoviruses, PLRV, BYDV, and BWYV. The size in kilodaltons (K) of the protein encoded by each open reading frame is shown. Numbers between brackets indicate the lengths of the RNAs in nucleotides. Similar shading indicates regions of high amino acid sequence homology among the open reading frames of the different viruses. Computer-assisted sequence comparisons revealed that the BYDV polymerase is homologous to that of carmovirus, whereas the polymerases of BWYV and PLRV are homologous to that of sobemovirus (Van der Wilk et al., 1989). ■, The conserved GDD domain characteristic of all plant viral RNA polymerases.

counterpart is found in the genomes of related animal viruses, Sindbis virus nsP3 and envelope protein genes, TRV 16K gene, etc.). The coupling of common genes to unique genes in various viral genomes suggests that, in addition to common ancestry, there is at least a second major evolutionary mechanism, i.e., interviral recombination by which sets of genes may be exchanged among viruses belonging to different taxonomic groups. This has led to the concept of modular evolution, i.e., the idea that viral RNA genomes are constructed by the mixing and joining of gene modules (Zimmern, 1987; Gibbs, 1987; Goldbach and Wellink, 1988). Indeed, recombination has been shown to occur for several virus families or groups, e.g., picornaviruses (Cooper, 1977), coronaviruses (Makino et al., 1986), and bromoviruses (Bujarski and Kaesberg, 1986). For poliovirus, evidence has been presented that this process occurs by a copy choice mechanism during RNA replication (Kirkegaard and Baltimore, 1986), and this may hold for the other viruses as well. For a considerable number of plant RNA viruses there is an accumulating amount of evidence that present-day RNA genomes have actually arisen by recombination events. For example, (i) the TRV RNA-2 segment may be extended at its 3′ end with sequences of RNA-1 of variable length (depending on the strain) (Angenent et al., 1986; Cornelissen et al., 1986); (ii) some TRV isolates appear to represent true, natural recombinants of TRV and pea early browning virus (Robinson et al., 1987); and (iii) some luteoviruses (e.g., PLRV and BWYV; Fig. 4) specify a polymerase homologous to that of sobemoviruses, while another luteovirus (e.g., BYDV; Fig. 4) specifies a very distinct polymerase homologous to that of carmoviruses (Van der Wilk et al., 1989).

Two important notes can be made with reference to these examples. First, recombination can occur within a single plant virus (example i), between viruses belonging to the same taxonomic group (example ii), and between viruses belonging to different taxonomic groups (example iii). Second, recombination not only occurs under the selective pressure of artificial, laboratory conditions but also occurs in nature, some recombinant genomes obviously being more advantageous under certain conditions than those of the parental viruses.

CONCLUDING REMARKS

With the current flood of sequence data, it has become possible to compare groups of viruses, both plant and animal, in terms of nucleotide and protein sequence. This has led to super-grouping of plant and animal viruses that were previously not suspected to be genetically related. The data available so far indicate that RNA virus evolution is based on two major mechanisms: divergency from common ancestor viruses and interviral recombination. They also illustrate the evolutionary dynamics and flexibility of viral RNA genomes which ensure that, despite all attempts to eradicate viruses by active vaccination programs or resistance breeding, RNA viruses will always retain their position among the most successful parasites of eucaryotic organisms.

ACKNOWLEDGMENTS. I thank Joan Wellink for stimulating discussions and Annemieke van der Jagt for typing the manuscript.

REFERENCES

Angenent, G. C., H. J. M. Linthorst, A. F. Van Belkum, B. J. C. Cornelissen, and J. F. Bol. 1986. RNA2 of tobacco rattle virus strain TCM encodes an unexpected gene. *Nucleic Acids Res.* **14**:4673–4682.

Bujarski, J. J., and P. Kaesberg. 1986. Genetic recombination between RNA components of a multipartite plant virus. *Nature* (London) **321**:528–531.

Cooper, P. D. 1977. Genetics of picornaviruses. *Compr. Virol.* **9**:133–207.

Cornelissen, B. J. C., H. J. M. Linthorst, F. T. Bredero, and J. F. Bol. 1986. Analysis of the genome structure of tobacco rattle virus strain PSG. *Nucleic Acids Res.* **14**:2157–2169.

Franssen, H., J. Leunissen, R. Goldbach, G. Lomonossoff, and D. Zimmern. 1984. Homologous sequences in non-structural proteins from cowpea mosaic virus and picornaviruses. *EMBO J.* **3**:855–861.

Gibbs, A. 1987. Molecular evolution of viruses: 'trees,' 'clocks' and 'modules.' *J. Cell Sci.* **7**:319–337.

Gildow, F. E., and C. J. D'Arcy. 1988. Barley and oats as reservoirs for an aphid virus and the influence on barley yellow dwarf virus transmission. *Phytopathology* **78**:811–816.

Goldbach, R. 1987. Genomic similarities between plant and animal RNA viruses. *Microbiol. Sci.* **4**:197–202.

Goldbach, R., and A. Van Kammen. 1985. Structure, replication and expression of the bipartite genome of cowpea mosaic virus, p. 83–120. In J. W. Davies (ed.), *Molecular Plant Virology*, vol. 2. CRC Press, Inc., Boca Raton, Fla.

Goldbach, R., and J. Wellink. 1988. Evolution of plus-strand RNA viruses. *Intervirology* **29**:260–267.

Goldbach, R. W. 1986. Molecular evolution of plant RNA viruses. *Annu. Rev. Phytopathol.* **24**:289–310.

Gorbalenya, A. E., V. M. Blinov, and E. V. Koonin. 1985. Prediction of nucleotide-binding properties of virus-specific proteins from their primary structure. *Mol. Genet.* **11**:30–36.

Gorbalenya, A. E., E. V. Koonin, A. P. Donchenko, and V. M. Blinov. 1988. A novel superfamily of nucleoside triphosphate-binding motif containing proteins which are probably involved in duplex unwinding in DNA and RNA replication and recombination. *FEBS Lett.* **235**:16–24.

Haseloff, J., P. Goelet, D. Zimmern, P. Ahlquist, R. Dasgupta, and P. Kaesberg. 1984. Striking similarities in amino acid sequence among nonstructural proteins encoded by RNA viruses that have dissimilar genomic organization. *Proc. Natl. Acad. Sci. USA* **81**:4358–4362.

Hodgman, T. C. 1988a. A new superfamily of replicative proteins. *Nature* (London) **333**:22–23.

Hodgman, T. C. 1988b. A new superfamily of replicative proteins. *Nature* (London) **333**:578.

Kamer, G., and P. Argos. 1984. Primary structural comparison of RNA-dependent polymerases from plant, animal and bacterial viruses. *Nucleic Acids Res.* **12**:7269–7282.

King, A. M. Q., D. McCahon, W. R. Slade, and J. W. I. Newman. 1982. Recombination in RNA. *Cell* **29**:921–928.

Kirkegaard, K., and D. Baltimore. 1986. The mechanism of RNA recombination in poliovirus. *Cell* **47**:433–443.

Lain, S., J. L. Riechmann, M. T. Martin, and J. A. Garcia. Homologous potyvirus and flavivirus proteins belonging to a superfamily of helicase-like proteins. Submitted for publication.

Makino, S., J. G. Keck, S. A. Stohlman, and M. M. C. Lai. 1986. High-frequency RNA recombination of murine coronaviruses. *J. Virol.* **57**:729–733.

Matthews, R. E. F. 1981. *Plant Virology*, 2nd ed. Academic Press, Inc., New York.

Moore, N. F., B. Reavy, and L. A. King. 1985. General characteristics, gene organisation and expression of small RNA viruses of insects. *J. Gen. Virol.* **66**:647–659.

Murant, A. F., S. K. Hemida, and M. A. Mayo. 1987. Plant viruses that resemble picornaviruses. *Abstr. 7th Int. Congr. Virol.*, R24.2.

Robinson, D., W. D. O. Hamilton, B. D. Harrison, and D. C. Baulcombe. 1987. Two anomalous tobravirus isolates: evidence for RNA recombination in nature. *J. Gen. Virol.* **68**:2551–2561.

Stauffacher, C., R. Usha, T. Schmidt, M. Harrington, and J. E. Johnson. 1987. The structure of cowpea mosaic virus at 3.0 A resolution. *Abstr. 7th Int. Congr. Virol.*, R16.35.

Strauss, E. G., and J. H. Strauss. 1986. Structure and replication of the alphavirus genome, p. 35–90. *In* S. Schlesinger and M. J. Schlesinger (ed.), *The Togaviridae and Flaviviridae*. Plenum Publishing Corp., New York.

Takegami, T., R. J. Kuhn, C. W. Anderson, and E. Wimmer. 1983. Membrane-dependent uridylation of the genome-linked protein VPg of poliovirus. *Proc. Natl. Acad. Sci. USA* **80**:7447–7451.

Van der Wilk, F., M. J. Huisman, B. J. C. Cornelissen, H. Huttinga, and R. Goldbach. 1989. Nucleotide sequence and organization of the potato leafroll virus genomic RNA. *FEBS Lett.* **245**:51–56.

Zimmern, D. 1987. Evolution of RNA viruses, p. 211–240. *In*: J. Holland, E. Domingo, and P. Ahlquist (ed.), *RNA Genetics*. CRC Press, Inc., Boca Raton, Fla.

Chapter 2

Comparative and Evolutionary Aspects of Coronaviral, Arteriviral, and Toroviral Genome Structure and Expression

Willy J. M. Spaan, Johan A. den Boon, Peter J. Bredenbeek, Ewan D. Chirnside, Ans F. H. Noten, Eric J. Snijder, Antonius A. F. de Vries, and Marian C. Horzinek

Positive-stranded RNA viruses display a great variety in genome organization and expression. In this chapter, the replication strategy and its evolutionary implications are discussed for three groups of positive-stranded RNA viruses: the corona-, toro-, and arteriviruses.

VIRUS STRUCTURE, GENOME ORGANIZATION, AND EXPRESSION

Coronaviruses

Coronaviruses cause infections in humans, other mammals, and birds. Most experimental data have been obtained from studies of mouse hepatitis virus (MHV) and infectious bronchitis virus of chickens (IBV). Additional representatives of the family discussed in this chapter are the human coronavirus OC43 and the bovine coronavirus (BCV). Coronaviruses are enveloped RNA viruses that possess three major structural proteins: a phosphorylated nucleocapsid protein (N), a small integral membrane glycoprotein (M), and a large spike glycoprotein (S). Although all coronaviruses contain these three proteins, a subgroup has now been shown to possess an additional spike glycoprotein (gp65), which has probably been obtained by recombination (see below, RNA Recombination: Importance for Viral Evolution).

The coronavirus nucleocapsid, which contains the N protein and the genomic RNA, displays helical symmetry. The M protein is largely embedded in the membrane, whereas the S protein forms the characteristic elongated spikes of the virion.

The genome of coronaviruses is about 30 kilobases (kb) long and contains multiple open reading frames (ORFs). Several subgenomic mRNAs are synthesized to position internal genes at the unique 5' end of each mRNA (Fig. 1). A typical feature of the coronavirus mRNAs is the presence of a common leader sequence of about 72 nucleotides, which is derived from the 5' end of the genome.

For an extensive review of coronavirus structure and genome expression, the reader is referred to a recent review by Spaan et al. (1988).

Toroviruses

Recently, Horzinek et al. (1987) have proposed a new family of positive-stranded RNA viruses, the *Toroviridae*. Initially, these viruses were considered to be coronaviruslike, but an extensive analysis of torovirus structure and replication has clearly distinguished the *Toroviridae* from the *Coronaviridae*. Toroviruses cause enteric infections and are probably transmitted by the fecal-oral route. So far, equine (Berne virus [BEV]), bovine (Breda virus), and human toroviruses have been described. Most experimental data have been obtained from studies on the family prototype, BEV.

Toroviruses have been characterized as enveloped, peplomer-bearing particles containing

Willy J. M. Spaan, Johan A. den Boon, Peter J. Bredenbeek, Ewan D. Chirnside, Ans F. H. Noten, Eric J. Snijder, Antonius A. F. de Vries, and Marian C. Horzinek, Institute of Virology, Veterinary Faculty, State University of Utrecht, Yalelaan 1, 3584 CL Utrecht, The Netherlands.

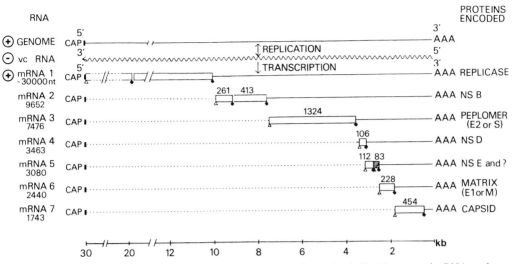

FIGURE 1. Genomic organization of the coronavirus MHV strain A-59. The genomic RNA and subgenomic mRNAs are represented by solid lines; the RNA lengths (in nucleotides) are indicated on the left. Open boxes show the positions of ORFs; numbers indicate their amino acid-coding capacity. Initiation and termination codons for translation are indicated by △ and ◆, respectively.

an elongated tubular capsid with helical symmetry. The capsid may bend into an open torus, conferring a biconcave disk or kidney-shaped morphology to the virion, or remain straight, resulting in a rod-shaped particle. The capsid contains a nucleocapsid protein with a molecular size of 19 kilodaltons (kDa) (Horzinek et al., 1985; Snijder et al., 1989) and a positive-stranded genomic RNA of about 25 kb (Snijder et al., 1988; Snijder et al., 1990). The BEV peplomer (P) consists of glycoproteins with molecular sizes of 80 to 100 kDa, which have been shown to be cleavage products of an intracellular precursor protein with a molecular size of 200 kDa (Horzinek et al., 1986). In addition to the P proteins, two other membrane-associated polypeptides (M and E) have been described.

The genome organization and expression of BEV are comparable to those of coronaviruses. Sequence analysis revealed at least five ORFs on the genome of BEV (Snijder et al., 1990). These ORFs are expressed from a nested set of subgenomic RNAs, each containing one of the ORFs at its 5' end. The BEV mRNAs are 3' coterminal, but in contrast to the coronaviral mRNAs, no common leader sequence has been detected at their 5' ends (Fig. 2; Snijder et al., 1990).

Arteriviruses

Equine arteritis virus (EAV) is the only virus assigned to the novel genus *Arterivirus* of the family *Togaviridae* (Westaway et al., 1986). Its classification as a togavirus was mainly based on morphological criteria. However, EAV synthesizes six subgenomic RNAs in infected cultures (Van Berlo et al., 1982), whereas other members of the *Togaviridae* (alpha- and rubiviruses) direct the synthesis of only one subgenomic RNA in infected cells (Strauss and Strauss, 1983).

cDNA cloning and sequence analysis have revealed that the EAV genome is 12.7 kb in length and, like corona- and toroviral genomes, contains multiple ORFs. So far, six ORFs have been identified (Fig. 3). The mRNAs again form a 3'-coterminal nested set and, as concluded from sequence analysis and Northern (RNA) blot hybridizations, they possess a common leader sequence of about 200 nucleotides at their 5' ends. This sequence is also present at the 5' end of the genome (A. A. F. de Vries et al., submitted for publication).

Genome Expression

On the basis of their translation strategy, the genomes of positive-stranded animal RNA viruses can be divided into three main groups (Strauss and Strauss, 1988): (i) genomes with only one ORF (e.g., picornaviruses, pestiviruses, and flaviviruses), (ii) genomes that contain two ORFs (e.g., alphaviruses and rubiviruses), and (iii) genomes that contain multiple

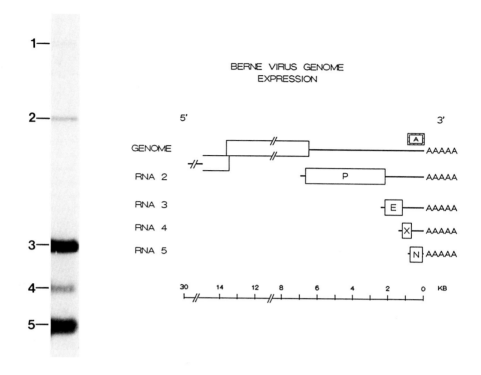

FIGURE 2. Genomic organization of the torovirus BEV. Genomic RNA and subgenomic mRNAs are represented by solid lines and indicated by number. Open boxes represent ORFs. The autoradiograph shows the result when a probe from the 3′ end of the BEV genome (depicted as a boxed A in the drawing) is hybridized to RNA from BEV-infected cells. This Northern blot hybridization proves that the BEV RNAs form a 3′-coterminal nested set.

ORFs (e.g., coronaviruses, toroviruses, and arteriviruses).

As described in the previous section, all members of the third group express their genetic information by the synthesis of multiple subgenomic mRNAs. Their transcription mechanism has been studied by analysis of the UV inactivation kinetics of mRNA synthesis (Jacobs et al., 1981; Stern and Sefton, 1982; Van Berlo et al., 1982; Snijder et al., 1990) and by sequence analysis of the mRNAs and the genome.

UV transcription mapping demonstrated that the coronavirus mRNAs do not result from the processing of a precursor RNA (Jacobs et al., 1981; Stern and Sefton, 1982). Most of the ORFs on the coronavirus genome are separated by a conserved junction sequence. This junction sequence is also present on free leader RNAs that can be detected in infected cells. Current data suggest a mechanism in which transcription of the viral mRNA is primed by leader transcripts (transcribed from the 3′ end of the minus-strand RNA [Fig. 1]) which bind to one of the complementary junction sequences on the negative-stranded template and function as a primer for the mRNA body transcription (reviewed by Lai [1986] and Spaan et al. [1988]).

The BEV RNAs do not possess a common leader sequence. Nevertheless, UV mapping showed that the RNAs are not derived from processing but are transcribed independently. Near transcription initiation sites, conserved regions have been identified that are postulated to function as promoter sequences (Snijder et al., 1990). Therefore, the transcription mechanism of the BEV RNAs is thought to show similarity to the mechanism of subgenomic RNA synthesis of alpha- and rubiviruses.

In contrast to the results obtained for corona- and toroviruses, the UV transcription mapping of EAV mRNAs indicated the presence of a genome-length precursor (Van Berlo et al., 1982). Since the RNAs have an identical leader sequence, these data suggest that splicing of a genome-length precursor is involved in the synthesis of the subgenomic RNAs of this virus. It

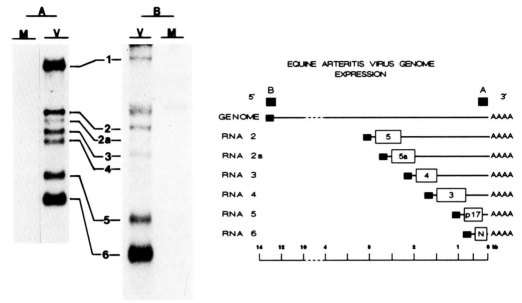

FIGURE 3. Genomic organization of the arterivirus EAV. Genomic RNA and subgenomic mRNAs are indicated by number and are represented by solid lines. Open boxes represent ORFs. The autoradiographs show the results of Northern blot hybridizations using RNA from EAV-infected (V) or mock-infected (M) cell cultures and probes from the 3′ end (probe A) or 5′ end (probe B) of the EAV genome. The positive reaction of all EAV RNAs with probe A proves that they are 3′ coterminal. The pattern obtained with probe B demonstrates that the RNAs also contain a common sequence at their 5′ ends, which is derived from the 5′ end of the genome. In the drawing, the EAV leader sequence is represented as a closed box at the 5′ end of each mRNA.

is not known whether the nucleus is involved in the replication of EAV. The sequences at the junction sites of EAV RNA leader and body are not compatible with sequences involved in premRNA splicing, splicing of tRNAs, and self-splicing of RNAs that contain group II introns. However, the sequence at the 3′ end of the EAV leader and that of an area 43 nucleotides downstream are almost identical to the *Tetrahymena* pre-rRNA 5′ splice site and internal guiding sequences, respectively (Cech, 1988).

RNA RECOMBINATION: IMPORTANCE FOR VIRAL EVOLUTION

Two major mechanisms are involved in the evolution of eukaryotic RNA viruses (see chapters by R. Goldbach and J. Wellink et al., this volume): (i) divergence from a common ancestor and (ii) interviral recombination.

The gene order polymerase (POL)-S protein-M protein-N protein is shared by coronaviruses belonging to different antigenic clusters. However, the number and position of additional

reading frames can vary significantly (Spaan et al., 1988, and references therein), indicating that, possibly because of recombination, ORFs can be gained or lost during coronaviral evolution. This is illustrated by the presence of two extra ORFs between the M and N genes in IBV and by the identification of an additional ORF downstream of the N gene in transmissible gastroenteritis virus (TGEV).

Another well-documented example of this variation is the presence of two ORFs between the POL gene and the S gene in MHV. Though the second of these reading frames was shown to be a pseudogene in MHV strain A-59 (Luytjes et al., 1988), its equivalent in the JHM strain of MHV was shown to encode the gp65 spike protein (see chapter by S. G. Siddell et al., this volume), which has been detected in only a limited number of coronaviruses. Recently, the presence of a homologous ORF has also been established on the BCV genome (Parker et al., 1989).

Comparison of the gp65 amino acid sequence with the entries in protein libraries revealed significant sequence similarity between this additional spike protein of MHV and the hemagglutinin subunit 1 of influenza C virus (Luytjes et al., 1988). The hemagglutinin protein of influenza C virus recognizes modified sialic acids as receptors. It also possesses a receptor-destroying activity, which removes the acidyl groups. The same biological activities have now been described for the coronaviruses BCV (Vlasak et al., 1988a; Vlasak et al., 1988b) and MHV strain JHM (see chapter by Siddell et al., this volume). For human coronavirus OC43, the region of the genome expected to contain the gp65 ORF has not yet been sequenced. However, the presence of the gene can be predicted from the fact that this virus also possesses the characteristic biological activities described above (Vlasak et al., 1988b).

To our surprise, the similarity between the gp65 of coronaviruses and the hemagglutinin subunit 1 of influenza C virus is not unique. Reading frame X of the torovirus BEV (Fig. 2) shares significant amino acid sequence similarity with the carboxy-terminal part of the proteins mentioned above. The percentages of identity between all three sequences are about 30% (E. J. Snijder et al., unpublished results).

Homologous recombination has been described to occur with relatively high frequency in coronaviruses (Makino et al., 1986; Keck et al., 1988). To explain the striking similarity between the coronaviruses and influenza C virus, it was postulated that nonhomologous RNA recombination has led to the incorporation of the hemagglutinin subunit 1 into the coronaviral genome (Luytjes et al., 1988). Because the extra spike protein in coronaviruses has a biological function similar to that of the influenza C virus protein, it is likely that this recombination has resulted in a coronavirus with a different cell tropism (see chapters by K. V. Holmes et al. and Siddell et al., this volume).

Since the corresponding ORF in BEV is located at an entirely different position on the genome, it is assumed that this similarity is the result of a second independent nonhomologous recombination event.

SUPERFAMILIES OF POSITIVE-STRANDED RNA VIRUSES

A number of RNA viruses have been grouped into two superfamilies, the alphaviruslike superfamily and the picornaviruslike superfamily (Strauss and Strauss, 1988). This classification was based on similarities in organization and expression of the genome and on amino acid

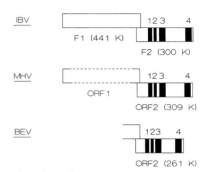

FIGURE 4. Schematic representation of the organization of the POL regions of IBV, MHV, and BEV. Open boxes represent ORFs; the nucleotide sequences of only small parts of the 5' and 3' regions of ORF1 of MHV and a part of the 3' region of ORF1 of BEV have been determined. Filled boxes indicate the position of conserved domains: 1, GDD motif; 2, potential zinc-binding finger; 3, helicase or GKS/T motif; 4, carboxy-terminal domain.

sequence homologies in regions of the polymerase or replicase proteins.

As described above, the genome organization and expression of corona-, toro-, and arteriviruses are very similar. Comparison of the deduced amino acid sequences of their structural proteins did not reveal any similarity. The sequence of the POL gene of the coronavirus IBV has been published by Boursnell et al. (1987), and recently about half of the polymerase-encoding regions of the coronavirus MHV and the torovirus BEV have been sequenced in our laboratory. Sequence data of the POL gene of EAV are not yet available.

Figure 4 shows what is known about the organization of the POL region of the coronaviruses MHV and IBV and the torovirus BEV. Several interesting conclusions can be drawn from these data. (i) The POL regions on the genomes of these viruses are about 20 kb long. (ii) For IBV, two ORFs have been identified in this region, encoding proteins of 441 and 300 kDa. (iii) The MHV and BEV POL regions also consist of two ORFs. The size of the second ORF of MHV is almost identical to the size of the IBV second ORF; the second ORF of BEV is somewhat smaller. (iv) In all three viruses, the first and second ORFs have a small overlap; the second ORF is in a −1 reading frame in comparison with the first ORF.

When the deduced amino acid sequences of the second ORFs of IBV and MHV were compared, it appeared that the amino acid sequence is highly similar for IBV and MHV, with a positional identity of 56% calculated. Remark-

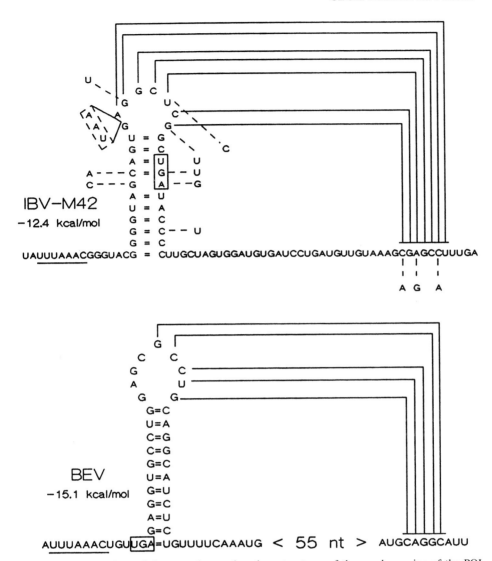

FIGURE 5. Comparison of the secondary and tertiary structures of the overlap region of the POL reading frames in IBV and BEV. The nucleotides (nt) potentially involved in frameshifting are underlined, and the termination codons of ORF1 (BEV and MHV) and frameshift 1 (IBV) are boxed. Solid lines illustrate the predicted pseudoknots. Differences between the sequences involved in the predicted RNA structures of MHV and IBV are indicated in the IBV structure. The complementary mutations illustrate the observed covariation described in the text. Dashed lines indicate point mutations; solid lines indicate insertions.

ably, amino acid similarities were also detected when the second ORF of BEV was included in the comparison. Positional identities of about 19% between the corresponding ORFs of BEV and both coronaviruses were calculated.

During recent years, comparisons of the RNA polymerases of different RNA viruses have revealed several interesting domains (Fig. 4). These domains include the so-called GDD motif (Argos, 1988; Gorbalenya and Koonin, 1988) and the nucleotide triphosphate-binding or GKS/T motif (Gorbalenya et al., 1988; Hodgman, 1988), which is also present in several *Escherichia coli* proteins with helicase activity. These two domains are present at comparable positions in the second ORFs of MHV, IBV, and BEV. Two additional conserved domains could also be identified: one containing conserved cysteine and histidine residues, which could potentially form a zinc-binding finger, and a second domain located at the carboxy terminus of the protein, previously unreported in viral RNA polymerases (E. J. Snijder et al., submitted for publication).

Apart from the conservation of amino acid sequences between the coronaviral polymerases and the BEV polymerase, the mechanism of expression of the second ORF of the POL gene is likely to be the same. For IBV, it has been shown that the second POL ORF is probably expressed by a mechanism of ribosomal frameshifting (Brierly et al., 1987). The involvement of a predicted stem-loop structure and RNA pseudoknot in this process has been demonstrated (Brierly et al., 1989).

A similar stem-loop structure and pseudoknot are predicted in the overlapping regions of the two POL ORFs of MHV and BEV (Fig. 5). When the ORF1-ORF2 overlapping regions of IBV and MHV are compared, it is noteworthy that mutations in the stem, as well as mutations in the loop and in the corresponding downstream region of the pseudoknot (Fig. 5), are complementary. When the BEV frameshift region was compared with its coronaviral equivalents, no nucleotide sequence similarity was detected. Nevertheless, the structural features necessary for ribosomal frameshifting seem to be well conserved in the toroviral sequence. Both observations stress the importance of the stem-loop structure and pseudoknot in ribosomal frameshifting.

Recently evidence was obtained that the ORF1-ORF2 overlapping regions of both MHV and BEV are able to direct ribosomal frameshifting in vitro and in vivo (Bredenbeek et al., in press; Snijder et al., submitted).

CONCLUSIONS

Although there are several important differences between the coronaviruses, toroviruses, and arteriviruses, striking similarities at the level of genome organization and expression have been observed. This finding implies that these viruses may be ancestrally related.

The existence of such a relationship between toro- and coronaviruses is underlined by the presence of several domains of amino acid sequence homology in the product of the second ORF of the POL gene. It has now been generally accepted that sequence similarities in the POL region indicate common ancestry (Strauss and Strauss, 1988).

On the basis of these similarities and homologies, the corona- and toroviruses may be classified into a third superfamily of positive-stranded RNA viruses, the coronaviruslike superfamily. Conclusions as to an evolutionary relationship between toro-, corona- and arteriviruses must await sequence analysis of the arteriviral POL gene.

Acknowledgments. P. J. Bredenbeek was supported by the Dutch Foundation for Chemical Research, with financial aid from the Dutch Organization for the Advancement of Pure Research. A. A. F. de Vries was supported by a research grant from Duphar B. V., Weesp, The Netherlands. E. D. Chirnside is the recipient of a European Molecular Biology Organization long-term fellowship. E. J. Snijder was supported by the Division for Health Research in cooperation with the Foundation for Medical Research (project 900-502-081).

REFERENCES

Argos, P. 1988. A sequence motif in many polymerases. *Nucleic Acids Res.* **16:**9909–9916.

Boursnell, M. E., T. D. Brown, I. J. Foulds, P. F. Green, F. M. Tomley, and M. M. Binns. 1987. Completion of the sequence of the genome of the coronavirus avian infectious bronchitis virus. *J. Gen. Virol.* **68:**57–77.

Bredenbeek, P. J., C. J. Pachuk, J. F. M. Noten, J. Charité, W. Loytjes, S. R. Weiss, and W. J. M. Spaan. The primary structure and expression of the second open reading frame of the polymerase gene of the coronavirus MHV-A5g. *Nucleic Acids Res.,* in press.

Brierly, I., M. E. G. Boursnell, M. M. Binns, B. Bilimoria, V. C. Blok, T. D. K. Brown, and S. C. Ingles. 1987. An efficient ribosomal frameshifting signal in the polymerase-encoding region of the coronavirus IBV. *EMBO J.* **6:**3779–3785.

Brierly, I., P. Digard, and S. C. Inglis. 1989. Characterization of an efficient coronavirus ribosomal frameshifting signal: requirement for an RNA pseudoknot. *Cell* **57:**537–547.

Cech, T. R. 1988. Conserved sequences and structures of group I introns: building an active site for RNA catalysis—a review. *Gene* **73:**259–271.

Gorbalenya, A. E., and E. V. Koonin. 1988. Birnavirus RNA polymerase is related to polymerases of positive strand RNA viruses. *Nucleic Acids Res.* **16:** 7735.

Gorbalenya, A. E., E. V. Koonin, A. P. Donchenko, and V. M. Blinov. 1988. A novel superfamily of nucleoside triphosphate-binding motif containing proteins which are probably involved in duplex unwinding in DNA and RNA replication and recombination. *FEBS Lett.* **235:**16–24.

Hodgman, T. C. 1988. A new superfamily of replicative proteins. *Nature* (London) **335:**22–23.

Horzinek, M. C., J. Ederveen, B. Kaeffer, D. de Boer, and M. Weiss. 1986. The peplomers of Berne virus. *J. Gen. Virol.* **67**:2475–2483.

Horzinek, M. C., J. Ederveen, and M. Weiss. 1985. The nucleocapsid of Berne virus. *J. Gen. Virol.* **66**:1287–1296.

Horzinek, M. C., T. H. Flewett, L. J. Saif, W. J. M. Spaan, M. Weiss, and G. N. Woode. 1987. A new family of vertebrate viruses: Toroviridae. *Intervirology* **27**:17–24.

Jacobs, L., W. J. M. Spaan, M. C. Horzinek, and B. A. M. van der Zeijst. 1981. Synthesis of subgenomic mRNA's of mouse hepatitis virus is initiated independently: evidence from UV transcription mapping. *J. Virol.* **39**:401–406.

Keck, J. G., G. K. Matsushima, S. Makino, J. O. Fleming, D. M. Vannier, S. A. Stohlman, and M. M. C. Lai. 1988. In vivo RNA-RNA recombination of coronavirus in the mouse brain. *J. Virol.* **62**:1810–1813.

Lai, M. M. C. 1986. Coronavirus leader-RNA primed transcription: an alternative mechanism to RNA splicing. *Bioessays* **5**:257–260.

Luytjes, W., P. J. Bredenbeek, A. F. H. Noten, M. C. Horzinek, and W. J. M. Spaan. 1988. Sequence of mouse hepatitis virus A-59 mRNA 2: indications for RNA recombination between coronaviruses and influenza C virus. *Virology* **166**:415–422.

Makino, S., J. G. Keck, S. A. Stohlman, and M. M. C. Lai. 1986. High frequency RNA recombination of murine coronaviruses. *J. Virol.* **57**:729–737.

Parker, M. D., G. J. Cox, D. R. Deregt, D. R. Fitzpatrick, and L. A. Babiuk. 1989. Cloning and in vitro expression of the E3 haemagglutinin glycoprotein of bovine coronavirus. *J. Gen. Virol.* **70**:155–164.

Snijder, E. J., J. A. den Boon, W. J. M. Spaan, G. M. G. M. Verjans, and M. C. Horzinek. 1989. Identification and primary structure of the gene encoding the Berne virus nucleocapsid protein. *J. Gen. Virol.* **70**:3363–3370.

Snijder, E. J., J. Ederveen, W. J. M. Spaan, M. Weiss, and M. C. Horzinek. 1988. Characterization of Berne virus genomic and messenger RNAs. *J. Gen. Virol.* **69**:2135–2144.

Snijder, E. J., M. C. Horzinek, and W. J. M. Spaan. 1990. A 3'-coterminal nested set of independently transcribed mRNAs is generated during Berne virus replication. *J. Virol.* **64**:331–338.

Spaan, W., D. Cavanagh, and M. C. Horzinek. 1988. Coronaviruses: structure and genome expression. *J. Gen. Virol.* **69**:2939–2952.

Stern, D. F., and B. M. Sefton. 1982. Synthesis of coronavirus mRNAs: kinetics of inactivation of infectious bronchitis virus RNA synthesis by UV light. *J. Virol.* **42**:755–759.

Strauss, E. G., and J. H. Strauss. 1983. Replication strategies of the single stranded RNA viruses of eukaryotes. *Curr. Top. Microbiol. Immunol.* **105**:1–98.

Strauss, J. H., and E. G. Strauss. 1988. Evolution of RNA viruses. *Annu. Rev. Microbiol.* **42**:657–683.

Van Berlo, M. F., M. C. Horzinek, and B. A. M. van der Zeijst. 1982. Equine arteritis virus-infected cells contain six polyadenylated virus-specific RNAs. *Virology* **118**:345–352.

Vlasak, R., W. Luytjes, J. Leider, W. J. M. Spaan, and P. Palese. 1988a. The E3 protein of bovine coronavirus is a receptor-destroying enzyme with acetylesterase activity. *J. Virol.* **62**:4686–4690.

Vlasak, R., W. Luytjes, W. J. M. Spaan, and P. Palese. 1988b. Human and bovine coronaviruses recognize sialic acid-containing receptors similar to those of influenza C viruses. *Proc. Natl. Acad. Sci. USA* **85**:4526–4529.

Westaway, E. G., M. A. Brinton, S. Y. Gaidamovich, M. C. Horzinek, A. Igarashi, L. Kääriäinen, D. K. Lvov, J. S. Porterfield, P. K. Russell, and D. W. Trent. 1986. Togaviridae. *Intervirology* **24**:125–139.

Chapter 3

Human Hepatitis Delta Virus: Unique or Not Unique?

John Taylor, Mei Chao, Mark Kuo, Lamia Sharmeen, and Sen-Yung Hsieh

This chapter begins with a brief review of certain aspects of hepatitis delta virus (HDV) and then moves to evaluate analogies between HDV and (i) certain pathogenic RNAs of plants, (ii) 7S L RNA, (iii) infectious introns, and finally (iv) maturases and (v) negative-stranded viruses. The discussion shows that in many ways HDV is not unique and that much has been gained, and more may still be gained, by examination of such analogies.

Human HDV is actually subviral; it carries out its replication only in the presence of a hepadnavirus, which for humans is hepatitis B virus (HBV) (Ponzetto et al., 1984; Rizzetto et al., 1977; Rizzetto et al., 1980). The virions of HDV are about 38 nm in diameter (He et al., 1989). The outside surface is a lipid-containing membrane substituted with the same HBV proteins that package the HBV virions (Bonino et al., 1984; Bonino et al., 1986). Inside the HDV virion are the RNA genome and the only known HDV-encoded protein, a 24-kilodalton species known as the delta antigen (Bonino et al., 1986; Weiner et al., 1988). HBV thus provides the coat for HDV and thereby controls the passage of HDV genomes to and from hepatocytes, which are the natural sites of replication of both viruses.

The structure of the HDV RNA genome is unique relative to known animal viruses; it is a single-stranded RNA that is joined to form a covalently closed circle (Chen et al., 1986; Kos et al., 1986; Kuo et al., 1988a; Makino et al., 1987; Wang et al., 1986); in addition, the nucleotide sequence is such as to allow the RNA to fold into an unbranched rod structure, with about 70% of the bases paired. This structure not only is predicted to occur (Kuo et al., 1988a; Wang et al., 1986) but has also been visualized by electron microscopy (Kos et al., 1986; Wang

et al., 1986). The sequence of the RNA genome of HDV shows no striking relationship to the genome of HBV, and so HDV is not a defective interfering variant of HBV but rather a subviral satellite agent.

HBV replicates by a pathway involving reverse transcription of RNA into DNA (Mason et al., 1987). In contrast, HDV replicates its genome by RNA-directed RNA synthesis. It has been shown by transfection studies with cloned HDV sequences that this genome replication can be carried out strictly independently of any HBV functions (Kuo et al., 1989). Also, the genome replication has been demonstrated to occur not only in hepatocytes (Taylor et al., 1987b) but also in other cell types, such as kidney and ovary cell lines (Kuo et al., 1989; unpublished observations). The normal genome replication involves the synthesis of a complementary RNA, referred to as the antigenome. In the liver, the majority of both genomic and antigenomic RNAs are unit length (about 1,700 bases), and more than half of these are circular. A small but significant amount of these RNAs are also greater than unit length, some behaving as if they were dimers and even trimers of the genome. There are usually 5 to 20 times more molecules of genomic than antigenomic RNA. The level of replication can accumulate to large amounts, for example, as much as 300,000 molecules of genomic RNA per average liver hepatocyte (Chen et al., 1986; Taylor et al., 1987b). This may be a large amount relative to that achieved by other animal viruses, but at the same time it is puzzlingly a relatively slow process; in animals the replication takes at least 5 weeks to reach this peak, and even in cultured hepatocytes and transfected cells the genome replication takes about 2 weeks.

The only known protein of HDV is the delta

John Taylor, Mei Chao, Mark Kuo, Lamia Sharmeen, and Sen-Yung Hsieh, Fox Chase Cancer Center, Philadelphia, Pennsylvania 19111.

antigen (Bergmann and Gerin, 1986; Bonino et al., 1984; Bonino et al., 1986; Chang et al., 1988; Rizzetto et al., 1977; Rizzetto et al., 1980; Weiner et al., 1988). The coding region for this protein of about 195 amino acids (Kuo et al., 1988a) is located on the antigenomic RNA. Thus, in the absence of other virus-coded proteins, HDV could be considered as a negative-stranded agent (Baltimore, 1971). This protein has been shown to be essential for replication of the HDV genome (Kuo et al., 1989); however, it has not yet been established as to how it is necessary. The protein is a phosphoprotein (Chang et al., 1988), and consistent with its predicted charge of +12 (unpublished observations), it behaves as an RNA-binding protein (Chang et al., 1988).

As will be explained in the subsequent sections, there are some potentially valuable analogies between known biological systems and the HDV genome and its replication. These analogies can provide valuable insights into questions relating to HDV and can provide clues as to evolutionary relatedness.

ANALOGY WITH PATHOGENIC RNAs OF PLANTS

It was probably first noted by Diener and Prusiner (1985) that HDV had certain similarities to certain known pathogenic RNAs of plants, specifically the viroids, virusoids, satellite RNAs, and satellite viruses. (This topic has been reviewed elsewhere [Taylor et al., 1988; Taylor et al., 1987a; Taylor, in press], and it should be noted that some workers now prefer to consider virusoids as a subclass of satellite RNAs.) What Diener and Prusiner especially noted as similar was that (i) HDV was a satellite of HBV and (ii) the genome of HDV was a small single-stranded RNA. This analogy was later extended by the findings mentioned above, that (iii) the HDV RNA genome was circular, (iv) it was able to fold into an unbranched rod structure, and (v) the RNA intermediates found inside infected cells seemed to be consistent with a rolling-circle model of genome replication (Taylor, in press). This overall analogy was so strong as to successfully predict that certain self-catalytic reactions observed for certain of the plant agents might also apply for HDV. More specifically, (vi) the self-cleavage ability of certain plant agent RNAs (Buzayun et al., 1986; Hutchins et al., 1986) was found for HDV (Kuo et al., 1988b; Sharmeen et al., 1989; Wu et al., 1989), and (vii) the self-ligation ability demonstrated for one plant agent (Buzayun et al., 1986; Prody et al.,

1986) was also found for HDV (Sharmeen et al., 1988). HDV has one such site on both the genomic and antigenomic RNAs, and the roles of the cleavage and ligation events have been incorporated into a model of the life cycle of the genome (Taylor et al., 1988). As with the plant agents, it is important to stress that even though these reactions may in the test tube be strictly RNA catalyzed, inside the infected cell they probably are codependent on host- and/or virus-encoded proteins.

In the foregoing discussion, it has been stressed that there are some striking similarities between the plant agents and HDV. Other similarities may yet be found; a potential example is that of RNA-directed transcription. For two of the plant agents, there is some evidence (Rackwitz et al., 1981; Semacnik and Harper, 1984) that an α-amanatin-sensitive host RNA polymerase is apparently redirected to copy RNA rather than DNA. No comparable data for HDV have yet been published.

ANALOGY WITH 7S L RNA

Recently, Haas and co-workers (1989) have pointed out that the sequence of the plant viroids shows a patch of complementarity to plant 7S L RNA. The latter RNA is the nucleic acid component of the signal recognition particle, which functions in the transport of nascent proteins across cellular membranes (Walter and Blobel, 1982). Haas et al. (1989) and subsequently Symons (1989) have pointed out that this similarity might be the basis for the pathogenicity of viroids.

Therefore, since there are so many features of similarity between certain of the plant agents and HDV (see above), we attempted a similar sequence analysis to compare HDV with human 7S L RNA. We did find some patches of complementarity. However, we also found more identity than complementarity. And if we scrambled one of the sequences (simply by reversing the order of the bases), the amount of relatedness increased. Thus, we are skeptical about the possibility of a significant relatedness between HDV and either 7S L RNA or its complement. However, one has to keep an open mind because there are cases where molecular mimicry, of cell proteins by viral proteins, has been proven to have a role in viral pathogenesis (Oldstone and Notkins, 1986).

ANALOGY WITH INFECTIOUS INTRONS

The ability of certain of the RNA plant agents to undergo self-cleavage and self-ligation is in many ways similar to the self-splicing that has been reported for various intron-containing RNAs (Bass and Cech, 1986). This finding has led others to carry out computer-based sequence comparisons. One such analysis has provided an indication that there are sequence similarities between a subclass of the plant agents, namely, the viroids, and introns that belong to group I (Dinter-Gottlieb, 1986; Hadidi, 1986). The similarities noted are in terms of (i) certain short sequence elements that are known to be shared among group I introns and (ii) the fact that both are known to be able to self-process. This observation has led to the speculation that certain of the viroids may be "escaped" introns (Dinter-Gottlieb, 1986; Hadidi, 1986). Whether or not the model is meaningful even for viroids, it certainly has not yet been demonstrated for HDV.

ANALOGY WITH MATURASES

Recently, another possible source of relationship between HDV and introns has been raised. Lambowitz (1989) has reviewed what he has termed infectious introns. Others have shown that some "introns" encode proteins. Some of these proteins have been shown to have a specific role in facilitating the processing of their own "introns" and for this reason have been named maturases. It is possible that there are valuable analogies between the delta antigen and such maturases, such as at the levels of function and gene expression and regulation.

ANALOGY WITH NEGATIVE-STRANDED VIRUSES

Even though HDV must be considered subviral, it is still negative stranded; only the antigenome codes for a functional protein. As a negative-stranded agent, it has some of the same problems as the negative-stranded RNA viruses and, as will be now explained, it probably uses a similar strategy.

Since the HDV RNA that enters the cell is actually the antisense strand, the new HDV-encoded protein, the delta antigen, must be made from a newly synthesized antigenomic RNA transcript. What is the mRNA for this? Such an mRNA would be expected to span the appropriate region of the antigenomic RNA and also to be cytoplasmic, polyadenylated, and capped. For several reasons, such a species cannot in turn be a template for the transcription of new genomic RNA. With the negative-

stranded viruses there is a regulatory mechanism, controlled by a virus-encoded protein, that facilitates the decision making as to whether to make an mRNA or an RNA transcript suitable for further genome replication (Krug et al., in press). Since HDV encodes one protein, it seems reasonable to speculate that one role of this protein is in such regulation of transcription.

CONCLUSIONS

HDV has been revealed as a subviral agent with a negative-stranded RNA genome. As a satellite of HBV, it apparently needs help only in getting packaged by HBV so as to allow exit and entry of its genome from cell to cell. It is sobering to think that this packaging requirement might readily be provided by other viruses so as to produce other tissue tropisms. Also, it is not unreasonable to expect that we will find other viral diseases of humans that are caused by other subviral agents in association with other viruses. And given what we know about that group of subviral agents of plants known as viroids, which have no recognized helper virus, it remains possible that subviral agents of humans, like HDV, might also be able to cause diseases in the absence of a helper virus.

The major aim of this chapter has been to evaluate various analogies between HDV and other systems with the intent of revealing additional potential relationships that may be tested experimentally. In this respect, the analogies presented here have in most cases already proven themselves to be extremely valuable.

ACKNOWLEDGMENTS. We thank William Mason and Robert Krug for critical comments on the manuscript and Robert Krug for pointing out the relevance to HDV of known mechanisms of regulation of transcription of negative-stranded viruses. J.T. is supported by Public Health Service grants AI-26522, CA-06927, and RR-05539 from the National Institutes of Health, grant MV-7M from the American Cancer Society, and an appropriation from the Commonwealth of Pennsylvania.

REFERENCES

Baltimore, D. 1971. Expression of animal virus genomes. *Bacteriol. Rev.* **35**:235–241.

Bass, B., and T. R. Cech. 1986. Biological catalysis by RNA. *Annu. Rev. Biochem.* **55**:599–629.

Bergmann, K., and J. L. Gerin. 1986. Antigens of hepatitis delta virus in the liver and serum of humans and animals. *J. Infect. Dis.* **154**:702–706.

Bonino, F., K. H. Heermann, M. Rizzetto, and W. H.

Gerlich. 1986. Hepatitis delta virus: protein composition of delta antigen and its hepatitis B virus-derived envelope. *J. Virol.* **58**:945–950.

Bonino, F., B. Hoyer, J. W.-K. Shih, M. Rizzetto, R. H. Purcell, and J. L. Gerin. 1984. Delta hepatitis agent: structural and antigenic properties of the delta-associated particles. *Infect. Immun.* **43**:1000–1005.

Buzayun, J. M., W. L. Gerlach, and G. Bruening. 1986. Non-enzymatic cleavage and ligation of RNAs complementary to a plant satellite RNA. *Nature* (London) **323**:349–353.

Chang, M.-F., S. C. Baker, L. H. Soe, T. Kamahora, J. G. Keck, S. Makino, S. Govindarajan, and M. M. C. Lai. 1988. Human hepatitis delta antigen is a nuclear phosphoprotein with RNA-binding activity. *J. Virol.* **62**:2403–2410.

Chen, P.-J., G. Kalpana, J. Goldberg, W. Mason, B. Werner, J. Gerin, and J. Taylor. 1986. The structure and replication of the genome of the hepatitis delta virus. *Proc. Natl. Acad. Sci. USA* **83**:8774–8778.

Diener, T. O., and S. B. Prusiner. 1985. The recognition of subviral pathogens, p. 3–20. *In* K. Maramorosch and J. J. McKelvey (ed.), *Subviral Pathogens of Plants and Animals: Viroids and Prions.* Academic Press, Inc., Orlando, Fla.

Dinter-Gottlieb, G. 1986. Viroids and virusoids are related to group I introns. *Proc. Natl. Acad. Sci. USA* **83**:6250–6254.

Haas, B., A. Klanner, K. Ramm, and H. L. Sanger. 1989. The 7S RNA from tomato leaf tissue resembles a signal recognition particle RNA and exhibits a remarkable sequence complementarity to viroids. *EMBO J.* **7**:4063–4074.

Hadidi, A. 1986. Relationships of viroids and certain other plant pathogenic nucleic acids to group I and II introns. *Plant Mol. Biol.* **7**:129–142.

He, L. F., E. Ford, R. H. Purcell, W. T. London, J. Phillips, and J. L. Gerin. 1989. The size of the hepatitis delta agent. *J. Med. Virol.* **27**:31–33.

Hutchins, C. J., P. D. Rathjen, A. C. Forster, and R. H. Symons. 1986. Self-cleavage of plus and minus RNA transcripts of avocado sunblotch viroid. *Nucleic Acids Res.* **14**:3627–3640.

Kos, A., R. Dijkema, P. H. van der Meide, and H. Schelleken. 1986. The hepatitis delta virus possesses a circular RNA. *Nature* (London) **322**:558–560.

Krug, R. M., F. V. Alonso-Caplen, I. Julkunen, and M. G. Katze. Expression and replication of the influenza virus genome. *In The Viruses*, in press.

Kuo, M. Y.-P., M. Chao, and J. Taylor. 1989. Initiation of replication of the human hepatitis delta virus genome from cloned DNA: role of delta antigen. *J. Virol.* **63**:1945–1950.

Kuo, M. Y.-P., J. Goldberg, L. Coates, W. Mason, J. Gerin, and J. Taylor. 1988a. Molecular cloning of hepatitis delta virus RNA from an infected woodchuck liver: sequence, structure, and applications. *J. Virol.* **62**:1855–1861.

Kuo, M. Y.-P., L. Sharmeen, G. Dinter-Gottlieb, and J. Taylor. 1988b. Characterization of self-cleaving RNA sequences on the genome and antigenome of human hepatitis delta virus. *J. Virol.* **62**:4439–4444.

Lambowitz, A. 1989. Infectious introns. *Cell* **56**:323–326.

Makino, S., M.-F. Chang, C.-K. Shieh, T. Kamahora, D. V. Vannier, S. Govindarajan, and M. M. C. Lai. 1987. Molecular cloning and sequencing of a human hepatitis delta virus RNA. *Nature* (London) **329**:343–346.

Mason, W. S., J. M. Taylor, and R. Hull. 1987. Retroid virus genome replication. *Adv. Virus Res.* **32**:35–96.

Oldstone, M. B., and A. L. Notkins. 1986. Molecular mimicry, p. 195–202. *In* A. L. Notkins and M. B. A. Oldstone (ed.), *Concepts in Viral Pathogenesis II.* Springer-Verlag, New York.

Ponzetto, A., P. J. Cote, H. Popper, B. H. Boyer, W. T. London, E. C. Ford, F. Bonino, R. H. Purcell, and J. L. Gerin. 1984. Transmission of the hepatitis B-associated delta agent to the eastern woodchuck. *Proc. Natl. Acad. Sci. USA* **81**:2208–2212.

Prody, G. A., J. T. Bakos, J. M. Buzayun, I. R. Schneider, and G. Bruening. 1986. Autolytic processing of dimeric plant virus satellite RNA. *Science* **231**:1577–1580.

Rackwitz, H.-R., W. Rohde, and H. L. Sanger. 1981. DNA-dependent RNA polymerase II of plant origin transcribes viroid RNA into full-length copies. *Nature* (London) **291**:297–301.

Rizzetto, M., M. G. Canese, J. Arico', O. Crivelli, F. Bonino, C. G. Trepo, and G. Verme. 1977. Immunofluorescence detection of a new antigen-antibody system (delta-antidelta) associated to the hepatitis B virus in the liver and in the serum of HBsAg carriers. *Gut* **18**:997–1003.

Rizzetto, M., B. Hoyer, M. G. Canese, J. W.-K. Shih, R. H. Purcell, and J. L. Gerin. 1980. Delta agent: association of delta antigen with hepatitis B surface antigen and RNA in serum of delta-infected chimpanzees. *Proc. Natl. Acad. Sci. USA* **77**:6124–6128.

Semacnik, J., and K. L. Harper. 1984. Optimum conditions for cell-free synthesis of citrus exorcitis viroid and the question of specificity of RNA polymerase activity. *Proc. Natl. Acad. Sci. USA* **81**:4429–4433.

Sharmeen, L., M. Y.-P. Kuo, G. Dinter-Gottlieb, and J. Taylor. 1988. The antigenomic RNA of human hepatitis delta virus can undergo self-cleavage. *J. Virol.* **62**:2674–2679.

Sharmeen, L., M. Y.-P. Kuo, and J. Taylor. 1989. Self-ligating sequences on the antigenome of human hepatitis delta virus. *J. Virol.* **63**:1945–1950.

Symons, R. 1989. Pathogenesis by antisense. *Nature* (London) **338**:542–543.

Taylor, J., M. Kuo, P.-J. Chen, G. Kalpana, J. Goldberg, C. Aldrich, L. Coates, W. Mason, J. Summers, J. Gerin, B. Baroudy, and E. Gowans. 1987a. Replication of hepatitis delta virus. *UCLA Symp. Cell. Mol. Biol. New Ser.* **70:**541–548.

Taylor, J., W. Mason, J. Summers, J. Goldberg, C. Aldrich, L. Coates, J. Gerin, and E. Gowans. 1987b. Replication of human hepatitis delta virus in primary cultures of woodchuck hepatocytes. *J. Virol.* **61:** 2891–2895.

Taylor, J., L. Sharmeen, M. Kuo, and G. Dinter-Gottlieb. 1988. The self-cleaving RNAs of human hepatitis delta virus. *UCLA Symp. Mol. Cell. Biol.* **94:**99–108.

Taylor, J. M. Human hepatitis delta virus: structure and replication of the genome. *Curr. Top. Microbiol. Immunol.*, in press.

Walter, P., and G. Blobel. 1982. Signal recognition particle contains a 7S RNA essential for protein translocation across the endoplasmic reticulum. *Nature* (London) **288:**691–698.

Wang, K.-S., Q.-L. Choo, A. J. Weiner, H.-J. Ou, R. C. Najarian, R. M. Thayer, G. T. Mullenbach, K. J. Denniston, J. L. Gerin, and M. Houghton. 1986. Structure, sequence and expression of the hepatitis delta viral genome. *Nature* (London) **323:** 508–513.

Weiner, A. J., Q.-L. Choo, K.-S. Wang, S. Govindarajan, A. G. Redeker, J. L. Gerin, and M. Houghton. 1988. A single antigenic open reading frame of the hepatitis delta virus encodes the epitope(s) of both hepatitis delta antigen polypeptides p24$^\delta$ and p27$^\delta$. *J. Virol.* **62:**594–599.

Wu, H.-N., Y.-J. Lin, F.-P. Lin, S. Makino, M.-F. Chang, and M. M. C. Lai. 1989. Human hepatitis δ virus RNA subfragments contain an autocleavage activity. *Proc. Natl. Acad. Sci. USA* **86:**1831–1835.

Chapter 4

Insertion of Ubiquitin-Coding Sequence Identified in the RNA Genome of a Togavirus

Gregor Meyers, Tillmann Rümenapf, and Heinz-Jürgen Thiel

Pestiviruses, currently classified as togaviruses, are positive-stranded RNA viruses that represent causative agents of severe animal epidemics, including bovine viral diarrhea and hog cholera (Westaway et al., 1985; Darbyshire, 1960; Baker, 1987). Bovine viral diarrhea virus (BVDV) is recognized as having worldwide distribution in cattle (Baker, 1987). Its genome consists of a 12.5-kilobase (kb) RNA molecule with positive polarity which is considered to be translated into a polyprotein, giving rise to mature viral proteins after subsequent processing (Renard et al., 1987; Collett et al., 1988b). Propagation of BVDV in tissue culture allows discrimination between cytopathic (cp-BVDV) and noncytopathic (noncp-BVDV) strains, both of which can be pathogenic in cattle (Baker, 1987). The most severe disease resulting from BVDV infections is mucosal disease (MD), with mortality rates approaching 100%. MD develops sporadically in cattle between the ages of 6 months and 2 years (Baker, 1987). A prerequisite for MD is a transplacental infection with noncp-BVDV during the first 3 months of gestation, resulting in birth of persistently infected animals immunotolerant to the respective BVDV strain (Baker, 1987). Remarkably, development of MD always coincides with the appearance of cp-BVDV in these animals from which noncp-BVDV can still be isolated (Baker, 1987; Corapi et al., 1988; Pocock et al., 1987). Several groups have explained these findings by superinfection of the animal with cp-BVDV, whereas others suggested a mutation of the existing noncp-BVDV (Baker, 1987). Noncp-BVDV and cp-BVDV isolated from the same animal suffering from MD show a high degree of serological relationship (Corapi et al., 1988; Pocock et al.,

1987), whereas antigenic variation among field isolates of BVDV is commonly observed (Baker, 1987; Corapi et al., 1988). This finding favors the idea that a cytopathic mutant of the transplacentally transmitted noncytopathic virus strain is responsible for development of MD (Corapi et al., 1988).

One remarkable difference between cp-BVDV and noncp-BVDV concerns the processing of a nonstructural 120-kilodalton (kDa) virus-encoded protein of unknown function to a product of about 80 kDa. This cleavage can be detected only in tissue culture cells infected with cp-BVDV, whereas the 120-kDa protein is not processed after infection with noncp-BVDV (Donis and Dubovi, 1987; Purchio et al., 1984). Comparison of the genomic sequences of two cp-BVDV strains (Osloss [Renard et al., 1987] and NADL [Collett et al., 1988b], respectively) and one hog cholera virus strain (Alfort [Meyers et al., 1989]) revealed insertions within the BVDV gene coding for the above-mentioned 120-kDa protein (Meyers et al., 1989; Collett et al., 1989). The Osloss insertion of 228 nucleotides and the NADL insertion of 270 nucleotides are located in the same genomic region (Fig. 1A). The two inserted sequences exhibit no nucleotide or deduced amino acid sequence homology to each other. Whereas similarity between the NADL insertion and known sequences could not be found, a data bank search revealed homology of 97% between the amino acid sequence deduced from the Osloss insertion and animal ubiquitin (Fig. 1B). In fact, the 228 nucleotides identified in the RNA genome of the BVDV Osloss strain code for a complete ubiquitin protein of 76 amino acids. In comparison with the ubiquitin sequence, conserved in all

---

Gregor Meyers, Tillmann Rümenapf, and Heinz-Jürgen Thiel, Federal Research Centre for Virus Diseases of Animals, D-7400 Tübingen, Federal Republic of Germany.

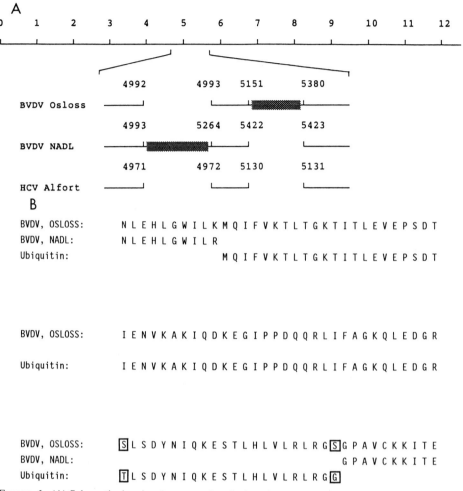

FIGURE 1. (A) Schematic drawing demonstrating the location of insertions identified in the genomes of two BVDV strains with respect to the hog cholera virus genome. A pestivirus genome is indicated in the upper line. Below, the regions containing insertions (dotted bars) are enlarged. Numbers refer to the published sequences (Renard et al., 1987; Collett et al., 1988b; Meyers et al., 1989) and indicate the nucleotides flanking the insertions or the respective positions in the sequences without insertions. (B) Amino acid sequence comparison of parts of the polyproteins encoded by the genomes of BVDV strains Osloss (Renard et al., 1987) and NADL (Collett et al., 1988b) and the animal ubiquitin (Rechsteiner, 1987). The amino acids differing between animal ubiquitin and the protein encoded by the Osloss insertion are boxed.

animals examined so far (Rechsteiner, 1987), two amino acid exchanges were detected in the Osloss ubiquitin (Fig. 1B).

To examine whether the ubiquitin-coding sequence is specific for the Osloss strain, a *Bgl*II fragment derived from a porcine polyubiquitin cDNA clone (pCL208 [Einspanier et al., 1987]) was hybridized to total RNA of Madin-Darby bovine kidney (MDBK) cells infected with five different strains of cp-BVDV. The ubiquitin probe recognized only genomic RNA from the Osloss strain (Fig. 2A, lane 1), whereas viral RNA of all strains was clearly detectable with a pestivirus-specific probe (Meyers et al., 1989) (Fig. 2B). Hybridization of the ubiquitin probe to poly(A)$^+$ RNA of noninfected MDBK cells indicated that the three bands visible in all lanes of Fig. 2A represent the bovine ubiquitin mRNAs (data not shown). Ubiquitin mRNA species varying in number and size have been

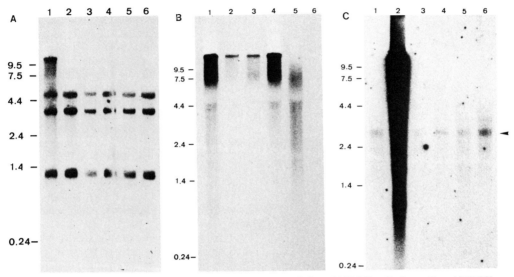

FIGURE 2. Northern blot analyses of total RNA from noninfected MDBK cells (lanes 6) and MDBK cells infected with BVDV strains Osloss (lanes 1), NADL (lanes 2), Oregon (lanes 3), Danmark (lanes 4), and Singer (lanes 5). The blots were hybridized with a 0.2-kb *Bgl*II fragment isolated from the porcine polyubiquitin cDNA clone pCL208 (Einspanier et al., 1987) (A), a 2.3-kb *Sal*I fragment derived from the hog cholera virus cDNA clone 4.2 (Meyers et al., in press) (B), or two oligonucleotides complementary to nucleotides 4994 to 5093 and 5094 to 5182 of the BVDV NADL genome (Collett et al., 1988b) (C). A 5-μg (A and B) or 10-μg (C) amount of glyoxylated RNA was separated in phosphate-buffered 1% agarose gels containing 5.5% formaldehyde and transferred to Duralon membranes (Stratagene) (Rümenapf et al., 1989). Hybridization with probes labeled with ^{32}P by nick translation (nick translation kit; Amersham Corp.) (A and B) or polynucleotide kinase (New England BioLabs, Inc.) (C) was performed in 0.5 M sodium phosphate (pH 6.8)–1 mM EDTA–7% SDS at 68°C (A) or 54°C (B and C). Posthybridization washes were carried out with 40 mM sodium phosphate (pH 6.8)–1 mM EDTA–5% SDS and 40 mM sodium phosphate (pH 6.8)–1 mM EDTA–1% SDS two times for 30 min each at hybridization temperature.

described for other mammalian species (Rechsteiner, 1987; Einspanier et al., 1987; Wiborg et al., 1985).

To find out whether the insertion in BVDV strain NADL is also homologous to cellular sequences, hybridization experiments with a mixture of two oligonucleotides complementary to this insertion were carried out. A cellular RNA of about 2.9 kb could be detected on the resulting Northern (RNA) blots (arrow in Fig. 2C). After longer exposure times, two additional bands were observed (data not shown). All three RNA species bind to oligo(dT)-cellulose (see below) and thus probably represent mRNAs. The hybridization experiments also showed that the NADL insertion is strain specific, at least with regard to the five strains tested here (Fig. 2C).

For further investigation, a cDNA library was constructed from poly(A)$^+$ RNA of noninfected MDBK cells and screened with the NADL insertion-specific oligonucleotide mixture; one positive clone (pcINS) containing an insert of about 1 kb hybridized to genomic RNA of BVDV NADL and to three species of poly-(A)$^+$ RNA of noninfected MDBK cells (Fig. 3). Nucleotide sequencing revealed that the respective cDNA fragment contains a sequence highly homologous to the insertion identified in the genome of BVDV strain NADL (Fig. 4). The complete sequence spanning the insertion is conserved except for two nucleotide exchanges, whereas the flanking regions show almost no homology (Fig. 4).

These data show that both the Osloss and NADL strains of BVDV have integrated cellular sequences into their genomes by recombination. In the Osloss strain, this sequence represents exactly the coding region of one ubiquitin gene. Togaviruses are considered to replicate in the cytoplasm, and reverse transcription has never been reported for these viruses. Thus, recombi-

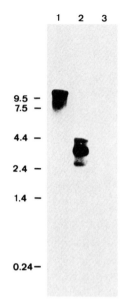

1   2   3

9.5 —
7.5 —

4.4 —

2.4 —

1.4 —

0.24 —

FIGURE 3. Northern hybridization analysis of 0.5 μg of total RNA of MDBK cells infected with BVDV NADL (lane 1) and 10 μg of poly(A)$^+$ (lane 2) or 20 μg of poly(A)$^-$ (lane 3) RNA of noninfected MDBK cells, using the insert of pcINS as a probe. RNA electrophoresis, transfer, and hybridization were done as described for Fig. 2A.

nation most likely happens at the RNA level. Because there is no homology between the viral sequences surrounding the two insertions, a recombination mechanism based on obvious signal sequences can be excluded. Since the viral sequences flanking the insertions exhibit no significant similarity to the corresponding cellular RNAs, a mechanism like that proposed for recombinations between homologous viral RNAs (King et al., 1987) appears unlikely. Nonhomologous recombinations have also been reported for other RNA viruses, but the mechanism is still speculative (King et al., 1987). It might be that the recombination process proposed for BVDV in this chapter depends on sequence and structural elements involved in pestivirus genome replication.

The two insertions are located in close proximity in the genomic region coding for the 120-kDa protein (Fig. 1A). The 80-kDa products resulting from the above-mentioned cp-BVDV-specific processing of this protein exhibit similar migration rates upon sodium dodecyl sulfate (SDS)-polyacrylamide gel electrophoresis for all BVDV strains examined so far, indicating similar positions at which cleavage occurs (Donis and Dubovi, 1987; Purchio et al., 1984). For the

NADL strain, this position was vaguely mapped around amino acid 1750 (Collett et al., 1988a), which is close to the carboxy-terminal serine residue of the Osloss ubiquitin (position 1662 [corrected; Collett et al., 1989]). It is known that monomeric ubiquitin is released from oligomeric forms as well as from fusion products by specific cleavage at the carboxy-terminal amino acid (Rechsteiner, 1987; Butt et al., 1988). Thus, for the Osloss strain, processing of the 120-kDa protein might be dependent on the insertion. The carboxy-terminal end of the NADL insertion is located at position 1626, which is also close to the suggested processing site. It is therefore possible that cleavage of the 120-kDa protein from both BVDV strains is influenced by the insertions.

The expression of mutated cellular sequences within the viral polyprotein could be responsible for the cytopathic effect of cp-BVDV. For the Osloss strain of BVDV, it has to be considered that the ubiquitinlike sequence encoded by the insertion exhibits two amino acid exchanges with regard to the conserved animal ubiquitin. Yeast cells overexpressing an in vitro-mutagenized ubiquitin with the carboxy-terminal glycine replaced by alanine stopped to divide (Butt et al., 1988). At this position, the Osloss ubiquitin contains a serine instead of the glycine residue. Uncontrolled viral expression of such a mutated protein might have deleterious effects on the host cell.

At this point, we suggest a novel model for pathogenesis of MD. Accordingly, the recombination event described above represents the mutation that changes a given noncp-BVDV strain into a cytopathic strain. This hypothesis explains the isolation of serologically closely related noncp-BVDV and cp-BVDV from one animal that has contracted MD. In contrast to point mutations, which generally are considered to be frequent in RNA viruses (Strauss and Strauss, 1988; Steinhauer and Holland, 1987), the process of RNA recombination should be a rare event and thus could explain why MD develops only sporadically after long periods of persistent infections characterized by a high degree of viremia. Once this recombination has occurred in one animal, horizontal transmission of the resulting cp-BVDV strain to other persistently infected animals harboring antigenically similar noncp-BVDV follows. This process explains the spread of MD within one herd. The hypothesis presented here accounts for all typical features for development of MD, which has been a mystery since its first description about 40 years ago.

```
BVDV:  ACTTTGAGGGT | ATGTGCAGCCGATGCCAGGGAAAGCATAGGAGGTTTGAAATGGACCGGGAAC
cINS:  TGAATACGATG | ATGTGCAGCCGATGCCAGGGAAAGCATAGGAGGTTTGAAATGGACCGGGAAC


       CTAAGAGTGCCAGATACTGTGCTGAGTGTAATAGGCTGCATCCTGCTGAGGAAGGTGACTTTTG
       CTAAGAGTGCCAGATACTGTGCTGAGTGTAATAGGCTGCATCCTGCTGAGGAAGGTGACTTTTG


                   *
       GGCAGAGTCGAGCATGTTGGGCCTCAAAATCACCTACTTTGCGCTGATGGATGGAAAGGTGTAT
       GGCAGAGTCAAGCATGTTGGGCCTCAAAATCACCTACTTTGCGCTGATGGATGGAAAGGTGTAT
                   *


       GATATCACAGAGTGGGCTGGATGCCAGCGTGTGGGAATCTCCCCAGATACCCACAGAGTCCCTT
       GATATCACAGAGTGGGCTGGATGCCAGCGTGTGGGAATCTCCCCAGATACCCACAGAGTCCCTT


       *
       GTCACATCTCATTTGGTTCACGGATG | CCTTTCAGGCAGGAA
       ATCACATCTCATTTGGTTCACGGATG | CCAGGCACCAGTGGG
       *
```

FIGURE 4. Comparison of the nucleotide sequences of the region of the BVDV strain NADL genome (Collett et al., 1988b) containing the insertion (upper line) and of part of a cDNA clone isolated from an MDBK poly(A)$^+$ cDNA library (cINS; lower line). The region identified as an insertion in the viral genome and the corresponding part of the cellular sequence are boxed. Nucleotide exchanges in this region are indicated by asterisks. cDNA synthesis (Rümenapf et al., 1989), cloning in lambda ZAPII bacteriophages (Stratagene) (Rümenapf et al., 1989), screening of the library with oligonucleotides (Meyers et al., 1989) (see legend to Fig. 2), and nucleotide sequencing (Meyers et al., 1989) were performed as described before.

It has been speculated that acquisition of new properties from the host cell by recombination represents an important force in RNA virus evolution (Strauss and Strauss, 1988). The recombination events proposed in this chapter actually resulted in formation of altered viruses by integration of host cellular sequences into viral genomes. These large-scale evolutionary jumps might be responsible for newly acquired properties of the mutated viruses such as development of novel features in tissue culture and of a lethal disease in the natural host. Because of the characteristics of persistent BVDV infection (Baker, 1987), rare processes like this kind of RNA recombination are probably easier to detect than in other RNA virus systems. However, such events may also play a role in the evolution of other RNA viruses.

ACKNOWLEDGMENTS. We thank K. H. Scheit, Max-Planck-Institut für Biophysikalische Chemie, Göttingen, Federal Republic of Germany, for providing the ubiquitin probe and C. Rein and P. Ulrich for excellent technical assistance.

REFERENCES

Baker, J. C. 1987. Bovine viral diarrhea virus: a review. *J. Am. Vet. Med. Assoc.* **190**:1449–1458.

Butt, T. R., M. I. Khan, J. Marsh, D. J. Ecker, and S. T. Crooke. 1988. Ubiquitin-metallothionein fusion protein expression in yeast. *J. Biol. Chem.* **263**: 16364–16371.

Collett, M. S., R. Larson, S. K. Belzer, and E. Retzel. 1988a. Proteins encoded by bovine viral diarrhea virus: the genomic organization of a pestivirus. *Virology* **165**:200–208.

Collett, M. S., R. Larson, C. Gold, D. Strinck, D. K. Anderson, and A. F. Purchio. 1988b. Molecular cloning and nucleotide sequence of the pestivirus bovine viral diarrhea virus. *Virology* **165**:191–199.

Collett, M. S., V. Moennig, and M. C. Horzinek. 1989. Recent advances in pestivirus research. *J. Gen. Virol.* **70**:253–266.

Corapi, W. V., R. C. Donis, and E. J. Dubovi. 1988. Monoclonal antibody analyses of cytopathic and noncytopathic viruses from fatal bovine viral diarrhea virus infections. *J. Virol.* **62**:2823–2827.

Darbyshire, J. H. 1960. A serological relationship

between swine fever and mucosal disease of cattle. *Vet. Rec.* **72:**331–333.

Donis, R. O., and E. J. Dubovi. 1987. Molecular specificity of the antibody responses of cattle naturally and experimentally infected with cytopathic and noncytopathic bovine diarrhea virus biotypes. *Am. J. Vet. Res.* **48:**1549–1554.

Einspanier, R., H. S. Sharma, and K. H. Scheit. 1987. An mRNA encoding polyubiquitin in porcine corpus luteum: identification by cDNA cloning and sequencing. *DNA* **6:**395–400.

King, A. M. Q., S. A. Ortlepp, J. W. I. Newman, and D. McCahon. 1987. Genetic recombination in RNA viruses, p. 129–152. *In* D. J. Rowlands, M. A. Mayo, and B. W. J. Mahy (ed.), *The Molecular Biology of the Positive Strand RNA Viruses*. Academic Press, Inc. (London), Ltd., London.

Meyers, G., T. Rümenapf, and H.-J. Thiel. 1989. Molecular cloning and nucleotide sequence of the genome of hog cholera virus. *Virology* **171:**555–567.

Pocock, D. H., C. J. Howard, M. C. Clarke, and J. Brownlie. 1987. Variation in the intracellular polypeptide profiles from different isolates of bovine virus diarrhoea virus. *Arch. Virol.* **94:**43–53.

Purchio, A. F., R. Larson, and M. S. Collett. 1984.

Characterization of bovine viral diarrhea viral proteins. *J. Virol.* **50:**666–669.

Rechsteiner, M. 1987. Ubiquitin-mediated pathways for intracellular proteolysis. *Rev. Cell Biol.* **3:**1–30.

Renard, A., D. Dino, and J. Martial. 1987. J. European Patent Application number 86870095.6. Publication number 0208672.

Rümenapf, T., G. Meyers, R. Stark, and H.-J. Thiel. 1989. Hog cholera virus—characterization of specific antiserum and identification of cDNA clones. *Virology* **171:**18–27.

Steinhauer, D. A., and J. J. Holland. 1987. Rapid evolution of RNA viruses. *Annu. Rev. Microbiol.* **41:**409–433.

Strauss, J. H., and E. G. Strauss. 1988. Evolution of RNA viruses. *Annu. Rev. Microbiol.* **42:**657–683.

Westaway, E. G., M. A. Brinton, S. Y. A. Gaidamovich, M. C. Horzineck, A. Igarashi, L. Kääriäinen, D. K. Lvov, J. S. Porterfield, P. K. Russel, and D. W. Trent. 1985. Togaviridae. *Intervirology* **24:**125–139.

Wiborg, O., M. S. Pedersen, A. Wind, L. E. Berglund, K. A. Marcker, and J. Vuusi. 1985. The human ubiquitin multigene family: some genes contain multiple directly repeated ubiquitin coding sequences. *EMBO J.* **4:**755–759.

Chapter 5

Organization of the Beet Yellows Closterovirus Genome

Valerian V. Dolja, Alexander V. Karasev, and Alexei A. Agranovsky

Closteroviruses may be the last well-known group of plant riboviruses whose molecular biology is little studied. This group includes viruses with a genome size of up to 20 kilobase pairs (kb), which is relatively large among single-stranded RNA genomes of plant viruses. The group itself seems rather diverse. It combines traditionally flexuous filamentous viruses differing substantially in particle length and mode of transmission. According to the accepted classification (Bar Joseph and Murant, 1982), closteroviruses are subdivided into subgroups A, B, and C. The best-studied representatives of the subgroups contain single positive-sense RNAs of about 7.5 kb for apple chlorotic leaf spot virus (ACLSV; subgroup A), 14.5 kb for beet yellows virus (BYV; subgroup B), and 20 kb for citrus tristeza virus (subgroup C). There is little doubt that these sharp differences in genome size reflect differences in genome structure in the subgroups. In fact, even preliminary papers on the molecular organization of the closterovirus genome clearly support this suggestion. For example, it has shown that ACLSV RNA is polyadenylated (Yoshikawa and Takahashi, 1988), whereas BYV RNA is devoid of a poly(A) tail (Karasev et al., 1989). Taken together, these data give grounds for believing that ACLSV and BYV belong to two different groups of plant viruses.

The subject of our work, BYV, is considered a type member of the closterovirus group. The virus is transmitted semipersistently by aphids, has a wide host range, and causes an economically important disease in sugar beets. BYV particles contain a single type of protein and RNA molecule. Our estimates of their sizes, 24 kilodaltons (kDa) for coat protein and 14.5 kb for virion RNA, coincide well with previous data (Bar Joseph and Hull, 1974; Carpenter et al., 1977). This chapter summarizes our recent findings on the structure and mode of expression of the BYV genome.

IN VITRO TRANSLATION OF BYV VIRION RNA

In vitro translation studies showed that BYV RNA was a rather inefficient messenger; its translation resulted only in a two- to fourfold stimulation of labeled amino acid incorporation. Thus, in our hands, the messenger activity of BYV RNA was at least 10 times less than that of potato virus X (PVX) or tobacco mosaic virus RNA. In these experiments, we used two cell-free systems, rabbit reticulocyte lysate and Krebs-2 cell extract. Both systems showed the same level of stimulation by BYV RNA. The analysis of resulting products revealed a set of polypeptides of up to 250 kDa in Krebs-2 cell extracts and up to 230 kDa in reticulocyte lysates. The Krebs-2 system was chosen for further experiments.

To test the translation of the whole spectrum of probable BYV-specific messengers present in the purified virus preparation, total virion RNA was fractioned in sucrose density gradients. Translation of the genomic-size RNA yielded a large major product, p250, as well as some smaller polypeptides (Fig. 1). Among the latter, the p65 band seemed the most prominent, discrete, and reproducible. In some gradients, the maximum translation of p65 was in BYV RNA fractions smaller than that for p250. It cannot be excluded that p65 is encoded by a specific subgenomic BYV messenger. On the other hand, other RNA gradient fractions directed no evident virus-specific translation. Time course experiments showed that p250 ap-

Valerian V. Dolja, Alexander V. Karasev, and Alexei A. Agranovsky, A. N. Belozersky Laboratory at Moscow State University, Moscow 119899, USSR.

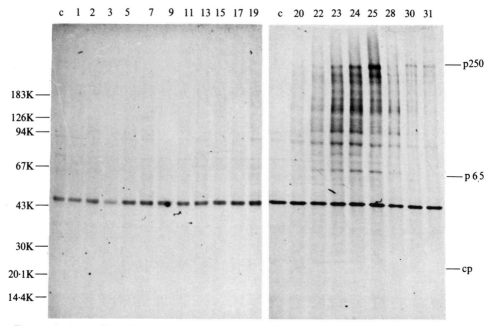

FIGURE 1. Autoradiograph of the in vitro translation products directed by separate sucrose density gradient fractions of BYV RNA. Each numbered gel lane corresponds to a gradient fraction. In lane c (control), no RNA was added. Molecular size markers used (from top to bottom) were two major TMV RNA translation products (183 and 126 kDa [K]), phosphorylase *b* (94 kDa), bovine serum albumin (67 kDa), egg white ovalbumin (43 kDa), bovine carbonic anhydrase (30 kDa), soybean trypsin inhibitor (20.1 kDa), and bovine milk alpha-lactalbumin (14.4 kDa). The positions of the viral p250, p65, and coat proteins are indicated. (Adapted from Karasev et al. [1989].)

peared after 2 h of incubation, with no appreciable changes in the ratio of p250 to smaller products observed for up to 4 h of incubation. These results clearly indicated that p250 synthesis was directed by the full-length BYV RNA. Most of the other translation products could arise from premature termination of p250 translation. Obviously, the data obtained are not sufficient to exclude the possibility of in vivo processing of p250.

It should be noted that none of BYV RNA in vitro translation products coincided with the coat protein in electrophoretic mobility. Moreover, no in vitro products were found to be immunoprecipitable with anti-BYV immunoglobulins.

In the other experiment, we checked the influence of a cap analog, m7Gpp, on in vitro translation of BYV RNA (Fig. 2). It was found that m7Gpp blocked the synthesis of all the BYV-specific polypeptides (Fig. 2, lanes 2 to 4). In controls, the cap analog completely inhibited translation of PVX RNA bearing the 5' cap structure (Fig. 2, lanes 5 to 7) and had no effect on translation of TMV intermediate-length sub-

genomic RNA I2, which is not capped (Fig. 2, lanes 8 and 9). These data indicate that the BYV genome is capped.

THE 3' REGION OF THE BYV GENOME

Affinity chromatography on oligo(dT)-cellulose revealed that about 98% of BYV RNA was not bound to immobilized oligo(dT). The same level of nonspecific adsorption (about 2%) was obtained for TMV RNA. At the same time, binding of the polyadenylated PVX RNA amounted to 90%. The fact that the addition of an oligo(dT) primer did not significantly stimulate the efficiency of reverse transcription of BYV RNA in vitro was also consistent with the absence of either terminal or internal poly(A) sequences in BYV genome.

The heteropolymeric nature of the 3' end of BYV RNA was confirmed by direct sequencing experiments. The virion RNA was labeled with 5'-[32P]pCp and RNA ligase. After complete digestion with RNase T2 followed by two-dimensional thin-layer chromatography, it was demon-

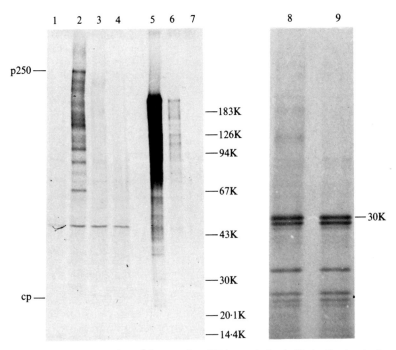

FIGURE 2. Effect of the cap analog m^{7}Gpp on virus RNA translation in Krebs-2 extracts. Lanes: 1, no added RNA; 2 to 4, BYV RNA translated in extract containing no (lane 2), 1 mM (lane 3), or 2 mM (lane 4) m^{7}Gpp; 5 to 7, PVX RNA translated with no (lane 5), 1 mM (lane 6), or 2 mM (lane 7) m^{7}Gpp; 8 and 9, TMV RNA I2 with no (lane 8) or 2 mM (lane 9) m^{7}Gpp. The volumes of translation mixture loaded on the gel were 8 and 2.5 μl for lanes 1 to 4 and 5 to 9, respectively. The molecular size markers were as for Fig. 1. (From Karasev et al. [1989].)

strated that the BYV genome ends in cytosine. Furthermore, we obtained 3'-labeled RNA fragments of about 100 nucleotides by partial RNase T$_1$ hydrolysis. These fragments were sequenced by the method of Peattie (1979). Unfortunately, the sequence could not be demonstrated here because some sequencing positions needed further verification. One could note, however, that the 3'-terminal region in BYV RNA was enriched in short oligo(A) and oligo(U) blocks and contained two putative stem-loop structures. It showed no obvious resemblance to any 3'-terminal sequence in other positive ribovirus genomes published to date.

BYV-SPECIFIC SUBGENOMIC RNAs

The first indication on possible expression of BYV genome via subgenomic RNAs was obtained upon analysis of double-stranded RNA (dsRNA) isolated from infected plant tissue. Besides the double-stranded counterpart of the full-length BYV RNA, at least four additional bands of shorter dsRNA were revealed (Fig. 3). The virus-specific nature of these dsRNAs was confirmed by Northern (RNA) hybridization with random-primed BYV-specific cDNA as a probe. In addition to the genomic and four subgenomic dsRNAs mentioned above, an additional short dsRNA was detected on Northern blots (Fig. 4). The rather dark picture in Fig. 4 was due to overexposure, which was the only way to make some bands visible. Nevertheless, all bands indicated were clearly visible on the original X-ray films. It is important that the counterparts for all of these five subgenomic dsRNAs were present in preparations of total single-stranded RNA (ssRNA) from BYV-infected leaves (Fig. 4, lanes DS and SS).

Some blots of virion BYV RNA displayed an additional band corresponding to subgenomic RNA-5 (data not shown). This result was obtained when the labeled cDNA clone representing a central part of the virus genome was used as a probe. It is tempting to speculate that the 6.3-kb subgenomic RNA-5 codes for p65 mentioned above.

ds RNAs
BYV TMV BSMV

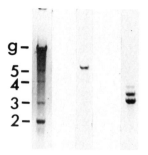

FIGURE 3. Analysis of viral dsRNAs in 1% agarose gel. The markers were TMV (6.4 kb) and barley stripe mosaic virus (BSMV; 4.2, 3.2, and 2.8 kb) dsRNAs. Designations at the left: g, genomic RNA; 2 to 5, subgenomic RNA-1 to -5. The gel was stained with ethidium bromide and photographed under a UV trans-illuminator.

Some additional, possibly artificial bands were visible in the ssRNA but not in the dsRNA pattern (Fig. 4, lanes SS and DS). These bands were also present in the control lane representing the artificial mixture of virion RNA and

ssRNA from noninfected plants (lane ni+V). At least one of these bands was visible in another control lane representing pure ssRNA from non-infected plants (lane n.i.). This band was even more prominent when Northern blots were hybridized to another labeled probe, a cDNA clone homologous to about 1 kb in the 5′ half of the BYV genome (data not shown). The cDNA clones not overlapping this particular clone did not hybridize with ssRNA from healthy plants. These data could indicate that the appearance of additional bands was due to some homology between the viral and plant RNAs (McClure and Perrault, 1985, 1986) in addition to some blotting technique artifacts (Dougherty, 1983).

Obviously, the 3′-coterminal nature of sub-genomic RNAs needs further confirmation. However, the overall impression is that expression of the BYV genome includes the formation of at least five subgenomic RNAs ranging in size from 1 to 6.3 kb.

To demonstrate the functional nature of the RNAs revealed, the three shortest dsRNAs were isolated from the agarose gel, followed by denaturation with methyl mercuric hydroxide. The RNAs thus prepared were translated in the rabbit reticulocyte lysate system; for unknown reasons, translation of the denatured RNAs in the Krebs-2 system was strongly inhibited. The shortest RNA codes for the 19-kDa polypeptide (Fig. 5, lane 1), the second codes for the 24-kDa

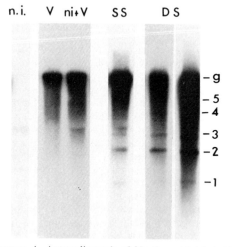

FIGURE 4. Autoradiograph of Northern blot hybridization of glyoxal-denatured RNA preparations. Lanes: n.i., total ssRNA from noninfected plants; V, BYV virion RNA; ni+V, artificial mixture of noninfected and virion RNA preparations; SS, total ssRNA from BYV-infected plants; DS, dsRNA from BYV-infected plants (exposed for 2 [left] and 12 [right] days). BYV-specific cDNA labeled with [α-^{32}P]CTP was used as a probe.

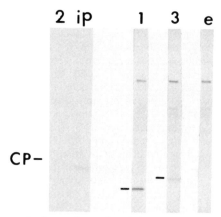

FIGURE 5. Autoradiograph of in vitro translation products of individual double-stranded forms of BYV subgenomic RNAs. Before translation, dsRNAs were denatured by methyl mercuric hydroxide. Lanes: 2, 1, and 3, translation products of BYV dsRNA-2, -1, and -3, respectively; ip, immunoprecipitation of BYV dsRNA-2 products with anti-BYV immunoglobulins; e, no RNA added. CP, Position of an unlabeled BYV coat protein marker.

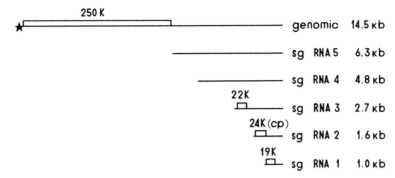

FIGURE 6. Tentative scheme of BYV genome structure and expression. The asterisk denotes the 5' cap structure. Open boxes are the in vitro translation products detected. sg, Subgenomic; K, kilodaltons.

polypeptide (lane 2), and the last codes for the 22-kDa polypeptide (lane 3). Preliminary experiments showed that the 24-kDa protein coded for by the major, 1.6-kb subgenomic RNA can be immunoprecipitated by antibodies to BYV, thus suggesting its relatedness to the BYV coat protein (lane ip).

CONCLUSIONS

The most important results of this work are summarized in Fig. 6. The BYV genome is presumably capped and ends in a heteropolymeric sequence at its 3' terminus. BYV virion RNA codes for a protein not smaller than 250 kDa. At least five subgenomic RNAs can be found in BYV-infected plants. All of these can be isolated in both single-stranded and double-stranded forms. The double-stranded forms of subgenomic RNA-1, -2, and -3 code in vitro for proteins of 19, 24 (presumably the coat protein), and 22 kDa, respectively. Preliminary data indicate that subgenomic RNA-5 codes for a 65-kDa protein.

In a search of positive-strand riboviruses for resemblance with the BYV genome, one could see some resemblance to coronaviruses. For example, expression of their 28- to 30-kb genomes also requires formation of five to seven subgenomic RNAs (Spaan et al., 1988). In addition, both the BYV and coronavirus genomic RNAs are translated in vitro into very large proteins. The obvious differences between BYV and coronaviruses are in genome size and the absence of a poly(A) tail in the BYV genome. Elucidation of the relationships of the BYV

genome with other ribovirus genomes requires sequencing, which is now in progress.

REFERENCES

Bar Joseph, M., and R. Hull. 1974. Purification and partial characterization of sugar beet yellows virus. *Virology* **62**:552–562.

Bar Joseph, M., and A. F. Murant. 1982. Closterovirus group. *CMI/AAB Descriptions of Plant Viruses*, no. 260.

Carpenter, J. M., B. Kassanis, and R. F. White. 1977. The protein and nucleic acid of beet yellows virus. *Virology* **77**:101–109.

Dougherty, W. G. 1983. Analysis of viral RNA isolated from tobacco leaf tissue infected with tobacco etch virus. *Virology* **131**:473–481.

Karasev, A. V., A. A. Agranovsky, V. V. Rogov, N. A. Miroshnichenko, V. V. Dolja, and J. G. Atabekov. 1989. Virion RNA of beet yellows closterovirus: cell-free translation and some properties. *J. Gen. Virol.* **70**:241–245.

McClure, M. A., and J. Perrault. 1985. Poliovirus genome RNA hybridizes specifically to higher eukaryotic rRNAs.

McClure, M. A., and J. Perrault. 1986. RNA virus genomes hybridize to cellular rRNAs and to each other. *J. Virol.* **57**:917–921.

Peattie, D. A. 1979. Direct chemical method for sequencing RNA. *Proc. Natl. Acad. Sci. USA* **76**:1760–1764.

Spaan, W., D. Cavanagh, and M. C. Horzinek. 1988. Coronaviruses: structure and genome expression. *J. Gen. Virol.* **69**:2939–2952.

Yoshikawa, N., and T. Takahashi. 1988. Properties of RNAs and proteins of apple stem grooving and apple chlorotic leaf spot viruses. *J. Gen. Virol.* **69**:241–245.

Chapter 6

Sequence Analysis of the Genome RNA of Lactate Dehydrogenase-Elevating Virus

E. K. Godeny, D. W. Speicher, and M. A. Brinton

Mainly on the basis of morphological criteria, lactate dehydrogenase-elevating virus (LDV) is currently listed as an unclassified togavirus (Westaway et al., 1985a). Recent sequence data have indicated that a number of the original members of the togavirus family differ significantly from the prototype alpha togaviruses. The flaviviruses have already been reclassified into a new family, *Flaviviridae* (Westaway et al., 1985b), and the pestiviruses may soon be included as members of this family (Collett et al., 1988). Equine arteritis virus is sufficiently unique (see chapter by W. J. M. Spaan et al., this volume) to justify its reclassification into a new family. So far, only rubella virus has been found to have a genome structure and replication strategy similar to those of the alpha togaviruses.

Too little is known about two additional togaviruses, LDV and simian hemorrhagic fever virus, to accurately classify them at this time. As a first step toward the molecular characterization of LDV, we have sequenced the 3' terminus of the viral genome RNA.

BIOLOGICAL CHARACTERISTICS OF LDV

LDV replicates only in mice, causing a persistent infection characterized by continuous virus production and a lifelong viremia (Brinton, 1986; Rowson and Mahy, 1985). Infection elicits antiviral antibody, but virus-immune complexes continue to infect a susceptible subpopulation of macrophages via Fc receptors. Infected mice display elevated levels of certain serum enzymes. The elevated level of serum lactate de-

hydrogenase in mice was responsible for the original fortuitous discovery of the virus by Riley et al. (1960) and is currently used as the endpoint for assay of virus infectivity.

Although LDV replicates to high titers in vivo, it is difficult to grow in vitro; LDV replicates in only 5 to 20% of the cells in primary murine cultures containing macrophages (Brinton, 1986; Rowson and Mahy, 1985). Established cell lines have not as yet been found to be permissive for LDV replication. Although most LDV isolates cause no illness in the mice they infect, one isolate, LDV-C, efficiently induces poliomyelitis in susceptible C58 and AKR mice (Martinez et al., 1980).

THE 3'-TERMINAL SEQUENCE OF THE GENOMIC RNA

The size of the LDV genome RNA has been estimated to be 14 kilobases (Brinton, 1986; Rowson and Mahy, 1985). By direct sequencing of end-labeled LDV genomic RNA, we have previously demonstrated the presence of a poly(A) tract of about 50 nucleotides at the 3' terminus of the LDV genome (Brinton et al., 1986b). cDNA synthesis from LDV genomic RNA was therefore primed with oligo(dT). The RNA template was then digested with RNase H, and the cDNA was made double stranded with DNA polymerase (Gubler and Hoffman, 1983). The resulting cDNAs were inserted into a pUC13 plasmid vector by using *Eco*RI linkers. *Escherichia coli* JM103 was transfected with the recombinant plasmids, and the resulting colonies were screened for inactivation of β-galactosidase gene expression (Close et al., 1983).

E. K. Godeny and M. A. Brinton, Department of Biology, Georgia State University, Atlanta, Georgia 30303. *D. W. Speicher,* Wistar Institute, 3601 Spruce Street, Philadelphia, Pennsylvania 19104.

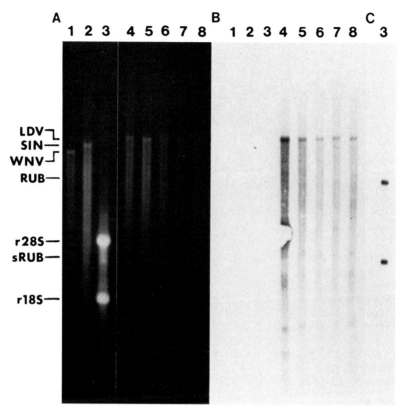

FIGURE 1. Northern blot analysis of various viral genomic RNAs and cellular RNAs. For each lane, approximately 0.5 to 2 µg of each of the sodium dodecyl sulfate-sucrose gradient-purified RNAs was electrophoresed in duplicate on a 1% agarose gel. The gel was stained with ethidium bromide (A), the electrophoresed RNA was transferred to Nytran paper and probed with radiolabeled cDNA synthesized from a 3'-terminal LDV-specific cloned DNA (B), and a crude extract of rubella virus RNA obtained from infected Vero cells was electrophoresed, transferred to Nytran, and probed with a radiolabeled cDNA synthesized from a rubella virus-specific cloned DNA (C). Lanes: 1, West Nile virus (WNV) genomic RNA; 2, Sindbis virus (SIN) genomic RNA; 3, Vero cellular rRNA and rubella virus genomic (RUB) and subgenomic (sRUB) mRNAs; 4, LDV-C genomic RNA; 5, LDV-2 genomic RNA; 6, LDV-3 genomic RNA; 7, LDV-4 genomic RNA; 8, LDV-5 genomic RNA.

The virus specificity of the cloned DNAs was determined by Southern blot analysis of miniprep plasmid DNA (Southern, 1975), using a radiolabeled cDNA probe reverse transcribed from purified LDV-C genomic RNA. Selected virus-specific, cloned DNAs were sequenced by the dideoxy-chain termination method (Sanger et al., 1977).

Clones that contained a terminal poly(A) region had overlapping identical sequences (Godeny et al., in press). Northern (RNA) blot analysis (Thomas, 1980) showed that these clones hybridized to the 14-kilobase LDV-C genome and to the genomes of four other isolates of LDV (Fig. 1), each having an RNA fingerprint pattern distinct from that of LDV-C (Brinton et al., 1986b). However, these clones did not hybridize with genomic RNA from an alpha togavirus (Sindbis virus), a flavivirus (West Nile virus), or rubella virus, nor did they hybridize with cellular RNA (Fig. 1). Comparison of the LDV 3' genomic sequence with those of flavivirus, alpha togavirus, and rubella virus genomes showed that none of the conserved sequences or secondary structures characteristic of the flaviviruses (Brinton et al., 1986a) or alpha togaviruses (Strauss and Strauss, 1986) were found within the LDV sequence (Godeny et al., 1989), and no homology to rubella virus (Frey et al., 1986) was found (Fig. 2).

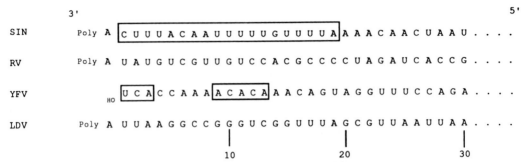

FIGURE 2. Comparison of the 3'-terminal genomic sequences of an alpha togavirus (Sindbis virus [SIN]), a togavirus (rubella virus [RV]), and a flavivirus (yellow fever virus [YFV]) with that of LDV-C. Boxes denote highly conserved 3' sequences characteristic of alpha togaviruses (Strauss and Strauss, 1986) and flaviviruses (Brinton et al., 1986a). (Reproduced with permission from Godeny et al. [1989].)

MAP POSITION OF THE LDV CAPSID PROTEIN, VP1

The virions of LDV are composed of at least three structural proteins. VP1 (14 kilodaltons) is the capsid protein, VP2 (18 kilodaltons) is a nonglycosylated envelope-associated protein, and VP3 (25 to 43 kilodaltons) is a glycosylated, envelope-associated protein that migrates heterogeneously on polyacrylamide gels. It is not yet known whether VP3 represents one or more proteins. The LDV structural proteins, VP1, VP2, and VP3, are present in virions in ratios of (3 to 5):1:1, respectively (Darnell and Plagemann, 1972; Michaelides and Schlesinger, 1973).

In vitro, LDV can be grown only in primary murine cell cultures containing macrophages. Since only a small subpopulation of cells within such cultures replicates LDV (Tong et al., 1977), it has not been possible to detect intracellular nonstructural viral proteins against the background of cell proteins radiolabeled in the presence of actinomycin D.

The LDV structural proteins were subjected to N-terminal amino acid sequencing. Purified virions obtained from mouse plasma were disrupted, and the viral structural proteins were separated on polyacrylamide gels and electrophoretically transferred to polyvinylidine difluoride-paper. Individual viral proteins were located by Coomassie blue staining and autoradiography. The viral capsid protein was then subjected to automated Edman degradation on an Applied Biosystems 470A gas-phase sequencer (Hewick et al., 1981). The following N-terminal sequence of 33 amino acids was obtained for VP1: N terminus-SQNKKKGGQN KGANQQLNQLISALLRNAGQNKG...

Analysis of open reading frames (ORFs) in an extended 3' sequence obtained from several overlapping clones revealed an ORF starting at nucleotide 432 and ending 80 nucleotides from the poly(A) tract (Godeny et al., in press). When this ORF was translated, it was found to contain the 33 VP1 N-terminal amino acids determined by amino acid sequencing at its 5' end. Since no other ORF was found between VP1 and the poly(A) tract, the virion capsid protein is therefore the 3'-terminal protein of the LDV genomic coding region. VP1 is a basic protein. The actual molecular size of VP1 (12.2 kilodaltons) is close to that (14 kilodaltons) predicted by its migration in polyacrylamide gels. The amino acid composition of VP1 is compared with those of a flavivirus, an alpha togavirus, and a rubella capsid protein in Table 1.

The next upstream ORF was found to encode an as yet unidentified protein of 170 amino acids. This ORF overlapped the VP1 ORF by 11 nucleotides and was in a different reading frame from the VP1 ORF.

TABLE 1
Compositions of Togavirus, Flavivirus, and LDV Capsid Proteins

Amino acids	% of total amino acids in capsid protein of:			
	Sindbis virus	Rubella virus	Yellow fever virus	LDV
Nonpolar	42	47	32	51
Positive charge	21	15	24	16
Negative charge	8	10	2	3
Uncharged	30	27	42	31

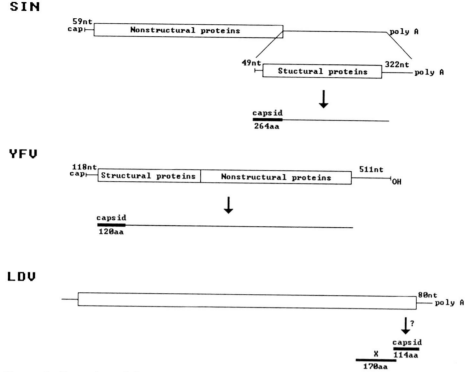

FIGURE 3. Comparison of the genome organization and mode of protein expression of togaviruses (Sindbis virus [SIN]), flaviviruses (yellow fever virus [YFV]), and LDV. Symbols: ——, untranslated regions of the genomic RNAs; ☐, translated regions. The Sindbis virus subgenomic mRNA is shown below the genomic RNA. The translated polyproteins are indicated by single lines, with the capsid protein in boldface. ? indicates that it is not yet known whether the LDV capsid protein is translated directly from LDV genomic RNA or from a subgenomic mRNA. X designates an overlapping ORF encoding an unidentified LDV protein. nt, Nucleotides; aa, amino acids.

CONCLUSIONS

The map position of the LDV capsid protein differs from that of both the flaviviruses and the togaviruses (Fig. 3). The picornaviruses are similar to the flaviviruses in that their structural genes are located at the 5′ end of the genome coding region. A common feature of these viruses is that all of their proteins are translated as polyproteins.

In addition to the recently defined picornavirus and alpha togavirus superfamilies of positive-strand RNA viruses (see chapter by R. Goldbach, this volume), Spaan et al. (this volume) have proposed a third superfamily consisting of the coronaviruses, toraviruses, and arteriviruses. The data so far obtained for LDV indicate that it may be similar to equine arteritis virus. Both viruses have a polyadenylated genome of approximately 14 kilobases, and the capsid protein is the 3′-terminal protein of the coding region. Furthermore, although no subgenomic mRNAs have yet been identified for LDV, the fact that VP1 and the next upstream protein (as yet unidentified) are translated from overlapping ORFs that are in different frames strongly suggests that some of the LDV proteins may be translated from separate subgenomic mRNAs. However, if LDV subgenomic mRNAs are produced, they are present in infected macrophage cultures in too low an abundance to be detectable by Northern (RNA) blot analysis. Completion of the LDV genomic sequence and analysis of its homology with the as yet unpublished equine arteritis virus sequence will be necessary before any final conclusions can be drawn about the classification of LDV.

REFERENCES

Brinton, M. A. 1986. Lactate dehydrogenase-elevating virus, p. 389–420. *In* P. N. Bhatt, R. O. Jacoby, H. C. Morse, and A. E. New (ed.), *Viral and*

Mycoplasma Infections of Laboratory Rodents. Academic Press, Inc., San Diego, Calif.

Brinton, M. A., A. V. Fernandez, and J. H. Dispoto. 1986a. The 3'-nucleotides of flavivirus genomic RNA form a conserved secondary structure. *Virology* **153**:113–121.

Brinton, M. A., E. I. Gavin, and A. V. Fernandez. 1986b. Genetic variation among strains of lactate dehydrogenase-elevating virus (LDV). *J. Gen. Virol.* **67**:2673–2684.

Close, T. J., J. L. Christmann, and R. L. Rodriguez. 1983. M13 bacteriophage and pUC plasmids containing DNA inserts but still capable of β-galactosidase α-complementation. *Gene* **23**:131–136.

Collett, M. S., D. K. Anderson, and E. J. Retzel. 1988. Comparisons of the pestivirus bovine viral diarrhoea virus with members of the Flaviviridae. *J. Gen. Virol.* **69**:2637–2643.

Darnell, M. B., and P. G. W. Plagemann. 1972. Physical properties of lactic dehydrogenase-elevating virus and its ribonucleic acid. *J. Virol.* **10**:1082–1085.

Frey, T. K., L. D. Marr, M. L. Hemphill, and G. Dominquez. 1986. Molecular cloning and sequencing of the region of the rubella virus genome coding for glycoprotein E1. *Virology* **154**:228–232.

Godeny, E. K., D. W. Speicher, and M. A. Brinton. Map location of lactate dehydrogenase-elevating virus (LDV) capsid protein (Vp1). *Virology*, in press.

Godeny, E. K., M. R. Werner, and M. A. Brinton. 1989. The 3' terminus of lactate dehydrogenase-elevating virus genome RNA does not contain togavirus or flavivirus conserved sequences. *Virology* **172**:647–650.

Gubler, U., and B. J. Hoffman. 1983. A simple and very efficient method for generating cDNA libraries. *Gene* **25**:263–269.

Hewick, R. M., M. W. Hunkapilar, L. E. Hood, and W. J. Dreyer. 1981. A gas-liquid solid phase peptide and protein sequenator. *J. Biol. Chem.* **256**:7990–7997.

Martinez, D., M. A. Brinton, T. G. Tachovsky, and A. H. Phelps. 1980. Identification of lactate dehydrogenase-elevating virus as the etiologic agent of genetically restricted age-dependent polioencephalomyelitis of mice. *Infect. Immun.* **27**:979–987.

Michaelides, M. C., and S. Schlesinger. 1973. Structural proteins of lactic dehydrogenase virus. *Virology* **55**:211–217.

Riley, V., F. Lilly, E. Huerto, and D. Bardell. 1960. Transmissible agent associated with 26 types of experimental mouse neoplasms. *Science* **132**:545–547.

Rowson, K. E. K., and B. W. J. Mahy. 1985. Lactate dehydrogenase-elevating virus. *J. Gen. Virol.* **66**:2297–2312.

Sanger, F., S. Nicklen, and A. R. Coulson. 1977. DNA sequencing with chain-terminating inhibitors. *Proc. Natl. Acad. Sci. USA* **74**:5463–5467.

Southern, E. 1975. Detection of specific sequences among DNA fragments separated by gel electrophoresis. *J. Mol. Biol.* **98**:503–517.

Strauss, E. G., and J. H. Strauss. 1986. Structure and replication of the alphavirus genome, p. 35–90. *In* S. Schlesinger and M. J. Schlesinger (ed.), *The Togaviridae and Flaviviridae.* Academic Press, Inc., San Diego, Calif.

Thomas, P. S. 1980. Hybridization of denatured RNA and small DNA fragments transferred to nitrocellulose. *Proc. Natl. Acad. Sci. USA* **77**:5201–5205.

Tong, S. L., J. Stueckemann, and P. G. W. Plagemann. 1977. An autoradiographic method for detection of lactate dehydrogenase-elevating virus (LDV)-infected cells in primary mouse macrophage cultures. *J. Virol.* **22**:219–227.

Westaway, E. G., M. A. Brinton, S. Y. Gaidamovich, M. C. Horzinek, A. Igarashi, L. Kääriäinen, D. K. Lvov, J. S. Porterfield, P. K. Russell, and D. W. Trent. 1985a. Togaviridae. *Intervirology* **24**:125–139.

Westaway, E. G., M. A. Brinton, S. Ya Gaidamovich, M. C. Horzinek, A. Igarashi, L. Kääriäinen, D. K. Lvov, J. S. Porterfield, P. K. Russell, and D. W. Trent. 1985b. Flaviviridae. *Intervirology* **24**:183–192.

Chapter 7

Characterization of Sequence Variation among Isolates of RNA Viruses: Detection of Mismatched Cytosine and Thymine in RNA-DNA Heteroduplexes by Chemical Cleavage

Peter J. Wright and R. G. H. Cotton

Genetic variability among isolates of RNA viruses is currently determined by one or more of several methods, such as determination of the mobility of genomic molecules upon electrophoresis (Palese and Schulman, 1976), partial denaturation of RNA-RNA or DNA-RNA heteroduplexes (Smith et al., 1986), RNase T₁ oligonucleotide fingerprinting (Trent et al., 1983), restriction enzyme cleavage of cDNA (Walker et al., 1988), cleavage by RNase A at mismatches in RNA-RNA heteroduplexes (Lopez-Galindez et al., 1988), and nucleotide sequencing of RNA or cDNA. In this chapter, we present a powerful new technique for detecting variation in viral RNA by chemical cleavage at mismatched C and T in DNA-RNA heteroduplexes, using hydroxylamine and osmium tetroxide, respectively (Lamande et al., 1989; Cotton and Wright, 1989). The technique is applicable to both positive- and negative-strand viruses and allows a specific region of the RNA to be targeted for analysis by selection of a suitable end-labeled cDNA probe. Potential uses for the technique include epidemiological surveys and diagnosis of new isolates. Because single-base changes are detected, it may be possible to use the method in defining the genetic basis of phenotypic changes in tropism, pathogenicity, and resistance to neutralizing monoclonal antibodies. Other potential uses are the locating of crossover points in recombinant viruses, the checking or proofreading of two known sequences against each other, and the screening of cloned cDNA against the viral RNA population from which it was derived in order to confirm that an unrepresentative molecule was not cloned.

Three isolates of dengue virus type 2 were used in our experiments. The New Guinea C strain (NGC) and PUO-218 isolate were selected because we possessed sequenced cDNA for these viruses (Gruenberg et al., 1988). D80-100 had not been sequenced. NGC is the prototype dengue virus type 2 strain isolated in 1944 by Sabin and Schlesinger (1945). It has been extensively passaged through suckling mice. PUO-218 and D80-100 were isolated during 1980 in Bangkok by D. S. Burke from patients with dengue fever and dengue hemorrhagic fever, respectively. Both have received very limited passaging in LLC-MK2 and C6/36 cells. These three strains reflect a situation that often occurs in virus laboratories, when strains have different passage histories and different places of isolation and are associated with different disease symptoms, and one requires some information on their genetic relatedness.

ANALYSIS OF MISMATCHES

The dengue viruses contain a single-stranded RNA genome of 11 kilobases and of positive sense. In the initial experiments reported here, we analyzed a small region of the genome (450 nucleotides) encoding the carboxy terminus of protein M and the amino terminus of protein E (Fig. 1). An end-labeled probe of negative-sense PUO-218 cDNA was prepared by

Peter J. Wright, Department of Microbiology, Monash University, Clayton, Victoria 3168, Australia.   *R. G. H. Cotton,* Olive Miller Protein Laboratory, Murdoch Institute, Royal Children's Hospital, Parkville, Melbourne, Victoria 3052, Australia.

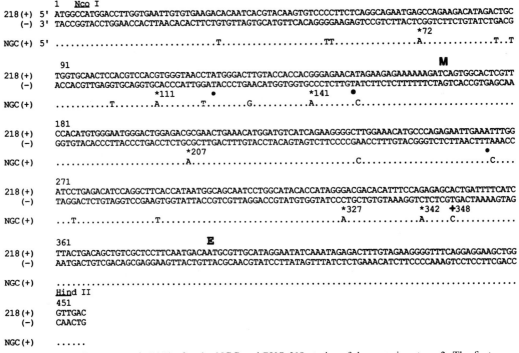

FIGURE 1. Sequences of cDNAs for the NGC and PUO-218 strains of dengue virus type 2. The first nucleotide corresponds to nucleotide 471 in the NGC sequence reported by Gruenberg et al. (1988). The start points of the coding regions for the virion proteins M and E are indicated. The positions of mismatched C (*) and T (+) in the PUO-218 negative-sense probe are designated. Additional T's susceptible to chemical cleavage are also marked (●).

[32]P labeling at the NcoI site. When this probe was hybridized with NGC nucleic acid of positive sense, mismatched C's in the probe were at positions 72, 111, 141, 207, 327, and 342; a mismatched T was at position 348. The probe was first hybridized with NGC cDNA, and by using the published procedure for analysis of DNA-DNA heteroduplexes (Cotton et al., 1988), the six mismatched C's and one mismatched T were detected by reactivity with hydroxylamine and osmium tetroxide, respectively (not shown). In addition, some bases correctly paired but within one or two bases of known mismatches also showed some reactivity to the modifying reagents. This reactivity in DNA-DNA heteroduplexes has been extensively studied for the steroid 21-hydroxylase gene (Cotton and Campbell, 1989).

The more significant experiments were those using the same PUO-218 cDNA probe hybridized with viral RNA prepared from purified virions (Biedrzycka et al., 1987) or contained in total infected-cell RNA (Khan and Wright, 1987) (Fig. 2). DNA-RNA heterodu-

plexes were formed in 80% formamide–40 mM piperazine-N,N'-bis(2-ethanesulfonic acid) (PIPES; pH 6.5)–1 mM EDTA–400 mM NaCl at 90°C for 5 min and at 55°C for 60 min, followed by a reduction to 45°C over 60 min, and finally at 45°C for 60 min. For each reaction, either 0.34 μg of purified viral RNA with 3.9 μg of sonicated salmon sperm DNA as carrier or 5 μg of total infected-cell RNA was used. Base modification and base displacement reactions were the same as those used for DNA-DNA heteroduplexes (Cotton et al., 1988).

Lanes 4 to 9 in Fig. 2 show the results of the hybridizations using viral RNA from purified virions. Hybridization of the probe with NGC RNA (lanes 4 and 5) gave results similar to those obtained with the DNA-DNA hybridization described above. True mismatches at positions 72, 111, 141, 207, 327, 341, and 348 were detected. The faint bands at positions 122, 149, and 265 in lane 4 indicated T's that were within one or two bases of a known mismatch and had some reactivity with osmium tetroxide. The hybridization between the PUO-218 cDNA probe and homol-

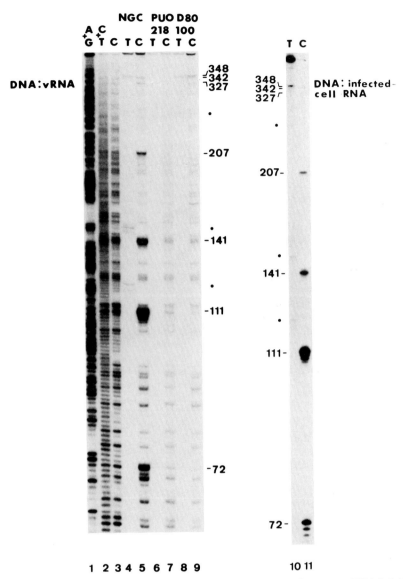

FIGURE 2. Analysis of mismatches, using end-labeled PUO-218 negative-sense cDNA hybridized to unlabeled viral RNA from purified virions (lanes 4 to 9) or from NGC virus-infected Vero cells (lanes 10 and 11). Lanes 1 to 3 show the probe treated with Maxam and Gilbert reagents. Mismatched T was detected by modification with osmium tetroxide for 5 min; mismatched C was detected by modification with hydroxylamine for 60 min.

ogous PUO-218 viral RNA resulted in a background ladder, particularly with hydroxylamine (lane 7). Nevertheless, the bands representing true mismatched bases were much stronger for NGC (lane 5). For D80-100 we had no sequence information, yet it was clear from lanes 8 and 9 that D80-100 more closely resembled PUO-218 than did NGC. Three differences were detected

in the region of positions 265 and 342 to 348. This example suggested the potential usefulness of the method in virus typing.

It is tedious and not always possible to purify virus for RNA extraction, so the PUO-218 probe was hybridized with total cell RNA extracted from NGC-infected Vero cells (Fig. 2, lanes 10 and 11). Again, all known mismatches were detected, and the gel pattern was enriched by cleavages at bases adjacent to mismatches. These additional cleavages, which enrich the gel pattern, provide extra information on the sequence under consideration and are useful in the comparative fingerprinting of isolates. Since total infected-cell RNA can be used for the hybridizations, rapid screening of a large number of isolates is feasible.

At present, reagents specific for mismatched A and G are not available. However, use of probes of both senses enables one to locate all point mutations, even if the actual base changes cannot be determined. In the case of a positive-strand virus such as dengue virus, this requires using end-labeled cDNA probes of both senses against positive- and negative-strand RNA extracted from infected cells.

SUMMARY

Chemical cleavage at mismatched C and T in DNA-RNA heteroduplexes is a new approach for detecting genetic variation in viral RNA. Advantages of the technique are the ability to detect single-base changes in defined areas of the genome and the potential to screen large numbers of isolates by analysis of total infected-cell RNA.

ACKNOWLEDGMENTS. This work was supported by the National Health and Medical Research Council of Australia and by the World Health Organization as part of its Programme for Vaccine Development.

REFERENCES

Biedrzycka, A., M. Cauchi, A. Bartholomeusz, J. J. Gorman, and P. J. Wright. 1987. Characterization of protease cleavage sites involved in the formation of the envelope glycoprotein and three non-structural proteins of dengue virus type 2, New Guinea C strain. J. Gen. Virol. 68:1317–1326.

Cotton, R. G. H., and R. D. Campbell. 1989. Chemical reactivity of matched cytosine and thymine bases near mismatched and unmatched bases in a heteroduplex between DNA strands with multiple differences. Nucleic Acids Res. 17:4223–4333.

Cotton, R. G. H., N. R. Rodrigues, and R. D. Campbell. 1988. Reactivity of cytosine and thymine in single-base-pair mismatches with hydroxylamine and osmium tetroxide and its application to the study of mutations. Proc. Natl. Acad. Sci. USA 85:4397–4401.

Cotton, R. G. H., and P. J. Wright. 1989. Rapid chemical mapping of dengue virus variability using RNA isolated directly form cells. J. Virol. Methods 26:67–76.

Gruenberg, A., W. S. Woo, A. Biedrzycka, and P. J. Wright. 1988. Partial nucleotide sequence and deduced amino acid sequence of the structural proteins of dengue virus type 2, New Guinea C and PUO-218 strains. J. Gen. Virol. 69:1391–1398.

Khan, A. M., and P. J. Wright. 1987. Detection of flavivirus RNA in infected cells using photobiotin-labelled hybridization probes. J. Virol. Methods 15:121–130.

Lamande, S. R., H.-H. Dahl, W. G. Cole, and J. B. Bateman. 1989. Characterization of point mutations in the collagen COL1A1 and COL1A2 genes causing lethal perinatal osteogensis imperfecta. J. Biol. Chem. 264:15809–15812.

Lopez-Galindez, C., J. A. Lopez, J. A. Melero, L. de la Fuente, C. Martinez, J. Ortin, and M. Perucho. 1988. Analysis of genetic variability and mapping of point mutations in influenza virus by the RNase A mismatch method. Proc. Natl. Acad. Sci. USA 85:3522–3526.

Palese, P., and J. L. Schulman. 1976. Differences in RNA patterns of influenza viruses. J. Virol. 17:876–884.

Sabin, A. B., and R. W. Schlesinger. 1945. Production of immunity to dengue with virus modified by propagation in mice. Science 101:640–642.

Smith, F. I., J. D. Parvin, and P. Palese. 1986. Detection of single base substitutions in influenza virus RNA molecules by denaturing gradient gel electrophoresis of RNA-RNA or DNA-RNA heteroduplexes. Virology 150:55–64.

Trent, D. W., J. A. Grant, L. Rosen, and T. P. Monath. 1983. Genetic variation among dengue 2 viruses of different geographic origin. Virology 128:271–284.

Walker, P. J., E. A. Henchal, J. Blok, P. M. Repik, L. S. Henchal, D. S. Burke, S. J. Robbins, and B. M. Gorman. 1988. Variation in dengue type 2 viruses isolated in Bangkok during 1980. J. Gen. Virol. 69:591–602.

II. GENOME REPLICATION

Chapter 8

Replication, Repair, and Recombination of Brome Mosaic Virus RNA

Timothy C. Hall, A. L. N. Rao, Gregory P. Pogue, Clayton C. Huntley, and Loren E. Marsh

Brome mosaic virus (BMV) is a single-stranded, positive-sense RNA virus possessing three genomic RNAs and a single subgenomic RNA (Fig. 1) that serve as mRNAs encoding polypeptides involved in replication, transmission, and encapsidation. The viral RNAs also function as templates for replication and processes such as nucleotidyl transferase and aminoacylation that involve the participation of host-encoded factors. Here, we review the characteristics of the three distinct promoter functions exhibited by BMV RNAs, the repair functions associated with the aminoacylatable 3′ terminus of the positive strands, and the apparent frequency of recombination of the BMV genome. We have used the term repliscription (Hall et al., 1987; Marsh et al., 1988) rather than transcription to denote the process by which new viral RNA is generated from an RNA template by the mediation of an RNA-dependent RNA polymerase (replicase). Use of this term provides differentiation from the production of RNAs from DNA templates and avoids confusion with the synthesis in vitro (usually by SP6 or T7 polymerase) of full-length (infectious) or partial viral RNA transcripts from cDNA templates.

Although the 3′-terminal 200 nucleotides of BMV RNAs are nearly identical (Ahlquist et al., 1981; Ahlquist et al., 1984a; Ahlquist et al., 1984b), the few sequence differences present (Fig. 2) may be important in maintaining the constant ratios observed between the genomic RNAs. The 3′ terminus of each of the RNAs of all bromoviruses and cucumoviruses can be tyrosylated with kinetics and specificity similar to those for tRNA^{Tyr} (Hall et al., 1972; Kohl and Hall, 1974). Examples of 3′ aminoacylation by other groups of plant viruses include the accep-

tance of valine by tymovirus RNAs and histidine by many tobamovirus RNAs (Haenni et al., 1982; Hall, 1979; Kohl and Hall, 1974). The tRNA-like structure shown in Fig. 2 is derived from that proposed for BMV by Rietveld et al. (1983) and Joshi et al. (1983).

REPLICATION

Synthesis of Negative Strands

Development of an in vitro template-dependent and template-specific BMV replicase system (Bujarski et al., 1982; Hardy et al., 1979; Miller and Hall, 1983), combined with a hybrid-arrested replication assay, indicated that the highly structured 3′ region of the BMV positive-strand RNAs provided signals for recognition by the replicase and initiation of negative-strand synthesis (Ahlquist et al., 1984a). Dreher et al. (1984) placed a cDNA containing the 3′ 200 nucleotides of BMV RNA-3 downstream from an SP6 promoter sequence and inserted a *Tth*111I site that permitted runoff transcription of RNAs that correctly terminate in $-CCA_{OH}$. Initially, single- and double-nucleotide substitutions were introduced by site-directed mutagenesis into the putative anticodon loop and 3′ terminus. The aminoacylation and replication template activities of these mutants in vitro were compared with those of corresponding wild-type RNA sequences. The anticodon mutants showed little difference in tyrosylation characteristics but were poor (less than 15% of wild type) templates for repliscription. Subsequently, over 40 mutations within the 3′ region were constructed, and their aminoacylation, nucleotidyl transferase, and replicase functions were

Timothy C. Hall, A. L. N. Rao, Gregory P. Pogue, Clayton C. Huntley, and Loren E. Marsh, Biology Department, Texas A&M University, College Station, Texas 77843-3258.

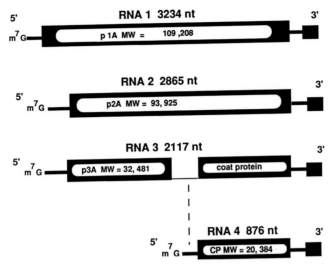

FIGURE 1. Genomic organization of BMV. BMV has a tripartite genome, RNA-1, -2, and -3. RNA-4 is a subgenomic RNA derived from the dicistronic RNA-3. Open boxes represent the protein-coding regions; small boxes at the 3' end represent the tRNA-like structure. nt, Nucleotides; MW, molecular weight.

characterized (Dreher and Hall, 1988a, 1988b; Table 1). Although these functions are distributed over multiple domains, we have identified certain mutations that specifically debilitate each function. Thus, negative-strand promoter activity in vitro is very low for mutants 67-GUA and Δknob; Δ5' and 5'AGA are deficient in aminoacylation, and ψGG is a poor substrate for nucleotidyl transferase. Mutants M2 and M4 are interesting in having elevated levels in vitro for negative-strand initiation; 5'+3'PsK and SSA$_{12}$ are two to four times better templates than is the wild-type sequence for nucleotidyl transferase. The effects of these interesting mutations are now under analysis in vivo and are further discussed below in regard to repair and recombination functions.

Synthesis of Subgenomic Positive-Strand RNA-4

In contrast to the complex structure of the 3' negative-strand promoter, the region containing the subgenomic promoter has minimal secondary structure (Fig. 3). Although there is no extensive sequence similarity to the negative-strand promoter, three short blocks of putative homology with subgenomic promoter regions of alphavirus RNAs are evident (Marsh et al., 1987). As noted below, these sequences may contribute to the production of genomic as well as subgenomic RNAs.

Experiments in which negative-strand RNA-3 sequences having differing 3' termini or bearing various deletions were supplied to the in vitro replicase system have delimited the core sequence to 20 bases immediately upstream of, and including, the initial nucleotide of subgenomic RNA-4 (Marsh et al., 1987, 1988; Miller et al., 1985; Miller et al., 1986). The complete subgenomic promoter is considerably larger, encompassing at least 62 bases and containing three modulating domains in addition to the core sequence (Marsh et al., 1987, 1988). The domain immediately downstream of the core sequence is essential for correct initiation and overlaps the 5' untranslated A/U-rich leader of subgenomic RNA-4. Two domains upstream of the core sequence, the internal poly(A) tract and the sequence UUAUUAUU, are required for high levels of activity. The internal poly(A) may serve as a non-base-paired spacer that facilitates access of the replicase to the promoter core sequence. The in vitro results were complemented by parallel studies in vivo of the subgenomic promoter (French and Ahlquist, 1987, 1988) that revealed additional enhancement from sequences 5' of the internal poly(A) tract. The fact that regions upstream of the subgenomic RNA initiation site are essential for its replication explains in part its inability to replicate independently.

Synthesis of Infectious Positive-Strand RNAs

Although the synthesis of negative-strand RNAs is intrinsic to the initial establishment of

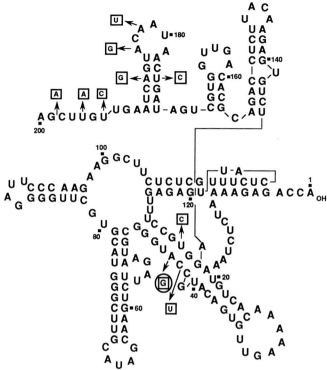

FIGURE 2. Secondary structure of the 3' homologous region of BMV RNAs. The sequence shown represents the 3'-terminal nucleotides of RNA-3. The RNA-2 sequence differs from that of RNA-3 in having a G at position 44 (boxed and circled). The RNA-1 sequence differs from the RNA-3 sequence at the following positions (boxed): 43 (U), 44 (G), 131 (G), 176 (C), 182 (U), 184 (G), and 187 (G).

infection within the host cell, it is the positive strands that are the infectious agents. Despite this importance, and also the fact that synthesis of positive strands often exceeds that of negative strands in infected cells by several hundredfold, remarkably little is known about the molecular mechanisms involved in generation of positive strands from negative strands. We have been unable to synthesize positive strands from correctly terminated genomic negative strands by using our in vitro replicase system (T. W. Dreher and T. C. Hall, unpublished observations), suggesting the existence of differences in processes for positive- and negative-strand synthesis.

Examination of the 5' end of BMV positive-strand RNAs initially revealed the presence of sequences similar to the short domains of putative homology to the alphavirus subgenomic promoter region noted above. More detailed consideration of these sequences led to the exciting realization that they closely resemble the A and B boxes (internal control regions 1 and 2 [ICR1 and -2]) characteristic of polymerase III (Pol III) promoters (Fig. 4). On the basis of these similarities, we (Marsh and Hall, 1987) speculated that ICR-like sequences might form the core of the positive-strand promoter and that Pol III transcription factors might be involved in the asymmetric generation of viral positive strands. Pol III is involved in the transcription of tRNA genes, and the presence of tRNA characteristics at both ends of the BMV genome implies an evolutionary relationship between viral replication and the generation of tRNAs (Marsh et al., 1989). Indeed, the fact that Pol III promoters lie internal to the sequence to be transcribed can be considered a primitive feature. Motifs similar to ICR1 and -2 sequences also exist within the intercistronic region of RNA-3 (Marsh and Hall, 1987), and it is known (French and Ahlquist, 1987) that the intercistronic region of BMV RNA-3 contributes significantly to replication of this genomic component; i.e., the intercistronic motifs promote initiation (distantly) upstream. These observations further explain the impor-

TABLE 1
Properties of BMV RNA 3' Mutants In Vitro[a]

| Mutant | Sequence alteration | Tyrosylation activity | | | | Adenylation activity | | Minus-strand promoter activity |
| | | Wheat germ | | Yeast | | Wheat germ | Escherichia coli | |
		Initial rate	Final level	Initial rate	Final level			
Wild type	None	100	100	100	100	100	100	100
3'9	-CAC$_{OH}$ and U97		0					5
3'64	-CCU$_{OH}$ and U97		2					34
3'95	-CUU$_{OH}$ and U97		1					5
3'98	-CCC$_{OH}$ and U97		1					32
5'PsK	-115-AGA-113	0	0	7	10	<1	<1	8
5'+3'PsK	-115-AGA-113 and 8-UCU-6	10	54	9	9	282	256	71
A12	A12	100	96					87
A14	A14	100	79					93
G14	G14	100	93					84
SSA$_{12}$	16-AAAAA-12	96	67	82	64	364	400	39
ΔB	Δ32-27	70	77	94	96	85	81	42
A46/C44	A46/C44	11	30	28	52	6	5	20
C46/C44	C46/C44	13	48	33	48	1	7	17
5-1i	A46/G44 and 42-UGUAC-41		48					11
5-2i	C46/G44 and 42-UGUAC-41		54					12
X1	A132 and U97		51			12	10	80
67-ACA	A67/C66					115	105	
67-CCA	C67/C66					100	89	
67-CUA	C67/U66				101	83		
67-GUA	G67/U66	92	104	98	99	100	101	6
67-UUA	U67/U66				101	122		
A32	U67/A66 and U97	100	100			45	14	
A48	G67/A66 and U97	110	108	112	94	47	9	7
A50	C67/U66 and U97	100	82			57	29	
A60	C67/A66 and U97	100	94			53	24	
A74	C67/C66 and U97	100	99			54	10	7
Δknob	Δ56-53	104	98			8	10	6
U97	U97	100	100			69	21	
M2	Δ94-88 and U97							200
M4	Δ100-81	100	100		65	18	220	
ψAG	A103/G102	90	94	122	108	2		48
ψGA	G103/A102	100	89			1		46
ψGG	G103/G102	100	91			3	15	39
G107	G107	106	78		103	78	75	38
U107	U107	100	89			123		90
G125	G125	100	95			5	12	22
G126	G126	100	95			38	17	21
G126/G125	G126/125	102	100	102	92	3	11	19
Δ5'	Δ156-135	5	19	33	56	112	122	90
5'AGA	137-AGA-135	3	17	22	35	137	130	62

[a] Data (Bujarski et al., 1985; Dreher et al., 1984; Dreher and Hall, 1988a, 1988b) represent activities obtained for transcripts of the 3'-terminal 200 nucleotides of BMV RNA-3 containing the indicated sequence alterations (see also Fig. 2).

tance of the modulating region present in the BMV RNA-4 subgenomic promoter in determining the downstream initiation site of this component (Marsh et al., 1988). The lack of conserved ICR-like sequences at the 5' terminus of RNA-4 provides further insight as to why it cannot replicate independently.

Recent experiments in which parts of the

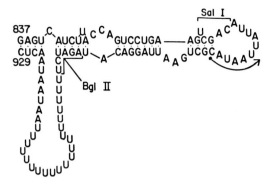

FIGURE 3. Secondary structure for RNA-3 (bases 5'-837 to 929-3' of the negative strand) in the region of the subgenomic promoter generated with a program based on the algorithm of Zucker and Stiegler (1981) Symbols: →, direction of RNA-4 repliscription; *, base complementary to the initial base of RNA-4. Brackets indicate bases corresponding to restriction sites for *Bgl*II and *Sal*I in the cDNA. (From Marsh et al. [1988], with permission.)

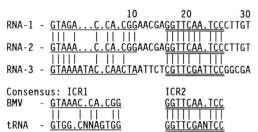

FIGURE 4. ICR-like sequences of the 5' termini of BMV genomic RNAs. Sequence similarities are indicated by alignment (|) between the 5' ends of BMV RNAs and consensus ICR1 (single underline) and -2 (double underline) regions of tRNA genes (Marsh and Hall, 1987; Sharp et al., 1985). Spaces introduced to maximize alignment to the BMV RNA consensus sequence are marked (.). (Modified from Marsh and Hall [1987].)

BMV RNA-2 5' ICR region were deleted have confirmed that its presence is integral to positive-strand repliscription (L. E. Marsh et al., manuscript in preparation). Indeed, substitution of individual bases known to be highly conserved within the ICR2 consensus sequence of tRNA promoters has resulted in a 55 to 80% decrease in positive-strand repliscription (G. P. Pogue et al., manuscript in preparation).

The occurrence of the ICR motifs appears to be frequent among plant viruses (Marsh et al., 1989). Indeed, the intrinsic similarity of BMV and alphaviruses (Ahlquist et al., 1985) and the functional homologies among the major groups of plant, insect, and animal viruses (Goldbach, 1987) suggest that general similarities in the strategy for positive-strand replication exist. Consequently, it appears important that additional research be undertaken to verify the possibility that Pol III factors are widely associated with the replication of positive-strand RNA viruses.

REPAIR

The 3'-terminal adenosine of BMV positive-strand RNAs was found to have no complement in the negative-strand template and to be absent from double-stranded replicative-form RNAs (Miller et al., 1986). To elucidate the possible involvement of host tRNA nucleotidyl transferase in adding the terminal $-A_{OH}$ and maintaining an intact 3'-CCA_{OH} terminus (Deutscher,

1982) to the viral RNA, four mutations of the 3'-CCA_{OH} terminus (3'CAC, 3'CCU, 3'CUU, and 3'CCC) of BMV RNA-3 that were shown in vitro to be defective in aminoacylation and negative-strand promoter functions (Dreher et al., 1984) were selected. These were substituted for the wild-type sequence in a full-length cDNA clone of BMV RNA-3 (Dreher et al., 1989), and transcripts were assayed in vivo for the ability to replicate in barley protoplasts and plants (Rao et al., 1989). In protoplasts, all four mutants were fully viable, the levels of both positive- and negative-strand RNA-3 progeny for each mutant being similar (relative to RNA-1 and -2) to wild type 2.5 h after inoculation (Fig. 5). This experiment also revealed the asymmetric nature of BMV RNA replication. Although similar levels of positive- and negative-strand RNAs were obtained at 2.5 h postinoculation, evidenced by the similar exposure times used (24 h) to detect progeny RNAs, by 5 h some 10-fold more positive- than negative-strand RNA was produced, and exposure time was reduced to 2 h. Inoculation of barley plants with these mutants resulted in phenotypic symptoms and viral yields that were similar to those from wild-type infections. Analysis of each mutant progeny RNA-3 confirmed that the altered sequence at the 3' terminus was restored to that of wild type (Rao et al., 1989). The rapid turnover and correction of the 3' termini of BMV RNAs observed in this study provide functional evidence for the concept that the tRNA-like structure found in BMV and other RNA viruses functions analogously to telomeres of chromosomal DNA in protecting the end of the genomic information from degradation.

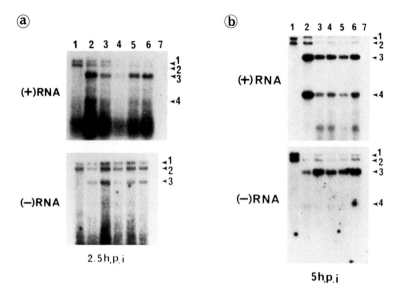

FIGURE 5. Rapid repair of mutagenized BMV RNA 3'-CCA$_{OH}$ termini. Autoradiograph showing Northern (RNA) hybridization of progeny RNA samples isolated from protoplasts 2.5 (a) or 5 (b) h postinoculation. In each case, inocula contained wild-type RNA-1 and -2 transcripts and no RNA-3 (lane 1) or the following RNA-3 transcripts: wild type (lane 2), 3'CAC (lane 3), 3'CCU (lane 4), 3'CUU (lane 5), or 3'CCC (lane 6). No RNA (mock inoculation) was added for the reaction shown in lane 7. RNAs extracted from barley protoplasts were electrophoresed in 1% agarose gels after denaturation with glyoxal and transferred to nylon membranes. The blots were then hybridized with ^{32}P-labeled RNA probes of negative and positive sense that represented the homologous 3' region present on each of the four BMV RNAs. Similar exposure times (24 h) were used to detect positive- and negative-strand progeny RNAs for the 2.5-h-postinoculation samples; by 5 h, >10-fold more positive- than negative-strand RNA was produced, and a shorter exposure (2 h) was used for the positive-strand blots. (From Rao et al. [1989], with permission.)

RECOMBINATION

Early evidence for recombination between wild-type BMV RNA-1 and -2 and the M4 deletion mutant of BMV RNA-3 (Bujarski and Kaesberg, 1986) in barley plants suggested that this was a frequent event. In contrast, studies on RNA-3 mutants ψGG and 5'PsK that are severely debilitated in replication yielded no evidence for recombination (Dreher et al., 1989). Since these mutants exhibit only 40 and 20%, respectively, of wild-type progeny RNA-3 in protoplasts, and virtually no RNA-3 in barley plant infections, strong selection pressure for rescue of functionally restored sequences by recombination was anticipated. Although the absence of recombination in these latter experiments calls into question the frequency of recombination events, we have recently obtained further evidence for high-frequency recombination through the use of RNA-2 mutants in protoplast inoculations. In these experiments, replication of RNA-2, vital in providing an mRNA

encoding replicase function(s) (French et al., 1986; Kiberstis et al., 1981), was eliminated by deletion of the 3' 200 nucleotides, thereby removing the entire negative-strand promoter. Nevertheless, in two of eight experiments, restoration of the 3' sequence and of normal replication of RNA-2 (and consequently of the other RNAs) appeared to occur within 24 h after inoculation (Rao and Hall, 1990). Consequently, the frequency with which recombination occurs may be subject to unknown parameters that could include the physiological state of the host cell. Since recombination can potentially occur between viral genomes infecting animals, plants, and insects, elucidation of conditions favoring its occurrence is of great importance.

SUMMARY

The ability to construct specific mutations in the genomes of RNA viruses and analyze their biological effects provides powerful new approaches to understanding processes involved in infection. Using such techniques, we have obtained considerable insight into promoter functions that generate progeny negative strands and subgenomic RNAs. Innovative concepts concerning the origin of infectious positive strands, possibly through the mediation of host Pol III transcription factors, have been adduced that are amenable for testing. It appears that viral genomes can use host functions such as nucleotidyl transferase to maintain their integrity. Under conditions yet to be elucidated, viral genomes can rapidly recombine to restore infectivity, thereby creating pathogens that are either novel or enhanced in virulence. Understanding these processes is likely to reveal new approaches to blocking viral functions, heralding a new era in preventing, and perhaps ameliorating, diseases of viral etiology.

ACKNOWLEDGMENTS. Support from the National Institutes of Health (Public Health Service grant AI22354) and the Texas Advanced Research Program is gratefully acknowledged.

REFERENCES

Ahlquist, P., J. J. Bujarski, P. Kaesberg, and T. C. Hall. 1984a. Localization of the replicase recognition site within brome mosaic virus by hybrid-arrested RNA synthesis. *Plant Mol. Biol.* **3**:37–44.

Ahlquist, P., R. Dasgupta, and P. Kaesberg. 1981. Near identity of 3′ RNA secondary structure in bromoviruses and cucumber mosaic virus. *Cell* **23**:183–189.

Ahlquist, P., R. Dasgupta, and P. Kaesberg. 1984b. Nucleotide sequence of brome mosaic virus genome and its implications for viral replication. *J. Mol. Biol.* **172**:369–383.

Ahlquist, P., E. G. Strauss, C. M. Rice, J. H. Strauss, J. Haseloff, and D. Zimmern. 1985. Sindbis virus proteins nsP1 and nsP2 contain homology to nonstructural proteins from several RNA plant viruses. *J. Virol.* **53**:536–542.

Bujarski, J. J., T. W. Dreher, and T. C. Hall. 1985. Deletions in the 3′-terminal tRNA-like structure of brome mosaic virus RNA differentially affect aminoacylation and replication *in vitro*. *Proc. Natl. Acad. Sci. USA* **82**:5636–5640.

Bujarski, J. J., S. F. Hardy, W. A. Miller, and T. C. Hall. 1982. Use of dodecyl-β-D-maltoside in the purification and stabilization of RNA polymerase from brome mosaic virus infected barley. *Virology* **119**:465–473.

Bujarski, J. J., and P. Kaesberg. 1986. Genetic recombination between RNA components of a multipartite plant virus. *Nature* (London) **321**:528–531.

Deutscher, M. P. 1982. tRNA nucleotidyl transferase and the -C-C-A terminus of transfer RNA, p. 159–

183. *In* S. T. Jacob (ed.), *The Enzymes of Nucleic Acid Synthesis and Modification*, vol. 2. CRC Press, Inc., Boca Raton, Fla.

Dreher, T. W., J. J. Bujarski, and T. C. Hall. 1984. Mutant viral RNAs synthesized *in vitro* show altered aminoacylation and replicase template activities. *Nature* (London) **311**:171–175.

Dreher, T. W., and T. C. Hall. 1988a. Mutational analysis of the sequence and structural requirements in brome mosaic virus RNA for minus strand promoter activity. *J. Mol. Biol.* **201**:31–40.

Dreher, T. W., and T. C. Hall. 1988b. Mutational analysis of the tRNA mimicry of brome mosaic virus RNA: sequence and structural requirements for aminoacylation and 3′ adenylation. *J. Mol. Biol.* **201**:41–55.

Dreher, T. W., A. L. N. Rao, and T. C. Hall. 1989. Replication *in vivo* of mutant brome mosaic virus RNAs defective in aminoacylation. *J. Mol. Biol.* **206**:425–438.

French, R., and P. Ahlquist. 1987. Intercistronic as well as terminal sequences are required for efficient amplification of brome mosaic virus RNA3. *J. Virol.* **61**:1457–1465.

French, R., M. Janda, and P. Ahlquist. 1986. Bacterial gene inserted in an engineered RNA virus: efficient expression in monocotyledonous plant cells. *Science* **231**:1294–1297.

French, R. P., and P. Ahlquist. 1988. Characterization and engineering of sequences controlling in vivo synthesis of brome mosaic virus subgenomic RNA. *J. Virol.* **62**:2411–2420.

Goldbach, R. 1987. Genomic similarities between plant and animal RNA viruses. *Microbiol. Sci.* **4**:197–202.

Haenni, A.-L., S. Joshi, and F. Chapeville. 1982. tRNA-like structures in the genome of RNA viruses. *Prog. Nucleic Acid Res. Mol. Biol.* **27**:85–104.

Hall, T. C. 1979. Transfer RNA-like structures in viral genomes. *Int. Rev. Cytol.* **60**:1–26.

Hall, T. C., L. E. Marsh, and T. W. Dreher. 1987. Analysis of brome mosaic virus replication and aminoacylation functions by site-specific mutagenesis. *J. Cell Sci. Suppl.* **7**:287–301.

Hall, T. C., D. S. Shih, and P. Kaesberg. 1972. Enzyme-mediated binding of tyrosine to brome mosaic virus ribonucleic acid. *Biochem. J.* **129**:969–976.

Hardy, S. F., T. L. German, L. S. Loesch-Fries, and T. C. Hall. 1979. Highly active template-specific RNA-dependent RNA polymerase from barley leaves infected with brome mosaic virus. *Proc. Natl. Acad. Sci. USA* **76**:4956–5960.

Joshi, R. L., S. Joshi, F. Chapeville, and A. L. Haenni. 1983. tRNA-like properties of plant viral RNAs: conformational requirements for adenylation and aminoacylation. *EMBO J.* **2**:1123–1127.

Kiberstis, P. A., L. S. Loesch-Fries, and T. C. Hall. 1981. Viral protein synthesis in barley protoplasts inoculated with native and fractionated brome mosaic virus RNA. *Virology* **112**:804–808.

Kohl, R. J., and T. C. Hall. 1974. Aminoacylation of RNA from several viruses: amino acid specificity and differential activity of plant, yeast and bacterial synthetases. *J. Gen. Virol.* **25**:257–261.

Marsh, L. E., T. W. Dreher, and T. C. Hall. 1987. Mutational analysis of the internal promoter for transcription of subgenomic RNA4 of BMV, p. 327–336. *In* M. Brinton and P. Kaesburg (ed.), *Positive Strand RNA Viruses.* Alan R. Liss, Inc., New York.

Marsh, L. E., T. W. Dreher, and T. C. Hall. 1988. Mutational analysis of the core and modulator sequences of the BMV RNA3 subgenomic promoter. *Nucleic Acids Res.* **16**:981–995.

Marsh, L. E., and T. C. Hall. 1987. Evidence implicating a tRNA heritage for the promoters of (+) strand RNA synthesis in brome mosaic virus and related viruses. *Cold Spring Harbor Symp. Quant. Biol.* **52**:331–341.

Marsh, L. E., G. P. Pogue, and T. C. Hall. 1989. Similarities among plant virus (+) and (−) RNA termini imply a common ancestry with promoters of eucaryotic tRNAs. *Virology* **172**:415–427.

Miller, W. A., J. J. Bujarski, T. W. Dreher, and T. C. Hall. 1986. Minus-strand initiation by brome mosaic virus replicase within the 3' tRNA-like structure of native and modified RNA templates. *J. Mol. Biol.* **187**:537–546.

Miller, W. A., T. W. Dreher, and T. C. Hall. 1985. Synthesis of brome mosaic virus subgenomic RNA *in vitro* by initiation on (−)-sense genomic RNA. *Nature* (London) **313**:68–70.

Miller, W. A., and T. C. Hall. 1983. Use of micrococcal nuclease in the purification of highly template-dependent RNA-dependent RNA polymerase from brome mosaic virus infected barley. *Virology* **125**:236–241.

Rao, A. L. N., T. W. Dreher, L. E. Marsh, and T. C. Hall. 1989. Telomeric function of the tRNA-like structure of brome mosaic virus RNA. *Proc. Natl. Acad. Sci. USA* **84**:7383–7387.

Rao, A. L. N., and T. C. Hall. 1990. Requirement for a viral *trans*-acting factor encoded by brome mosaic virus RNA-2 provides strong selection in vivo for functional recombinants. *J. Virol.* **64**, in press.

Rietveld, K., C. W. A. Pleij, and L. Bosch. 1983. Three-dimensional models of the tRNA-like 3' termini of some plant viral RNAs. *EMBO J.* **2**:1079–1085.

Sharp, S. J., J. Shaack, L. Cooley, D. J. Burke, and D. Söll. 1985. Structure and transcription of eucaryotic tRNA genes. *Crit. Rev. Biochem.* **19**:107–125.

Zucker, M., and P. Stiegler. 1981. Optimal computer folding of large RNA sequences using thermodynamics and auxiliary information. *Nucleic Acids Res.* **9**:133–148.

Chapter 9

Mechanism of Poliovirus Negative-Strand RNA Synthesis and the Self-Catalyzed Covalent Linkage of VPg to RNA

J. Bert Flanegan, Gregory J. Tobin, M. Steven Oberste, B. Joan Morasco, Dorothy C. Young, and Philip S. Collis

Poliovirus has a single-stranded RNA genome of positive polarity that contains a 3'-terminal poly(A) sequence and a 5'-terminal covalently linked protein, VPg (Lee et al., 1977; Flanegan et al., 1977). All newly synthesized viral RNAs including negative strands and nascent positive strands are covalently linked to VPg (Nomoto et al., 1977; Pettersson et al., 1978) by a phosphodiester bond between the 5'-terminal UMP residue in the RNA and the tyrosine residue in VPg (Ambros and Baltimore, 1978; Rothberg et al., 1978). Poliovirus RNA replicates in the cytoplasm of infected cells and requires a virus-specific RNA-dependent RNA polymerase ($3D^{pol}$). Unlike many other RNA polymerases, the poliovirus polymerase cannot initiate RNA synthesis de novo and requires either an oligonucleotide primer (Flanegan and Van Dyke, 1979; Dasgupta et al., 1979) or a host factor isolated from uninfected cells (Dasgupta et al., 1980; Dasgupta, 1983; Baron and Baltimore, 1982). Different host factor preparations that are associated with different enzymatic activities have been described (e.g., protein kinase, terminal uridylyltransferase, or endonuclease) (Morrow et al., 1985; Andrews et al., 1985; Andrews and Baltimore, 1986; Hey et al., 1987). The precise mechanism of action and the roles that different host factors play in viral RNA replication have not yet been determined. In addition, little information is available on the exact nature of the interactions that exist between the polymerase and its templates, primers, and host factor proteins.

We have proposed a template-priming model in which a 3'-terminal hairpin acts as a primer for negative-strand synthesis. This model was suggested by our finding that negative-strand RNA synthesized in vitro by purified polymerase and host factor was covalently linked at one end to the positive-strand RNA template (Young et al., 1985; Young et al., 1986). This model was further defined and extended by the observation that terminal uridylyltransferase stimulates the initiation of RNA synthesis (Andrews and Baltimore, 1986; Andrews et al., 1985). The direct uridylylation of the poly(A) sequence in positive-strand RNA templates may facilitate the formation of a 3'-terminal hairpin primer, which could then be elongated into negative-strand RNA by the polymerase (Fig. 1). Thus, the model predicts that negative-strand RNA is synthesized by a template elongation mechanism, with the initiation of RNA synthesis occurring at the 3' end of the poly(A) sequence. In addition, a mechanism must exist for the separation of the template and product RNAs and for the covalent linkage of VPg to the 5' end of negative-strand RNA. As discussed below, our results indicate that VPg-linked negative-strand RNA was formed in a reaction that required synthetic VPg, Mg^{2+}, and a replication intermediate synthesized in vitro on poliovirion RNA.

POLYMERASE-BINDING STUDIES

Previous genetic studies with poliovirus indicate that sequences and presumably the structures at the 3' and 5' ends of the viral genome are required for RNA replication in vivo (Racaniello

J. Bert Flanegan, Gregory J. Tobin, M. Steven Oberste, B. Joan Morasco, Dorothy C. Young, and Philip S. Collis, Department of Immunology and Medical Microbiology, University of Florida, College of Medicine, Gainesville, Florida 32610-0266.

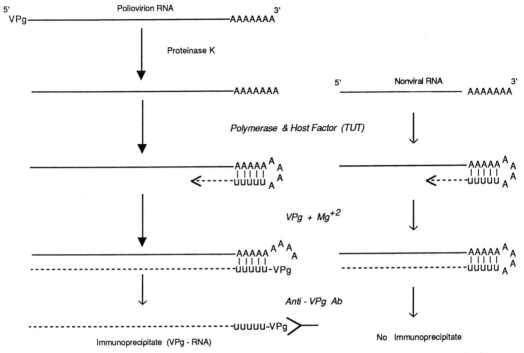

FIGURE 1. Model for the synthesis of VPg-linked RNA in vitro. Poliovirus RNA pretreated with proteinase K to remove VPg is added to a reaction with purified polymerase and host factor. Labeled negative-strand RNA is synthesized by a template-priming mechanism and contains a 5′ poly(U) sequence covalently linked to the 3′ poly(A) sequence in the template RNA. This forms a replication intermediate that is the substrate for the VPg linkage reaction. VPg is linked to the negative-strand RNA in a transesterification reaction (see Fig. 2). Labeled VPg-linked RNA is detected by immunoprecipitation with anti-VPg antibody (Ab). A similar replication intermediate is formed on nonviral polyadenylated RNA but was not active in the VPg linkage reaction. The exact site of the linkage to negative-strand RNA has not yet been determined.

and Meriam, 1986; Sarnow et al., 1986). In contrast, some internal regions in the viral genome are not required for replication. Kaplan and Racaniello (1988) reported that large in-frame deletions in the capsid-coding region of the genome do not inhibit RNA replication in cells transfected with RNA transcripts containing these deletions. We have confirmed this observation and are investigating what additional internal regions of the genome can be deleted without affecting RNA replication. To initiate RNA synthesis, the polymerase must bind to its template RNA, perhaps at specific sites. The binding of purified poliovirus polymerase to virion RNA, subgenomic RNA transcripts, and ribohomopolymers was characterized by using a filter-binding assay. The relative order of binding of the polymerase to ribohomopolymeric RNAs was poly(G) >>> poly(U) > poly(C) > poly(A) (Oberste

Flanegan, 1988). In competitive binding studies, the equilibrium association constant (K_a) for the binding of the polymerase to poly(A) was 200-fold less than the value for poly(G) and 17-fold less than the value for poly(U). This finding suggests that polymerase-binding sites in poliovirus RNA may contain G- and/or U-rich sequences. In addition, a 3′-terminal oligo(U) sequence synthesized by terminal uridylyltransferase as proposed above may be important in the binding of the polymerase at the initiation step of RNA synthesis on polyadenylated RNA templates. To quantitate the binding of the polymerase to poliovirion RNA, we determined the K_a by using previously described methods (Riggs et al., 1970). The K_a was found to be 10^9 M^{-1}, which is similar to the values found for other RNA-binding proteins (Wu et al., 1988). This value was within a factor of 2 for the average K_a obtained with 3′-terminal subge-

TABLE 1
RNA Transcribed by SP6 Polymerase from
Poliovirus-Specific 3' Sequences
Inserted into pGEM-1

Clone	Poliovirus bases	RNA size[a] (nucleotides)
1213	7210–7440	318
1205	6516–7440	1,012
1209	6012–7440	1,516
1211	4912–7440	2,616

[a] Includes 3'-terminal poly(A)$_{83}$ sequence.

nomic positive-strand transcripts (Table 1). In competitive binding studies with two nonviral RNAs, we found that the K_a for both 18S rRNA and globin mRNA was about fivefold lower (2×10^8 M^{-1}). Thus, there appears to be a relatively small but significant difference between polymerase binding to viral and nonviral RNAs. One could envision that the polymerase might first bind to a specific G/U-rich sequence in the 3'-terminal region and then translocate to a poly(A)-oligo(U) hairpin at the end of the poly(A) tract to initiate negative-strand RNA synthesis. In future studies, we will determine how factors such as sequence, structure, and other proteins affect polymerase binding.

NEGATIVE-STRAND RNA SYNTHESIS IN VITRO

To further test the template-priming model, we characterized the product RNA synthesized in vitro on small polyadenylated RNA templates. Subgenomic poliovirus-specific positive-strand RNAs were transcribed by SP6 polymerase from cDNA clones that represented the 3' end of the viral genome. The transcripts contained increasing amounts of virus-specific RNA and a 3'-terminal poly(A) sequence (Table 1). The individual RNA transcripts were used as templates in complete replication reactions that contained purified poliovirus RNA polymerase and host factor isolated as described previously (Young et al., 1985; Young et al., 1987). In each case, a band of labeled product RNA that was twice the size of the template RNA was recovered from each reaction. Time course experiments showed that the labeled product RNA was template size at early reaction times (30 s) and then increased in length with time to form a band of dimer-size product RNA. In other experiments, we showed that ^{32}P-labeled RNA templates could be chased into bands of dimer-size product RNA and that poly(U)$_{80}$ could be

isolated from the dimer-size product RNA digested with RNases T$_1$ and U$_2$. To ensure that the five vector-derived nucleotides at the 3' end of poly(A) sequence in these transcripts did not affect our results, we constructed a new series of cloned plasmid DNAs that had a MluI site in place of the EcoRI site. Digestion of these plasmids with MluI, treatment with mung bean nuclease, and transcription with SP6 polymerase resulted in the synthesis of 3'-terminal transcripts that terminated with a poly(A) sequence with no additional vector-derived nucleotides. These transcripts were efficiently copied in vitro and formed bands of labeled product RNA that were exactly twice the size of the template RNAs. On the basis of the results from these studies and others, we concluded that the polymerase can initiate RNA synthesis at the 3' end of polyadenylated RNAs, using a template-priming mechanism.

VPg LINKAGE REACTION

Synthetic VPg was prepared by using solid-phase methods (Young et al., 1986) and was used to study the molecular mechanisms involved in the covalent linkage of VPg to poliovirus RNA. To determine whether VPg-linked RNA was formed in vitro, VPg was added to reactions that contained ^{32}P-labeled product RNA synthesized by purified polymerase and host factor on proteinase K-treated poliovirion RNA. At the end this reaction, noncovalently linked VPg was removed by phenol-sodium dodecyl sulfate extraction, and the VPg-linked RNA was immunoprecipitated with affinity-purified anti-VPg antibody (Fig. 1). Labeled VPg-linked RNA immunoprecipitated from reactions that contained the template-linked replication intermediates synthesized in the host factor-dependent reaction. The largest VPg-linked RNA recovered from these reactions was about the same size as virion RNA. Under optimal conditions for this reaction (i.e., 42°C, 13 mM Mg^{2+}, and pH 7.5), 30% or more of the labeled product RNA was linked to VPg in a 1-h reaction. In other experiments, we showed that VPg-linked RNA was not recovered from reactions that contained poliovirus oligo(U)-primed negative-strand RNA or replication intermediates synthesized on two nonviral polyadenylated RNAs. The VPg linkage reaction did not require the addition of ribonucleoside triphosphates, poliovirus polymerase, host factor, or any protein other than VPg itself. Thus, the linkage reaction was specific for template-linked negative-strand RNA synthesized on poliovirion RNA.

STRUCTURE OF BOND LINKING VPg TO RNA

The VPg-linked RNA was characterized to determine the structure of the bond between VPg and the RNA. It was previously shown that the complete digestion of ^{32}P-labeled poliovirion RNA with RNase yields labeled VPg-pUp, which can be identified by its characteristic migration toward the cathode during high-voltage paper ionophoresis at pH 3.5 (Flanegan et al., 1977; Rothberg et al., 1978). Using the same experimental approach, we were able to isolate labeled VPg-pUp from [^{32}P]UMP-labeled VPg-linked product RNA prepared as described above. By labeling the product RNA with labeled ATP, GTP, or CTP, we were able to show that VPg was specifically linked to a UMP residue in the RNA. On the basis of the structure of the native bond between VPg and poliovirion RNA, we hypothesized that the tyrosyl-hydroxyl group in the synthetic VPg was the site of a phosphodiester bond with the RNA. This hypothesis was supported by our finding that a modified synthetic VPg which contained phenylalanine in place of tyrosine was totally inactive in the VPg linkage reaction. This hypothesis was confirmed by the isolation of phosphotyrosine from ^{32}P-labeled VPg-pUp. Thus, VPg was linked to the RNA by a phosphodiester bond between the tyrosine in VPg and the 5′ UMP residue in the RNA.

MODEL FOR VPg LINKAGE REACTION

The formation of a phosphodiester bond between VPg and poliovirus RNA in the absence of an exogenous energy source (e.g., ATP hydrolysis) was consistent with a transesterification reaction. The cleavage of the original phosphodiester bond in the RNA would provide the energy required to form the new bond. Transesterification reactions that involve the formation of a phosphodiester bond with tyrosine have been observed with certain topoisomerases and with the bacteriophage φX174 A protein (Rowe et al., 1984; Sanhuenza and Eisenberg, 1985; Champoux, 1981; Tse et al., 1980). Our current model proposes that the covalent bond between VPg and poliovirus RNA is formed by nucleophilic attack by the tyrosyl-hydroxyl group in VPg on a phosphate in the RNA replication intermediate (Fig. 2). The resulting transesterification reaction would cleave a phosphodiester bond in the RNA and form a new bond between VPg and the RNA. Because the VPg linkage reaction required the template-linked replication

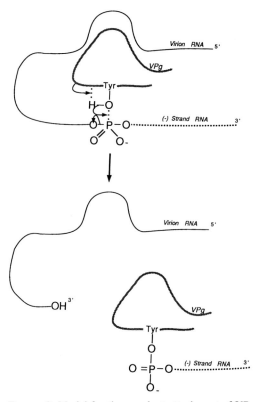

FIGURE 2. Model for the covalent attachment of VPg to poliovirus RNA. The tyrosyl-hydroxyl group in VPg acts as a nucleophile and attacks the phosphate in the bond linking the positive- and negative-strand RNAs. A new phosphodiester bond is formed as a result of this transesterification reaction. The catalytic center for this may be formed by the RNA, VPg, or a combination thereof.

intermediate and because we could isolate VPg-poly(U) from VPg-linked RNA, the linkage of VPg to the product RNA appears to take place near the terminal poly(A)-poly(U) hairpin. Thus, the VPg linkage reaction would serve two important functions in poliovirus RNA replication: (i) cleavage of the bond linking the template and product RNAs and (ii) linkage of VPg to negative-strand RNA.

Our results indicate that the linkage of VPg to poliovirus RNA is a self-catalyzed reaction. It appears that the catalytic center for the reaction is provided by either the RNA, the protein, or some combination thereof. Numerous examples of RNA-catalyzed transesterification reactions that require Mg^{2+} as a cofactor are now well documented in the literature (Cech and Bass, 1986; Cech, 1987). The VPg linkage reaction, however, would be one of the first examples of an RNA-catalyzed reaction that directly involves a protein. An alternate possibility that must be considered is that VPg provides the catalytic activity for this reaction. This seems unlikely, considering its small size (22 amino acids) and the lack of activity that was found with nonviral RNAs. In addition, we have found that long (1.5 kilobases) but not short (0.35 kilobase) 3'-terminal positive-strand RNA transcripts are active in the linkage reaction. This finding suggests that a specific sequence in the 3'-terminal region of the viral genome is required for the VPg linkage reaction. This sequence might provide part of the catalytic center for the reaction, a specific binding site for VPg, or both (Fig. 2). Additional studies are now in progress to identify RNA sequences that are required for the VPg linkage reaction and to characterize the catalytic center for this reaction.

SUMMARY

We have measured the binding of the polymerase to RNA and have proposed a replication model to explain the initiation of negative-strand RNA synthesis and the covalent linkage of VPg to viral RNA. In binding studies, the polymerase showed about a fivefold difference in its specificity for binding to poliovirion RNA relative to two nonviral RNAs. This difference in binding may provide some specificity for viral RNA replication but would not in itself appear sufficient to explain the specificity observed in vivo for the replication of viral RNAs. In other studies, we demonstrated that the polymerase and host factor initiated RNA synthesis at the 3' end of polyadenylated RNAs to form a template-linked replication intermediate that was active in the VPg linkage reaction. VPg was covalently linked to negative-strand RNA in a self-catalyzed reaction that required synthetic VPg, Mg^{2+}, and the replication intermediate synthesized on poliovirion RNA. This reaction did not require the polymerase or any other proteins and did not require a ribonucleoside triphosphate. VPg was linked to the RNA by a phosphodiester bond between the tyrosine in VPg and the 5' UMP in the RNA. Our model proposes that this bond forms in a self-catalyzed transesterification reaction that involves nucleophilic attack by the tyrosyl-hydroxyl group on a 5' phosphate in the terminal hairpin joining the template and product RNAs. Thus, either the RNA, VPg, or both form the catalytic center for this reaction.

ACKNOWLEDGMENTS. This work was supported by Public Health Service research grant AI15539 from the National Institute of Allergy and Infectious Diseases and grant MV-400 from the American Cancer Society.

REFERENCES

Ambros, V., and D. Baltimore. 1978. Protein is linked to the 5' end of poliovirus RNA by a phosphodiester linkage to tyrosine. *J. Biol. Chem.* **253:**5263–5266.

Andrews, N. C., and D. Baltimore. 1986. Purification of a terminal uridylyltransferase that acts as host factor in the *in vitro* poliovirus replicase reaction. *Proc. Natl. Acad. Sci. USA* **83:**221–225.

Andrews, N. C., D. Levin, and D. Baltimore. 1985. Poliovirus replicase stimulation by terminal uridylyl transferase. *J. Biol. Chem.* **260:**7628–7635.

Baron, M. H., and D. Baltimore. 1982. Purification and properties of a host cell protein required for poliovirus replication *in vitro*. *J. Biol. Chem.* **257:**12351–12358.

Cech, T. R. 1987. The chemistry of self-splicing RNA and RNA enzymes. *Science* **236:**1532–1539.

Cech, T. R., and B. L. Bass. 1986. Biological catalysis by RNA. *Annu. Rev. Biochem.* **55:**599–629.

Champoux, J. J. 1981. DNA is linked to rat liver DNA nicking-closing enzyme by a phosphodiester bond to tyrosine. *J. Biol. Chem.* **256:**4805–4809.

Dasgupta, A. 1983. Purification of host factor required for *in vitro* transcription of poliovirus RNA. *Virology* **128:**245–251.

Dasgupta, A., M. H. Baron, and D. Baltimore. 1979. Poliovirus replicase: a soluble enzyme able to initiate copying of poliovirus RNA. *Proc. Natl. Acad. Sci. USA* **76:**2679–2683.

Dasgupta, A., P. Zabel, and D. Baltimore. 1980. Dependence of the activity of the poliovirus replicase on a host cell protein. *Cell* **19:**423–429.

Flanegan, J. B., R. F. Pettersson, V. Ambros, N. J. Hewlett, and D. Baltimore. 1977. Covalent linkage of a protein to a defined nucleotide sequence at the 5'-terminus of virion and replicative intermediate RNAs of poliovirus. *Proc. Natl. Acad. Sci. USA* **74:**961–965.

Flanegan, J. B., and T. A. Van Dyke. 1979. Isolation of a soluble and template-dependent poliovirus RNA polymerase that copies virion RNA in vitro. *J. Virol.* **32:**155–161.

Hey, T. D., O. C. Richards, and E. Ehrenfeld. 1987. Host factor-induced template modification during synthesis of poliovirus RNA in vitro. *J. Virol.* **61:**802–811.

Kaplan, G., and V. R. Racaniello. 1988. Construction and characterization of poliovirus subgenomic replicons. *J. Virol.* **62:**1687–1696.

Lee, Y. F., A. Nomoto, B. M. Detjen, and E. Wimmer. 1977. A protein covalently linked to poliovirus genome RNA. *Proc. Natl. Acad. Sci. USA* **74:**59–63.

Morrow, C. D., G. F. Gibbons, and A. Dasgupta. 1985.

The host protein required for *in vitro* replication of poliovirus is a protein kinase that phosphorylates eukaryotic initiation factor-2. *Cell* **40**:913–921.

Nomoto, A., B. Detjen, R. Pozzatti, and E. Wimmer. 1977. The location of the polio genome protein in viral RNAs and its implication for RNA synthesis. *Nature* (London) **268**:208–213.

Oberste, M. S., and J. B. Flanegan. 1988. Measurement of poliovirus RNA polymerase binding to poliovirion and nonviral RNAs using a filter-binding assay. *Nucleic Acids Res.* **16**:10339–10352.

Pettersson, R. F., V. Ambros, and D. Baltimore. 1978. Identification of a protein linked to nascent poliovirus RNA and to the polyuridylic acid of negative-strand RNA. *J. Virol.* **27**:357–365.

Racaniello, V. R., and C. Meriam. 1986. Poliovirus temperature-sensitive mutant containing a single nucleotide deletion in the 5′-noncoding region of the viral RNA. *Virology* **155**:498–507.

Riggs, A. D., H. Suzuki, and S. Bourgeois. 1970. *Lac* repressor-operator interaction. I. Equilibrium studies. *J. Mol. Biol.* **48**:67–83.

Rothberg, P. G., T. J. Harris, A. Nomoto, and E. Wimmer. 1978. The genome-linked protein of picornaviruses. V. O⁴-(5′-uridylyl)tyrosine is the bond between the genome-linked protein and the RNA of poliovirus. *Proc. Natl. Acad. Sci. USA* **75**:4868–4872.

Rowe, T. C., K. M. Tewey, and L. F. Liu. 1984. Identification of the breakage-reunion subunit of T4 DNA topoisomerase. *J. Biol. Chem.* **259**:9177–9181.

Sanhuenza, S., and S. Eisenberg. 1985. Bacteriophage ϕX174 A protein cleaves single stranded DNA and binds to it covalently through a tyrosyl-dAMP phosphodiester bond. *J. Virol.* **53**:695–697.

Sarnow, P., H. D. Bernstein, and D. Baltimore. 1986. A poliovirus temperature-sensitive RNA synthesis mutant in a noncoding region of the genome. *Proc. Natl. Acad. Sci. USA* **83**:571–575.

Tse, Y. C., K. Kirkegaard, and J. C. Wang. 1980. Covalent bonds between protein and DNA. Formation of phosphotyrosine linkage between certain DNA topoisomerases and DNA. *J. Biol. Chem.* **255**:5560–5565.

Wu, H.-N., K. A. Kastelic, and O. C. Uhlenbeck. 1988. A comparison of two phage coat protein-RNA interactions. *Nucleic Acids Res.* **16**:5055–5066.

Young, D. C., B. M. Dunn, G. J. Tobin, and J. B. Flanegan. 1986. Anti-VPg antibody precipitation of product RNA synthesized in vitro by the poliovirus polymerase and host factor is mediated by VPg on the poliovirion RNA template. *J. Virol.* **58**:715–723.

Young, D. C., G. J. Tobin, and J. B. Flanegan. 1987. Characterization of product RNAs synthesized in vitro by poliovirus RNA polymerase purified by chromatography on hydroxylapatite or poly(U) Sepharose. *J. Virol.* **61**:611–614.

Young, D. C., D. M. Tuschall, and J. B. Flanegan. 1985. Poliovirus RNA-dependent RNA polymerase and host cell protein synthesize produce RNA twice the size of poliovirion RNA in vitro. *J. Virol.* **54**:256–264.

Chapter 10

Functions of the 5'-Terminal and 3'-Terminal Sequences of the Sindbis Virus Genome in Replication

James H. Strauss, Richard J. Kuhn, Hubert G. M. Niesters, and Ellen G. Strauss

The alphaviruses consist of about 25 closely related viruses, many of which are important human pathogens (Calisher and Karabatsos, 1988). The viruses contain a single-stranded RNA genome of 11.7 kilobases that serves as messenger for four nonstructural proteins which are all believed to be components of the viral RNA replicase (reviewed by Strauss and Strauss [1986]). A subgenomic messenger of 4.1 kilobases of the same polarity encodes the virus structural proteins, consisting of a nucleocapsid protein and two membrane glycoproteins. Both the nonstructural and structural proteins are produced as polyproteins that are processed posttranslationally (reviewed by Strauss et al. [1987]). In nature, alphaviruses replicate in their mosquito vectors as well as in a wide selection of vertebrates. Members of this group have been isolated from amphibians, reptiles, birds, and mammals, but the host range of any particular virus is more restricted and includes certain genera of mosquito vectors as well as particular vertebrates.

The 5' and 3' ends of viral genomic RNAs are thought to contain *cis*-acting regulatory sequences that are required for viral RNA replication (reviewed by Strauss and Strauss [1983]). From comparative sequencing studies of a number of alphaviruses, we hypothesized some time ago that conserved RNA sequence elements located at the 5' and 3' ends of the genome were important for alphavirus RNA replication (reviewed by Strauss and Strauss [1986]). In particular, the 5'-terminal 40 nucleotides of the virus RNA could form a structure that was postulated to serve as an essential binding element during initiation of transcription of plus strands from minus-strand templates (Ou et al.,

1983), and a conserved sequence element of 19 nucleotides at the 3' end of the RNA was hypothesized to be required for transcription of these minus strands from the genomic RNA (Ou et al., 1981; Ou et al., 1982b). A conserved double-hairpin structure of 51 nucleotides found between nucleotides 155 and 205 in Sindbis virus RNA was also postulated to play a role in RNA replication, possibly in the initiation of minus strands from a plus-strand template (Ou et al., 1983). In addition, repeated sequence elements approximately 60 nucleotides long were found in the 3' untranslated region, whose function is unknown but which could also be involved in minus-strand synthesis (Ou et al., 1982b). Finally, a conserved sequence element 21 nucleotides long was found at the junction region between the structural and nonstructural protein-coding domains which contained the start of the subgenomic mRNA, and this was postulated to be the promoter for transcription of the subgenomic RNA (Ou et al., 1982a). Support for this last hypothesis has come from the work of Levis (1988), and the importance of the 19-nucleotide 3'-terminal element for RNA replication was shown by Levis et al. (1986).

The construction of a full-length cDNA clone of Sindbis virus from which infectious RNA can be transcribed in vitro (Rice et al., 1987) has made it possible to examine the functions of these *cis*-acting regulatory elements by site-specific mutagenesis. We have begun a program to isolate and characterize a large number of site-specific mutants in the 5' and 3' conserved domains. With these mutants, it should be possible to show whether our original hypotheses as to the functions of these sequence elements are correct and, in turn, to use these

---

James H. Strauss, Richard J. Kuhn, Hubert G. M. Niesters, and Ellen G. Strauss, Division of Biology, California Institute of Technology, Pasadena, California 91125.

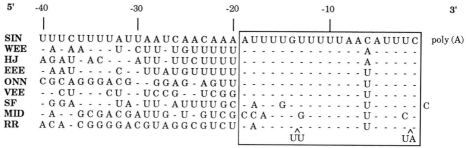

FIGURE 1. Sequences at the 3′ terminus of several alphaviruses. The genomic RNA is shown 5′ to 3′; nucleotides are numbered 3′ to 5′ from the poly(A). Virus abbreviations: SIN, Sindbis; WEE, Western equine encephalitis; HJ, Highlands J; EEE, Eastern equine encephalitis, ONN, O'Nyong-nyong; VEE, Venezuelan equine encephalitis; SF, Semliki Forest; MID, Middelburg; RR, Ross River. The 19-nucleotide conserved element is boxed. A dash indicates that the nucleotide is the same as that in the top line.

mutants to isolate second-site revertants in proteins that interact with these sequence elements. Mapping of such revertants would allow us to identify the *trans*-acting viral proteins that interact with each sequence element and possibly to assess the participation of host cell factors as well.

MUTATIONS IN THE 3′ NONTRANSLATED REGION

The conserved nucleotide sequence element present at the 3′ terminus of alphavirus RNAs is illustrated in Fig. 1. The 19 terminal nucleotides adjacent to the poly(A) tail are absolutely conserved among most of these viruses with the exception of nucleotide 6 from the 3′ end. In Semliki Forest, Ross River, and Middelburg viruses, the element diverges somewhat more; in the case of Ross River virus, the element is longer.

We have made 14 different nucleotide substitutions within this 19-nucleotide sequence element in Sindbis virus. These mutants are all viable, and most of the mutations have only modest effects on the virus growth rate. Representative growth curves of three of these mutants in chicken cells at 40°C are shown in Fig. 2. These three mutants grow to final titers equal to that of wild-type virus under these conditions, but virus production is delayed by about 1 h (notice that the growth curves are differential curves in which the yield of virus in the last hour is plotted, not the cumulative yield of virus at a given time point). Thus, all of the substitution mutations do in fact have deleterious effects on virus growth; moreover, the effects are generally more pronounced for growth in mosquito cells.

For example, the three mutants shown in Fig. 2 produce an order of magnitude less virus in mosquito cells than does the wild-type virus. The opposite has also been found, however. Some mutants grow as well as or even better than wild-type virus in mosquito cells but grow poorly in chicken cells in comparison with the parental virus.

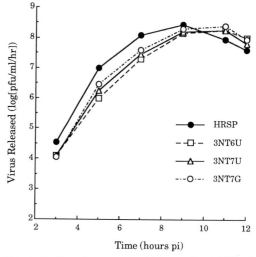

FIGURE 2. Growth curves in chicken cells at 40°C of Sindbis virus strain HRSP and of three mutants with substitutions in the 3′ nontranslated region. The medium over monolayers of infected cells was replaced every hour, and the samples from the times shown were assayed for virus titer by plaque assay on monolayers of chicken cells. The name of the mutant indicates the substitution made; e.g., mutant 3NT6U has a substitution of U for the parental C at nucleotide 6 in the 3′ nontranslated region. pi, Postinfection.

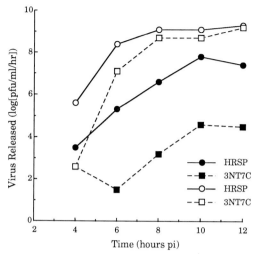

FIGURE 3. Growth curves of Sindbis virus strain HRSP and substitution mutant 3NT7C in mosquito cells (solid symbols) and chicken cells (open symbols) at 30°C. The experimental protocol was the same as for Fig. 2.

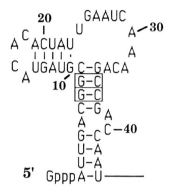

FIGURE 4. Conserved stem-loop structure at the 5' end of Sindbis virus. Deletion of any of the boxed residues gives mutants with temperature-sensitive phenotypes.

Mutant 3NT7C (containing a substitution of C for the parental A at position 7 from the 3' end) had a more profound effect on virus growth. It was temperature sensitive in chicken cells, producing plaques at 30°C but not at 40°C, and grew very poorly in mosquito cells at any temperature tested. Growth curves of this mutant in chicken cells and in mosquito cells at 30°C are shown in Fig. 3. In chicken cells at 30°C the virus grows almost as well as wild type but with a delay in virus production (as noted above, the virus is temperature sensitive and at 40°C produces much less virus than does the wild type in these cells). In mosquito cells at 30°C, however, the virus grows very poorly and produces some 3 orders of magnitude less virus than does the wild type.

We have also made 17 deletion mutants in which from 1 to 8 nucleotides within the 3' 25 nucleotides were deleted. Deletions involving the 3'-terminal C were not viable, but most of the other deletions tested were viable. In general, these mutants had phenotypes similar to those of the point (substitution) mutations in that they were viable, most grew well in chicken cells but more slowly than wild type, and most were more affected for growth in mosquito cells than in chicken cells.

We have also made six constructs with much larger deletions (31 to 358 nucleotides long) in the 3' nontranslated region, of which three were viable. The largest viable deletion

was 3NTd(25-318), in which all of the untranslated region but the 3'-terminal 24 nucleotides, and 1 nucleotide following the stop codon that terminates the open reading frame for the structural proteins, was deleted. This mutant grows reasonably well in chicken cells but very poorly in mosquito cells.

MUTANTS IN THE 5' NONTRANSLATED REGION

The 5' nontranslated region of Sindbis virus is 59 nucleotides long. The first 44 nucleotides of the sequence are shown in Fig. 4, drawn as a theoretically stable stem-loop structure. Similar structures can be drawn for the 5' termini of other alphaviruses, although the linear sequences are not conserved. We have made 25 deletion mutants in the 5' nontranslated region in which from 1 to 15 nucleotides were deleted and in addition constructed 2 substitution mutants that change the theoretical stability of the stem structure. All but two of these mutants were viable, and most grew fairly well in chicken cells. Growth curves of four mutants in chicken cells at 30°C are shown in Fig. 5; these data illustrate that although these mutants in general grow well, they are delayed in growth by 1 to 2 h relative to wild-type virus, and some never reach the titers obtained with wild-type virus. As was the case for the mutants in the 3' nontranslated region, most of the 5' mutations had a more severe effect in mosquito cells than they did in chicken cells.

One interesting mutant that gave unusual results was 5NTd(41-55). This virus threw off both large and small plaques whose growth properties appeared to be almost identical. Be-

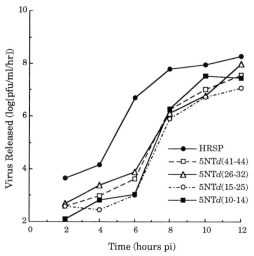

FIGURE 5. Growth curves of mutants with deletions in the 5′ nontranslated region in chicken cells at 30°C. The experimental protocol was the same as for Fig. 2.

cause neither virus seemed to have a growth advantage, it is strange that both would be present. Furthermore, the virus grows better in mosquito cells than does the wild-type virus, as illustrated in Fig. 6 for the small-plaque virus. In chicken cells, the virus grows almost identically to wild-type virus, with a slight reduction in yield late in infection. The differences in the large-plaque and small-plaque variants and their origin are presently unknown.

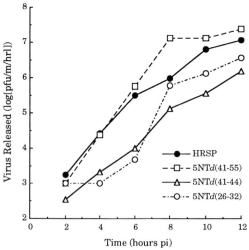

FIGURE 6. Growth curves of mutants with deletions in the 5′ nontranslated region in mosquito cells at 30°C. The experimental protocol was the same as for Fig. 2.

One set of deletions gave rise to temperature-sensitive virus that could form plaques and grow fairly well in chicken cells at 30°C but which grew poorly and did not form plaques at 40°C. These involve deletions in the G · C base pairs indicated by the boxes in the stem-loop structure (Fig. 4). Deletion of either G7 or G8 (which gives the same sequence) or of either of the corresponding nucleotides C37 and C36, or deletion of the G · C base pair by deleting both G7 and C37, has a drastic effect on virus replication, giving rise to a temperature-sensitive phenotype in chicken cells. In mosquito cells these mutants grow poorly, and the double deletion is barely able to replicate. Since none of the other deletions tested was temperature sensitive, this result suggests that the G · C base pair is indeed formed and that the structure is important for virus replication.

DOUBLE MUTANTS

We have constructed several double mutants in which one of the defined mutations at the 5′ end, described above, has been combined with one of the characterized mutations from the 3′ end. These double mutants grow very poorly, much more poorly than would have been predicted from the phenotypes of the individual mutants. Thus, in the double mutants the effects of these mutations on replication are cooperative in some way.

THE 51-NUCLEOTIDE-SEQUENCE ELEMENT

The sequence of Sindbis virus RNA between nucleotides 155 and 205 is shown in Fig. 7. As illustrated, this sequence is capable of forming two hairpin structures with a calculated thermal stability sufficiently high that these hairpins probably exist in solution under physiological conditions. Because this sequence encodes part of nonstructural protein 1 of Sindbis virus, the number of nucleotide changes that could be made in this structure without changing the protein sequence was limited (Fig. 7). Nevertheless, we have constructed a number of nucleotide substitution mutants (utilizing mostly third-codon-position silent changes) that would be expected to destabilize the structure shown. In every case where a silent change was made, the mutant was viable. Two of these mutations had only minor effects on virus growth rate, but all others had a pronounced effect, with virus yields being depressed by 1 to 4 orders of magnitude.

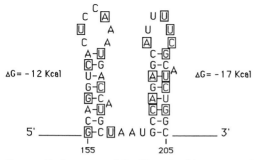

FIGURE 7. Sequence of the 51-nucleotide conserved sequence element in Sindbis virus. The sequence is shown as a double stem-loop structure. The stabilities of the proposed hairpins are indicated. Nucleotides at which it is possible to make silent changes are boxed; 20 silent mutants were made and tested.

Once again, differences were noted between chicken cells and mosquito cells. These results suggest that these stem-loop structures are in fact important for replicase binding and RNA replication and that changes which destabilize the conformation or change the sequence are viable but the virus grows at reduced rates, presumably because of weakened interactions of the replicase with its template.

REPLACEMENT OF SINDBIS SEQUENCE ELEMENTS WITH THOSE FROM OTHER ALPHAVIRUSES

We have constructed a Sindbis virus in which the entire 3′ nontranslated region was replaced with that from Ross River virus. This virus grows essentially as wild-type Sindbis virus. Little or no difference in the growth rates of the mutant and wild type can be detected in chicken cells or in mosquito cells, and the plaques formed by the two viruses appear identical. Sindbis and Ross River viruses differ at several positions in the conserved 3′-terminal sequence element, and, as noted above, the element is longer by four nucleotides in Ross River than in Sindbis virus. The lengths of the 3′ nontranslated regions and the sequence and number of repeated elements are also different in the two viruses. However, replacement of the entire 3′ nontranslated region of Sindbis virus with that of Ross River virus has little effect. Thus, even though individual substitutions in the 3′-sequence element are deleterious, replacement of the whole unit with another functional unit that differs at many positions has little effect in this case. It may be that the sequence and

structure of the entire 3′ nontranslated domain as a whole, and not just the 3′-terminal conserved element, must be considered when one contemplates its function in RNA replication.

CONCLUSIONS

The alphavirus genome appears to be remarkably plastic. Even though in nature sequence elements are rigorously conserved among different alphaviruses and therefore must play an important role in virus replication, large changes in these elements are tolerated in that viable virus is obtained and in many cases the virus grows quite well, at least in the laboratory. However, almost every change that we have made in this study has led to virus that grows more slowly, and often to lower yield, than the parental virus in at least one of the cell lines tested. The differential effects of many of the mutations on virus growth in chicken versus mosquito cells, including the observation that some mutants grow relatively better in chicken cells whereas others grow better in mosquito cells, implies that host factors may be involved in viral RNA replication. The conserved consensus sequences that have emerged in alphavirus genomes thus appear to constitute a compromise between that sequence which would best adapt the virus to mosquitoes and that which would best adapt the virus to vertebrates. It is also important to keep in mind the difference between virus competent for transmission in tissue culture cells in the laboratory and virus capable of being transmitted in nature. Changes that have only modest effects on the fitness of the virus for passage in culture may have profound effects on the ability of the virus to be maintained in nature.

The changes at nucleotide 7 in the 3′ nontranslated region are of particular interest and illustrate the fact that at the current time it is impossible to predict the results of simple substitutions or deletions in these viral sequence elements. Nucleotide 7 is an A in the parental virus. Changing this nucleotide to U or to G, as well as deleting this nucleotide, has only modest effects on the virus growth rate. However, change of this nucleotide to C has a dramatic effect on virus replication, giving rise to virus that is rigorously temperature sensitive in chicken cells and grows very poorly in mosquito cells. It is also noteworthy that this mutation does not readily revert, even though change of this C to any other nucleotide would lead to temperature-insensitive virus. This region may possess an overall structure that is important for replication and that also leads to a slow rate of nucleotide substitution.

Acknowledgments. We are grateful to Zhang Hong for competent technical assistance.

The work reviewed here was supported by National Science Foundation grant DMB-8617372.

REFERENCES

Calisher, C. H., and N. Karabatsos. 1988. Arbovirus serogroups: definition and geographic distribution, p. 19–57. *In* T. P. Monath (ed.), *The Arboviruses: Epidemiology and Ecology*. CRC Press, Inc., Boca Raton, Fla.

Levis, R. 1988. Sindbis virus cis-acting sequences. Ph.D. thesis, Washington University School of Medicine, St. Louis, Mo.

Levis, R., B. G. Weiss, M. Tsiang, H. Huang, and S. Schlesinger. 1986. Deletion mapping of Sindbis virus DI RNAs derived from cDNAs defines the sequences essential for replication and packaging. *Cell* **44:**137–145.

Ou, J.-H., C. M. Rice, L. Dalgarno, E. G. Strauss, and J. H. Strauss. 1982a. Sequence studies of several alphavirus genomic RNAs in the region containing the start of the subgenomic RNA. *Proc. Natl. Acad. Sci. USA* **79:**5235–5239.

Ou, J.-H., E. G. Strauss, and J. H. Strauss. 1981. Comparative studies of the 3′-terminal sequences of several alphavirus RNAs. *Virology* **109:**281–289.

Ou, J.-H., E. G. Strauss, and J. H. Strauss. 1983. The 5′-terminal sequences of the genomic RNAs of several alphaviruses. *J. Mol. Biol.* **168:**1–15.

Ou, J.-H., D. W. Trent, and J. H. Strauss. 1982b. The 3′-non-coding regions of alphavirus RNAs contain repeating sequences. *J. Mol. Biol.* **156:**719–730.

Rice, C. M., R. Levis, J. H. Strauss, and H. V. Huang. 1987. Production of infectious RNA transcripts from Sindbis virus cDNA clones: mapping of lethal mutations, rescue of a temperature sensitive marker, and in vitro mutagenesis to generate defined mutants. *J. Virol.* **61:**3809–3819.

Strauss, E. G., and J. H. Strauss. 1983. Replication strategies of the single stranded RNA viruses of eukaryotes. *Curr. Top. Microbiol. Immunol.* **105:**1–98.

Strauss, E. G., and J. H. Strauss. 1986. Structure and replication of the alphavirus genome, p. 35–90. *In* S. Schlesinger and M. J. Schlesinger (ed.), *The Togaviridae and Flaviviridae*. Plenum Publishing Corp., New York.

Strauss, J. H., E. G. Strauss, C. S. Hahn, Y. S. Hahn, R. Galler, W. R. Hardy, and C. M. Rice. 1987. Replication of alphaviruses and flaviviruses: proteolytic processing of polyproteins. *UCLA Symp. Mol. Cell. Biol.* **54:**209–225.

Chapter 11

Murine Coronavirus RNA Synthesis

Julian L. Leibowitz, Phillip W. Zoltick, Kathryn V. Holmes, Emilia L. Oleszak,
and Susan R. Weiss

The coronaviruses are large, pleomorphic, enveloped viruses that bud exclusively from intracellular membranes (Robb and Bond, 1979). The coronavirus genome is a single-stranded, nonsegmented RNA molecule of positive polarity, 27 to 35 kilobases (kb) in length, making it the largest RNA genome described to date (Lai and Stohlman, 1978; Boursnell et al., 1987; Pachuk et al., 1989). The strategy coronaviruses utilize to express their genetic information entails the synthesis of a series of five to seven subgenomic mRNAs, an RNA species that is indistinguishable from the virion RNA, and a single negative-strand molecule that has the same apparent molecular size as virion RNA (Stern and Kennedy, 1980; Lai et al., 1981; Lai et al., 1982; Leibowitz et al., 1981; Spaan et al., 1981, 1982; Wege et al., 1981; Weiss and Leibowitz, 1981; Dennis and Brian, 1982). The subgenomic mRNAs together with the virion RNA make up a 3'-coterminal nested set in which each successively larger RNA species contains all of the sequences present in the smaller RNA species and additional sequences at its 5' end consistent with its molecular size (Stern and Kennedy, 1980; Lai et al., 1981; Leibowitz et al., 1981; Weiss and Leibowitz, 1981, 1983). This is shown schematically in Fig. 1. In addition, each mRNA contains a leader sequence of approximately 70 bases at its 5' terminus (Lai et al., 1981; Lai et al., 1982; Lai et al., 1984; Leibowitz et al., 1981; Spaan et al., 1982; Spaan et al., 1983; Armstrong et al., 1983; Budzilowicz et al., 1985). The leader sequence is identical to the sequence of the 5' terminus of the virion RNA and contains a short sequence of 7 to 11 bases very similar to sequences preceding each gene in genomic RNA.

The genetic organization of the coronavirus genome has been determined by two approaches. Cell-free translation studies of individual mRNAs, either purified from infected cells or synthesized by in vitro transcription of cDNA clones, have allowed investigators to determine which mRNAs encode particular gene products (Rottier et al., 1981; Leibowitz et al., 1982b; Siddell, 1983). These experiments have been complemented by extensive cloning and sequencing studies. Taken together, these results have allowed the construction of the genetic maps of several coronaviruses. The map of the extensively studied mouse hepatitis virus (MHV) is shown in Fig. 1. Several features of this map are worth noting. In general, only the first open reading frame of each mRNA is thought to be translated. There are two exceptions to this. The second open reading frame of gene 5 (ORF b) is translated both in vitro and in the infected cell, presumably by internal initiation of protein synthesis on mRNA5 (Skinner et al., 1985; Budzilowicz and Weiss, 1987; Leibowitz et al., 1988). Perhaps interestingly, the second open reading frame in gene 1 appears to be expressed by a ribosomal frameshifting mechanism (Boursnell et al., 1987; Brierley et al., 1987). This results in potentially allowing the expression of MHV gene 1, the putative RNA-dependent RNA polymerase gene, as a primary translation product greater than 700 kilodaltons in size (Boursnell et al., 1987; Brierley et al., 1987). This primary translation product is

Julian L. Leibowitz and Emilia L. Oleszak, Department of Pathology and Laboratory Medicine, University of Texas Medical School, P.O. Box 20708, Houston, Texas 77225.   *Phillip W. Zoltick and Susan R. Weiss,* Department of Microbiology, University of Pennsylvania Medical School, Philadelphia, Pennsylvania 19104-6076.   *Kathryn V. Holmes,* Department of Pathology, Uniformed Services University of the Health Sciences, Bethesda, Maryland 20814-4799.

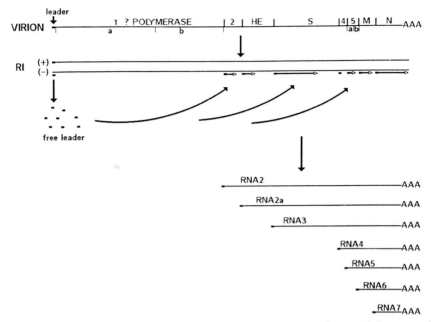

FIGURE 1. Leader-primed model of coronavirus mRNA synthesis. A genetic map of the coronavirus MHV is shown at the top. Genes for which functions have been assigned are indicated by letters; N is the nucleocapsid gene, M (formerly E1) is the virion membrane-associated protein, HE is the hemagglutinin/esterase protein, and S (formerly E2) is the peplomer protein. Nonstructural genes with unknown or unproven functions are designated by numbers. Genes 1 and 5 contain two potential open reading frames (a and b), each of which is indicated.

thought to undergo proteolytic processing to yield functional gene products.

Several alternative models can be envisioned for generating coronavirus subgenomic mRNAs from a single genome-size negative-strand template. In the first model, the subgenomic RNAs are derived from a large precursor molecule. This model is unlikely to be correct, since UV transcriptional mapping demonstrates that the UV target size of each subgenomic mRNA is consistent with its physical size (Jacobs et al., 1981). A second possible model proposes that the subgenomic mRNAs are generated by looping out of the negative-strand template RNA after synthesis of the leader sequence, followed by transcription of the body of the mRNAs by jumping across the loop. There is little evidence for or against this model. The most likely model for synthesizing coronavirus subgenomic RNA is the leader-primed model. In this model (Fig. 1), leader RNAs are synthesized and released from the template RNA and subsequently reassociate with template RNA molecules at intragenic regions, presumably via the 7 to 11 bases contained in the leader RNAs that can base pair with the intragenic sequences on

negative-strand RNAs. Evidence in support of this model includes the association of leader RNAs with partially double-stranded transcription complexes, the presence of free small virus-specific leader-related RNA molecules, and genetic and biochemical evidence suggesting that leader RNAs can freely reassort (Baric et al., 1983; Baric et al., 1985; Baric et al., 1987; Shieh et al., 1987; Makino et al., 1986). A fourth model, also consistent with most of the data, would invoke transplicing of free leader RNA to the body of subgenomic mRNAs, rather than the free leader serving a priming function, although the finding of leader RNA associated with replicative intermediates implies that if transplicing occurs, it occurs cotranscriptionally.

Thus, the mechanism for synthesizing MHV RNAs is quite complex. We have taken two approaches to understanding the mechanisms of MHV replication. The first of these is genetic. We are characterizing the phenotypes of temperature-sensitive mutants unable to induce the synthesis of virus-specific mRNAs, with the eventual goal of associating individual genetic lesions in different locations along the MHV genome with particular defects in MHV

TABLE 1
Effect of Time of Temperature Shift on Virus Yield

Mutant	% Yield at time (h) of shift given[a]						
	34°C → 39°C				39°C → 34°C		
	2	4	6	8	2	4	6
tsA204	<1	<1	<1	<1	42	6.6	<1
tsB105	<1	<1	<1	<1	36	<1	<1
tsC60	<1	<1	<1	<1	18	3.8	<1
tsD264	<1	<1	<1	<1	7.4	<1	<1
tsE248	<1	<1	<1	1.2	38	3.8	<1
tsF203	25	34	9.4	7.1	<1	<1	<1

[a] Expressed as percentage of virus yield obtained after incubation at 34°C for 16 h. Values for all mutants after incubation for 12 h at 39°C were <1.

RNA synthesis. The second approach has been to develop antibodies directed against polypeptide sequences encoded in the putative MHV polymerase region of the genome, gene 1, and biochemically characterize the functions and structure of proteins containing these sequences.

GENETIC STUDIES

Our initial genetic studies of two panels of MHV strain JHM (MHV-JHM) mutants allowed us to divide these mutants into seven complementation groups (Leibowitz et al., 1982a). We had previously demonstrated that mutants from six of these complementation groups (A to F) were unable to induce the synthesis of MHV virion or mRNAs in cells incubated continuously under restrictive conditions (39°C). Initial blot hybridization experiments with strand-specific probes determined that cells infected with mutants representing complementation groups A to F did not synthesize any detectable MHV negative-strand RNA (data not shown). When a duplicate blot was probed for positive-strand RNA, similar results were obtained.

We subsequently performed a series of temperature shift experiments with our MHV-JHM mutants to determine when in the replicative cycle the gene products altered by mutation were required (Table 1). Mutant tsF203, representing complementation group F, appeared to be defective in a gene that was required very early in the infectious cycle. Incubation at the permissive temperature (34°C) for as little as 2 h was sufficient for significant virus replication to occur. Conversely, incubating tsF203-infected cells at 39°C during the first 2 h of infection was sufficient to reduce virus replication to less than

1% of that observed under permissive conditions. Mutants representing complementation groups A to E also were sensitive to 39°C during the early phase of infection, although at somewhat later times than tsF203. Incubation of mutants representing groups A to E at 39°C until 4 to 6 h postinfection prevented viral replication. Cultures infected with these mutants and incubated at 34°C until 8 h postinfection and then shifted to 39°C yielded little or no infectious virus. Extending the time of incubation at 34°C to 10 h postinfection, a time when mRNA synthesis was established, resulted in the production of small amounts of infectious virus (data not shown). These results suggested that the functions represented in our panel of RNA mutants could be classified as an extremely early function, defective in our group F mutant, and functions that are required slightly later in infection, represented by complementation groups A to E.

To better understand the possible gene functions needed for virus-specific mRNA synthesis, we performed a series of temperature shift experiments, using mutants representing complementation groups A, B, and D. Replicate cell cultures were infected with either wild-type or mutant MHV-JHM and incubated at 34°C for 3 to 6 h. At the times indicated in Fig. 2, one set of cultures was shifted to 39°C and incubated for 1 h with $^{32}P_i$ in the presence of actinomycin D; a second set of cultures was maintained at 34°C and similarly labeled for 1 h at that temperature. At the end of the labeling period, the intracellular RNA was extracted and analyzed for MHV-specific mRNA synthesis by agarose gel electrophoresis (Fig. 2). In cultures infected with wild-type virus and maintained at 34°C, MHV-specific RNA synthesis was first observed during the labeling period at 6 to 7 h postinfection. At that time, all of the MHV-specific RNA species were present. It is possible that MHV-specific RNA would have been detected 1 h earlier at 34°C, but the RNA from that particular time point appears to have been somewhat degraded. Temperature shift to 39°C increased the amount of MHV-specific RNA synthesized but did not alter the species observed. This finding is in contrast to what occurred in cultures infected with the temperature-sensitive mutants. For all three mutants, although most obviously for ts201, MHV genome-size RNA synthesis was first detected by labeling at 34°C from 5 to 6 h postinfection. Shiftup to 39°C at 6 h postinfection, a time when genome-size RNA synthesis had started but mRNA synthesis was not yet established, prevented the cells from synthesiz-

MOCK   TS201   TS216   TS129   WT

<u>5 - 6   6 - 7</u> <u>5 - 6   6 - 7</u> <u>5 - 6   6 - 7</u> <u>5 - 6   6 - 7</u> <u>5 - 6   6 - 7</u>

a   b   a   b   a   b   a   b   a   b   a   b   a   b   a   b   a   b   a   b

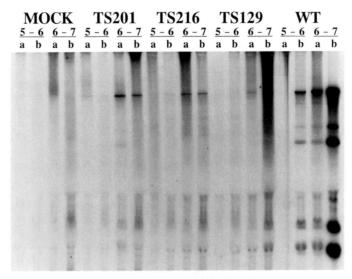

FIGURE 2. Effect of temperature shiftup on MHV-specific RNA synthesis. Replicate cell cultures were infected with either wild-type MHV-JHM (WT), tsA201 (TS201), tsB216 (TS216), or tsD129 (TS129) or were mock infected and were then incubated at 34°C. For each time point, actinomycin D (5 μg/ml) was added 15 min before labeling, and one set of cultures (set b) was shifted to 39°C; the second set (set a) remained at 34°C. Cells were labeled with 100 μCi of $^{32}P_i$ for 1 h, and total intracellular RNA was extracted, denatured, and resolved by electrophoresis on 0.8% agarose gels. The times (in hours) of labeling for each sample are indicated.

ing detectable amounts of MHV-specific subgenomic mRNAs, although genome RNA synthesis continued.

To determine whether the apparent inability of tsA201, tsB216, and tsD129 to direct the synthesis of MHV subgenomic mRNAs after temperature shiftup could be related to defects in leader RNA synthesis, we performed a series of temperature shift experiments similar in design to those depicted in Fig. 2. In these experiments, the cultures were not metabolically labeled with $^{32}P_i$, since several abundant host-derived RNA products of approximately the same size (70 bases) as leader RNAs were labeled even in the presence of high doses of actinomycin D, interfering with our ability to identify leader RNA by direct labeling in the presence of actinomycin D. However, the accumulation of leader RNA could be monitored by polyacrylamide gel electrophoresis (PAGE) followed by RNA blot hybridization with leader-specific probes. The result of this analysis is shown in Fig. 3. Mutants ts216 and ts129 demonstrated similar phenotypes. Cells infected with either of these mutants contained much smaller amounts of leader-related RNAs than did cells infected with wild-type MHV. This result contrasts with that for cells infected with tsA201, which accumulated leader-related

RNAs after shiftup to nonpermissive temperatures to at least as great an extent as did cells infected with wild-type MHV-JHM. A mutant of MHV-A59 similar in many respects to tsA201 has been identified by Baric et al. (1985). These results suggest that the defect of subgenomic RNA synthesis of group A mutants is likely to involve a step subsequent to the synthesis of leader RNAs, whereas the defects in group B and D mutants are in functions that are necessary for leader RNA synthesis. A possible alternative explanation for the results observed with group B and D mutants is that these mutations result in leader RNA turning over more rapidly.

The data described above suggest that our temperature-sensitive mutants exhibit several different phenotypes under restrictive conditions. One complementation group, group F, appears to encode a function that is required very early in the replicative cycle. The biochemical function of the gene represented by this group remains to be elucidated. A biochemical characterization of mutants representing three of the five remaining RNA complementation groups has begun. Two of these mutants, tsB216 and tsD129, have similar phenotypes, synthesizing reduced amounts of leader RNA (as compared with wild type) when shifted to restrictive conditions. Although a more detailed biochemi-

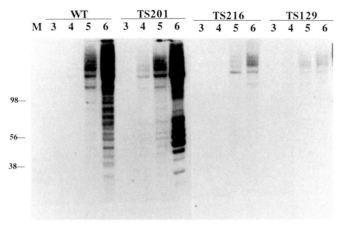

FIGURE 3. Effect of temperature shiftup on MHV leader RNA synthesis. Replicate cell cultures were infected with either wild-type MHV-JHM (WT), *ts*A201 (TS201), *ts*B216 (TS216), or *ts*D129 (TS129) or were mock infected (M) and were then incubated at 34°C. At the times (in hours) indicated, one set of cultures was shifted to 39°C and incubated for 1 h. Total RNA was extracted, electrophoresed on a 12% polyacrylamide gel, and electrophoretically transferred to a nylon membrane. The membranes were probed with a leader-specific probe and autoradiographed. The positions of in vitro RNA transcripts used as size markers (98, 56, and 38 bases) are shown on the left.

cal characterization of these mutants is needed, these preliminary data suggest that at least two genes may be involved in leader RNA synthesis. The function represented by complementation group A appears to be needed subsequent to leader RNA synthesis. Again, a more detailed biochemical knowledge of MHV mRNA synthesis is needed to determine the precise functions that are defective in group A mutants. If the generally accepted leader-primed model of coronavirus mRNA synthesis is correct, this gene could be involved in either priming or elongation of subgenomic mRNA synthesis.

BIOCHEMICAL AND IMMUNOLOGICAL STUDIES

The second approach we have taken to understanding MHV RNA synthesis has been to develop immunologic reagents that recognize gene products encoded by the region of the MHV genome thought to encode polypeptides involved in virus-specific RNA synthesis. Portions of putative MHV polymerase proteins were expressed in *Escherichia coli* and used to raise antibodies (Zoltick et al., in press). These reagents can then be used in biochemical studies to identify the polypeptide products of particular regions of the MHV genome and, in appropriate assays, to identify the functions carried out by these polypeptides.

The large size, about 23 kb, of the putative polymerase region of the MHV genome (Fig. 1) made it desirable to use a strategy for the expression of putative polymerase proteins that did not require detailed and complete sequence information of the gene being expressed. For this purpose, we have adapted an expression system that we have previously used for expressing MHV-A59 gene 5 ORF b. The vector we used contains the 5' end of the E. coli recA gene (35 codons) located upstream of and out of frame with the lacZ coding sequence. Bacteria containing this plasmid do not synthesize β-galactosidase, since recA and lacZ are out of frame. Viral sequences are inserted between the recA and lacZ genes by homopolymer tailing. This serves to randomize the junctions between the viral sequences and the flanking procaryotic sequences, resulting in about 5% of transformants synthesizing tripartite proteins consisting of 35 amino acids of RecA protein fused to the desired viral protein sequences fused in turn to β-galactosidase. These transformants are easily selected by virtue of their LacZ[+] phenotype.

This expression system was used to express several portions of the putative polymerase region of MHV-A59. Results obtained by using antibodies raised against protein sequences encoded 5.3 to 6.7 kb (clone 1533) and 3 to 4.5 kb (clone 1410) from the 5' end of the genome are presented in Fig. 4. Previous work has demon-

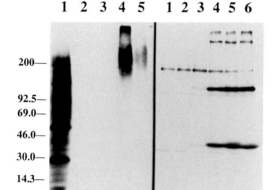

FIGURE 4. Radioimmunoprecipitation of putative polymerase gene products. (A) Virion RNA was translated in an mRNA-dependent reticulocyte lysate, and the translation products were immunoprecipitated with anti-1533 antibodies and analyzed by SDS-PAGE. Lanes: 1, translation products analyzed without immunoprecipitation; 2, translation products immunoprecipitated with preimmune serum A; 3, translation products immunoprecipitated with preimmune serum B; 4, translation products immunoprecipitated with anti-1533 serum A; 5, translation products immunoprecipitated with anti-1533 serum B. (B) Cells were infected with MHV-A59 (lanes 2 to 6) or mock infected (lane 1) and labeled with [^{35}S]methionine at 4 (lane 2), 8 (lane 3), 12 (lane 4), 16 (lane 5), or 20 (lane 6) h postinfection. Samples were immunoprecipitated with anti-1410 antibodies and analyzed by SDS-PAGE. The positions of molecular size markers (indicated in kilodaltons) are shown on the left.

strated that cell-free translation of viral genome results in several related proteins of 200 to 250 kilodaltons (Leibowitz et al., 1981; Dennison and Perlman, 1986) that correspond to the open reading frame initiated at the 5' end of the genome. As expected, antisera raised against the sequences encoded in clone 1533 immunoprecipitated a series of large polypeptides synthesized in rabbit reticulocyte lysates programmed with MHV virion RNA (Fig. 4A). Antisera directed against clone 1410 immunoprecipitated similar bands when reacted with cell-free translation products of genome RNA. Radioimmunoprecipitation of infected cell lysates with antibody directed against sequences encoded in clone 1410 demonstrated that this antibody reacted with a series of proteins, the largest of which had an estimated molecular size greater than 400 kilodaltons (Fig. 4B). This protein could potentially correspond to the gene 1 ORF a (Fig. 1).

A biochemical approach to identify the

TABLE 2
Antibody Inhibition of MHV-Specific RNA
Synthesis in Permeabilized Cells[a]

Antibody	% Activity	
	Expt 1	Expt 2
1.16-1 (α-N)	67	ND[b]
Anti-MHV	ND	31
Anti-1533	72	64
Anti-1410	90	64

[a] Measured by incorporation of [^{3}H]UTP into actinomycin D-resistant, trichloroacetic acid-precipitable radioactivity and expressed as a percentage of that in permeabilized MHV-infected cultures to which no antibody was added. Control cultures to which irrelevant monoclonal or polyclonal antibodies were added incorporated [^{3}H]UTP at from 98 to 84% of the levels observed in cultures not treated with antibody.
[b] ND, Not done.

functions required for MHV-specific RNA synthesis required the development of a reliable assay for MHV polymerase activity. Toward that end, we have developed a permeabilized cell system for assaying RNA-dependent RNA polymerase activity (Leibowitz and DeVries, 1988). This system has much of the flexibility of the truly cell-free systems that have been described but is more active and more reproducible in our hands. Most importantly, permeabilized cells allow antibodies to enter the cell and have access to polypeptides that are involved with MHV RNA synthesis. We have tested four antibodies for their ability to interfere with virus-specific RNA synthesis, anti-1533, anti-1410, a polyvalent anti-MHV antibody raised against purified virions, and a monoclonal antinucleocapsid antibody. Both antinucleocapsid and antivirion antibodies inhibited MHV RNA polymerase activity, as measured by incorporation of [^{3}H]UTP into trichloroacetic acid-precipitable material in the presence of actinomycin D (Table 2). This result was expected, since nucleocapsid protein binds to leader RNA (Baric et al., 1988; Stohlman et al., 1988) and antinucleocapsid antibody inhibits UTP incorporation in a cell-free RNA polymerase assay (Compton et al., 1987). Similarly, both antibodies raised against putative polymerase polypeptides expressed in E. coli also inhibited UTP incorporation. These results, although very preliminary, provide the first direct biochemical evidence for a role for the proteins encoded in clones 1533 and 1410 in MHV RNA synthesis. It should be pointed out that the products synthesized in permeabilized cells late in infection, when these experiments were performed, are largely subgenomic mRNAs. The inhibition we observed most likely reflects an inhibition of the synthesis of these RNAs. This conclusion will need to be confirmed by biochemical analysis of the products synthesized in the presence of these antibodies. It is our hope that the application of this approach, using antibodies raised to proteins encoded in other regions of MHV gene 1, and further biochemical characterization of the effects of these antibodies on MHV RNA synthesis will allow us to more precisely define the role of various portions of MHV gene 1 in this process.

ACKNOWLEDGMENTS. This work was supported by Public Health Service grants AI 17418, NS 21954, and NS 20834 from the National Institutes of Health and by Biomedical Research Support Grant 05745. P.W.Z. was supported in part by Public Health Service training grant NS 07180 from the National Institutes of Health.

We thank Lee Erickson, Ed Murray, and Jim DeVries for technical assistance.

REFERENCES

Armstrong, J., S. Smeekens, and P. J. M. Rottier. 1983. Sequence of the nucleocapsid gene from murine coronavirus MHV-A59. *Nucleic Acids Res.* **11**:883–891.

Baric, R. S., G. W. Nelson, J. O. Fleming, R. J. Deans, J. G. Keck, N. Casteel, and S. A. Stohlman. 1988. Interactions between coronavirus nucleocapsid protein and viral RNAs: implications for viral transcription. *J. Virol.* **62**:4280–4287.

Baric, R. S., C. K. Shieh, S. A. Stohlman, and M. M. C. Lai. 1987. Analysis of intracellular small RNAs of mouse hepatitis virus: evidence for discontinuous transcription. *Virology* **156**:342–354.

Baric, R. S., S. A. Stohlman, and M. M. C. Lai. 1983. Characteristic of replicative intermediate RNA of mouse hepatitis virus: presence of leader RNA sequences on nascent chains. *J. Virol.* **48**:633–640.

Baric, R. S., S. A. Stohlman, M. K. Razavi, and M. M. C. Lai. 1985. Characterization of leader-related small RNAs in coronavirus-infected cells: further evidence for leader-primed mechanism of transcription. *Virus Res.* **3**:19–33.

Boursnell, M. E. G., T. D. Brown, I. J. Foulds, P. F. Green, F. M. Tomley, and M. M. Binns. 1987. Completion of the sequence of the genome of the coronavirus avian infectious bronchitis virus. *J. Gen. Virol.* **68**:57–77.

Brierley, I., M. E. G. Boursnell, M. M. Binns, B. Bilimoria, V. C. Blok, T. D. K. Brown, and S. C. Inglis. 1987. An efficient ribosomal frame-shifting signal in the polymerase-encoding region of the coronavirus IBV. *EMBO J.* **6**:3779–3785.

Budzilowicz, C. J., and S. R. Weiss. 1987. In vitro synthesis of two polypeptides from a nonstructural gene of coronavirus mouse hepatitis virus strain A59. *Virology* **157**:509–515.

Budzilowicz, C. J., S. P. Wilczynski, and S. R. Weiss. 1985. Three intergenic regions of coronavirus mouse hepatitis virus strain A59 genome RNA contain a common nucleotide sequence that is homologous to the 3′ end of the viral mRNA leader sequence. *J. Virol.* **53**:834–840.

Compton, S. R., D. B. Rodgers, K. V. Holmes, D. Fertsch, J. Remenick, and J. J. McGowan. 1987. In vitro replication of mouse hepatitis virus strain A59. *J. Virol.* **61**:1814–1820.

Dennis, D. E., and D. A. Brian. 1982. RNA-dependent RNA polymerase activity in coronavirus-infected cells. *J. Virol.* **42**:153–164.

Dennison, M. R., and S. Perlman. 1986. Translation and processing of mouse hepatitis virus virion RNA in a cell-free system. *J. Virol.* **60**:12–18.

Jacobs, L., W. J. M. Spaan, M. C. Horzinek, and B. A. M. Van Der Zeijst. 1981. Synthesis of subgenomic mRNAs of mouse hepatitis virus is initiated independently: evidence from UV transcription mapping. *J. Virol.* **39**:401–406.

Lai, M. M., R. S. Baric, P. R. Brayton, and S. A. Stohlman. 1984. Characterization of leader RNA sequences on the virion and mRNAs on mouse hepatitis virus, a cytoplasmic RNA virus. *Proc. Natl. Acad. Sci. USA* **81**:3626–3630.

Lai, M. M., P. R. Brayton, R. C. Armen, C. D. Patton, C. Pugh, and S. A. Stohlman. 1981. Mouse hepatitis virus A59: mRNA structure and genetic localization of the sequence divergence from hepatotropic strain MHV-3. *J. Virol.* **39**:823–834.

Lai, M. M., C. D. Patton, and S. A. Stohlman. 1982. Further characterization of mRNA's of mouse hepatitis virus: presence of common 5′-end nucleotides. *J. Virol.* **41**:557–565.

Lai, M. M., and S. A. Stohlman. 1978. RNA of mouse hepatitis virus. *J. Virol.* **26**:236–242.

Leibowitz, J. L., and J. R. DeVries. 1988. Synthesis of virus-specific RNA in permeabilized murine coronavirus-infected cells. *Virology* **166**:66–75.

Leibowitz, J. L., J. R. DeVries, and M. V. Haspel. 1982a. Genetic analysis of murine hepatitis virus strain JHM. *J. Virol.* **42**:1080–1087.

Leibowitz, J. L., S. Perlman, G. Weinstock, J. R. DeVries, C. Budzilowicz, J. M. Weissemann, and S. R. Weiss. 1988. Detection of a murine coronavirus nonstructural protein encoded in a downstream open reading frame. *Virology* **164**:156–164.

Leibowitz, J. L., S. R. Weiss, E. Paavola, and C. W. Bond. 1982b. Cell-free translation of murine coronavirus RNA. *J. Virol.* **43**:905–913.

Leibowitz, J. L., K. C. Wilhelmsen, and C. W. Bond.

1981. The virus-specific intracellular RNA species of two murine coronavirus: MHV-A59 and MHV-JHM. *Virology* **114**:39–51.

Makino, S., S. A. Stohlman, and M. M. C. Lai. 1986. Leader sequences of murine coronavirus RNAs can be freely reassorted: evidence for the role of free leader RNA in transcription. *Proc. Natl. Acad. Sci. USA* **83**:4204–4208.

Pachuk, C. J., P. J. Bredenbeek, P. W. Zoltick, W. J. Spaan, and S. R. Weiss. 1989. Molecular cloning of the gene encoding the putative polymerase of mouse hepatitis coronavirus, strain A59. *Virology* **171**:141–148.

Robb, J. A., and C. W. Bond. 1979. Coronaviridae. *Compr. Virol.* **14**:193–247.

Rottier, P. J. M., W. J. M. Spaan, M. C. Horzinek, and B. A. M. Van Der Zeijst. 1981. Translation of three mouse hepatitis virus strain A59 subgenomic RNAs in *Xenopus laevis* oocytes. *J. Virol.* **38**:20–26.

Shieh, C. K., L. Soe, S. Makino, M. F. Chang, S. A. Stohlman, and M. M. C. Lai. 1987. The 5′ end sequence of the murine coronavirus genome: implications of multiple fusion sites in leader primed transcription. *Virology* **156**:321–330.

Siddell, S. G. 1983. Coronavirus JHM: coding assignments of subgenomic mRNAs. *J. Gen. Virol.* **64**:113–125.

Skinner, M. A., D. Ebner, and S. G. Siddell. 1985. Coronavirus MHV-JHM mRNA 5 has a sequence arrangement which potentially allows translation of a second, downstream open reading frame. *J. Gen. Virol.* **66**:581–592.

Spaan, W. J., P. J. Rottier, M. C. Horzinek, and B. A. Van der Zeijst. 1982. Sequence relationships between the genome and the intracellular RNA species 1, 3, 6, and 7 of mouse hepatitis virus strain A59. *J. Virol.* **42**:432–439.

Spaan, W. J. M., J. Delius, M. A. Skinner, J. Armstrong, P. J. W. Rottier, S. Smeekens, B. A. M. Van Der Zeust, and S. G. Siddell. 1983. Coronavirus mRNA synthesis involves the fusion of non-contiguous sequences. *EMBO J.* **2**:1839–1844.

Spaan, W. J. M., P. J. M. Rottier, M. C. Horzinek, and B. A. M. Van Der Zeijst. 1981. Isolation and identification of virus-specific mRNAs in cells infected with mouse hepatitis virus (MHV-A59). *Virology* **108**:424–434.

Stern, D. F., and S. I. T. Kennedy. 1980. Coronavirus multiplication strategy. I. Identification and characterization of virus-specified RNA. *J. Virol.* **34**:665–674.

Stohlman, S. A., R. S. Baric, G. N. Nelson, L. H. Soe, L. M. Welter, and R. J. Deans. 1988. Specific interaction between coronavirus leader RNA and nucleocapsid protein. *J. Virol.* **62**:4288–4295.

Wege, H., S. Siddell, M. Sturm, and V. ter Meulen. 1981. Coronavirus JHM: characterization of intracellular viral RNA. *J. Gen. Virol.* **54**:213–217.

Weiss, S. R., and J. L. Leibowitz. 1981. Comparison of the RNAs of murine and human coronaviruses, p. 245–259. *In* V. ter Meulen, S. G. Siddell, and H. Wege (ed.), *Biochemistry and Biology of Coronaviruses*. Plenum Publishing Corp., New York.

Weiss, S. R., and J. L. Leibowitz. 1983. Characterization of murine coronavirus RNA by hybridization with virus-specific cDNA probes. *J. Gen. Virol.* **64**:127–133.

Zoltick, P. W., J. L. Leibowitz, J. R. DeVries, G. M. Weinstock, and S. R. Weiss. A general method for the induction and screening of antisera for cDNA-encoded polypeptides: antibodies specific for a coronavirus putative polymerase-encoding gene. *Gene*, in press.

Chapter 12

Association of Alphavirus Replication with the Cytoskeletal Framework and Transcription In Vitro in the Absence of Membranes

David J. Barton, Dorothea L. Sawicki, and Stanley G. Sawicki

Replication complexes formed in vivo by the positive-strand RNA alphaviruses Sindbis virus (SIN) and Semliki Forest virus (SFV) are found associated with cytoplasmic membranes and are capable of synthesizing viral RNA in vitro (Friedman et al., 1972; Martin, 1969; Martin and Sonnabend, 1967; Michel and Gomatos, 1972; Sreevalsan and Yin, 1969). Clewley and Kennedy (1976) reported the purification of SFV RNA polymerase after solubilization of the mitochondrial pellet (P15) fraction with Triton X-100. Although detergent treatment of the membrane-associated replication complexes released 25S replication complexes, there was at least a 5- to 10-fold loss of the overall amount of RNA polymerase activity. Ranki and Kääriäinen (1979) and Gomatos et al. (1980) also reported obtaining solubilized replication complexes from SFV-infected cells, although the RNA polymerase activity of the solubilized complexes was greatly diminished compared with the activity present in the P15 fraction (Gomatos et al., 1980). We (Barton, 1989) observed during our attempts to purify active replication complexes from alphavirus-infected cells that 25S replication complexes could be obtained by detergent treatment of the P15 fraction from SFV-infected cells as reported by others (Clewley and Kennedy, 1976; Gomatos et al., 1980; Ranki and Kääriäinen, 1979), but we could not solubilize the replication complexes from SIN-infected cells. Rather, we found that the SIN RNA polymerase activity was detergent insoluble and pelleted with low-speed (15,000 × g) centrifugation after detergent treatment. We report here that both SIN and SFV replication complexes

are associated with the detergent-insoluble cytoplasmic fraction, i.e., the cytoskeletal framework, and that detergent treatment of the replication complexes does not diminish RNA polymerase activity.

ALPHAVIRUS IN VITRO TRANSCRIPTION

We have reported previously the details of our methods of cell fractionation and in vitro transcription (Barton, 1989; Barton et al., 1988). Figure 1 is a flow chart explaining the derivation of the different subcellular fractions that we used to determine RNA polymerase activity: the P15 fraction, the detergent-insoluble fraction, the 1.25-g/ml cytoskeletal framework fraction, and the 1.16-g/ml smooth membrane fraction. In vitro RNA polymerase assays were performed by using an amount of each subcellular fraction that came from an equivalent number of cells. Figure 2 demonstrates the RNA polymerase activity found in the P15 fraction from SIN- and SFV-infected cells, using $[\alpha\text{-}^{32}P]CTP$ as the radiolabel. The labeled CTP is incorporated first after 1 min of incubation into the RI (replicative intermediate) RNA and then with increasing length (15, 30, and 60 min) of the incubation period into the 26S mRNA and the 49S genome RNA. P15 fractions from SFV-infected cells contain approximately twice as much RNA polymerase activity as do P15 fractions from SIN-infected cells. Nuclease treatment of the samples generates the expected RF (replicative-form) RNA, RFI, RFII, and RFIII. There are two major problems with the in vitro assay for

David J. Barton, Dorothea L. Sawicki, and Stanley G. Sawicki, Department of Microbiology, Medical College of Ohio, Toledo, Ohio 43699.

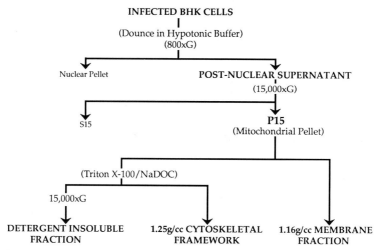

FIGURE 1. The subcellular fractions that contain RNA polymerase activity in vitro. The P15 fraction was prepared as described by Barton et al. (1988) from BHK cells that were infected with SFV or SIN and either (i) layered over linear 30 to 60% sucrose gradients in 10 mM Tris hydrochloride (pH 7.8)–10 mM NaCl (RS buffer) and centrifuged in an SW28 rotor for 20 h at 150,000 × g to obtain the 1.16-g/ml smooth membrane fraction or (ii) suspended in 1% Triton X-100–0.5% NaDOC in RS buffer. The detergent-treated P15 either was centrifuged for 15 min at 15,000 × g to obtain the detergent-insoluble pellet fraction, which was resuspended in RS buffer, or was layered over linear 30 to 60% sucrose gradients in RS buffer and centrifuged in a TLS 55 rotor for 20 h at 200,000 × g to obtain the 1.25-g/ml cytoskeletal fraction.

RNA polymerase activity that uses subcellular fractions of alphavirus-infected cells. The first is the presence of endogenous RNase that degrades the single-stranded products. The second is the presence of endogenous phosphatases that degrade the nucleoside triphosphate precursors to diphosphates (Barton, 1989).

One way of dealing with these two problems is to run the in vitro assays to observe only the synthesis of RI RNA that is resistant to RNase. We call this kind of reaction an instantaneous labeling reaction. The labeled nucleoside triphosphate, which in this case is CTP, is used undiluted and at a high specific activity. This low concentration of CTP prevents extensive elongation and keeps the incorporated label in an RNase-protected, double-stranded form. Therefore, by using instantaneous labeling conditions that limit RNA synthesis to only a short burst, we can find and quantitate RNA polymerase activity in the presence of endogenous RNase. Whether we use low specific activity, and therefore a high concentration of CTP, or a high specific activity, and therefore a low concentration of CTP, in our in vitro assays, the RNA polymerase is found associated with the mitochondrial pellet or P15 fraction, as reported previously by several other laboratories (Clew-ley and Kennedy 1976; Friedman et al., 1972; Martin, 1969; Martin and Sonnabend, 1967; Michel and Gomatos, 1972; Ranki and Kääriäinen, 1979; Sreevalsan and Yin, 1969).

Treating the P15 fraction with increasing concentrations of Triton X-100 and sodium deoxycholate (NaDOC) as high as 1% Triton X-100–0.5% NaDOC does not release the RNA polymerase activity in a detergent-soluble form, but it remains pelletable by centrifugation at 15,000 × g (Fig. 3). These detergent treatments solubilize, in addition to the lipids, more than 80% of the protein in the P15 (data not shown; Barton, 1989). Because detergent treatment also releases endogenous RNase that degrades the single-stranded 26S and 49S RNA products of the RNA polymerase activity, we had to use instantaneous labeling conditions to demonstrate that there was essentially no loss of RNA polymerase activity after detergent treatment. Detergent treatment of P15 from both SFV- and SIN-infected cells gave similar results except that the SFV extracts were more active than the SIN extracts (D. J. Barton et al., manuscript in preparation). Thus, alphavirus RNA replication complexes do not require membranes for elongation activity.

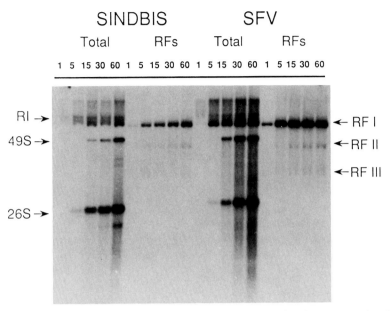

FIGURE 2. SIN and SFV RNA polymerase activity in vitro by the P15 fraction. The P15 fraction from SIN- or SFV-infected BHK cells was incubated at 30°C for 1, 5, 15, 30, and 60 min with [α-^{32}P]CTP (3 μCi/μl; 0.2 mM [final concentration] CTP) as described elsewhere (Barton et al., 1988). The samples were phenol-chloroform extracted and ethanol precipitated. Half of the samples were treated with RNase A, and both the RNase A-treated (RFs) and the untreated (total) samples were electrophoresed on 0.8% agarose gels containing morpholinepropanesulfonic acid (MOPS) buffer and 2.2 M formaldehyde.

DENSITY OF REPLICATION COMPLEXES AFTER DETERGENT TREATMENT

We then determined the density of the detergent-treated replication complexes. The P15 fraction from SFV- and SIN-infected cells was either detergent treated (1% Triton X-100–0.5% NaDOC) or not treated and layered on 30 to 60% sucrose gradients and spun at 200,000 × g for 20 h. Every other fraction across the gradients was assayed by using instantaneous labeling conditions (Fig. 4). Without detergent treatment, the polymerase activity was in the 1.16-g/ml fraction, which is at the top of this type of gradient. With detergent treatment, the density of the polymerase activity was shifted to 1.25 g/ml (Barton, 1989). There was the same amount of activity in the 1.25-g/ml fraction obtained after detergent treatment as in the 1.16-g/ml fraction obtained without detergent treatment. Thus, detergent treatment did not reduce the RNA polymerase activity of the alphavirus replication complexes.

Another way of identifying replication complexes besides looking for RNA polymerase activity is to prelabel in vitro the viral replication

complexes in the P15 fraction by using instantaneous labeling conditions. When the prelabeled replication complexes in the P15 fraction were solubilized with detergents and banded isopycnically on 30 to 60% sucrose gradients, the prelabeled complexes were found also now in the 1.25-g/ml fraction (Barton, 1989; Barton et al., in preparation). Negligible amounts of the prelabeled replication complexes were found in the pellet or at the top of the gradient. The detergent-treated SFV replication complexes were found to be very large; they pelleted under conditions that left the 260S virions at the top of the glycerol gradient. Protein gels of the detergent-insoluble fraction from both mock-infected and alphavirus-infected cells showed the presence of components of the cytoskeletal framework, such as actin, myosin, and tubulin (Barton, 1989; Barton et al., in preparation).

DISTRIBUTION OF THE VIRAL NONSTRUCTURAL PROTEINS

Monospecific, polyclonal antibodies to each of the four SIN nonstructural proteins were obtained from Reef Hardy in Jim Strauss's lab.

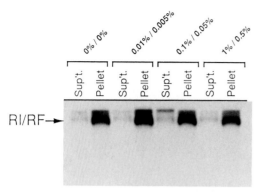

FIGURE 3. Treatment of the P15 fraction with detergent in RS buffer does not solubilize the RNA polymerase activity. The P15 fraction from SFV-infected BHK cells was incubated with the indicated concentrations of Triton X-100/NaDOC (respectively) in RS buffer on ice for 10 min and centrifuged at 10,000 × g in a microcentrifuge for 10 min. The pellet was resuspended in the same volume as the supernatant of RS buffer containing the appropriate concentration of detergents. An equal volume of 2× reaction mix (Barton et al., 1988) that lacked added unlabeled CTP was added to each sample together with [α-³²P]CTP (3 μC/μl). The reaction products, i.e., RI/RF RNA, were separated on 0.8% agarose gels in Tris-borate buffer.

These antibodies detect the four processed nonstructural proteins, nsP1, nsP2, nsP3, and nsP4, as well as their precursor polyproteins, nsP12, nsP34, nsP123, and nsP1234 (Hardy and Strauss, 1989). All four nonstructural proteins were in the total extract, before removal of the nuclei, and in both the S15 and the P15 fractions from SIN-infected cells (Barton, 1989; Barton et al., in preparation). Approximately 85% of the RNA polymerase activity and prelabeled replication complexes were found in the P15. Because the distribution of the four nonstructural proteins did not coincide directly with the distribution of polymerase activity, the presence of nonstructural proteins does not of itself demonstrate SFV and SIN replication complexes. The 1.25-g/ml fraction obtained after detergent treatment of the P15 contained all four of the nonstructural proteins. Therefore, all four nonstructural proteins are found in the detergent-treated fraction that contains complexes possessing RNA polymerase activity.

CONCLUSIONS

Alphavirus transcription in vitro did not require membranes. Both the SIN and SFV

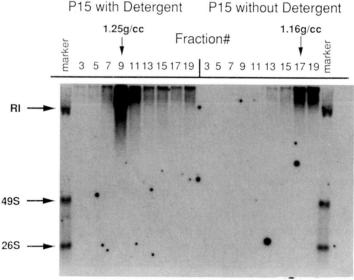

FIGURE 4. Detergent treatment of the P15 fraction shifted the density of the alphavirus RNA polymerase activity from 1.16 to 1.25 g/ml. The P15 fraction in RS buffer from SFV-infected cells was made 1% Triton X-100–0.5% NaDOC or was not treated with detergent and was layered over linear 30 to 60% sucrose gradients in RS buffer. After centrifugation for 20 h at 200,000 × g, the gradients were fractionated into 20 fractions, and every other fraction was assayed for RNA polymerase activity under high-specific-activity labeling conditions. The products of the reaction, i.e., RI/RF RNA, were resolved on 0.8% agarose gels in Tris-borate buffer.

replication complexes were associated with the detergent-insoluble fraction, which had a density of 1.25 g/ml. This fraction contained elements of the cytoskeletal framework. All four viral nonstructural proteins were found in the detergent-insoluble fraction that retained polymerase activity. Recently, Froshauer et al. (1988), using electron microscopy of immunogold labeling in semipermeabilized cells, showed that the monospecific antibodies against nsP3 and nsP4 reacted with fibrous, granular material in the cytosol of cells infected with SIN. Although they interpreted their results as suggesting that the cytopathic vacuoles found in alphavirus-infected cells (Friedman et al., 1972) are the site of viral replication, we propose that their results demonstrate the association of the viral nonstructural proteins with the cytoskeletal framework to which is attached the cytoplasmic vesicles and cytopathic vacuoles. Furthermore, our studies indicated that the viral nonstructural proteins are overproduced and that fractions lacking polymerase activity contain viral nonstructural proteins, as did fractions containing polymerase activity.

REFERENCES

Barton, D. J. 1989. In vitro analysis of alphavirus RNA synthesis. Ph.D. thesis, Medical College of Ohio, Toledo.

Barton, D. J., S. G. Sawicki, and D. L. Sawicki. 1988. Demonstration in vitro of temperature sensitive elongation of RNA in Sindbis virus mutant ts6. *J. Virol.* **62**:3597–3602.

Clewley, J. P., and S. I. T. Kennedy. 1976. Purification and polypeptide composition of Semliki Forest virus RNA polymerase. *J. Gen. Virol.* **32**:413–430.

Friedman, R. M., J. G. Levin, P. M. Grimly, and I. K. Berzesky. 1972. Membrane-associated replication complex in arbovirus infection. *J. Virol.* **10**:504–514.

Froshauer, S., J. Kartenbeck, and A. Helenius. 1988. Alphavirus RNA replicase is located on the cytoplasmic surface of endosomes and lysosomes. *J. Cell Biol.* **107**:2075–2086.

Gomatos, P. J., L. Kääriäinen, S. Keränen, M. Ranki, and D. L. Sawicki. 1980. Semliki Forest virus replication complex capable on synthesizing 42S and 26S nascent RNA chains. *J. Gen. Virol.* **49**:61–69.

Hardy, W. R., and J. R. Strauss. 1989. Processing the nonstructural polypeptides of Sindbis virus: study of the kinetics in vivo by using monospecific antibodies. *J. Virol.* **62**:998–1007.

Martin, M. E. 1969. Studies on the RNA polymerase of some temperature-sensitive mutants of Semliki Forest virus. *Virology* **39**:107–117.

Martin, M. E., and J. A. Sonnabend. 1967. Ribonucleic acid polymerase catalyzing synthesis of double-stranded arbovirus ribonucleic acid. *J. Virol.* **1**: 97–109.

Michel, M. R., and P. J. Gomatos. 1972. Semliki Forest virus-specific RNAs synthesized in vitro by enzyme from infected BHK cells. *J. Virol.* **11**: 900–914.

Ranki, M., and L. Kääriäinen. 1979. Solubilized RNA replication complex from Semliki Forest virus infected cells. *Virology* **98**:298–307.

Sreevalsan, T., and F. H. Yin. 1969. Sindbis virus-induced ribonucleic acid polymerase. *J. Virol.* **3**: 599–604.

III. DEFECTIVE INTERFERING RNAs AND INFECTIOUS CLONES

Chapter 13

Molecular Genetic Analysis of Poliovirus RNA Replication by Mutagenesis of a VPg Precursor Polypeptide

Cristina Giachetti and Bert L. Semler

Although replication of poliovirus RNA has been studied for a number of years, many of the mechanistic details of RNA synthesis are not understood (Rueckert, 1985; Koch and Koch, 1985; Kuhn and Wimmer, 1987; Semler et al., 1988; Wimmer et al., 1988). In vivo, all poliovirus RNA synthesis appears to proceed in a membranous environment. How template RNA and replication proteins are sequestered in this complex is not known. Moreover, since nearly all viral proteins are found in the membranous replication complex (Caliguiri and Mosser, 1971; Roder and Koschel, 1975), the nature and significance of specific interactions between template RNA and various viral gene products are difficult to assess. Viral proteins 3D^{pol}, 2BC, 2C, and 3AB (a precursor of VPg) are specifically associated with membranes (Tershak, 1984). All of these proteins, along with 3C^{pro} (which is not membrane associated), are believed to participate in viral RNA replication.

Polypeptide 3D^{pol} has been isolated from infected cells and characterized as the virus-encoded RNA polymerase (Lundquist et al., 1974). After the isolation of 3D^{pol}, a soluble, membrane-free in vitro RNA synthesis system was developed (Flanegan and Baltimore, 1977, 1979; Van Dyke and Flanegan, 1980). With use of this system, the viral polymerase was characterized as a template- and primer-dependent enzyme. 3D^{pol} catalyzes the synthesis of double-stranded RNA when viral RNA (as template) and oligo(U) (as primer) are added to the in vitro reaction (Flanegan and Baltimore, 1979). The product of this reaction resembles the replicative-form RNA found in vivo in infected cells.

Polypeptide 3D^{pol} is thought to be the only viral protein capable of elongating the nascent viral RNA strand (Dasgupta et al., 1980). Further studies using the soluble in vitro system demonstrated that the priming activity of oligo(U) can be replaced by a cell protein called host factor (molecular size, 67,000 daltons [Baron and Baltimore, 1982c; Dasgupta, 1983; Young et al., 1985; Kaplan et al., 1985]). A host factor activity has been reported to copurify with a terminal uridylyltransferase activity from uninfected cells (Andrews and Baltimore, 1986). Since the product of the host factor-stimulated reaction is usually very heterogeneous in chain length and sometimes is covalently linked to the template RNA, several models for initiation of RNA synthesis and for the role of host factor were proposed. These include the hairpin model (Young et al., 1985; Andrews and Baltimore, 1986) and one involving endonuclease(s), with the fortuitous formation of snapback structures during the in vitro reaction (Lubinski et al., 1986; Hey et al., 1987).

Recent work reported by several groups (Morrow et al., 1987; Rothstein et al., 1988; Plotch et al., 1989) involved the expression of cDNA clones coding for 3D^{pol} in bacteria. The purified bacterially expressed polymerase shows properties similar to those of the intracellular virus-induced protein: the enzyme exhibits poly(A)-dependent, oligo(U)-primed poly(U) polymerase activity as well as RNA polymerase activity. Interestingly, Plotch et al. (1989) reported that in the absence of added primer or host factor, the bacterially expressed polymerase synthesizes RNA products up to twice the

Cristina Giachetti and Bert L. Semler, Department of Microbiology and Molecular Genetics, College of Medicine, University of California, Irvine, California 92717.

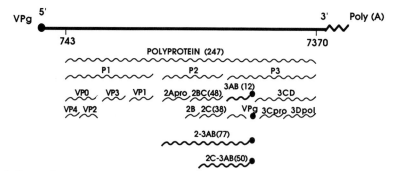

FIGURE 1. Processing map of poliovirus polypeptides. Numbers in parentheses indicate molecular weights (in thousands). Polypeptide precursors of VPg are indicated by bold lines.

length of the template, and in the presence of a single nucleoside triphosphate, [α-^{32}P]UTP, the enzyme catalyzes the incorporation of ^{32}P into template RNA. The mechanism involved in the generation of the dimer-sized product and whether the terminal uridylyltransferase activity is catalyzed solely by the viral polymerase are aspects of the in vitro RNA synthesis reactions that remain to be determined.

Since authentic poliovirus RNA (i.e., 3′ polyadenylated and 5′ linked to VPg) was never found by using the above-mentioned soluble in vitro RNA replication system, it appears that the system is unable to initiate RNA synthesis with concomitant VPg linkage, and therefore it is likely that it does not reflect the complete in vivo mechanism(s) for RNA synthesis. To better explain the role of VPg as part of the polymerase-host factor-mediated poliovirus RNA synthesis, a new model was proposed by Flanegan et al. (1987). According to this model, in the presence of host factor, the polymerase uses a template priming mechanism to initiate RNA synthesis. The generation of unit-length product RNA linked to VPg is the result of a second step, in which VPg (or a precursor of VPg) is responsible for a concerted cleavage and linkage reaction.

On the basis of the observation that in vivo RNA synthesis seems to be membrane associated, studies in E. Wimmer's laboratory followed a different approach to develop another in vitro system for replication of poliovirus RNA (Takeda et al., 1986; Takegami et al., 1983a). The system uses a membrane fraction isolated from infected cells (named crude replication complex [CRC]). This system supports the synthesis of viral RNA (linked to VPg), as well as replicative-form and replicative intermediate RNA. Interestingly, addition of nonionic detergents to the CRC abolishes single-strand RNA synthesis without altering synthesis of replica-

tive-form RNA (Takegami et al., 1983a). This finding was interpreted to indicate that detergent interferes with initiation of RNA, perhaps because of an alteration of the membranous environment required for viral RNA synthesis. Using this system, Takegami et al. (1983a) demonstrated the synthesis of VPg-pU and VPg-pUpU (the 5′-terminal moiety of both plus- and minus-strand RNA). Takeda et al. (1986) demonstrated that preformed VPg-pU can be chased into VPg-pUpU and nucleotidyl proteins containing nine or more of the poliovirus 5′-proximal nucleotides. The synthesis of VPg-pUpU (as in the case of full-length viral RNA) is strictly dependent on the presence of intact membranes (Takegami et al., 1983a). Formation in vitro of VPg-pUpU is sensitive to pretreatment of the CRC with nuclease, suggesting that the uridylylation of VPg is a template-dependent reaction (Takeda et al., 1987). In addition, the following observations support a model in which VPg-pU can function as primer for poliovirus plus-strand synthesis within a membranous environment: (i) VPg is attached to the 5′ end of the RNA genome of poliovirus, to the 5′ end of the intracellular plus- (except mRNA associated with polyribosomes) and minus-strand RNA, and to the nascent RNAs (Nomoto et al., 1977; Pettersson et al., 1978; Flanegan et al., 1977); (ii) antibodies against VPg immunoprecipitate VPg as well as VPg-pUpU from infected cells (Crawford and Baltimore, 1983); and (iii) in in vitro replication reactions, anti-VPg antibody specifically inhibits initiation of viral RNA synthesis (Baron and Baltimore, 1982b). However, since VPg is a strongly polar (basic) protein, it must be delivered to the hydrophobic site by a lipophilic carrier.

Three polypeptide precursors of VPg, P2-3AB, 2C-3AB, and 3AB (formerly called 3b/9, X/9, and P3-9, respectively) (Fig. 1), are associ-

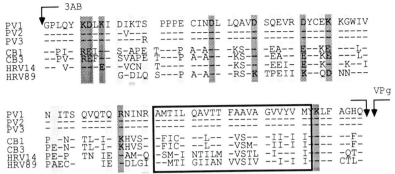

FIGURE 2. Alignment of the amino acid sequences of the 3AB regions of poliovirus types 1, 2, and 3 (PV1, 2, and 3), coxackieviruses B1 and B3 (CB1 and CB3), and human rhinoviruses 14 and 89 (HRV14 and HRV89). Dashes indicate homology; darkly shaded areas indicate the presence of charged amino acids in all of the sequences compared; lightly shaded areas indicate the presence of charged amino acids in some of the sequences compared. Boxed sequences indicate the uncharged amino acid domain within 3AB.

ated with membrane fractions of poliovirus-infected cells (Semler et al., 1982; Takegami et al., 1983b; Baron and Baltimore, 1982a), presumably because they contain a nonpolar sequence of 22 amino acids just amino terminal to the amino acid sequence of VPg (Semler et al., 1982; Fig. 2). Proteolytic degradation assays of the membrane-associated viral polypeptides confirmed the specific interaction of this hydrophobic domain with membranes (Takegami et al., 1983b). These studies indicated that whereas polypeptide 2C and the amino-terminal region of 3AB are susceptible to proteolytic attack, the hydrophobic region of 3AB is protected. The protection can be abolished by the addition of mild nonionic detergents (Takegami et al., 1983b). The smallest VPg-containing polypeptide, 3AB (molecular size, 12,000 daltons), is the most likely candidate for a donor of VPg to the membrane-associated poliovirus RNA-synthesizing complex because it is an abundant viral product of proteolytic processing that appears to turn over very slowly (not more than 5% of it would be needed to initiate all newly synthesized RNA strands) (Semler et al., 1982; Takegami et al., 1983b). In addition, recent studies on a cold-sensitive mutant bearing a defined lesion in 3A (insertion of an extra serine at position 15 of 3A [Bernstein and Baltimore, 1988]) indicated that this protein is involved in RNA replication. Giving further support for a role of 3A or 3AB in poliovirus RNA replication were in vitro mutagenesis experiments in which the spacer region between the 3A-3B (VPg) cleavage site was disrupted by insertion of five extra amino acids, resulting in nonviable mutations (Kuhn et al., 1988b).

It is also possible that other VPg-containing polypeptides (P2-3AB or 2C-3AB) are involved in the formation of the membrane-bound replication complex. Protein 2C (contained in these two precursors) is thought to be involved in RNA replication because mutations to guanidine resistance (that lead to an altered RNA synthesis phenotype) map in 2C (Pincus et al., 1986). In vitro mutagenesis of 2C also generated mutants defective in RNA synthesis (Li and Baltimore, 1988). However, the role of 2C in viral replication is unclear (Pincus et al., 1986; Li and Baltimore, 1988; Bienz et al., 1987). The common 22-amino-acid stretch of hydrophobic residues preceding VPg seems to be crucial for the proposed role of these polypeptides (3AB, 2C-3AB, and P2-3AB) in poliovirus RNA synthesis. Moreover, sequence analysis of polypeptide 3AB of the three strains of poliovirus (Toyoda et al., 1984) and of other closely related picornaviruses, coxsackievirus (Iizuka et al., 1987; Lindberg et al., 1987) and human rhinovirus (Duechler et al., 1987; Stanway et al., 1984), indicates the presence of a similar, nonpolar amino acid sequence preceding VPg (Fig. 2). Conservation of such a hydrophobic domain among different picornaviruses adds further support for its proposed role in viral RNA replication.

The existence of poliovirus cDNA clones that produce infectious virus upon transfection of primate cells (Racaniello and Baltimore, 1981; Semler et al., 1984) makes it possible to genetically manipulate poliovirus RNA. By using this technology, mutations within coding (Ypma-Wong and Semler, 1987; Dewalt and Semler, 1987; Kuhn et al., 1988a; Nicklin et al., 1987;

Kean et al., 1988; Bernstein et al., 1986; Bernstein and Baltimore, 1988; Kuhn et al., 1988b; Li and Baltimore, 1988) and noncoding (Sarnow et al., 1986; Semler et al., 1986; Racaniello and Meriam, 1986; Trono et al., 1988; Dildine and Semler, 1989) regions of the genome have been made. Previous work by Dewalt and Semler (1987) and Kuhn et al. (1988a; Kuhn et al., 1988b) demonstrated that the use of synthetic mutagenesis cassettes permits a rapid and extensive mutational analysis of defined regions of the poliovirus genome. In the work described here, we used this approach to introduce amino acid substitutions into the hydrophobic domain of 3AB. Among the mutants we isolated, mutant Se1-3AB-310/4 (hereafter referred to as 310/4) showed a strong temperature-sensitive phenotype. We show here that a biochemical analysis combined with a genetic analysis of this mutant provides evidence for a direct role of polypeptide 3AB in poliovirus RNA synthesis.

GENERATION OF AMINO ACID REPLACEMENT MUTANTS IN THE HYDROPHOBIC DOMAIN CONTAINED IN POLYPEPTIDE 3AB

Because there is only one unique restriction site in the cDNA encoding the hydrophobic domain of 3AB (*Bgl*I, at position 5318), we engineered a new *Pst*I restriction site at position 5297, that is, 21 nucleotides upstream of the *Bgl*I site. The new *Pst*I site was introduced by changing the A at N5299 to a G and the A at N5302 to a G. These changes resulted in the sequence CTGCAG, which corresponds to the recognition site for *Pst*I. Since CTA (original) and CTG (mutant) codons specify Leu and CAA (original) and CAG (mutant) codons specify Gln, the introduced changes did not result in amino acid substitutions. The mutations were created by using oligonucleotide-directed, site-specific mutagenesis according to the methodology of Inouye and Inouye (1987). The new restriction site allowed us to insert one mutagenesis cassette directed to mutate Thr-67 within the sequence of 3AB. Briefly, a synthetic pair of oligonucleotides, corresponding to the region of the poliovirus genome between nucleotides 5297 and 5323, was prepared. These oligonucleotides contained nucleotides randomized at position 5310 and sequences that allowed direct cloning back into the *Pst*I-*Bgl*I region within 3AB. This changed the original codon ACA to AGA, AUA, or AAA, thus changing the original amino acid Thr at position 67 to Arg, Ile, or Lys, respectively.

TABLE 1
Mutations Generated in Polypeptide 3AB

Randomized nucleotide[a]	Amino acid	Virus or plasmid designation	Growth characteristic
5310	67		
* ACA	Thr	Wild type	Wild type
AAA	Lys	pPVA55-310/5	Nonviable
AUA	Ile	Se1-3AB-310/4	Temperature sensitive
AGA	Arg	pPVA55-310/7	Nonviable

[a] Asterisk indicates randomized nucleotide.

The full-length plasmids containing these mutations, as well as wild-type poliovirus cDNA (pPVA55 [Kuhn et al., 1987]), were transfected into HeLa cell monolayers at different temperatures (33, 37, and 39°C). Transfections were made by using the calcium phosphate coprecipitation technique (Graham and van der Eb, 1973). Two independent transfection assays with mutant cDNAs 310/5 and 310/7 (which specify Lys and Arg, respectively, at position 67) did not result in recovery of infectious virus at any of the temperatures assayed. We conclude that these cDNAs generate nonviable mutant plasmids.

Transfection of HeLa cells with mutant cDNA 310/4 resulted in low levels of virus at 33°C, and no virus was recovered from transfections incubated at 37 or 39°C. Supernatants from the transfected cultures at 33°C were assayed for plaque production on fresh HeLa cell monolayers at 33°C, and individual plaques were isolated. One plaque was used to generate a virus stock (passage two [P2]) of mutant 310/4. RNA was extracted from P2 mutant virus, and the presence of the introduced mutation was confirmed by dideoxynucleotide sequence analysis. Table 1 summarizes the infectivity characteristics of the plasmids containing mutations in polypeptide 3AB.

GROWTH CHARACTERISTICS OF MUTANT 310/4

Virus mutant 310/4 was subjected to further biochemical analysis. First we determined the ability of virus 310/4 to produce plaques at 33, 37, and 39°C. This analysis indicated that the mutant grew to titers that were almost 2 log units per ml lower at 37 than at 33°C and 4.5 log units per ml lower at 39 than at 33°C (Table 2). In

TABLE 2
Reproductive Capacity of Mutant[a] 310/4 Assayed
at Different Temperatures

Virus type	Titer (log PFU/ml)			Log difference (titer at 33°C/ titer at 39°C)
	33°C	37°C	39°C	
Wild type	9.60	9.60	9.45	0.05
310/4	9.55	7.67	5.00	4.55

[a] P2 stocks.

addition, the plaques were very small at all temperatures assayed, and they required 6 to 7 days of incubation to appear, instead of the 2 or 3 days usually required for wild-type virus, poliovirus type 1 (Fig. 3). One-step growth curves of mutant virus at 33 and 39°C showed a dramatic delay in the kinetics of viral growth at these temperatures. At 8 h postinfection (p.i.), very little progeny virus was detected in the mutant-infected cells, whereas wild-type virus yield at 33°C was already 90% of the maximum level reached at this temperature. Maximum wild-type virus production occurred at 10 h p.i. at 33°C and at 5.5 h p.i. at 39°C, whereas maximum virus production of the mutant was reached at 14 h p.i. at 33°C and at 12 h p.i. at

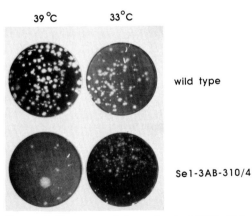

FIGURE 3. Plaque morphology of mutant 310/4. HeLa cells infected with dilutions of P2 stocks of wild-type or mutant virus were incubated under agar overlay at 39 or 33°C. After approximately 40 h of incubation at 39°C or 64 h at 33°C, wild-type virus-infected cells were fixed and stained. Mutant-infected cells were fixed and stained after 6 or 7 days of incubation at 39 or 33°C. Mutant-infected cells at 39°C were inoculated with 1,000 times more virus than mutant-infected cells at 33°C. Note that the large plaque in mutant-infected cells at 39°C is due to the presence of revertants in the mutant stock.

39°C. The maximum mutant virus yield at 33°C was similar to the wild-type yield at 33°C, but mutant virus yield was almost 2 log units lower at 39 than at 33°C. Plaque assays at 33 versus 39°C of the virus produced by mutant 310/4 at 39°C indicated that approximately 50% of the virus generated at 39°C was revertant virus. These results demonstrated that the mutant virus 310/4 is temperature sensitive for growth at 37 and 39°C.

PROTEIN SYNTHESIS OF MUTANT 310/4 AT PERMISSIVE AND NONPERMISSIVE TEMPERATURES

The kinetics of appearance of virus-specific polypeptides in HeLa cells infected with mutant 310/4 at 33 and 39°C were assayed. Wild-type- or 310/4-infected cells were pulse-labeled with [^{35}S]methionine for 1- or 2-h intervals beginning at 2 h p.i. After the labeling period, cells were harvested and the labeled extracts were subjected to sodium dodecyl sulfate-polyacrylamide gel electrophoresis (SDS-PAGE). The results of this experiment indicated that 310/4 showed delayed kinetics of viral protein synthesis at 33°C. Viral proteins were not detected before 7 h p.i., and the maximum level of viral protein synthesis occurred between 13 and 15 h p.i. (Fig. 4B), whereas the maximum level of protein synthesis induced by wild-type poliovirus at this temperature occurred between 7 and 9 h p.i. (Fig. 4A). At 15 h p.i., normal levels of viral proteins were present in the cells infected with the mutant virus at 33°C, and host cell protein synthesis was effectively shut off. In addition, protein processing appeared to be normal. The only polypeptide difference observed was an altered electrophoretic mobility of the 3AB and 3A polypeptides. At the nonpermissive temperature (39°C), no viral proteins were detected, nor was significant host cell shutoff observed, even as late as 21 h p.i. (Fig. 4B). When the extract from cells infected with the mutant virus at 39°C was immunoprecipitated with antisera against polypeptide 3AB before the SDS-PAGE analysis, we observed extremely reduced levels (at least 100 times lower than at 33°C) of virus-specific proteins (data not shown).

To determine whether the absence of protein synthesis at 39°C was due to a primary defect on protein synthesis or processing, we performed temperature shift experiments. In these experiments, we initiated virus infections at 33°C to allow synthesis of viral mRNA templates and proteins. After a 12-h incubation period at the permissive temperature, the in-

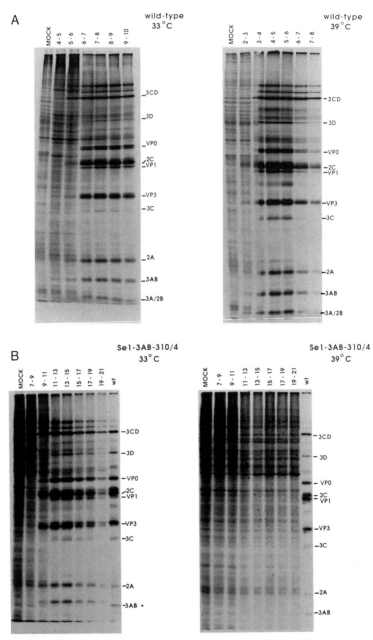

FIGURE 4. Kinetics of protein synthesis of wild-type and mutant 310/4 virus. HeLa cells were infected with wild-type virus (A) or with mutant virus 310/4 (B) at a multiplicity of infection of 50. Infected cells were incubated at 33 or 39°C and were pulse-labeled with [^{35}S]methionine for the indicated period of time at different times p.i. At the end of the labeling period, cells were suspended in Laemmli gel sample buffer and analyzed on 12.5% SDS-polyacrylamide gels. Protein gels were fluorographed and exposed to Kodak XAR film. Mock, Samples derived from mock-infected cells, labeled at 4.5 h p.i. at 33 or 39°C.

fected cells were shifted to 39°C. In one set of experiments, designed to analyze protein processing at the nonpermissive temperature, cells were pulse-labeled with [35S]methionine for 15 min (either immediately after the shiftup or after 1.5 h of incubation at 39°C) and chased in the presence of a 100-fold excess of nonradioactive methionine for up to 4 h after the pulse. Samples of the infected cells were withdrawn at different times and analyzed by SDS-PAGE. The results of this experiment showed normal protein processing of the viral polyprotein at 39°C (data not shown). In another set of experiments, cells were pulse-labeled for 1 h with [35S]methionine at different times after the shift. In this experiment, we observed that viral protein synthesis was normal at 39°C during the first hour after the shift from 33°C, but a clear reduction in the overall amount of viral protein synthesized at 39°C occurred afterwards. Our results indicated that the mRNA synthesized at 33°C was a good template for protein synthesis at 39°C. The reduction in the amount of protein synthesized at the nonpermissive temperature at late times after the shift was probably due to a reduction in the amount of mRNAs synthesized at 39°C. These results suggested that the lesion we introduced in 3AB does not directly affect protein synthesis.

VIRAL RNA SYNTHESIS INDUCED BY MUTANT 310/4 AT PERMISSIVE AND NONPERMISSIVE TEMPERATURES

We next examined the kinetics of viral RNA synthesis induced by mutant 310/4 during a single cycle of infection at the permissive and nonpermissive temperatures. HeLa cells were infected with wild-type virus or with mutant 310/4 at both temperatures and labeled with [3H]uridine in the presence of actinomycin D. At different times p.i., samples were removed and total RNA synthesis was determined by trichloroacetic acid precipitation. Mutant virus 310/4 induced extremely low levels of viral RNA synthesis at the nonpermissive temperature (Table 3). The amount of [3H]uridine-labeled trichloroacetic acid-precipitable material detected at different times p.i. was similar to the amount detected in uninfected cells (not shown). At 33°C, as in the case of viral growth and viral protein synthesis, the mutant showed delayed kinetics of viral RNA synthesis. Mutant-induced viral RNA was not detectable until 10 h p.i., and the maximum level was reached as late as 17 h p.i. On the other hand, wild-type virus maximum levels of RNA synthesis were reached at

TABLE 3
RNA Synthesis Levels at 33 and 39°C
for Mutant 310/4

Time (h) p.i.	% of maximum value for wild type		% of wild-type maximum incorporation[a] by 310/4	
	39°C	33°C	39°C	33°C
2.5	17.5	0	0	0
5.5	100	29.6	0	0
10.0	77	100	0	0
13.0		67	0.5	2.7
15.0			0	22
17.0			1.2	58
19.0			0	51

[a] Based on wild-type synthesis at the respective temperature.

5.5 h p.i. at 39°C and at 10 h p.i. at 33°C. The maximum level of viral RNA synthesized by the mutant at 33°C was lower than the maximum levels synthesized by the wild-type virus at this temperature. Clearly, these results indicated that the lesion we introduced in mutant 310/4 affects viral RNA production. Since the RNA synthesized at the permissive temperature is stable at 39°C (data not shown), it is likely that the absence of viral RNA in cell cultures infected with the mutant at 39°C was due to a defect in the synthesis of RNA at this temperature. Preliminary results from temperature shift experiments (from 33 to 39°C) showed that the mutant was able to synthesize RNA at a normal rate immediately after the shiftup, but longer incubation at 39°C resulted in loss of the RNA synthesis capability (data not shown). This result, and the results of the protein synthesis analysis, strongly suggested that the mutation we introduced in polypeptide 3AB has a detrimental effect on viral RNA synthesis.

ISOLATION AND CHARACTERIZATION OF NON-TEMPERATURE-SENSITIVE REVERTANTS OF MUTANT 310/4

We isolated revertants of mutant 310/4 to provide further genetic evidence that the mutation we introduced in polypeptide 3AB was responsible for the temperature-sensitive defect in RNA synthesis of mutant 310/4. Also, the possible isolation of non-temperature-sensitive pseudorevertants containing second-site mutations in the mutant RNA would provide new information regarding structure-function relationships within the hydrophobic domain of

TABLE 4

Reproductive Capacity of Revertant Virus 310/4-R6[a] Assayed at Different Temperatures

Virus type	Titer (log PFU/ml)			Log difference (titer at 33°C/ titer at 39°C)
	33°C	37°C	39°C	
Wild type	9.17	9.44	9.32	−0.15
310/4-R6	9.25	9.43	9.36	−0.11

[a] P2 stocks.

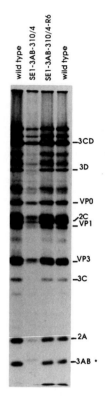

FIGURE 5. Virus protein synthesis induced by revertant virus 310/4-R6. HeLa cell monolayers were infected at a multiplicity of infection of 50 with wild-type, 310/4, or 310/4-R6 virus, pulse-labeled with [35S]methionine for 1.5 h, and analyzed by SDS-PAGE. Cells infected with wild-type virus or 310/4-R6 were incubated at 39°C and labeled at 4 h p.i. Cells infected with 310/4 were incubated at 33°C and pulse-labeled at 13 h p.i.

3AB. To isolate revertant viruses, dilutions of a P2 stock of 310/4 were incubated at 39°C under agar overlay for 40 h. At this time, a few viral plaques of a size similar to that of wild-type virus were observed on the monolayers infected with the lower dilutions. Two plaques were picked, and after plaque purification for a second time at 37°C, P1 and P2 viral stocks were prepared. One of the revertants we isolated, virus Se1-3AB-310/4-R6, was further characterized.

Revertant virus 310/4-R6 was able to produce plaques at 33, 37, and 39°C with the same efficiency as wild-type virus (Table 4). Also, the morphological characteristics of the plaques produced by the revertant virus were similar to those of wild-type virus. This finding confirmed that virus 310/4-R6 had completely reverted the temperature-sensitive growth defect of the original mutant 310/4. Sequence analysis of 310/4-R6 viral RNA indicated that the nucleotide change we had introduced to generate virus 310/4 reverted to the original wild-type sequence in virus 310/4-R6. Therefore, virus 310/4-R6 represents a true revertant of mutant 310/4. Proteins synthesized in cells infected with 310/4-R6 virus are shown in Fig. 5. Analysis of the data indicated that virus 310/4-R6 induced a normal pattern of viral proteins. In addition, as predicted by the sequence data, the electrophoretic mobilities of 3AB and 3A polypeptides of virus 310/4-R6 were shifted back to the normal position for wild-type virus (Fig. 5).

Finally, we examined the kinetics of viral RNA synthesis induced by virus 310/4-R6 during single-cycle infections at 33, 37, and 39°C. The kinetics of viral RNA synthesis induced by the revertant virus were nearly identical to the ones induced by the wild-type virus at all temperatures tested (Fig. 6). This finding demonstrated that virus 310/4-R6 had also reverted the temperature-sensitive RNA synthesis defect displayed by the original mutant, 310/4. The results of the revertant analysis clearly indicated that the amino acid change we introduced in polypeptide 3AB to generate mutant 310/4 was responsible for the temperature-sensitive viral growth and viral RNA synthesis exhibited by this virus.

SUMMARY

Our data indicate that changing the wild-type amino acid Thr-67 to Ile within the sequence of 3AB generated a temperature-sensitive virus. Viral growth, protein synthesis, and RNA synthesis of mutant 310/4 were dramatically reduced at 39°C. Temperature shift experiments (from 33 to 39°C) indicated that the lesion we introduced in 3AB did not directly affect protein synthesis or processing. Incubation of the mutant virus-infected cells at 33°C for 12 h before the shift to the nonpermissive temperature resulted in partial (temporary) rescue of the

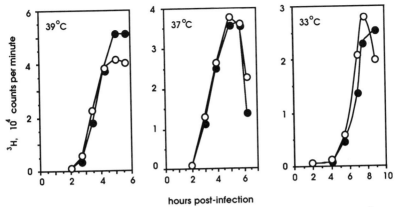

FIGURE 6. Kinetics of viral RNA synthesis induced by revertant virus 310/4-R6. HeLa cell monolayers were infected at a multiplicity of infection of 50 with wild-type (●) or 310/4-R6 (○) virus and incubated at 33, 37, or 39°C. After a 30-min adsorption period, medium containing 5 μg of actinomycin D per ml was added, cells were incubated for 30 min, and [³H]uridine (final concentration, 5 μCi/ml) was added. At different times p.i., samples were withdrawn, and incorporation of ³H into trichloroacetic acid-precipitable material was determined.

RNA synthesis capability at 39°C. These results suggested that the mutation we introduced in 3AB primarily affects RNA synthesis and also that the elongation function is probably not altered. The isolation and analysis of mutant 310/4 provide evidence for a direct role of polypeptide 3AB in poliovirus RNA synthesis. Further analysis of the temperature-sensitive defect of mutant 310/4 will provide a better understanding of the function of 3AB. Identification of the specific RNA synthetic process affected by the mutation will determine the precise role of polypeptide 3AB in poliovirus RNA synthesis.

ACKNOWLEDGMENTS. This work was supported by Public Health Service grant AI 22693 from the National Institutes of Health and by a grant from the Cancer Research Coordinating Committee of the University of California.

REFERENCES

Andrews, N. C., and D. Baltimore. 1986. Purification of a terminal uridylyltransferase that acts as host factor in the *in vitro* poliovirus replicase reaction. *Proc. Natl. Acad. Sci. USA* **83**:221–225.

Baron, M. H., and D. Baltimore. 1982a. Antibodies against the chemically synthesized genome-linked protein of poliovirus react with native virus-specific proteins. *Cell* **28**:395–404.

Baron, M. H., and D. Baltimore. 1982b. Anti-VPg antibody inhibition of the poliovirus replicase reaction and production of covalent complexes of VPg-related proteins and RNA. *Cell* **30**:745–752.

Baron, M. H., and D. Baltimore. 1982c. Purification

and properties of a host cell protein required for poliovirus replication *in vitro*. *J. Biol. Chem.* **257**:12351–12358.

Bernstein, H. D., and D. Baltimore. 1988. Poliovirus mutant that contains a cold-sensitive defect in viral RNA synthesis. *J. Virol.* **62**:2922–2928.

Bernstein, H. D., P. Sarnow, and D. Baltimore. 1986. Genetic complementation among poliovirus mutants derived from an infectious cDNA clone. *J. Virol.* **60**:1040–1049.

Bienz, K., D. Egger, and L. Pasamontes. 1987. Association of polioviral proteins of the P2 genomic region with the viral replication complex and virus-induced membrane synthesis as visualized by electron microscopic immunocytochemistry and autoradiography. *Virology* **160**:220–226.

Caliguiri, L. A., and A. G. Mosser. 1971. Proteins associated with the poliovirus RNA replication complex. *Virology* **46**:375–386.

Crawford, N. M., and D. Baltimore. 1983. Genome-linked protein VPg of poliovirus is present as free VPg and VPg-pUpU in poliovirus-infected cells. *Proc. Natl. Acad. Sci. USA* **80**:7452–7455.

Dasgupta, A. 1983. Purification of host factor required for *in vitro* transcription of poliovirus RNA. *Virology* **127**:245–251.

Dasgupta, A., P. Zabel, and D. Baltimore. 1980. Dependence of the activity of the poliovirus replicase on a host cell protein. *Cell* **19**:423–429.

Dewalt, P. G., and B. L. Semler. 1987. Site-directed mutagenesis of proteinase 3C results in a poliovirus mutant deficient in synthesis of viral RNA polymerase. *J. Virol.* **61**:2162–2170.

Dildine, S. L., and B. L. Semler. 1989. The deletion of

41 proximal nucleotides reverts a poliovirus mutant containing a temperature-sensitive lesion in the 5′ noncoding region of genomic RNA. *J. Virol.* **63**:847–862.

Duechler, M., T. Skern, W. Sommergruber, C. Neubauer, P. Guendler, I. Fogy, D. Blaas, and E. Kuechler. 1987. Evolutionary relationships within the human rhinovirus genus: comparison of serotypes 89, 2, and 14. *Proc. Natl. Acad. Sci. USA* **84**:2605–2609.

Flanegan, J. B., and D. Baltimore. 1977. Poliovirus-specific primer-dependent RNA polymerase able to copy poly(A). *Proc. Natl. Acad. Sci. USA* **74**:3677–3680.

Flanegan, J. B., and D. Baltimore. 1979. Poliovirus polyuridylic acid polymerase and RNA replicase have the same viral polypeptide. *J. Virol.* **29**:352–360.

Flanegan, J. B., R. F. Pettersson, V. Ambros, M. J. Hewlett, and D. Baltimore. 1977. Covalent linkage of a protein to a defined nucleotide sequence of the 5′-terminus of virion and replicative intermediate RNAs of poliovirus. *Proc. Natl. Acad. Sci. USA* **74**:961–965.

Flanegan, J. B., D. C. Young, G. J. Tobin, M. M. Stokes, C. D. Murphy, and S. M. Oberste. 1987. Mechanism of RNA replication by the poliovirus RNA polymerase, HeLa cell host factor, and VPg, p. 273–284. *In* M. A. Brinton and R. R. Rueckert (ed.), *Positive Strand RNA Viruses.* Alan R. Liss, Inc., New York.

Graham, F. L., and A. J. van der Eb. 1973. A new technique for the assay of infectivity of human adenovirus 5 DNA. *Virology* **52**:456–467.

Hey, T. D., O. C. Richards, and E. Ehrenfeld. 1987. Host factor-induced template modification during synthesis of poliovirus RNA in vitro. *J. Virol.* **61**:802–811.

Iizuka, N., S. Kuge, and A. Nomoto. 1987. Complete nucleotide sequence of the genome of coxsackievirus B1. *Virology* **156**:64–73.

Inouye, S., and M. Inouye. 1987. Oligonucleotide-directed site-specific mutagenesis using double-stranded plasmid DNA, p. 181–206. *In* S. A. Narang (ed.), *Synthesis and Applications of DNA and RNA.* Academic Press, Inc., New York.

Kaplan, G., J. Lubinski, A. Dasgupta, and V. R. Racaniello. 1985. *In vitro* synthesis of infectious poliovirus RNA. *Proc. Natl. Acad. Sci. USA* **82**:8424–8428.

Kean, K. M., H. Agut, O. Fichot, E. Wimmer, and M. Girard. 1988. A poliovirus mutant defective for self-cleavage at the COOH-terminus of the 3C protease exhibits secondary processing defects. *Virology* **163**:330–340.

Koch, F., and G. Koch. 1985. *The Molecular BIology of Poliovirus.* Springer-Verlag, New York.

Kuhn, R. J., H. Tada, M. F. Ypma-Wong, J. J. Dunn, B. L. Semler, and E. Wimmer. 1988a. Construction of a mutagenesis cartridge for poliovirus VPg: isolation and characterization of viable and non-viable mutants. *Proc. Natl. Acad. Sci. USA* **85**:519–523.

Kuhn, R. J., H. Tada, M. F. Ypma-Wong, B. L. Semler, and E. Wimmer. 1988b. Mutational analysis of the genome-linked protein VPg of poliovirus. *J. Virol.* **62**:4207–4215.

Kuhn, R. J., and E. Wimmer. 1987. The replication of picornaviruses, p. 17–51. *In* D. J. Rowlands, B. W. J. Mahy, and M. Mayo (ed.), *The Molecular Biology of Positive Strand RNA Viruses.* Academic Press, Inc., New York.

Kuhn, R. J., E. Wimmer, and B. L. Semler. 1987. Expression of the poliovirus genome from infectious cDNA is dependent upon arrangements of eukaryotic and prokaryotic sequences in recombinant plasmids. *Virology* **157**:560–564.

Li, J.-P., and D. Baltimore. 1988. Isolation of poliovirus 2C mutants defective in viral RNA synthesis. *J. Virol.* **62**:4016–4021.

Lindberg, A. M., P. O. K. Stalhandske, and U. Pettersson. 1987. Genome of coxsackievirus B3. *Virology* **156**:50–63.

Lubinski, J. M., G. Kaplan, V. R. Racaniello, and A. Dasgupta. 1986. Mechanism of in vitro synthesis of covalently linked dimeric RNA molecules by the poliovirus replicase. *J. Virol.* **58**:459–467.

Lundquist, R. E., E. Ehrenfeld, and J. V. Maizel. 1974. Isolation of a viral polypeptide associated with poliovirus RNA polymerase. *Proc. Natl. Acad. Sci. USA* **71**:4773–4777.

Morrow, C. D., B. Warren, and M. R. Lentz. 1987. Expression of enzymatically active poliovirus RNA-dependent RNA polymerase in *Escherichia coli. Proc. Natl. Acad. Sci. USA* **84**:6050–6054.

Nicklin, M. J. H., H. G. Krausslich, H. Toyoda, J. J. Dunn, and E. Wimmer. 1987. Poliovirus polypeptide precursors: expression *in vitro* and processing by exogenous 3C and 2A. *Proc. Natl. Acad. Sci. USA* **84**:4002–4006.

Nomoto, A., B. Detjen, R. Pozzatti, and E. Wimmer. 1977. The location of the polio genome protein in viral RNAs and its implication for RNA synthesis. *Nature* (London) **268**:208–213.

Pettersson, R. F., V. Ambros, and D. Baltimore. 1978. Identification of a protein linked to nascent poliovirus RNA and to the polyuridylic acid of negative-strand RNA. *J. Virol.* **27**:357–365.

Pincus, S. E., D. C. Diamond, E. A. Emini, and E. Wimmer. 1986. Guanidine-selected mutants of poliovirus: mapping of point mutations to polypeptide 2C. *J. Virol.* **57**:638–646.

Plotch, S. J., O. Palant, and Y. Gluzman. 1989. Purification and properties of poliovirus polymerase expressed in *Escherichia coli. J. Virol.* **63**:216–225.

Racaniello, V. R., and D. Baltimore. 1981. Cloned poliovirus complementary DNA is infectious in mammalian cells. *Science* **214**:916–919.

Racaniello, V. R., and C. Meriam. 1986. Poliovirus temperature-sensitive mutant containing a single nucleotide deletion in the 5′ noncoding region of the viral RNA. *Virology* **155**:498–507.

Roder, A., and K. Koschel. 1975. Virus-specific proteins associated with the replication complex of poliovirus RNA. *J. Gen. Virol.* **28**:85–98.

Rothstein, M. A., O. C. Richards, C. Amin, and E. Ehrenfeld. 1988. Enzymatic activity of poliovirus RNA polymerase synthesized in *Escherichia coli* from viral cDNA. *Virology* **164**:301–308.

Rueckert, R. R. 1985. Picornaviruses and their replication, p. 705–738. *In* B. N. Fields (ed.), *Virology*. Raven Press, New York.

Sarnow, P., H. S. Bernstein, and D. Baltimore. 1986. A poliovirus temperature-sensitive mutant located in a non-coding region of the genome. *Proc. Natl. Acad. Sci. USA* **83**:571–575.

Semler, B. L., C. W. Anderson, R. Hanecak, L. F. Dorner, and E. Wimmer. 1982. A membrane-associated precursor to poliovirus VPg identified by immunoprecipitation with antibodies directed against a synthetic heptapeptide. *Cell* **28**:405–412.

Semler, B. L., A. J. Dorner, and E. Wimmer. 1984. Production of infectious poliovirus from cloned cDNA is dramatically increased by SV40 transcription and replication signals. *Nucleic Acids Res.* **12**:5123–5141.

Semler, B. L., V. H. Johnson, and S. Tracy. 1986. A chimeric plasmid from cDNA clones of poliovirus and coxsackievirus produces a recombinant virus that is temperature-sensitive. *Proc. Natl. Acad. Sci. USA* **83**:777–781.

Semler, B. L., R. J. Kuhn, and E. Wimmer. 1988. Replication of poliovirus genome, p. 23–48. *In* E. Domingo, J. J. Holland, and P. Ahlquist (ed.), *RNA Genetics*, vol 1. CRC Press, Inc., Boca Raton, Fla.

Stanway, G., P. J. Hughes, R. C. Mountford, P. D. Minor, and J. W. Almond. 1984. The complete nucleotide sequence of a common cold virus: rhinovirus 14. *Nucleic Acids Res.* **12**:7859–7875.

Takeda, N., R. J. Kuhn, C.-F. Yang, T. Takegami, and E. Wimmer. 1986. Initiation of poliovirus plus-strand RNA synthesis in a membrane complex of infected HeLa cells. *J. Virol.* **60**:43–53.

Takeda, N., C.-F. Yang, R. J. Kuhn, and E. Wimmer. 1987. Uridylylation of the genome-linked protein of poliovirus *in vitro* is dependent upon an endogenous RNA template. *Virus Res.* **8**:193–204.

Takegami, T., R. J. Kuhn, C. W. Anderson, and E. Wimmer. 1983a. Membrane-dependent uridylylation of the genome-linked protein VPg of poliovirus. *Proc. Natl. Acad. Sci. USA* **80**:7447–7451.

Takegami, T., B. L. Semler, C. W. Anderson, and E. Wimmer. 1983b. Membrane fractions active in poliovirus RNA replication contain VPg precursor polypeptides. *Virology* **128**:33–47.

Tershak, D. R. 1984. Association of poliovirus proteins with the endoplasmic reticulum. *J. Virol.* **52**:777–783.

Toyoda, H., M. Kohara, Y. Kataoda, T. Suganuma, T. Omata, N. Imura, and A. Nomoto. 1984. Complete nucleotide sequences of all three poliovirus serotype genomes: implication for genetic relationship, gene function and antigenic determinants. *J. Mol. Biol.* **174**:561–585.

Trono, D., R. Andino, and D. Baltimore. 1988. An RNA sequence of hundreds of nucleotides at the 5′ end of poliovirus RNA is involved in allowing viral protein synthesis. *J. Virol.* **62**:2291–2299.

Van Dyke, T. A., and J. B. Flanegan. 1980. Identification of poliovirus polypeptide p63 as a soluble RNA-dependent RNA polymerase. *J. Virol.* **35**:732–740.

Wimmer, E., R. J. Kuhn, S. Pincus, C.-F. Yang, H. Toyoda, M. Nicklin, and N. Takeda. 1988. Molecular events leading to picornavirus genome replication, p. 251–276. *In* J. W. Davies, K. F. Chafer, T. H. N. Ellis, R. Hull, R. Townsend, and H. W. Woolhouse (ed.), *Virus Replication and Genome Interactions*. 7th John Innes Symposium. The Company Biologists Ltd., Cambridge.

Young, D. C., D. M. Tuschall, and J. B. Flanegan. 1985. Poliovirus RNA-dependent RNA polymerase and host cell protein synthesize product RNA twice the size of poliovirion RNA in vitro. *J. Virol.* **54**:256–264.

Ypma-Wong, M. F., and B. L. Semler. 1987. Processing determinants required for in vitro cleavage of the poliovirus P1 precursor to capsid proteins. *J. Virol.* **61**:3181–3189.

Chapter 14

Natural and Artificial Poliovirus Defective Interfering Particles

Kimiko Hagino-Yamagishi, Shusuke Kuge, and Akio Nomoto

Defective interfering (DI) particles are deletion mutants that are constantly generated at low levels by nearly all infectious viruses (Holland et al., 1980; Huang and Baltimore, 1977; Perrault, 1981). These particles contain normal viral structural proteins, are able to replicate only with the aid of a helper standard infectious virus, and interfere specifically with the replication of a homologous standard helper virus. The DI genomes, therefore, must retain genetic information that is essential for their replication and not provided by the genome of a helper virus. Thus, analysis of the interaction between DI and helper genomes may provide insight into the basic mechanisms of viral replication. Furthermore, these DI particles may be powerful tools for studying molecular mechanisms of rearrangement in RNA genomes (Perrault, 1981).

DI particles of the Mahoney strain of poliovirus type 1 were initially described by Cole et al. (1971). Kajigaya et al. (1985) observed the generation of DI particles [PV1(Sab)DI] from the Sabin strain of poliovirus type 1 [PV1(Sab)]. Both types of the DI particles were generated at very low rates in viral preparations after many serial high-multiplicity-of-infection passages. The purified DI particles are able to initiate a normal poliovirus replication cycle but fail to synthesize capsid proteins and therefore cannot produce progeny virions. These data suggest that the DI RNAs lack a genome region encoding viral capsid proteins.

SIZES AND LOCATIONS OF DELETIONS

Kajigaya et al. (1985) observed that a small amount of particles of lower density appeared upon CsCl density gradient centrifugation after more than 30 passages of PV1(Sab). The gradient profile of a mixture of the standard virus and DI particles obtained at passage 42 is shown in Fig. 1. The densities of the standard virus and DI particles were 1.34 and 1.32 g/cm^3, respectively. This finding suggested that the size of the deletion in the DI particle genome was similar to that in the Mahoney DI particle genome (Cole et al., 1971).

Both the standard RNA and DI RNA were labeled with ^{32}P and analyzed by fingerprint analysis. The fingerprints of the RNAs obtained after RNase T$_1$ digestion are shown in Fig. 2. As can be seen, spots 6, 25, 38, and 47 were missing in the DI RNA pattern. A similar observation was reported for the Mahoney DI RNA (Nomoto et al., 1979). This observation indicated that the location of the deletion was limited to a certain region of the genome. According to the known total nucleotide sequence of PV1(Sab) RNA (Nomoto et al., 1982) and the sequences of large RNase T$_1$-resistant oligonucleotides of PV1(Sab) RNA (Nomoto et al., 1981), the nucleotide positions of oligonucleotides in spots 6, 25, 38, and 47 were deduced to be 2070 to 2094, 1795 to 1811, 1986 to 2000, and 1653 to 1667, respectively. Therefore, almost all DI RNAs must lack the genome region corresponding to nucleotide positions 1653 to 2094, which includes the genome region coding for capsid proteins VP2 and VP3 (Kitamura et al., 1981; Nomoto et al., 1982). Thus, like Mahoney DI particle RNAs, PV1(Sab)DI RNAs lack the capsid protein-coding region, indicating that the presence of a helper virus is required to supply capsid proteins for the production of DI particles.

Kimiko Hagino-Yamagishi and Akio Nomoto, Department of Microbiology, The Tokyo Metropolitan Institute of Medical Science, Honkomagome, Bunkyo-ku, Tokyo 113, Japan. *Shusuke Kuge,* Gene Regulation Laboratory, Imperial Cancer Research Fund, Lincoln's Inn Fields, London WC2A 3PX, United Kingdom.

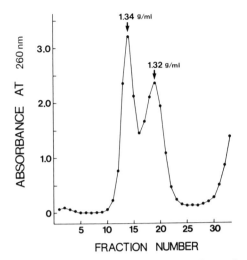

FIGURE 1. Profile of CsCl equilibrium density gradient of progeny virions produced at passage 42 (Kajigaya et al., 1985). PV1(Sab) was passaged in suspension-cultured HeLa S3 cells. Partially purified progeny virions were analyzed by CsCl density gradient. The gradient was fractionated from bottom to top.

electrophoresis under alkaline conditions. Five main DI-specific cDNA fragments (2a, 2b, 2c, 2d, and 2e) were observed on the gel. To determine the map locations of these DI-specific cDNA fragments, two different DNA probes, a and b, corresponding to the 5' and 3' ends of the genome of PV1(Sab), respectively, were used (Fig. 3). The three shorter bands (2c, 2d, and 2e) and the two other bands (2a and 2b) were specifically hybridized to probes a and b, respectively (Fig. 3). The population of genome species was estimated by the density of the corresponding bands observed on the autoradiogram and is indicated by the thickness of the filled bars in Fig. 3. Thus, the precise map locations of deletions and the population of DI RNAs were established as illustrated schematically in Fig. 3.

It is of interest that both ends of the deletions appear to be located in limited areas on the PV1(Sab) genome. Nucleotide sequences in these areas might contribute to the enhancement of DI generation. Therefore, determination of nucleotide sequences of many DI RNAs may provide important information about the mechanism of generation.

More precise map locations for the deletions were established by nuclease S1 mapping (Kuge et al., 1986). Genomic RNAs extracted from PV1(Sab)DI were hybridized with an infectious cDNA clone of PV1(Sab) that had been linearized with restriction endonuclease SalI and then treated with nuclease S1. Nuclease S1-resistant materials (cDNA-RNA hybrid molecules) were separated by 0.8% agarose gel

NUCLEOTIDE SEQUENCES OF REARRANGED GENOME REGIONS

Molecular cloning and rapid sequence analysis techniques were used to determine the nucleotide sequences of individual DI genomes. Double-stranded cDNAs of the DI genomes were synthesized from the virion RNA and digested with restriction endonuclease BanII.

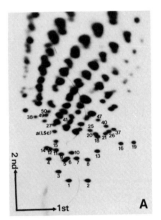
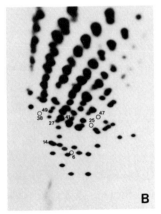

FIGURE 2. Fingerprints of RNase T_1-resistant oligonucleotides of [^{32}P]RNAs of the standard virus and lower-density particles (Kajigaya et al., 1985). Purified [^{32}P]RNAs of the standard virus (A) and lower-density particles (B) were digested with RNase T_1 and analyzed by two-dimensional polyacrylamide gel electrophoresis. Spots absent in panel B but present in panel A are indicated by open circles.

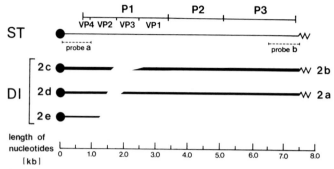

FIGURE 3. Results of nuclease S1 mapping (Kuge et al., 1986). Map positions of cDNA fragments are illustrated schematically. ST and DI, Standard PV1(Sab) and PV1(Sab)DI genomes, respectively; P1, viral capsid protein region; P2 and P3, viral replication protein regions; kb, kilobases. VPg (●) and poly(A) (∿) in PV1(Sab)DI genomes are drawn without experimental confirmation. The population of genome species is indicated by the thickness of the filled bars; thicker bars represent genome species of larger populations.

This enzyme is able to digest PV1(Sab) cDNA at only two sites, corresponding to nucleotide positions 909 and 3523 of PV1(Sab) RNA. Therefore, BanII-resistant internal fragments of DI cDNA should include all kinds of rearranged genome regions. BanII cDNA fragments were then inserted into cloning vector pBR322.

We obtained 80 independent cDNA clones. These cDNA clones were analyzed by digestion with restriction endonuclease RsaI, followed by polyacrylamide gel electrophoresis to determine the sizes and locations of deletions. By this analysis, locations of individual deletions as well as size distribution were estimated for individual

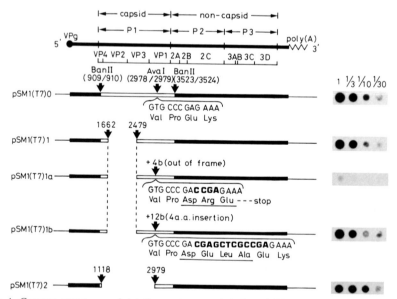

FIGURE 4. Genome structures of deletion mutants and their activities as RNA replicons (Hagino-Yamagishi and Nomoto, 1989). Closed and open bars represent segments from the Mahoney and Sabin 1 genomes, respectively. Numbers over the RNAs or in parentheses are the corresponding nucleotide positions from the 5' end of the genome of PV1(Sab). Results of dot blot hybridization experiments are shown on the right. Dilutions of the RNA preparation extracted from the transfected cells are indicated by magnification of dilution. Nick-translated pSM1(T7)0 was used as a probe.

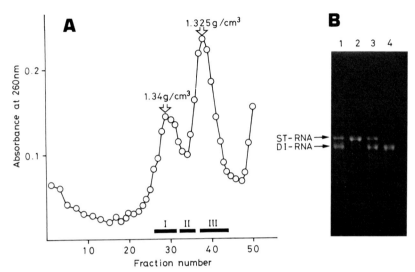

FIGURE 5. Identification of an artificial DI particle (Hagino-Yamagishi and Nomoto, 1989). (A) Profile of CsCl equilibrium density gradient of passage 6 viruses. The gradient was fractionated from bottom to top. Fractions indicated by bars were pooled and designated fractions I, II, and III. (B) RNAs extracted from a mixture of the standard virus and naturally occurring DI particles (lane 1), fraction I (lane 2), fraction II (lane 3), and fraction III (lane 4) were subjected to 0.8% agarose gel electrophoresis. ST-RNA, RNA of the standard virus; DI-RNA, RNA of DI particles.

cDNA clones (data not shown). On the basis of these data, we selected 13 cDNA clones of PV1-(Sab)DI RNAs that appeared to represent most different species of DI genome in order to analyze their nucleotide sequences.

Sequence analysis revealed precise map locations of individual clones (data not shown). The size distribution of the deletions was from 9.6 to 13.2% of the total genome length, except for one clone whose deletion size was only 4.2% of the entire genome. The deletions occurred between nucleotide positions 1226 and 2705, and the rearranged sites appeared to occur in limited areas of the genome. These data are compatible with the results obtained by S1 mapping (Fig. 3).

All of the deletions occurred in such a manner that the correct reading frame of the poliovirus polyprotein downstream of the deletion was maintained. This finding strongly suggests that certain nonstructural proteins must be provided by translation of the DI RNA for its own replication and not be supplied by helper standard virus, at least at a certain stage of viral replication. It appears, therefore, that the genetic information of the nonstructural proteins of the poliovirus genome cannot function in *trans*. To confirm this assumption, we constructed poliovirus DI particles in vitro.

CONSTRUCTION OF ARTIFICIAL DI PARTICLES

Recently, van der Werf et al. (1986) reported the synthesis of a large amount of infectious poliovirus RNA by using purified T7 RNA polymerase. The specific infectivity of the synthesized RNA is approximately 10^5 PFU/μg of RNA in HeLa cells. This infectivity is almost 100 times higher than that of the most efficient infectious cDNA clone of poliovirus (Kohara et al., 1986; Kuge and Nomoto, 1987). Availability of the synthesized RNA of high specific infectivity may make it possible to construct mutant polioviruses with low viability. It is also possible to examine the ability of an RNA as a replicon by measuring the amount of RNA replicated in the transfected cells even if the RNA is inactive in producing infectious particles. In fact, Kaplan and Racaniello (1988) produced poliovirus RNA containing in-frame deletions within the capsid-coding region to show that the entire capsid-coding sequence of the poliovirus genome is not required for translation or RNA replication. We took advantage of this highly infectious RNA synthesized in vitro to construct artificial poliovirus DI particles.

Kuge et al. (1986) have isolated a number of

cloned cDNAs of the DI genomes spontaneously generated by serial passagings of PV1(Sab). Of these, pVS(1)DI-213 contains a deletion of 816 bases from the corresponding nucleotide positions 1663 to 2478, which encodes parts of capsid proteins VP2 and VP3 (Kuge et al., 1986). The *Ban*II fragment of plasmid pSM1(T7)0 (Fig. 4) was replaced by the shorter *Ban*II DNA fragment of plasmid pVS(1)DI-213. The resulting plasmid was designated pSM1(T7)1.

To generate the DI particle derived from an RNA synthesized in vitro, semiconfluent HeLa S3 monolayer cells were transfected with 1 to 5 μg of RNA that had been transcribed from pSM1(T7)1 linearized by digestion with either *Eco*RI or *Pvu*I (Hagino-Yamagishi and Nomoto, 1989). After transfection, the cells were superinfected with PV1(M) as a helper standard virus. Viruses produced in the cells were passaged to amplify the DI particle. To confirm that a DI particle exists in the passaged virus preparation, the virus preparation of passage 6 was analyzed by equilibrium centrifugation in a CsCl density gradient (Hagino-Yamagishi and Nomoto, 1989). In addition to a peak of standard virus at a density of 1.34 g/cm³, a peak at a density of 1.325 g/cm³ was observed (Fig. 5A). The fractions of different densities were pooled separately into three classes (fractions 25 to 31, 32 to 36, and 37 to 44). RNAs were extracted from virus particles of each fraction and separated by agarose gel electrophoresis under denaturing conditions with formamide and Formalin. The RNA from fractions I (Fig. 5B, lanes 2) and III (lane 4) migrated to positions of standard RNA and DI RNA, respectively. The data clearly show that a DI particle exists in the passaged isolates (Hagino-Yamagishi and Nomoto, 1989).

Furthermore, nucleotide sequence analysis was carried out on the RNA of the recovered DI particle. The result indicated that the DI particle shown in Fig. 5A was derived from the RNA synthesized from the recombinant deletion plasmid pSM1(T7)1. Thus, we succeeded in constructing an artificial DI particle of poliovirus.

REQUIREMENTS FOR THE
DI PARTICLE GENOME

To investigate the possible requirement of in-frame deletions for the DI RNAs to be packaged in virion particles, two more recombinant plasmids, pSM1(T7)1a and pSM1(T7)1b, were constructed (Hagino-Yamagishi and Nomoto, 1989) (Fig. 4). Plasmid pSM1(T7)1a has an insertion sequence of 4 bases at the *Ava*I cleavage site of the corresponding nucleotide position

2978 in the nucleotide sequence of pSM1(T7)1, and therefore the RNA transcript from pSM1(T7)1a has an incorrect reading frame for viral protein synthesis in the nucleotide sequence downstream of position 2978. Plasmid pSM1(T7)1b has an insertion sequence of 12 bases at the same site of pSM1(T7)1. Thus, the RNA from pSM1(T7)1b has an extra nucleotide sequence encoding four amino acids, but the correct reading frame is retained.

RNAs transcribed from *Eco*RI-linearized plasmids pSM1(T7)0, pSM1(T7)1, pSM1(T7)1a, and pSM1(T7)1b were transfected into HeLa S3 cells, and their abilities as RNA replicons were examined by comparing amounts of RNAs rep-

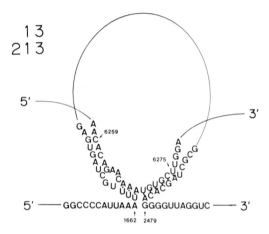

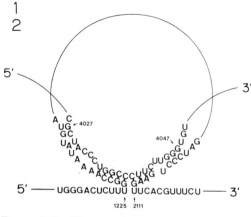

FIGURE 6. Possible secondary structure of rearranged genome regions (Kuge et al., 1986). Names of clones are indicated by numbers on the left; the prefix pVS(1)DI- is omitted. Numbers on the nucleotide sequences indicate nucleotide positions.

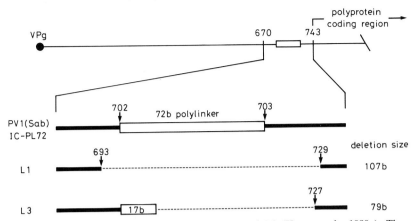

FIGURE 7. Genome structures of deletion variants L1 and L3 (Kuge et al., 1989a). The genome structure of virus PV1(Sab)IC-PL72 is indicated at the top. Symbols: ▢, sequence inserted at nucleotide position 702; – – –, deletion regions in the genomes of viruses L1 and L3. b, Bases.

licated in the transfected cells, using dot blot hybridization. Synthesized RNAs replicated well in the transfected cells except for an RNA from pSM1(T7)1a (Fig. 4). To determine whether all of the replication proteins were able to act in *trans*, the cells transfected with these RNAs were superinfected with the standard poliovirus. The DI RNA derived from plasmid pSM1(T7)1a was not packaged, whereas other RNAs containing in-frame deletions were easily packaged into virion particles. These observations support our previous notion that at least one replication protein of poliovirus has little or no activity to function in *trans* on a certain stage in viral replication (Hagino-Yamagishi and Nomoto, 1989).

Plasmid pSM1(T7)2 has an in-frame large deletion of 1,860 bases in the corresponding genome region encoding viral capsid proteins (Fig. 4). The RNA transcribed from this plasmid replicated well in the transfected cells (Fig. 4). However, this active RNA replicon was not able to be packaged under the conditions described above. The mechanisms underlying this observation are unclear at present. However, it is possible that a minimum length of RNA is required to be packaged into the poliovirus capsid. Alternatively, a specific packaging signal might exist within nucleotide sequences that are missing in the sequence of pSM1(T7)2-derived RNA.

A MODEL FOR DI GENERATION

Deletion mutants of single-stranded RNA viruses leading to DI particles are thought to have arisen as a result of an error during RNA replication. For this genetic event to occur, a copy choice model has been proposed. According to this model, the synthesis of a nascent daughter chain is interrupted during viral RNA synthesis and resumes on another template strand at sites selected by unknown mechanisms. However, the relationship between premature termination and heterologous reinitiation sites is obscure.

Recently, we found a secondary structure common to every deletion site, and we speculate that this structure may be involved in the generation of DI particles (Kuge et al., 1986) (Fig. 6). Typical examples are shown in Fig. 6. According to our model, deletion regions could be looped out from the remaining sequences in a possible common structure supported by sequences existing on the positive or negative RNA strand. Replicase may skip these transiently formed loop structures with certain frequencies. This model was designated the supporting sequence-loop model (Kuge et al., 1986). Statistical analysis, however, did not verify the model. Therefore, the probability of formation of the supporting sequence-loop may be low, which explains why the rate at which deletion mutants of poliovirus occur is very low.

To provide higher reliability to our supporting sequence-loop model, analysis of the genomes of deletion mutants that can easily be generated from the parent virus is essential. Among many viable in vitro mutants of poliovirus, an insertion mutant, PV1(Sab)IC-PL72, has a deleterious AUG in the insertion sequence in the 5' noncoding sequence of the genome (Fig. 7). Fortunately, the AUG was found to be elim-

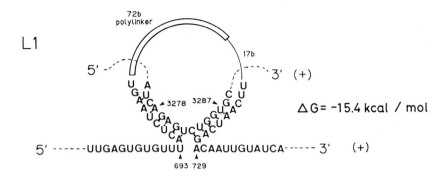

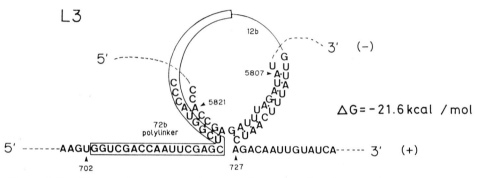

FIGURE 8. Possible secondary structure of rearranged genome regions (Kuge et al., 1989a). Names of virus clones are indicated on the left. Numbers on the nucleotide sequences indicate nucleotide numbers. (+) and (−), Positive and negative strands, respectively. The calculated free energies for the secondary structure are indicated on the right. b, Bases.

inated by point mutations or deletion mutations during viral replication (Kuge et al., 1989a, 1989b). The genome structures of variants L1 and L3 that eliminated the AUG by deletion mutations are shown in Fig. 7. This kind of deletion mutant can be isolated at fairly high frequency (Kuge et al., 1989a). Therefore, there may be strong supporting sequences by which deleted sequences are looped out from the remaining genome sequence.

As we expected, a computer-aided search detected fairly stable secondary structures at the deletion sites (Kuge et al., 1989a) (Fig. 8). This observation may support our DI generation model. Our sequence data for the deleted genomes did not contradict a recombination model proposed by Romanova et al. (1986). These kinds of genetic events mediated by the transient highly ordered RNA structures may have been involved in the evolution of poliovirus.

ACKNOWLEDGMENT. This work was supported in part by a grant from the Ministry of Education, Science and Culture of Japan.

REFERENCES

Cole, C. N., D. Smoler, E. Wimmer, and D. Baltimore. 1971. Defective interfering particles of poliovirus. I. Isolation and physical properties. J. Virol. 7:478–485.

Hagino-Yamagishi, K., and A. Nomoto. 1989. In vitro construction of poliovirus defective interfering particles. J. Virol. 63:5389–5392.

Holland, J. J., S. I. T. Kennedy, B. L. Semler, C. L. Jones, L. Roux, and E. A. Graban. 1980. Defective interfering RNA viruses and the host cell response. Compr. Virol. 16:137–192.

Huang, A. S., and D. Baltimore. 1977. Defective interfering animal viruses. Compr. Virol. 10:73–116.

Kajigaya, S., H. Arakawa, S. Kuge, T. Koi, N. Imura, and A. Nomoto. 1985. Isolation and characterization of defective-interfering particles of poliovirus Sabin 1 strain. *Virology* **142**:307–316.

Kaplan, G., and V. R. Racaniello. 1988. Construction and characterization of poliovirus subgenomic replicons. *J. Virol.* **62**:1687–1696.

Kitamura, N., B. L. Semler, P. G. Rothberg, G. R. Larsen, C. J. Adler, A. J. Dorner, E. A. Emini, R. Hanecak, J. J. Lee, S. van der Werf, C. W. Anderson, and E. Wimmer. 1981. Primary structure, gene organization and polypeptide expression of poliovirus RNA. *Nature* (London) **291**:547–553.

Kohara, M., S. Abe, S. Kuge, B. L. Semler, T. Komatsu, M. Arita, H. Itoh, and A. Nomoto. 1986. An infectious cDNA clone of the poliovirus Sabin strain could be used as a stable repository and inoculum for the oral polio live vaccine. *Virology* **151**:21–30.

Kuge, S., N. Kawamura, and A. Nomoto. 1989a. Genetic variation occurring on the genome of an in vitro insertion mutant of poliovirus type 1. *J. Virol.* **63**:1069–1075.

Kuge, S., N. Kawamura, and A. Nomoto. 1989b. Strong inclination toward transition mutation in nucleotide substitutions by poliovirus replicase. *J. Mol. Biol.* **207**:175–182.

Kuge, S., and A. Nomoto. 1987. Construction of viable deletion and insertion mutants of the Sabin strain of type 1 poliovirus: function of the 5′ noncoding sequence in viral replication. *J. Virol.* **61**:1478–1487.

Kuge, S., I. Saito, and A. Nomoto. 1986. Primary structure of poliovirus defective-interfering particle genomes and possible generation mechanisms of the particles. *J. Mol. Biol.* **192**:473–487.

Nomoto, A., A. Jacobson, Y. F. Lee, J. Dunn, and E. Wimmer. 1979. Defective interfering particles of poliovirus: mapping of deletion and evidence that the deletion in the genome of DI(1), (2) and (3) are located in the same region. *J. Mol. Biol.* **128**:179–196.

Nomoto, A., N. Kitamura, J. J. Lee, P. G. Rothberg, N. Imura, and E. Wimmer. 1981. Identification of point mutations in the genome of the poliovirus Sabin vaccine LSc 2ab, and catalogue of RNase T1- and RNase A-resistant oligonucleotides of poliovirus type 1 (Mahoney) RNA. *Virology* **112**:217–227.

Nomoto, A., T. Omata, H. Toyoda, S. Kuge, H. Horie, Y. Kataoka, Y. Genba, Y. Nakano, and N. Imura. 1982. Complete nucleotide sequence of the attenuated poliovirus Sabin 1 strain genome. *Proc. Natl. Acad. Sci. USA* **79**:5793–5797.

Perrault, J. 1981. Origin and replication of defective interfering particles. *Curr. Top. Microbiol. Immunol.* **93**:151–207.

Romanova, L. I., V. M. Blinov, E. A. Tolskaya, E. G. Viktorova, M. S. Kolesnikova, E. A. Guseva, and V. I. Agol. 1986. The primary structure of crossover regions of intertypic poliovirus recombinants: a model of recombination between RNA genomes. *Virology* **155**:202–213.

van der Werf, S., J. Bradley, E. Wimmer, F. W. Studier, and J. J. Dunn. 1986. Synthesis of infectious poliovirus RNA by purified T7 RNA polymerase. *Proc. Natl. Acad. Sci. USA* **83**:2330–2334.

Chapter 15

Defective Interfering Particles of Hepatitis A Virus in Cell Cultures and Clinical Specimens

Günter Siegl, Jürg P. F. Nüesch, and José de Chastonay

For an extended period of time, hepatitis A virus (HAV) resisted cultivation and characterization. Physicochemical parameters of the virus particle such as size, buoyant density, sedimentation coefficient, and structural proteins, as well as size and organization of the linear positive-stranded RNA genome, now support its classification among the picornaviruses (Gust et al., 1983). Specific features, however, preclude grouping of HAV within one of the genera accepted to date (enterovirus, rhinovirus, cardiovirus, or aphtovirus) of this virus family: HAV surpasses in stability known picornaviruses, and it retains its physical integrity and biological activity even after treatment at 60°C for 30 min (Siegl et al., 1984). At the level of nucleotide sequence, homologies are also low between HAV and accepted picornaviruses and amount at best to 28% of amino acid identity in regions of the genome supposedly coding for functions required to replicate viral RNA and to process viral translation products (Ticehurst et al., 1988). Last but not least, HAV shows a protracted replication behavior, with the replication cycle extending over 24 to 36 h. In the vast majority of HAV-cell systems, persistent infection rather than lysis of the infected cells ensues (Siegl, 1988).

Persistent HAV infection differs significantly from other persistently infected picornavirus-cell systems. In general, all cells of an in vitro culture are infected and remain so for a prolonged period of time of months or even years. The reasons for this virus-cell relationship may be sought in the lack of significant interference of HAV replication with host cell metabolism, in defects in the expression of virus-encoded functions which thus allow for a slow and inefficient replication of HAV only, in accumulation of temperature-sensitive mutants, or in the generation of defective interfering (DI) particles. In fact, evidence for contribution of all of these factors and functions to the establishment and maintenance of persistent HAV infection has been obtained: even in cytolytic HAV infection, shutoff of host cell macromolecular synthesis has so far not been observed (Siegl, 1988). Processing of translation products of the HAV genome obviously is rather inefficient. In the course of in vitro translation of HAV RNA, precursor polypeptides up to almost 200 kilodaltons in size were found to accumulate, yet they failed to be processed to the expected low-molecular-weight products (de Chastonay-Krech et al., in preparation). Experimental evidence for continued accumulation of temperature-sensitive variant viruses in the course of HAV infection is also available (Nüesch et al., in preparation). The significance of such variants in HAV persistent infections, however, awaits further investigation. Finally, we have good evidence for the presence in viral harvests and for continued generation in infected cultures of defective, subgenomic viral RNAs. These molecules have distinct and stable physicochemical characteristics (Nüesch et al., 1988). HAV particles harboring the very same defective genomes were also detected in clinical materials (Nüesch et al., 1989). Because the presence of such particles could be shown to interfere with replication of standard hepatitis A virions, the latter finding may provide the possibility of investigating the true role of DI particles in the course of a naturally occurring viral infection.

Günter Siegl, Jürg P. F. Nüesch, and José de Chastonay, Institute for Medical Microbiology, University of Bern, Friedbühlstrasse 51, CH-3010 Bern, Switzerland.

TABLE 1

Synthesis of Infectious HAV in MRC-5 Cells
Infected at Various MOIs

Day postinfection	Titer ($\log_{10}$/ml of total lysate) at given MOI		
	0.5	10^{-2}	10^{-4}
8	6.66	5.66	3.5
20	6.33	7.66	7.77

DETECTION AND CHARACTERIZATION OF DI GENOMES IN HAV-INFECTED CULTURE SYSTEMS

The presence of defective HAV particles in cell culture harvests and their ability to modulate (in other words, to interfere with) replication of standard mature hepatitis A virions can be demonstrated in the classical experiment in which cultures are infected at high and low multiplicities of infection (MOIs) and production of infectious standard virions is monitored. For example, infection of MRC-5 cells at an MOI of about 0.5 results in relatively rapid accumulation of maximal titers of infectious HAV. At the same time, a persistent virus-cell relationship is established, in the course of which the concentration of infectious virus in cell cultures remains more or less constant. Infection at an MOI of 10^{-2} or even 10^{-4}, on the other hand, leads to the expected delay in accumulation of infectious HAV. Titers finally recorded under these conditions, however, are higher by a factor of 10 to 100. This finding can be interpreted as the effect of reduction of DI particles in the inoculum by dilution. Nonetheless, persistent infection ensues as soon as cells in an infected culture become involved in HAV replication.

When particles contained in the inoculated seed virus or in harvests from the resulting persistently infected cultures are purified by buoyant density centrifugation in CsCl and/or sucrose sedimentation, Northern (RNA) analysis of encapsidated RNAs reveals, for example, the standard genome and additional subgenomic RNA species in viruses sedimenting in the peak supposed to harbor standard HAV (156S) (Fig. 1). The full-size viral genome can also be detected in particles sedimenting ahead (>160S) of standard virions and possibly derived from the high-density component (1.4 to 1.46 g/cm^3) of HAV. In addition, subgenomic and sometimes quite short RNA molecules seem to be present in particles at this location. Hybridization of such RNA blots with clones of HAV cDNA representing distinct regions of the genome

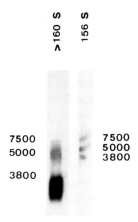

FIGURE 1. Northern analysis of RNA present in HAV particles propagated in BGM cells and sedimenting in a sucrose gradient around 156S and >160S. RNA was detected by hybridization with radiolabeled negative-stranded cRNA transcribed by SP6 polymerase from subcloned HAV cDNA.

pointed to the existence of internal deletions in the region of the genome coding for the HAV structural proteins as well as to the presence of truncations in the 3'-proximal part of viral RNA.

The exact locations of both deletions and truncations were determined by S1 mapping and confirmed by exonuclease VII mapping. The strategy and cDNA probes used for this purpose have been described in detail elsewhere (Nüesch et al., 1988). Figure 2 summarizes the findings. Three internal deletions were mapped. One of them is predominantly located within the region supposedly coding for HAV structural proteins. It spans from nucleotides (nt) 1370 to 3240. A second deletion extends from nt 1200 to 3820 and hence affects the information both for structural proteins VP2, VP3, and VP1 and for nonstructural proteins 2A and 2B. A third type of defective genome lacks nt 930 to 4380, i.e., coding sequences for both structural proteins VP2, VP3, and VP1 and nonstructural proteins 2A to 2C. Given the size of 7,450 nt of standard HAV RNA (Baroudy et al., 1985), these deletions result in molecules of 5,580, 4,830, and 4,000 nt, respectively. Such values are compatible with the size of the distinct species of subgenomic viral RNAs detected by Northern blotting (Fig. 1).

In addition to the described internal deletions, S1 mapping experiments in the 3'-terminal region of RNA derived from virus pools probably representing the high-density component of HAV revealed truncation points clustering around nt 4800, 5350, 5650, 6300, and 7250 of the

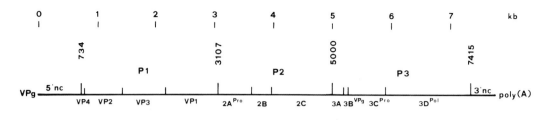

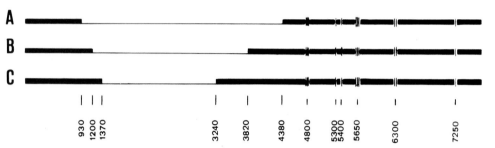

FIGURE 2. Tentative organization of the genome and locations of the three prominent internal deletions and the 3′-proximal truncations in HAV particles present in cell culture harvests. Endpoints of deletions and breakpoints of truncations are given as nucleotide positions.

HAV genome. In combination with the presence of the internal deletions, this multitude of truncations explains the apparently continuous size spectrum of short subgenomic RNA molecules present in the latter pool.

The existence of defective genomes in HAV particles and the ability of these specific particles to interfere with replication of standard HAV is not an unexpected finding. Major internal deletions affecting predominantly sequences coding for viral structural proteins have been described repeatedly for other picornaviruses, e.g., polio-, coxsackie-, echo-, and mengoviruses (Holland, 1986). However, the existence of defective genomes which, in addition to internal deletions, have truncations of variable length in their 3′-terminal regions is a unique feature of HAV. Such truncated molecules seem to be exclusively present in particles of the so-called high-density component of HAV. Particles of this component are characterized by aberrant physicochemical properties (higher buoyant density and faster sedimentation than the hepatitis A virion), including perhaps also a reduced stability of the viral capsid. Under these conditions, the viral genome may be more accessible to RNase than when located within the extraordinarily stable capsid of the mature virus. Contrary to what might be expected as the result of digestion with endogenous RNase, however, truncations were always restricted to regions of the HAV genome between nt 4800 and the 3′ terminus. Moreover, such molecules could be directly recovered from infected cells by phenol extraction, whereby digestion with RNase during purification of particles could be avoided. Hence, truncated viral RNA seems to be a genuine intermediate or side product of HAV RNA replication.

A final interesting finding concerns the constant presence of identical deletions and truncations in all HAV strains and in harvests from all HAV-cell systems so far analyzed. In addition, the internal deletions and 3′-proximal truncations mapped at identical locations with all of the viruses, whether virus isolates originated in Australia (strain HM 175), India (H141), North Africa (MBB 11/5), Switzerland (CLF), or Costa Rica (CR326). The type and number of deletions, however, varied. In general, deletions C and B were the most abundant. In some instances, they became detectable as early as passage 5 in cell culture of a certain HAV isolate. This is in striking contrast to the situation with the generation of DI particles in known picornavirus systems, which usually require multiple, consecutive, and undiluted passages to generate detectable amounts of DI particles (Holland, 1986).

In this context, it is also noteworthy that replication of defective RNA type C could be observed easily in cell culture both under conditions of high MOI (0.6) and after infection of cultures at an MOI of 10^{-2} or even 10^{-4}. In each instance, synthesis of defective molecules proceeded parallel to accumulation of total viral

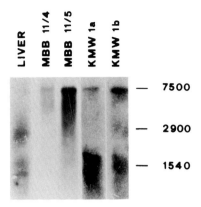

FIGURE 3. Northern analysis of HAV RNA extracted from the indicated human clinical specimens and from the liver of an experimentally infected marmoset monkey. These samples were later used for mapping of deletion breakpoints.

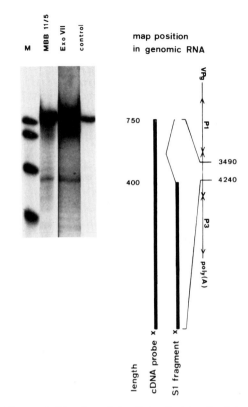

RNA but with a delay of 1 to 2 days (Nüesch and Siegl, in preparation).

PRESENCE OF DEFECTIVE HAV PARTICLES IN CLINICAL MATERIAL

The regular appearance of defective HAV particles in cell culture harvests early after primary isolation of the virus led to the hypothesis that these particles might also be present in the clinical materials from which the viruses originally had been recovered. To probe into this possibility, HAV particles were purified from human fecal samples (clinical specimens from which HAV isolates MBB 11/5, MBB 11/9, KMW1a, KMW1b, and CLF, respectively, originated), from viremic human plasma (Hollinger et al., 1983), and from the liver of a marmoset monkey experimentally infected with HAV strain CR326 (Provost et al., 1975). Northern analysis of RNA recovered from these virus samples then furnished evidence for the existence of subgenomic viral RNA in detectable yet usually low quantity (Fig. 3). When the endpoints of internal deletions A to C were searched for by S1 and exonuclease VII mapping, no evidence for the presence of molecules with deletion A (nt 930 to 4380) was obtained. For deletions B and C, however, the characteristic 3'-terminal endpoints located at nt 3820 and 3240, respectively, could be documented in RNA of most virus isolates (Fig. 4), although the autoradiographic signals related to the presence of these deletions were sometimes extremely faint. In contrast, only one 5' terminus at nt 1200

FIGURE 4. Mapping of the 3'-terminal breakpoint of deletion B (cf. Fig. 2). RNA was prepared from HAV particles isolated from clinical specimen MBB 11/5 and hybridized to a cDNA probe labeled at the 5' end of the minus strand at the HindII site at position 4240. Hybrids were digested with S1 nuclease or exonuclease VII, and protected cDNA fragments were resolved on a sequencing gel. During digestion with S1 nuclease, protection by standard viral RNA yielded a fragment 750 nt in length, whereas partial protection by the defective genome of type C resulted in a smaller fragment of 400 nt. The corresponding deletion breakpoint is located at nt 3820. Lane M, DNA size marker (pBR322/HpaII). x, Position of label on the S1 hybridization probe.

could be identified in RNA derived from three fecal human samples as well as from marmoset liver extracts. This finding indicates that defective HAV genomes with deletion B are definitely present in clinical materials. Although the 3'-terminal endpoint of deletion C could be readily mapped in HAV particles contained in human stool, marmoset liver, and even human viremic blood, the same RNA samples continuously failed to yield evidence for the presence of the previously observed, specific 5' breakpoint at nt

1370. Therefore, it must be assumed that deletion C in the form described for cell culture-derived HAV is not present in clinical material. Rather, a corresponding deletion must be supposed to start at nt 1200, like deletion B, and to end at nt 3240. Evidence for the presence of new additional breakpoints was never obtained. This underlines again the existence of some few breakpoints within the HAV genome that are preferentially active in formation of internally deleted, subgenomic HAV RNAs. Mapping of 3'-proximal truncations could not be performed because most of the clinical specimens were available in rather limited quantity.

CONCLUSIONS

As in many other aspects, HAV differs from its known relatives within the picornavirus family in the generation and characteristics of DI particles. One of the most prominent and so far unobserved features of these structures concerns conservation of distinct breakpoints within the viral genome, both during their generation in vitro in various HAV-cell systems and in the course of natural or experimental infection in vivo in humans or monkeys, respectively. Moreover, these preferential breakpoints seem to be present in the genomes of all HAV strains independent of geographical origin and passage history of the virus. How they are generated, however, remains to be shown. Inspection of published nucleotide sequences of HAV RNA (Najarian et al., 1985) revealed splicing consensus sequences of eucaryotic pre-mRNAs in the locations flanking the 5' termini of deletions A and C as well as the 3'-terminal breakpoint of deletion C. This finding suggests that a splicing mechanism comparable to the lariat model proposed by Grabowski et al. (1984) could be active.

A further distinguishing characteristic consists in the ease with which HAV DI particles are generated during passage of the virus in vitro. This phenomenon concerns predominantly particles containing the most prominent internal deletions, B and C. They became detectable either individually or in combination in sometimes considerable concentrations early after primary isolation of a HAV strain in cell culture. Deletion C (the only one specifically looked for so far) also appeared constantly during propagation of HAV even after cells were infected at a rather low MOI such as 10^{-2} or 10^{-4}. In contrast, viruses of the four established picornavirus genera usually require several consecutive undiluted passages for DI particles to accumulate in significant quantity (Holland, 1986). The latter viruses, however, also fail to regularly establish persistent infection in vitro, and if they do so at all, the resulting virus-cell relationship is clearly different from that characteristic for HAV. In consequence, generation of deleted genomes as well as interference of the resulting defective virus particles with replication and spread of standard infectious HAV may indeed be implicated in the establishment and maintenance of persistent infection in vitro.

The situation in vivo may be completely different. DI particles of several viruses were shown to influence the course of acute infection in animal model systems (Barrett and Dimmock, 1986). Even if present in barely detectable quantities after experimental infection, they then proved able to ameliorate the effects of acute virus replication, to favor host immune response, and to facilitate development of persistent infection. Under conditions of natural infection, however, DI particles of the very same viruses have not yet been observed. Hence, it is not known whether the predicted positive and/or negative effects of this specific class of particles play the expected role during natural infection. Hepatitis A is a benign, usually self-limiting disease with no tendency toward persistent infection. Yet according to the results presented here, defective and, by analogy with in vitro studies, interfering HAV particles seem to be present both early and late in natural hepatitis A and in all clinical materials (liver, blood, and feces) known to yield infectious HAV at one time or the other in the course of the disease. For the first time, this provides the possibility of carefully investigating the true role of DI particles in the establishment, maintenance, and resolution of a viral infection. These studies should be facilitated by the fact that HAV DI particles can be studied in an in vitro system and in an experimental animal model, as well as under conditions of natural infection.

Aknowledgments. This work was supported in part by grant 3.829-0.86 from the Swiss National Science Foundation.

We are also very grateful to M. Weitz (University of Bern) for helpful discussion and critical comments.

REFERENCES

Baroudy, B. M., J. R. Ticehurst, T. A. Miele, J. V. Maizel, Jr., R. H. Purcell, and S. M. Feinstone. 1985. Sequence analysis of hepatitis A virus cDNA coding for capsid proteins and RNA polymerase. *Proc. Natl. Acad. Sci. USA* **82:**2143–2147.

Barrett, A. D. T., and N. J. Dimmock. 1986. Defective interfering viruses and infections of animals. *Curr. Top. Microbiol. Immunol.* **128**:55–84.

de Chastonay-Krech, S., M. Weitz, and G. Siegl. Characterization of hepatitis A virus RNA as a template for in vitro translation in a rabbit reticulocyte lysate. Manuscript in preparation.

Grabowski, P. J., R. A. Padgett, and P. A. Sharp. 1984. Messenger RNA splicing in vitro: an excised intervening sequence and a potential intermediate. *Cell* **37**:415–427.

Gust, J., A. G. Coulepis, S. M. Feinstone, S. A. Locarnini, Y. Moritsugu, R. Najera, and G. Siegl. 1983. Taxonomic classification of hepatitis A virus. *Intervirology* **20**:1–7.

Holland, J. J. 1986. Generation and replication of defective viral genomes, p. 77–99. *In* B. N. Fields and D. M. Knipe (ed.), *Fundamental Virology.* Raven Press, New York.

Hollinger, F. B., N. C. Khan, P. E. Oefinger, D. H. Yawn, A. C. Schmulen, G. R. Dreesman, and J. L. Melnick. 1983. Posttransfusion hepatitis type A. *J. Am. Med. Assoc.* **250**:2313–2317.

Najarian, R., D. Caput, W. Gee, S. J. Potter, A. Renard, J. Merryweather, G. V. Nest, and D. Dina. 1985. Primary structure and gene organization of human hepatitis A virus. *Proc. Natl. Acad. Sci. USA* **82**:2627–2631.

Nüesch, J., S. Krech, and G. Siegl. 1988. Detection and characterization of subgenomic RNAs in hepatitis A virus particles. *Virology* **165**:419–427.

Nüesch, J. P. F., J. de Chastonay, and G. Siegl. 1989. Detection of defective genomes in particles of hepatitis A virus present in clinical specimens. *J. Gen. Virol.* **70**:3475–3480.

Nüesch, J. P. F., and G. Siegl. Production of hepatitis A virus RNAs during exponential virus replication and in persistently infected cells. Manuscript in preparation.

Nüesch, J. P. F., M. Weitz, and G. Siegl. Contribution of defective interfering particles and temperature permissive mutants to persistent hepatitis A virus infection in MRC-5 cells. Manuscript in preparation.

Provost, P. J., B. S. Wolanski, W. J. Miller, O. L. Ittensohn, W. J. McAleer, and M. R. Hilleman. 1975. *Proc. Soc. Exp. Biol. Med.* **148**:532–539.

Siegl, G. 1988. Virology of hepatitis A, p. 3–7. *In* A. J. Zuckerman (ed.), *Viral Hepatitis and Liver Disease.* Alan R. Liss, Inc., New York.

Siegl, G., M. Weitz, and G. Kronauer. 1984. Stability of hepatitis A virus. *Intervirology* **22**:218–226.

Ticehurst, J., J. I. Cohen, and R. H. Purcell. 1988. Analysis of molecular sequences demonstrates that hepatitis A virus is a unique picornaviurs, p. 33–35. *In* A. J. Zuckerman (ed.), *Viral Hepatitis and Liver Disease.* Alan R. Liss, Inc., New York.

Chapter 16

cis-Acting Sequence Responsible for Sindbis Virus Subgenomic RNA Synthesis

Arash Grakoui, Robin Levis, Ramaswamy Raju, Sondra Schlesinger,
Henry V. Huang, and Charles M. Rice

Sindbis virus, the type species of the alphavirus genus (togavirus family), and the closely related Semliki Forest virus have been widely studied (reviewed in Schlesinger and Schlesinger, 1986). The genome of Sindbis virus consists of a single molecule of single-stranded RNA, 11,703 nucleotides in length (Strauss et al., 1984). The genomic RNA is infectious, is capped at the 5' terminus and polyadenylated at the 3' terminus, and serves as mRNA, and is therefore by convention of plus polarity. The 5' two-thirds of the genomic 49S RNA is translated early during infection to produce two polyproteins that are processed by cotranslational and/or posttranslational cleavage (Hardy and Strauss, 1988) into four nonstructural proteins which are required for RNA replication. A full-length minus strand complementary to the genomic RNA is first synthesized, which then serves as template for the synthesis of new 49S genomic RNA molecules. The minus strand is also transcribed from an internal site to produce a 26S subgenomic mRNA that is 4,106 nucleotides long and colinear with the 3'-terminal one-third of the 49S genome. This subgenomic mRNA is capped and polyadenylated, and it encodes the three virion structural proteins.

Synthesis and packaging of Sindbis virus-specific RNAs in vertebrate cells requires *trans*-acting components (both virus encoded and host specified) that act on specific viral RNA structures or sequences. The viral proteins and *cis*-acting RNA sequences mediating these regulated steps in the virus life cycle and the molecular mechanisms involved have only partially been defined. Comparison of several different alphavirus sequences (Ou et al., 1981, 1983; Ou et al., 1982) revealed at least four conserved RNA domains that could act as *cis*-acting RNA sequences or structures. A conserved sequence of 19 nucleotides adjacent to the 3'-terminal poly(A) tail was hypothesized to be important for initiation of minus-strand synthesis (Ou et al., 1981). Deletion mutagenesis studies, using a defective interfering (DI) RNA cDNA clone that could produce biologically active RNA transcripts, showed that this sequence was important for replication or packaging of DI RNA (Levis et al., 1986), although direct evidence of its role in initiation of minus-strand synthesis has yet to be obtained. The other three conserved RNA features in the alphavirus genome include a 5'-terminal secondary structure (Ou et al., 1983), a 51-base conserved sequence near the 5' end of the genomic RNA (Ou et al., 1983), and a stretch of 21 bases encompassing the start of the subgenomic RNA in the junction region (Ou et al., 1982). The junction region is defined as the region of the genome immediately preceding and including the beginning of 26S RNA (Ou et al., 1982). For DI RNA replication and packaging, the 5'-terminal secondary structure, the 51-base conserved sequence, and the junction region are nonessential (Levis et al., 1986).

The conserved 21-base sequence in the junction region was proposed to function, in the minus strand, as the promoter for subgenomic RNA synthesis (Ou et al., 1982; Fig. 1A). This sequence is absent from naturally occurring DI

Arash Grakoui, Ramaswamy Raju, Sondra Schlesinger, Henry V. Huang, and Charles M. Rice, Department of Molecular Microbiology, Box 8230, Washington University School of Medicine, 660 South Euclid Avenue, St. Louis, Missouri 63110-1093. *Robin Levis,* National Cancer Institute, Building 41, Room D909, Bethesda, Maryland 20892.

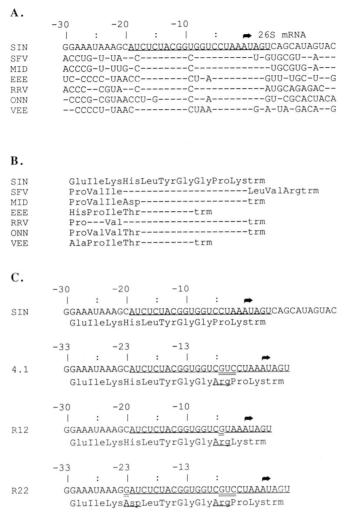

A.

```
         -30         -20         -10
          |    :      |    :      |    :      ➝ 26S mRNA
SIN   GGAAAUAAAGCAUCUCUACGGUGGUCCUAAAUAGUCAGCAUAGUAC
SFV   ACCUG-U-UA--C--------C----------U-GUGCGU--A---
MID   ACCCG-U-UUG-C--------C------------UGCGUG-A---
EEE   UC-CCCC-UAACC--------CU-A----------GUU-UGC-U--G
RRV   ACCC--CGUA--C--------C----------AUGCAGAGAC--
ONN   -CCCG-CGUAACCU-G-----C--A---------GU-CGCACUACA
VEE   --CCCCU-UAAC---------CUAA-------G-A-UA-GACA--G
```

B.

```
SIN   GluIleLysHisLeuTyrGlyGlyProLystrm
SFV   ProValIle--------------------LeuValArgtrm
MID   ProValIleAsp-----------------trm
EEE   HisProIleThr---------trm
RRV   Pro---Val-------------------trm
ONN   ProValValThr----------------trm
VEE   AlaProIleThr---------trm
```

C.

```
         -30         -20         -10
          |    :      |    :      |    :      ➝
SIN   GGAAAUAAAGCAUCUCUACGGUGGUCCUAAAUAGUCAGCAUAGUAC
      GluIleLysHisLeuTyrGlyGlyProLystrm


         -33         -23         -13
          |    :      |    :      |    :          ➝
4.1   GGAAAUAAAGCAUCUCUACGGUGGUCGUCCUAAAUAGU
      GluIleLysHisLeuTyrGlyGlyArgProLystrm


         -30         -20         -10
          |    :      |    :      |    :      ➝
R12   GGAAAUAAAGCAUCUCUACGGUGGUCGUAAAUAGU
      GluIleLysHisLeuTyrGlyGlyArgLystrm


         -33         -23         -13
          |    :      |    :      |    :          ➝
R22   GGAAAUAAAGGAUCUCUACGGUGGUCGUCCUAAAUAGU
      GluIleLysAspLeuTyrGlyGlyArgProLystrm
```

Fɪɢᴜʀᴇ 1. Alphavirus junction region sequences. (A) Alignment of the genomic sequences in the junction regions from several alphaviruses: Sindbis virus (SIN; Ou et al., 1982); Semliki Forest virus (SFV; Ou et al., 1982); Middelburg virus (MID; Ou et al., 1982); Eastern equine encephalitis virus (EEE; Chang and Trent, 1987); Ross River virus (RRV; Ou et al., 1982); O'Nyong-nyong virus (ONN; Strauss et al., 1988); and Venezuelan equine encephalitis virus (VEE; Kinney et al., 1986). The complement of these sequences in the negative-strand template was proposed to function as the promoter for subgenomic RNA synthesis. Sequence identity with respect to the Sindbis virus sequence is indicated by dashes. Nucleotides are numbered relative to the 26S mRNA start (shown by an arrow; see Ou et al., 1982). (B) Comparison of the deduced amino acid sequences of nsP4 in this region. (C) Nucleotide (double underline) and amino acid (underline) sequences of the 4.1 mutant and the R12 and R22 revertants that differ from the sequences of the parental virus, Toto1101. The sequence identified as the core promoter for subgenomic RNA synthesis (-19 to $+5$) is underlined (Levis et al., 1990). (Adapted from Grakoui et al. [1989].)

RNA genomes that have been sequenced (Lehtovaara et al., 1982; Monroe and Schlesinger, 1984), and such DI genomes do not make DI-derived subgenomic RNAs. The conclusions from two studies directed at understanding al-phavirus subgenomic RNA transcription are summarized below. In the first study, the extent of 5′ and 3′ sequences actually necessary for a functional Sindbis virus subgenomic promoter element in the context of a DI RNA have been

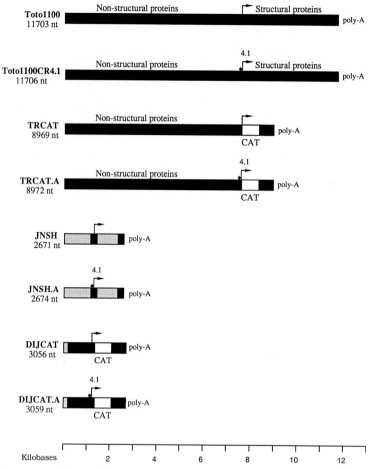

FIGURE 2. Structures of clones. Diagrams and sizes of genomic-length RNA transcripts derived by in vitro transcription of plasmid templates described in the text are shown. Sindbis virus (■■), the CAT gene coding (▢), and DI RNA (▨) sequences are indicated. Arrows mark locations in the genomic RNAs corresponding to the start sites of subgenomic RNAs. Constructs with the mutant junction (4.1) containing a three-base insertion (◆) are indicated. (Adapted from Grakoui et al. [1989].)

mapped (Levis et al., 1990). In the second study, the effects of a three-base insertion mutation in the putative subgenomic RNA promoter element were characterized in the context of both the full-length Sindbis virus genome and a DI genome (Grakoui et al., 1989).

DEFINITION OF THE MINIMAL SEQUENCE ELEMENT NECESSARY FOR SINDBIS VIRUS SUBGENOMIC RNA SYNTHESIS IN DI GENOMES

Since the putative promoter region overlaps with the C terminus of nsP4 (Fig. 1C), mutagenesis of this sequence in the context of full-length

viral genome is constrained (see below). Translocation of a segment of the viral genome containing the junction region into a DI genome leads to the production of DI-derived subgenomic RNA (JNSH; Fig. 2), thus allowing the minimal sequences required for subgenomic RNA synthesis to be mapped (Levis et al., 1990). By examining the ability of successively shorter sequences to promote subgenomic RNA synthesis, it was found that the minimal promoter for subgenomic RNA synthesis, in the context of the DI genome, extends from −18 or −19 to +5 nucleotides relative to the subgenomic RNA start (underlined in Fig. 1A and C). 5′ or 3′ deletions extending into this region abolished the accumulation of DI-derived subgenomic RNAs. Subgenomic RNAs were not observed when the orientation of the promoter was reversed, which suggests that the promoter may not be functional in the context of the positive strand.

GENETIC DEFINITION OF
TRANS-ACTING FACTORS

The definition of the *cis*-acting promoter element and our ability to manipulate the Sindbis virus genome allow the use of molecular genetics to help define viral *trans*-acting components important for alphavirus transcription. The initial step in this approach has been the creation of a deleterious mutation in the *cis*-acting element of interest, characterization of the mutant phenotype, and isolation of a set of preferably early-passage second-site revertants. Mapping and characterization of appropriate second-site revertants should be valuable in elucidating functional interactions of viral proteins with *cis*-acting regulatory sequences.

CHARACTERIZATION OF A DELETERIOUS MUTATION IN THE SUBGENOMIC RNA PROMOTER

A full-length cDNA clone of the Sindbis virus genome from which infectious transcripts can be generated by in vitro transcription (called Toto1100 or Toto1101; Rice et al., 1987) was used to construct a three-base insertion mutation in the junction region (called Toto1100CR4.1). The effects of this mutation on subgenomic RNA synthesis both in the Sindbis virus genome and in engineered DI RNA genomes have been characterized. The mutation is a three-base insertion (GUC) in the conserved sequence of the junction region, between nucleotides −4 and −5 relative to the subgenomic RNA start (Fig. 1A), and also results in the insertion of an arginyl residue between amino acid residues 608 and 609 of nsP4 (Fig. 1C) (Strauss et al., 1984).

The phenotype of Toto1100CR4.1 was studied after RNA transfection of chicken embryo fibroblast or baby hamster kidney cells. Plaque morphology variants were present in early-passage virus stocks, suggesting the presence of revertants in the population and emphasizing the need to study the phenotype as early as possible, i.e., immediately after transfection. Analysis of the mutant phenotype after RNA transfection showed that the mutation was deleterious, leading to a drastic reduction in the level of the subgenomic RNA without altering the start site of the RNA. Probably as a consequence of depressed structural protein synthesis, very few progeny virions were released and the mutant produced tiny or indistinct plaques even after prolonged incubation.

QUANTITATIVE EXAMINATION OF THE EFFECTS OF THE Toto1100CR4.1 MUTATION

Relative levels of subgenomic RNA synthesis were measured by using two packaging-defective but replication-competent Sindbis virus genomes: TRCAT (Xiong et al., 1989) and TRCAT.A (Fig. 2). In TRCAT, the Sindbis structural genes are replaced by the chloramphenicol acetyltransferase (CAT) gene such that a subgenomic RNA encoding CAT is produced by the Sindbis virus replication-transcription machinery (Xiong et al., 1989). TRCAT.A is identical to TRCAT except for the three-base insertion corresponding to the Toto1100CR4.1 mutation in the junction region. Since these genomes lack the structural protein genes, no virus will be released from transfected cells, and the levels of genomic and subgenomic RNAs in transfected cells can be more easily compared. Furthermore, translation of the subgenomic RNA leads to production of enzymatically active CAT, which can then be assayed to provide a more sensitive albeit indirect estimate of the relative subgenomic RNA levels. When the levels of CAT produced per cell by these constructs were compared at different times posttransfection, TRCAT.A produced between 200- and 500-fold less CAT per cell than did TRCAT. Examination of viral RNA levels in baby hamster kidney cells transfected with these constructs showed that whereas the three-base insertion drastically reduced subgenomic RNA synthesis, relatively little difference was seen in the level of genome-length RNAs.

EFFECTS OF THE THREE-BASE INSERTION ON SUBGENOMIC RNA PRODUCTION IN TRANS

Since the Toto1100CR4.1 mutation also affects the structure of nsP4 (Fig. 1C) by insertion of an arginyl residue two residues from the C terminus of the protein, the phenotypes of Toto1100CR4.1 and TRCAT.A may result from deleterious effects of the insertion mutation on the subgenomic RNA promoter or an effect of the inserted amino acid on the function of nsP4. The predicted carboxyl termini of alphavirus nsP4s are rather variable (Fig. 1B). The nsP4-coding region of Semliki Forest virus contains three additional carboxyl-terminal residues, whereas Venezuelan and Eastern equine encephalitis viruses contain a stop codon three residues upstream from the position homologous to the position of the Toto1100CR4.1 insertion mutation. Although these observations suggest that this region of nsP4 may not be functionally significant, and therefore that the Toto110CR4.1 insertion mutation acts primarily in *cis*, a direct test is clearly desirable.

The *cis*-acting effect of this mutation was demonstrated by incorporating either a wild-type or mutant junction region into a DI RNA and examining the relative synthesis of DI-derived subgenomic RNA in vivo in the presence of wild-type helper virus. DI RNAs that contained either the wild-type (JNSH) or the mutant (JNSH.A) junction region (Fig. 1) were tested for the ability to support subgenomic RNA synthesis when wild-type nonstructural proteins were supplied in *trans* by helper virus. The DI genome containing the junction region with the three-base insertion (JNSH.A) produced little, if any, subgenomic RNA as compared with the DI with the wild-type junction region (JNSH). Thus, the depressed synthesis of 26S RNA by Toto1100CR4.1 is primarily due to the deleterious effect of the three-base insertion on the efficiency of the genomic RNA promoter.

Considered together, these studies clearly demonstrate that the conserved sequence in the Sindbis virus junction region is indeed a *cis*-acting element involved in subgenomic RNA transcription. The mechanism by which the insertion mutation leads to depressed subgenomic RNA synthesis is unknown. One likely possibility is that the insertion of three bases alters the interaction of the promoter element with one or more *trans*-acting components (either viral or host) required for subgenomic RNA transcription. Since Toto1100CR4.1 subgenomic RNA transcripts apparently initiate at the same position as normal 26S RNA, the insertion of these three bases near the middle of the conserved sequence neither abolished transcription completely nor affected the subgenomic RNA start.

TRANS-ACTING TRANSCRIPTION FACTORS

Little is known about the identity of the factor(s) involved in initiation of subgenomic RNA synthesis. Minimally, promoter recognition and RNA polymerase activities are necessary. In addition, the 26S RNA is capped and methylated. Some clues have begun to emerge from mapping the causal lesions in Sindbis virus temperature-sensitive mutants defective in RNA replication (RNA⁻ mutants), which were previously grouped by genetic complementation into four groups (Burge and Pfefferkorn, 1966a, 1966b, 1968; Keränen and Kääriäinen, 1979; D. L. Sawicki et al., 1981; Sawicki and Sawicki, 1985; S. G. Sawicki et al., 1981; Strauss et al., 1976; reviewed by Strauss and Strauss [1980]). These data indicate that nsP1 may play a role in negative-strand synthesis (group B; Hahn et al., 1989b; D. L. Sawicki et al., 1981) and methyltransferase activity (Mi et al., 1989), that lesions in nsP2 affect the ratio of genomic to subgenomic RNA and polyprotein processing (groups A and G; Hahn et al., 1989b; Sawicki and Sawicki, 1985), and that lesions in nsP4 can exhibit generalized defects in RNA synthesis (group F) (Barton et al., 1988; Hahn et al., 1989a) and shutoff of negative-strand synthesis (S. G. Sawicki et al., 1981; Sawicki et al., 1990). Thus, at least one possibility is that nsP2 and nsP4 play major roles in initiation of subgenomic RNA synthesis. Indeed, genetic data exist which can be interpreted as evidence for a complex of these two proteins (Hahn et al., 1989a).

As mentioned above, second-site revertants may help to identify virus-specific proteins that participate in subgenomic RNA synthesis. When the Toto1100CR4.1 mutant was passaged in culture, plaque morphology variants readily arose. Of 24 independent revertants isolated, 16 have been characterized in detail. All revertants analyzed showed an increase in the level of subgenomic RNA synthesis. Sequence analysis of the junction region showed that all were pseudo-revertants, with only two (R12 and R22; Fig. 1C) containing potentially compensating changes in the junction region. The remaining revertants were identical to Toto1100CR4.1 in the region sequenced (from −43 to +58 relative to the subgenomic RNA start site). Thus, all of the revertants were second-site revertants, and many reversions involved changes outside of the junction region.

At least four formal classes of second-site revertants could compensate for the Toto1100CR4.1 mutation (Table 1). One class could involve *cis*-acting, secondary changes in the Toto1100CR4.1 junction region that restore or partially restore promoter function. A second class might involve changes in the structural proteins or RNA sequences involved in encapsidation, increasing the efficiency of that process

TABLE 1
Predicted Results of Transcomplementation Assay[a]

Helper virus	Level of CAT produced	
	DIJCAT	DIJCAT.A
Wild-type (Toto1101)	High	Low
Toto1100CR4.1 mutant	?	Low
Second-site revertants		
Class I (changes in junction region)	High	Low
Class II (increased packaging efficiency)	High	Low
Class III (changes suppressing nsP4 defect)	High	Low
Class IV (changes in *trans*-acting viral proteins)	?	Low-high

[a] Shown is the rationale used for the design of an assay to screen for revertants with potential changes in *trans*-acting viral proteins involved in subgenomic RNA synthesis. In the presence of a helper virus, the CAT activity produced was determined for DI RNAs (the structures are shown in Fig. 2) that expressed a CAT subgenomic RNA under the control of either the wild-type promoter (DIJCAT) or the promoter containing the three-base insertion mutation (DIJCAT.A). Listed are some of the possible classes of pseudorevertants that could compensate for the Toto1100CR4.1 insertion mutation. The results of this assay for several of the second-site revertants are discussed in the text. The behavior of the Toto1100CR4.1 mutant in this assay is unknown, since high-titered stocks of this virus, free of revertants, could not be obtained. If the insertion mutation does not affect the activity of nsP4, Toto1100CR4.1 should utilize the mutant and wild-type junction regions like the wild-type helper virus. (Adapted from Grakoui et al. [1989].)

even in the presence of low concentrations of the structural proteins. A third class could involve reversions that compensate for a possible effect of the insertion mutation on nsP4 function. A fourth class involves changes in the protein(s) that recognizes the promoter, increasing the efficiency of transcription initiation at the mutant junction region. Since large numbers of revertants are easily obtained, rapid methods for screening and classifying are clearly of value for identifying representatives in each of these classes.

TRANSCOMPLEMENTATION ASSAY TO IDENTIFY SECOND-SITE REVERTANTS IN *TRANS*-ACTING VIRAL COMPONENTS

Current emphasis in our laboratories is on the fourth class of revertants, and a rapid assay has been developed and used to identify sec-

ond-site revertants with potential compensating changes in virus-encoded *trans*-acting factors that allow better recognition of the Toto1100CR4.1 mutant junction region. The assay is based on the activity of the junction region when translocated to DI RNAs (Fig. 2 and Table 1). DI clones were constructed with the CAT gene downstream from a wild-type (DIJCAT) or mutant (DIJCAT.A) junction region (Fig. 2). These constructs do not encode functional nonstructural proteins and are inactive in the absence of helper virus. In the presence of Toto1101 helper virus, DIJCAT RNAs replicate, transcribe a subgenomic RNA, and express high levels of CAT enzyme. In the case of DIJCAT.A, much less CAT enzyme is made. Presumably, this is due to lower levels of subgenomic RNA production, as was found for JNSH.A. The Toto1100CR4.1 revertants were then tested as helpers, screening for those that show enhanced recognition of the mutant junction region. The ratio of CAT activity was used as a measure of the utilization of the mutant junction region relative to the wild-type junction region by each virus. Two revertants, R11 and R12, gave similar CAT expression with wild-type and mutant junction regions, suggesting that they contain changes in viral proteins which facilitate use of the mutant junction relative to the wild-type junction region in *trans*. Revertant R12 has a three-base deletion in the junction region, resulting in a single-base substitution at the -4 position of the putative promoter (Fig. 1C). This revertant makes nearly normal levels of subgenomic RNA. In addition to recognizing its own promoter, this revertant recognizes the wild-type promoter and the Toto1100CR4.1 mutant promoter, as shown by the transcomplementation assay. It may contain additional changes elsewhere in the genome. The other revertant, R11, retains the Toto1100CR4.1 mutation and recognizes both the mutant and the wild-type promoter. It too is likely to have changes in one or more *trans*-acting viral transcription factors.

Since care was taken to isolate early-passage revertants, relatively few changes in the revertants relative to the Toto1100CR4.1 mutant are expected. Once the compensating changes in the revertants are cloned, more thorough studies of the phenotypes of the reversions on the Toto1100CR4.1 and wild-type (Toto1101) backgrounds will be undertaken. It is hoped that mapping of these mutations, using essentially the strategies previously outlined (Hahn et al., 1989; Hahn et al., 1989a; Hahn et al., 1989b; Lustig et al., 1988; Mi et al., 1989; Polo et al.,

1988; Rice et al., 1987; Sawicki et al., 1990), will help to identify those viral proteins which are involved in recognizing the junction region for transcription of the subgenomic RNA.

THE IMPORTANCE OF FIRST-CYCLE ANALYSIS

An important observation emerging from the work summarized here and elsewhere (Hahn et al., 1989; Li and Rice, 1989) is the remarkable plasticity of the alphavirus genome as well as the facile generation of revertants. As has been found for other RNA viruses, deleterious mutations that allow RNA replication, even at low levels, coupled with high mutation rates, will strongly select for revertants. As shown by sequence analysis of pseudorevertants of Toto1100CR4.1, changes involving either substitutions or deletions arise even in early-passage stocks. Although the genome flexibility and high mutation frequency allow facile isolation of revertants, the results with Toto1100CR4.1 underscore the need to study mutant phenotypes as early as possible, ideally without generating virus stocks. Precautions that we have adopted include routine examination of the specific infectivity of mutant RNA transcripts and the characterization of the phenotype of a mutant immediately after primary transfection, which then serve as a standard for comparison with the phenotypes of any subsequent virus stocks generated by passaging. With these precautions, there is less risk of reporting a mutant phenotype that is actually that of a revertant.

ACKNOWLEDGMENTS. This work was supported by Public Health Service grants from the National Institute of Allergy and Infectious Disease, by the Pew Memorial Trust, and by a Monsanto/Washington University Biomedical Research contract. C.M.R. is a Pew Scholar in the Biomedical Sciences.

REFERENCES

Barton, D. J., S. G. Sawicki, and D. L. Sawicki. 1988. Demonstration in vitro of temperature-sensitive elongation of RNA in Sindbis virus mutant ts6. *J. Virol.* **62**:3597–3602.

Burge, B. W., and E. R. Pfefferkorn. 1966a. Isolation and characterization of conditional-lethal mutants of Sindbis virus. *Virology* **30**:204–213.

Burge, B. W., and E. R. Pfefferkorn. 1966b. Complementation between temperature-sensitive mutants of Sindbis virus. *Virology* **30**:214–223.

Burge, B. W., and E. R. Pfefferkorn. 1968. Functional defects of temperature-sensitive mutants of Sindbis virus. *J. Mol. Biol.* **35**:193–205.

Chang, G. J., and D. W. Trent. 1987. Nucleotide sequence of the genome region encoding the 26S mRNA of Eastern equine encephalitis virus and the deduced amino acid sequence of the viral structural proteins. *J. Gen. Virol.* **68**:2129–2142.

Grakoui, A., R. Levis, R. Raju, H. V. Huang, and C. M. Rice. 1989. A cis-acting mutation in the Sindbis virus junction region affecting subgenomic RNA synthesis. *J. Virol.* **63**:5216–5227.

Hahn, C. S., C. M. Rice, E. G. Strauss, E. M. Lenches, and J. H. Strauss. 1989. Sindbis virus ts103 has a mutation in glycoprotein E2 that leads to defective assembly of virions. *J. Virol.* **63**:3459–3465.

Hahn, Y. S., A. Grakoui, C. M. Rice, E. G. Strauss, and J. H. Strauss. 1989a. Mapping of RNA⁻ temperature-sensitive mutants of Sindbis virus: complementation group F mutants have lesions in nsP4. *J. Virol.* **63**:1194–1202.

Hahn, Y. S., E. G. Strauss, and J. H. Strauss. 1989b. Mapping of RNA⁻ temperature-sensitive mutants of Sindbis virus: assignment of complementation groups A, B, and G to nonstructural proteins. *J. Virol.* **63**:3142–3150.

Hardy, W. R., and J. H. Strauss. 1988. Processing of the nonstructural polyproteins of Sindbis virus: study of the kinetics in vivo using monospecific antibodies. *J. Virol.* **62**:998–1007.

Keränen, S., and L. Kääriäinen. 1979. Functional defects of RNA-negative temperature-sensitive mutants of Sindbis and Semliki Forest viruses. *J. Virol.* **32**:19–29.

Kinney, R. M., B. J. B. Johnson, V. L. Brown, and D. W. Trent. 1986. Nucleotide sequence of the 26S mRNA of the virulent Trinidad donkey strain of Venezuelan equine encephalitis virus and deduced sequence of the encoded structural proteins. *Virology* **152**:400–413.

Lehtovaara, P., H. Söderlund, S. Keränen, R. F. Pettersson, and L. Kääriäinen. 1982. 18S defective-interfering RNA of Semliki Forest virus contains a triplicated linear repeat. *Proc. Natl. Acad. Sci. USA* **78**:5353–5357.

Levis, R., S. Schlesinger, and H. V. Huang. 1990. Promoter for Sindbis virus RNA-dependent subgenomic RNA transcription. *J. Virol.* **64**:1726–1733.

Levis, R., B. G. Weiss, M. Tsiang, H. Huang, and S. Schlesinger. 1986. Deletion mapping of Sindbis virus DI RNAs derived from cDNAs defines the sequences essential for replication and packaging. *Cell* **44**:137–145.

Li, G., and C. M. Rice. 1989. Mutagenesis of the in-frame opal termination codon preceding nsP4 of Sindbis virus: studies of translational readthrough and its effect on virus replication. *J. Virol.* **63**:1326–1337.

Lustig, S., A. C. Jackson, C. S. Hahn, D. E. Griffin, E. G. Strauss, and J. H. Strauss. 1988. Molecular basis of Sindbis virus neurovirulence in mice. *J. Virol.* **62:**2329–2336.

Mi, S., R. Durbin, H. V. Huang, C. M. Rice, and V. Stollar. 1989. Association of the Sindbis virus RNA methyltransferase activity with the nonstructural protein nsP1. *Virology* **170:**385–391.

Monroe, S., and S. Schlesinger. 1984. Common and distinct regions of defective-interfering RNAs of Sindbis virus. *J. Virol.* **49:**865–872.

Ou, J.-H., C. M. Rice, L. Dalgarno, E. G. Strauss, and J. H. Strauss. 1982. Sequence studies of several alphavirus genomic RNA's in the region containing the start of subgenomic RNA. *Proc. Natl. Acad. Sci. USA* **79:**5235–5239.

Ou, J.-H., E. G. Strauss, and J. H. Strauss. 1981. Comparative studies of the 3′ terminal sequences of several alphavirus RNAs. *Virology* **109:**281–289.

Ou, J.-H., E. G. Strauss, and J. H. Strauss. 1983. The 5′ terminal sequences of the genomic RNAs of several alphaviruses. *J. Mol. Biol.* **168:**1–15.

Polo, J. M., N. L. Davis, C. M. Rice, H. V. Huang, and R. E. Johnston. 1988. Molecular analysis of Sindbis virus pathogenesis in neonatal mice using virus recombinants constructed in vitro. *J. Virol.* **62:** 2124–2133.

Rice, C. M., R. Levis, J. H. Strauss, and H. V. Huang. 1987. Production of infectious RNA transcripts from Sindbis virus cDNA clones: mapping of lethal mutations, rescue of a temperature-sensitive marker, and in vitro mutagenesis to generate defined mutants. *J. Virol.* **61:**3809–3819.

Sawicki, D. L., D. B. Barkhimer, S. G. Sawicki, C. M. Rice, and S. Schlesinger. 1990. Temperature sensitive shut-off of alphavirus minus strand RNA synthesis maps to a nonstructural protein, nsP4. *Virology* **174:**43–52.

Sawicki, D. L., and S. G. Sawicki. 1985. Functional analysis of the A complementation group mutants of Sindbis HR virus. *Virology* **144:**20–34.

Sawicki, D. L., S. G. Sawicki, S. Keränen, and L. Kääriäinen. 1981. Specific Sindbis virus coded function for minus strand RNA synthesis. *J. Virol.* **39:**348–358.

Sawicki, S. G., D. L. Sawicki, L. Kääriäinen, and S. Keränen. 1981. A Sindbis virus mutant temperature-sensitive in the regulation of minus-strand RNA synthesis. *Virology* **115:**161–172.

Schlesinger, S., and M. J. Schlesinger (ed.). 1986. *The Togaviridae and Flaviviridae.* Plenum Publishing Corp., New York.

Strauss, E. G., E. M. Lenches, and J. H. Strauss. 1976. Mutants of Sindbis virus. I. Isolation and partial characterization of 89 new temperature-sensitive mutants. *Virology* **74:**154–168.

Strauss, E. G., R. Levinson, C. M. Rice, J. Dalrymple, and J. H. Strauss. 1988. Nonstructural proteins nsP3 and nsP4 of Ross River and O'Nyong-nyong viruses: sequence and comparison with those of other alphaviruses. *Virology* **164:**265–274.

Strauss, E. G., C. M. Rice, and J. H. Strauss. 1984. Complete nucleotide sequence of the genomic RNA of Sindbis virus. *Virology* **133:**92–110.

Strauss, E. G., and J. H. Strauss. 1980. Mutants of alphaviruses: genetics and physiology, p. 393–426. *In* R. W. Schlesinger (ed.), *The Togaviruses.* Academic Press, Inc., New York.

Xiong, C., R. Levis, P. Shen, S. Schlesinger, C. M. Rice, and H. V. Huang. 1989. Sinbdis virus: an efficient, broad host range vector for gene expression in animal cells. *Science* **243:**1188–1191.

Chapter 17

Use of Full-Length DNA Copies in the Study of Expression and Replication of the Bipartite RNA Genome of Cowpea Mosaic Virus

J. Wellink, R. Eggen, J. Verver, R. Goldbach, and A. van Kammen

The genome of cowpea mosaic virus (CPMV), type member of the plant comoviruses, consists of two positive-stranded RNAs designated B RNA and M RNA (for a review, see Goldbach and Van Kammen [1985]). The RNAs are separately encapsidated in icosahedral particles with a diameter of 28 nm. The capsids are composed of 60 copies of each of two different coat proteins, VP37 and VP23. The RNAs are characterized by a small protein (VPg) at the 5' end, a poly(A) tail at the 3' end, and a single large open reading frame. Both RNAs are required for infectivity in plants, but B RNA can replicate independently from M RNA in cowpea protoplasts (Goldbach et al., 1980). B RNA must therefore carry information for viral RNA replication, whereas M RNA is thought to carry information for virus movement (Rezelman et al., 1982).

Expression of the CPMV RNAs has been studied extensively in the cowpea mesophyll protoplast system. Such studies showed that B RNA is translated into a polyprotein of molecular weight 200,000 (200K polyprotein) that is cleaved by a viral protease at specific Gln/Gly, Gln/Ser, and Gln/Met sites into several intermediate and five final cleavage products (Fig. 1; Rezelman et al., 1980; Wellink et al., 1986; Vos et al., 1988b). In vitro M RNA is translated into two overlapping polyproteins, 105K and 95K, as a result of initiation of translation at two different AUG codons (Vos et al., 1984). These polyproteins can be cleaved by a viral protease into 58K and 48K proteins and the 60K precursor to the capsid proteins (Fig. 1; Pelham, 1979; Franssen et al., 1982; Vos et al., 1988b). In protoplasts, the capsid proteins, the 60K precursor to the capsid proteins, and the 48K protein are the only M-RNA-encoded proteins detected so far (Wellink et al., 1987).

Comoviruses are very similar in genome organization to the plant nepo- and potyviruses and to the animal picornaviruses and use the same strategy for expression of their genetic information. Several proteins encoded by these viruses show considerable sequence homology and presumably have similar functions in viral RNA replication (Franssen et al., 1984; Goldbach and Wellink, 1988).

IMPACT OF TERMINAL SEQUENCES ON THE INFECTIVITY OF IN VITRO TRANSCRIPTS FROM cDNA CLONES

To study the expression and replication of the CPMV RNAs, full-length cDNA clones of both RNAs were constructed downstream of the bacteriophage T7 promoter (Vos et al., 1988a). By trimming the promoter and linker sequences, transcripts with varying numbers of additional nonviral sequences at the 5' end were obtained upon transcription with T7 RNA polymerase (Eggen et al., 1989a). The infectivity of the transcripts was tested in cowpea protoplasts. The numbers of infected protoplasts were determined by immunofluorescent staining with an antiserum against the B-RNA-encoded 24K protein to detect protoplasts in which B RNA is replicated and an antiserum against CPMV to detect protoplasts in which M RNA is replicated (Table 1).

J. Wellink, R. Eggen, J. Verver, R. Goldbach, and A. van Kammen, Department of Molecular Biology, Agricultural University, Dreijenlaan 3, 6703 HA Wageningen, The Netherlands.

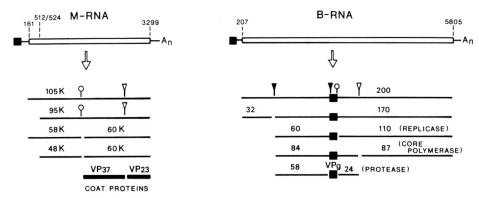

FIGURE 1. Model for expression of CPMV RNAs. The open reading frame on each RNA is indicated by an open bar, and positions of the translational start and stop codons are indicated on this bar. VPg is represented by a black square, and the other proteins are indicated by single lines. Both capsid proteins are indicated by thick lines. Cleavage sites: ○, Gln-Met; ▽, Gln-Gly; ▼, Gln-Ser.

The extensions at the 5' end greatly affected the specific infectivity of the RNA (Table 1). The highest level of infectivity was observed with transcripts containing only one extra G residue (M1G and B1G). With capped RNA, the percentage of infected cells was about two times higher than with uncapped RNA. We have not been able to prepare RNA without an extra G residue at the 5' end, as removal of the last G, which is the transcription start site, resulted in complete inactivation of transcription activity.

Extra nucleotides at the 3' end of the transcripts had no deleterious effect on the infectivity of the RNAs (B1G-A0).

The infectivity of B-RNA transcripts with one extra G at the 5' end was still low in comparison with viral RNA (~2%). The extra G

residue possibly disturbs the replication process. If this G residue is transcribed into the 3' end of the minus-strand RNA, the presence of the extra C residue may have a deleterious effect on the function of this RNA as a template for replication. This would also explain the great reduction in infectivity of transcripts with larger 5' extensions. Furthermore, transcripts lack a VPg, and perhaps this VPg protects the viral RNA against degradation by nucleases, thus contributing to the stability of the RNA during inoculation of protoplasts. The importance of protection of the 5' end of the RNA is substantiated by the higher infectivity of capped than of uncapped transcripts.

However, the infectivity of the transcripts is such that it is possible to infect cowpea plants

TABLE 1

Infectivity of Transcript and Wild-Type CPMV RNAs in Cowpea Protoplasts

Virus or clone	Sequence of 5' and 3' termini of RNA[a]	% Protoplasts[b] producing proteins encoded by:	
		B RNA	M RNA
CPMV RNA	VPg–UAUU------A$_{50-140}$	20	20
pTB35	X$_{35}$–UAUU------A$_{46}$UAUCG	0.2	
pTB2G	GG–UAUU------A$_{46}$UAUCG	5	
pTB2G-cap	m^{7}GpppGG–UAUU------A$_{46}$UAUCG	10	
pTB1G	G–UAUU------A$_{46}$UAUCG	25	
pTB1G-cap	m^{7}GpppG–UAUU------A$_{46}$UAUCG	40	
pTB1G-AO	G–UAUU------A$_{46}$X$_{440}$	25	
pTB1G + pTM2G	GG–UAUU------A$_{46}$UAUCG	25	0.3
pTB1G + pTM1G	G–UAUU------A$_{46}$UAUCG	25	10

[a] A$_x$ is the number of residues in the poly(A) tail, X$_x$ is the number of nonviral nucleotides at the termini.

[b] Cowpea protoplasts (2.5×10^6) were inoculated with 50 μg of transcripts or 2 μg of CPMV RNA. The infectivity of the RNA was determined by immunofluorescent staining; numbers shown are the results of several independent experiments.

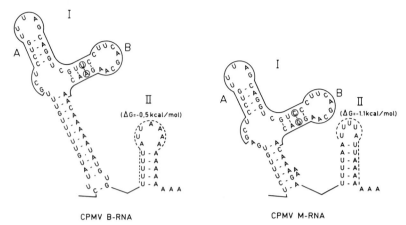

FIGURE 2. Secondary-structure predictions of the 3′ regions of CPMV B and M RNAs consisting of the noncoding regions immediately preceding the poly(A) tail and some residues of the poly(A) tail. The solid line in structure I represents an identical structure in the two RNAs allowing a UA-to-CG covariation in hairpin B (shown by open circles). The interrupted line in hairpin II represents a conserved stretch of 11 nucleotides in the 3′ ends of B and M RNAs which might represent a signal in viral RNA replication.

with these transcripts. In virus isolated from these plants, it was found that the RNA had not preserved the extra G at the 5′ termini (Eggen et al., 1989a). It is not known at which point the extra nucleotide is removed. It is possible that the extra G is removed by exonucleolytic attack or is lost during replication.

Analysis of the 3′ end of the RNA showed that the 40-residue poly(A) tail of the transcript was elongated during the multiple replication cycles in the plant and had regained the wild-type heterogeneous length distribution (Eggen et al., 1989a). Presumably either the replicase is slipping in synthesizing the poly(A) tail on a poly(U) template or one has to assume that the poly(A) tail can be extended by some terminal nucleotidyl transferase activity present in the cytoplasm of the plant cell.

ANALYSIS OF A SEQUENCE INVOLVED IN CPMV RNA REPLICATION

A series of mutants was generated in the full-length cDNA clone of B RNA to examine whether a sequence of 11 nucleotides, UUUUAUUAAAA, comprising nucleotides 5883 to 5893 in the 3′ noncoding region of B RNA, might represent a signal in replication (Eggen et al., 1989b). This sequence consists of the seven nucleotides preceding the poly(A) tail and the first four residues of the poly(A) tail. It is also present in the 3′ noncoding region of M RNA and, allowing one G-to-U transversion, in

a region close to the 3′ ends of the minus strands of both B and M RNAs (Eggen and Van Kammen, 1988).

The preserved sequence in the 3′ noncoding region of each RNA can be folded into a secondary structure which contains the homologous stretch of 11 nucleotides in a stem-loop structure (Fig. 2). The existence of this structure is supported both by a covariation of nucleotides in hairpin B of B and M RNA such that the base pairing potential is preserved (Fig. 2) and by the fact that a similar structure with two covariations in hairpin B can be deduced from the nucleotide sequence of the M RNA of another comovirus, red clover mottle virus.

By replacing the A residue at position 5887 in the B cDNA by a T, an AhaIII restriction site was created, which allowed the construction of mutants with insertions and deletions in this region (Eggen et al., 1989b; Table 2). Replication of B-RNA transcripts derived from these mutants was tested in cowpea protoplasts. Infectivity was determined by immunofluorescent staining and Western blot (immunoblot) analysis (Table 2). The U-to-A substitution in B RNA, which created the AhaIII site, decreased the infectivity of the transcripts by 40% (pTB1G-Aha; Table 2). Small insertions at this site resulted in nonviable transcripts (pTB1G-AS and pTB1G-AG). Successive deletions of one U upstream of the AhaIII site showed no further decrease in infectivity with removal of the first three nucleotides. Further deletions resulted in

TABLE 2
Infectivity of Transcripts of CPMV B-RNA cDNA Clones with Deletions in the 3' Region

B-RNA cDNA clone	Sequence of 3' region of transcript	Infectivity (%)[a]
pTB1G	$- \text{GUUUUUAUUA}_{46}\text{UAUCG}$	100
pTB1G-Aha	$- \text{GUUUUUUUUA}_{46}\text{UAUCG}$	60
pTB1G-Δ1	$- \text{GUUUUUUU A}_{46}\text{UAUCG}$	60
pTB1G-Δ2	$- \text{GUUUUUU A}_{46}\text{UAUCG}$	70
pTB1G-Δ3	$- \text{GUUUUU A}_{46}\text{UAUCG}$	70–80
pTB1G-Δ4	$- \text{GUUUU A}_{46}\text{UAUCG}$	20
pTB1G-Δ7	$- \text{GU A}_{46}\text{UAUCG}$	0.5
pTB1G-Δ8	$- \text{G A}_{46}\text{UAUCG}$	0
pTB1G-A28	$- \text{GUUUUUUUUA}_{28}\text{UAUCG}$	60
pTB1G-A19	$- \text{GUUUUUUUUA}_{19}\text{UAUCG}$	60
pTB1G-A4	$- \text{GUUUUUUUUA}_{4}\text{UAUCG}$	1
pTB1G-AS	$- \text{GUUUUUUUUCCCCCGGGGGA}_{46}\text{UAUCG}$	0
pTB1G-AS $\times$ *Sma*I	$- \text{GUUUUUUUUCCCCC}$	0
pTB1G-AG	$- \text{GUUUUUUUUGGGGGA}_{46}\text{UAUCG}$	0

[a] Cowpea protoplasts (2.5×10^6) were inoculated with 50 µg of RNA transcripts. The infectivity of each transcript was determined by immunofluorescent staining. The relative amounts of B-RNA-encoded protein present in extracts of these protoplasts were also determined by protein blot analysis. These amounts were in agreement with the numbers of infected cells as determined by immunofluorescent staining (data not shown). The infectivity of pTB1G transcripts was arbitrarily set at 100% (this corresponds to 30% infected cowpea protoplasts), and the values for the other transcripts were correlated to this value.

decreased infectivity, and when eight U residues were deleted, which has the effect that folding of this part of the RNA into a small stem-loop structure is no longer possible, infectivity was completely lost (Table 2 and Fig. 3). In addition, a decrease of infectivity was also observed when the pertinent part of the RNA was prevented from folding by shortening the poly(A) tail to four residues. These results indicate that the 11-nucleotide sequence is important for infectivity of the B RNA and probably has a function in viral RNA replication. The results furthermore are consistent with the existence of a small hairpin loop in this region. Nucleotide sequence analysis showed that the sequences of two deletion mutants, B1GΔ2 and B1GΔ3, ultimately reversed to the wild-type sequence after passaging in cowpea plants (Eggen et al., 1989b), which

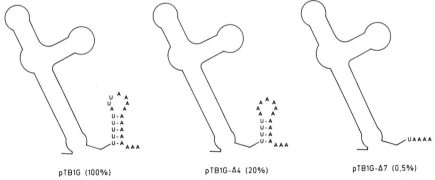

FIGURE 3. Predicted effects of mutations in CPMV B-RNA transcripts on hairpin II (see Fig. 2). The effects of the mutations have been classified into three groups, shown along with percentages indicating the specific infectivities of the transcripts (see also Table 2). In pTB1G, which also represents pTB1G-Δ1, pTB1G-Δ2, and pTB1G-Δ3, the stem length is not disturbed. A further deletion (pTB1G-Δ4) diminishes the stem length, which is even more pronounced in the last group, pTB1G-Δ7, which also represents mutant pTB1G-Δ8. The solid line indicates structure I from Fig. 2.

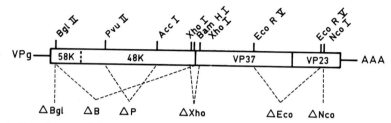

FIGURE 4. Genetic map of CPMV M RNA showing the positions of the cleavage sites, used for construction of some of the M-RNA mutants, above the open bar that represents the open reading frame of M RNA (see Table 3).

further supported the importance of this sequence in virus multiplication.

ROLE OF THE M-RNA-ENCODED PROTEINS IN CELL-TO-CELL TRANSPORT OF CPMV

Using sensitive immunological techniques, we have been able to show that, just as in vitro, two M-RNA-encoded polyproteins are produced in CPMV-infected protoplasts (Fig. 1; Rezelman et al., 1989). In addition, their respective N-terminal cleavage products could be detected; the 58K protein was present in the cytoplasmic fraction, whereas the 48K protein was found in both the cytoplasmic and membrane fractions (Rezelman et al., 1989). Mutations were introduced in the AUG codons at positions 161, 512, and 524 in a full-length cDNA clone of M RNA.

Transcripts derived from these clones were translated in reticulocyte lysates. The translations showed that in vitro, the AUG codons at 161 and 512 of the M RNA are used to produce the 105K and 95K polyproteins, respectively. Because the same proteins are formed in vitro and in vivo, it is very likely that these two start codons are used in vivo also.

To study the role of the M-RNA-encoded proteins in virus movement, two other sets of mutants were generated by using the full-length cDNA of M RNA (Wellink and Van Kammen, 1989). One set contained deletions in the coding region of the overlapping 58K and 48K proteins, and the other set contained deletions and small insertions in the coding region of the capsid proteins (Fig. 4 and Table 3). The transcripts derived from these clones were tested by in vitro translations. All mutants produced one (MΔBgl

TABLE 3
Properties of M-RNA Mutants

RNA	Mutation	Region	% Fluorescent cells with[a]:		Infectivity[b] in plants
			Anti-48K	Anti-CPMV	
M1G			100	100[c]	+
MΔBgl	4-nt[d] insertion	58K	0	0	−
MLac	141-nt insertion	58K	0	50	−
MΔP	486-nt deletion	58K/48K	100	100	−
MΔB	135-nt deletion	58K/48K	0.5	0.5	−
MΔA1	3-nt insertion	58K/48K	ND	100	−
MΔEco	774-nt deletion	VP37/VP23	50	0	−
MΔNco	4-nt insertion	VP23	ND	50	−
MΔA2	3-nt insertion	VP37	ND	100	−
MΔA3	3-nt insertion	VP23	ND	50	−

[a] Cowpea protoplasts (2.5 × 10⁶) were inoculated with 25 μg of B-RNA transcript and 25 μg of M1G or mutant M-RNA transcript. The infectivity of each transcript was determined by immunofluorescent staining.
[b] Plants were inoculated with 5 μg of B-RNA transcript and 5 μg of M-RNA or mutant M-RNA transcript on both primary leaves. Infectivity was determined by visual inspection and immunological analysis of extracts of leaves by Western blotting.
[c] Infectivity of the M1G transcripts was arbitrarily set at 100%, and values for the other transcripts were correlated to this value. ND, Not determined.
[d] nt, Nucleotide.

and MΔB) or two polyproteins, as expected. To check whether the mutant M RNAs could still be replicated, protoplasts were inoculated with these mutants in the presence of transcripts of the B-RNA cDNA clone. With the exception of MΔBgl RNA, all mutants still produced M-RNA-specific proteins, as detected by immunofluorescent staining (Table 3). Northern (RNA) blot analysis confirmed that the mutant RNAs were replicated in protoplasts and further revealed that MΔBgl RNA was also replicated, although at a very low level. From these results, it can be concluded that it is possible to delete considerable parts of the M RNA coding region without destroying the ability of the RNA to be replicated. Apparently, the sequences at the termini of the M RNA contain sufficient signals to allow replication.

If cowpea plants were inoculated with any of the mutant M RNAs combined with B-RNA transcripts, infection did not occur (Table 3). Visible symptoms did not develop, nor was it possible to detect virus-specific proteins in the inoculated and upper leaves. These results indicate that the M-RNA-encoded 58K and 48K proteins are involved in cell-to-cell transport of CPMV and that the virus can spread only if the RNA is encapsidated in particles (Wellink and Van Kammen, 1989). The involvement of 58K and/or 48K proteins in virus transport was further substantiated by immunogold labeling experiments of sections of infected leaves with anti-48K serum. Gold label was found to be associated with tubular structures protruding from the cell wall (Van Lent et al., 1990). These structures are present only in infected cells and are probably involved in transport of the virus from cell to cell. Also, many of these structures appear to contain virus particles. These results suggest a possible role for the 58K and 48K proteins in the induction or formation of the tubular structures in leaf cells to facilitate transport of CPMV.

With respect to CPMV M RNA, the following three questions remain to be answered. (i) Are both the 58K and 48K proteins essential for transport, and if so, what different functions do these largely overlapping protein have in this process? In protoplasts, the 58K protein is found mainly in the cytoplasmic fraction and the 48K protein is found predominantly in the membrane fraction. Unfortunately, despite numerous efforts, we have not been able to detect the 58K protein in infected plants. Their different locations in protoplasts indicate that the 58K and 48K proteins may have different functions. The analyses with mutants have shown that the 58K protein is essential for transport, but it has not yet been possible to establish whether the 48K protein also has a function in this process. In view of the prominent presence of the 48K protein in infected cells as well as the fact that all comoviruses seem to produce two M-RNA-encoded proteins (Goldbach and Krijt, 1982), it is very likely that the 48K protein also has an important role. (ii) It remains to be elucidated which elements on the RNA, if any, are responsible for the efficient internal initiation of translation on this RNA. Does the RNA possess an internal ribosome entry site (Jang et al., 1988; Pelletier and Sonnenberg, 1988), or do about half of the ribosomes just ignore the AUG codon at position 161 of the M RNA? Experiments are in progress to determine whether the M RNA possesses an internal ribosome entry site between positions 161 and 512. (iii) We have to think of an explanation for the results with MΔBgl RNA. Upon translation of this RNA in vitro, the 105K protein was no longer produced because of a four-nucleotide insertion behind the start codon of this protein. On the other hand, as might be expected, production of the 95K protein was not affected in vitro. To our surprise, no translation products could be detected in protoplasts inoculated with this mutant, although its RNA is replicated, as revealed by Northern blotting. These results suggest that translation of the M RNA in vivo differs from the situation in vitro.

ACKNOWLEDGMENTS. We thank Kees Pley for providing us with RNA structures and Gré Heitkönig for preparation of the manuscript.

This work was supported by the Netherlands Foundation for Chemical Research, with financial aid from the Netherlands Organisation for Scientific Research.

REFERENCES

Eggen, R., and A. Van Kammen. 1988. RNA replication in comoviruses, p. 49–70. In E. Domingo, J. J. Holland, and P. Ahlquist (ed.), RNA Genetics, vol. 1. CRC Press, Inc., Boca Raton, Fla.

Eggen, R., J. Verver, J. Wellink, A. De Jong, R. Goldbach, and A. Van Kammen. 1989a. Improvements of the infectivity of in vitro transcripts from cloned cowpea mosaic virus cDNA: impact of terminal nucleotide sequences. Virology 173:447–455.

Eggen, R., J. Verver, J. Wellink, K. Pley, A. Van Kammen, and R. Goldbach. 1989b. Analysis of sequences involved in cowpea mosaic virus RNA replication using site specific mutants. Virology 173:456–464.

Franssen, H., R. Goldbach, M. Broekhuysen, M. Moerman, and A. Van Kammen. 1982. Expression of middle-component RNA of cowpea mosaic virus: in vitro generation of a precursor to both capsid proteins by a bottom-component RNA-encoded protease from infected cells. *J. Virol.* **41**:8–17.

Franssen, H., J. Leunissen, R. Goldbach, G. Lomonossoff, and D. Zimmern. 1984. Homologous sequences in non-structural proteins from cowpea mosaic virus and picornaviruses. *EMBO J.* **3**:855–861.

Goldbach, R., and J. Krijt. 1982. Cowpea mosaic virus-encoded protease does not recognize primary translation products of M RNAs from other comoviruses. *J. Virol.* **43**:1151–1154.

Goldbach, R., G. Rezelman, and A. Van Kammen. 1980. Independent replication and expression of B-component RNA of cowpea mosaic virus. *Nature* (London) **286**:297–300.

Goldbach, R., and A. Van Kammen. 1985. Structure, replication and expression of the bipartite genome of cowpea mosaic virus, p. 83–120. *In* J. W. Davies (ed.), *Molecular Plant Virology*, vol. 2. CRC Press, Inc., Boca Raton, Fla.

Goldbach, R., and J. Wellink. 1988. Evolution of plus strand RNA viruses. *Intervirology* **29**:260–267.

Jang, S. K., H. G. Kräusslich, M. J. H. Nicklin, G. M. Duke, A. C. Palmenberg, and E. Wimmer. 1988. A segment of the 5' nontranslated region of encephalomyocarditis virus RNA directs internal entry of ribosomes during in vitro translation. *J. Virol.* **62**:2636–2643.

Pelham, H. R. B. 1979. Synthesis and proteolytic processing of cowpea mosaic virus proteins in reticulocyte lysates. *Virology* **96**:463–477.

Pelletier, J., and N. Sonnenberg. 1988. Internal initiation of translation of eukaryotic mRNA directed by a sequence derived from poliovirus RNA. *Nature* (London) **334**:320–325.

Rezelman, G., H. J. Franssen, R. W. Goldbach, T. S. Ie, and A. Van Kammen. 1982. Limits to the independence of bottom component RNA of cowpea mosaic virus. *J. Gen. Virol.* **60**:335–342.

Rezelman, G., R. Goldbach, and A. Van Kammen. 1980. Expression of bottom-component RNA of cowpea mosaic virus in cowpea protoplasts. *J. Virol.* **36**:366–373.

Rezelman, G., A. Van Kammen, and J. Wellink. 1989. Expression of cowpea mosaic virus M RNA in cowpea protoplasts. *J. Gen. Virol.* **70**:3043–3050.

Van Lent, J. W. M., J. Wellink, and R. Goldbach. 1990. Evidence for the involvement of the 58K and 48K proteins in the intercellular movement of cowpea mosaic virus. *J. Gen. Virol.* **71**:219–223.

Vos, P., M. Jaegle, J. Wellink, J. Verver, R. Eggen, A. Van Kammen, and R. Goldbach. 1988a. Infectious RNA transcripts derived from full-length DNA copies of the genomic RNAs of cowpea mosaic virus. *Virology* **165**:33–41.

Vos, P., J. Verver, M. Jaegle, J. Wellink, A. Van Kammen, and R. Goldbach. 1988b. Two viral proteins involved in the proteolytic processing of the cowpea mosaic virus polyproteins. *Nucleic Acids Res.* **16**:1967–1985.

Vos, P., J. Verver, P. Van Wezenbeek, A. Van Kammen, and R. Goldbach. 1984. Study of the genetic organisation of a plant viral RNA genome by in vitro expression of a full-length DNA copy. *EMBO J.* **3**:3049–3053.

Wellink, J., M. Jaegle, H. Prinz, A. Van Kammen, and R. Goldbach. 1987. Expression of the middle component RNA of cowpea mosaic virus in vivo. *J. Gen. Virol.* **68**:2577–2585.

Wellink, J., G. Rezelman, R. Goldbach, and K. Beyreuther. 1986. Determination of the proteolytic processing sites in the polyprotein encoded by the B-RNA of cowpea mosaic virus. *J. Virol.* **59**:50–58.

Wellink, J., and A. Van Kammen. 1989. Cell-to-cell transport of cowpea mosaic virus requires both the 58K/48K proteins and the capsid proteins. *J. Gen. Virol.* **70**:2279–2286.

Chapter 18

Defective Interfering RNAs Associated with Plant Virus Infections

T. J. Morris and D. A. Knorr

Laboratory cultures of viruses can often develop defective genomes that interfere with normal viral functions. These defective interfering viruses (DIs) have been identified in association with members of all major RNA virus families and in virtually every virus in which a systematic search has been made (Perrault, 1981). DIs generally consist of portions of a parental virus genome, but through rearrangements, deletions, or recombination, they have lost functions essential for independent existence. As such, DIs depend on the parental helper virus genome to supply those missing components. In most cases, competition between DIs and the helper virus, usually for *trans*-acting factors, leads to specific interference with normal replication processes of the helper virus (Schlesinger, 1988). Although there is some debate about the importance of defective viral genomes in natural infections, there is sufficient indication to suggest that they may be important in modulating virus disease processes (Huang, 1988).

For plant viruses, the phenomena of symptom attenuation and virus interference as it relates to disease modulation have not been as well documented as in animal systems. The majority of reports of these phenomena in plants have involved interactions between a helper virus and an associated component which has been identified as either a satellite virus or a satellite RNA (Francki, 1985). Satellite RNAs are similar to DIs in that they are dependent on a helper virus for replication and, in most cases, for encapsidation as well. However, satellite RNAs differ from DIs in that they share no sequence homology with their helper virus. This is true for all but the chimeric satellite RNA C of turnip crinkle virus (TCV) (Simon and Howell, 1986). Among the plant viruses, satellite RNAs are relatively common, having been found associated with at least 24 viruses in six groups. Interestingly, satellites may either attenuate or intensify normal disease symptoms expressed by the helper virus (Simon, 1988).

In contrast, there have been few reports of defective viruses or DI RNAs associated with plant virus infections. These include DI-like particles associated with negative-strand plant rhabdoviruses (Adam et al., 1983; Ismail and Milner, 1988) and the bunyaviruslike tomato spotted wilt virus (Verkleij and Peters, 1983). DI RNA components have also been demonstrated for a plant reovirus, wound tumor virus (Nuss, 1988). Definitive reports of DIs among plant viruses with positive-stranded single-stranded RNA genomes include only tomato bushy stunt virus (TBSV) (Hillman et al., 1987) and TCV (Li et al., 1989). The latter two reports represent the best characterized of the plant virus-associated DI systems because the DIs as well as both helper virus genomes have been completely sequenced (Carrington et al., 1989; Hillman et al., 1989; Hearne et al., in preparation). The value of both TCV and TBSV is further augmented by our ability to produce infectious RNA transcripts from cDNA clones of these two viruses (Heaton et al., 1989; Hearne et al., in press). TBSV and TCV are members from two groups of related small RNA viruses, the tombusviruses and the carmoviruses (Morris and Carrington, 1988). Viruses in these groups share several common features of structure and composition, including a monopartite genome of 4 to 5 kilobases and virions composed of 180 capsid subunits of about 40 kilodaltons each. Comparison of the genomic sequences of these two viruses suggests that the putative polymerase genes are closely related evolutionarily although the ge-

---

T. J. Morris and D. A. Knorr, Department of Plant Pathology, University of California, Berkeley, California 94720.

nomes have distinct organizational differences. Identification of similar DI RNAs in association with both of these viruses makes them valuable subjects for study. The remainder of this chapter discusses our current results on the characterization of DI RNAs associated with members of these virus groups.

DISCOVERY OF DI RNAs IN TOMBUSVIRUSES

We first described an interference phenomenon while studying the cherry strain of TBSV in which inocula containing highly concentrated virus tended to cause a less severe and more persistent disease syndrome in plants than did diluted inocula. The interference was found to be associated with a novel low-molecular-size (400-base) RNA species present in the virus isolate (Hillman et al., 1985). This system was intriguingly similar to the DI-mediated autointerference observed in animal virus systems (Huang, 1988). Gallitelli and Hull (1985) also reported the presence of a satellite RNA of about 700 bases in association with cymbidium ringspot virus, a close relative of TBSV, which they demonstrated by hybridization to be unrelated to the helper virus genome. In contrast, Northern (RNA) hybridizations performed on our TBSV strain consistently indicated that the small RNA was related to the helper virus genome. Subsequent cloning and sequencing of TBSV-associated small RNAs in conjunction with genomic sequencing clearly demonstrated that they were colinear deletion mutants consisting of a mosaic of helper virus sequence derived from 5′-proximal, internal, and 3′-proximal regions of the genome (Hillman et al., 1987). Plant infection studies demonstrated that DI RNA was responsible for symptom attenuation as well as reduced viral replication and virus accumulation (Hillman, 1986). More recently, we have shown that TBSV DIs specifically interfere with helper virus replication in plant protoplasts (Jones et al., in press). These features are all consistent with the definition of DIs and establish the first definitive proof for the existence of DIs in association with plant RNA viruses.

ORIGIN AND EVOLUTION OF TBSV DI RNAs

It was suspected that the presence of DIs might be a common feature of the tombusviruses. In preliminary experiments, DI-like RNA species could be readily detected after serial passage of several different tombusviruses (artichoke mottle crinkle virus, pelargonium leaf curl virus, petunia asteroid mosaic virus, and a field isolate from tomatoes closely related to the BS3 strain of TBSV) in *Nicotiana clevelandii* (T. J. Morris, unpublished data). In addition, Burgyan et al. (1989) recently reported that a 400-base DI-like RNA appeared in cymbidium ringspot virus after host passage of purified genomic RNA. Curiously, there is also a previous report claiming generation of a satellite RNA upon passage of cymbidium ringspot virus in *Nicotiana benthamiana* (Gallitelli and Hull, 1985). Skepticism about this observation and curiosity about the possible de novo origin of DIs prompted a further set of passage experiments. Sixteen local lesion isolates of the cherry strain of TBSV were passed on two *Nicotiana* hosts at 7-day intervals either at high multiplicity of infection (MOI) (undiluted sap) or low MOI (buffer-diluted sap). The appearance of symptom attenuation along with the presence of novel, virus-specific RNAs of distinct size (ranging from 500 to 600 bases) in each of the high-MOI isolates but in none of the low-MOI isolates was interpreted as strong evidence for de novo generation of DIs. More important, continued passage of DI-containing isolates resulted in further evolution of the DIs, as judged by size reduction (Morris and Hillman, 1989).

MOLECULAR CHARACTERIZATION OF A DE NOVO DI

One of the DIs (DI-B10) generated by high-MOI passage was selected for further characterization because it was larger and more efficiently encapsidated than the DIs originally isolated. Six nearly full-length cDNA clones of DI-B10 made from sucrose gradient-fractionated B10 virion RNA were completely sequenced (Knorr and Morris, in preparation). Alignment of the DI-B10 sequences and the previously characterized DI-1 sequence with a consensus sequence of the TBSV genome illustrates a consistent pattern of genomic sequences retained in both DI populations (Fig. 1). Regions conserved in the DIs are the 5′ leader sequence (including the first AUG), followed by a variable region of approximately 200 bases from the putative polymerase domain. This is fused to a small region from the 3′ terminus of the two small overlapping open reading frames, followed by the final 130 nucleotides from the nontranslated 3′ terminus. The three domains are conserved in both DI populations, but several substitutions and deletions within the domains of the smaller DI-1

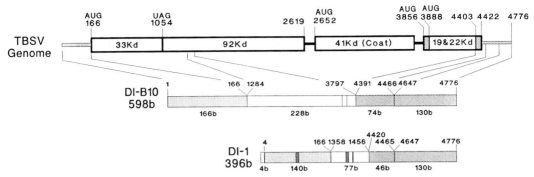

FIGURE 1. Comparison of the sequences of two independently isolated DI RNAs with each other and with the parental genome of TBSV from which they were derived. (Top) Complete TBSV genome; open boxes represent the open reading frames encoding polypeptides of the size indicated in kilodaltons (Kd). (Middle) Regions of sequence maintained by DI-B10, which was generated through high-MOI passage. (Bottom) Origin of sequences in DI-1, which was identified in association with the original TBSV isolate. Common regions of sequence are identified by similar shading, and the breakpoints for deletions corresponding to nucleotide positions along the TBSV genome are indicated above each DI. The numbers below each DI indicate the number of nucleotides maintained in each region. The additional lines and cross-hatched areas denote additional deletions or minor sequence variations within the larger blocks.

suggest that it may have evolved from a larger DI species.

To confirm biological activity of the DI sequences and also to begin addressing the functional significance of the regions of sequence, the DI-B10 cDNA clones were modified to allow in vitro production of RNA transcripts. This was accomplished using the polymerase chain reaction to simultaneously remove oligonucleotide tails added during cloning and install a bacteriophage T7 promoter sequence immediately adjoining the 5′ DI terminus. Transcripts produced from the polymerase chain reaction-amplified product were of homogeneous size consistent with that of the parental DI-B10 RNA population. Each of the in vitro transcripts was replicated when inoculated onto plants with helper virus, and each induced a level of symptom attenuation similar to that of the original DI RNA population (Knorr and Morris, in preparation). We have also assembled full-length cDNA clones of TBSV behind the bacteriophage T7 promoter which, when transcribed, produce infectious RNA (Hearne et al., in press); thus, we can now initiate studies of DI generation and evolution by using cloned inocula from each component.

IDENTIFICATION OF DI RNAs ASSOCIATED WITH TCV

TCV, a member of the carmovirus group, has both structural and genetic relatedness to

TBSV (Morris and Carrington, 1988). We have completed the cloning and sequencing of TCV (Carrington et al., 1989) and are currently studying its structure and gene expression, using inocula transcribed from full-length cDNA clones of the viral genome (Heaton et al., 1989). Previous workers have shown that TCV contains a family of satellite RNAs, one of which was a chimeric molecule composed of both satellite and TCV sequences (Simon and Howell, 1986). This species, termed TCV RNA C, consisted of 184 bases of a satellite RNA at the 5′ terminus, a 14-base region of internal TCV sequence, and 144 bases of the extreme 3′ terminus of TCV. We subsequently identified a satellite-sized RNA species associated with the Berkeley isolate of TCV and suspected that it was similar to this previously reported RNA C. This species was further characterized in collaboration with Anne Simon of the University of Massachusetts; surprisingly, the small RNA present in our TCV isolate has features consistent with those of a DI and not a satellite RNA. This RNA associated with the Berkeley isolate has been designated DI RNA G. A summary of preliminary sequence analysis in which the origins of the sequences of DI RNA G are given relative to the genomic RNA is depicted in Fig. 2. All of DI RNA G except for a block of 21 nucleotides at the extreme 5′ end was derived from the TCV genomic sequence. The virus-derived sequences consisted of a domain of about 140 bases from the viral 5′ region and a

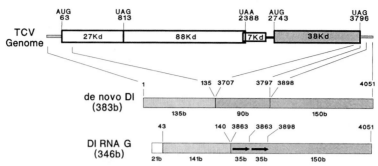

FIGURE 2. Comparison of two DIs with each other and with the parental genome of TCV. Symbols and numbering are as in Fig. 1. (Top) TCV genome. (Middle) Origins of the sequences in the de novo DI generated from cloned virus inoculum. (Bottom) Sequences in DI RNA G, which is the DI associated with the original virus isolate. The 21 bases of non-TCV sequence at the 5′ terminus is shown in white, and the region of the 35-base direct repeat is indicated by arrows. These diagrams represent a preliminary scheme of sequence origins available prior to publication and do not necessarily represent the precise models of the sequence blocks and their respective origins.

domain of 185 bases derived from the extreme 3′ terminus. At the beginning of this latter region of sequence is a duplication of 35 bases, followed by the remaining 115 bases of the 3′ terminus. Further experiments to explore the possible origins of the small RNAs from TCV were initiated by using in vitro transcripts from our TCV clone. Additional DI RNAs derived de novo from the cloned viral inoculum were identified (Fig. 2). They contained sequences entirely derived from the genomic RNA consisting of 135 bases of 5′ sequence, an internal region of about 90 bases from the 3′ terminus of the coat protein gene, and 150 bases of the viral 3′ terminus. In preliminary experiments, the TCV DI RNAs were shown to interfere with the accumulation of TCV in plants when present in excess of the genomic RNA in the inoculum. Curiously, the presence of DIs also tended to increase the severity of symptoms caused by the helper virus.

DISCUSSION

While DIs have been widely reported in animal virus systems, they have gone largely unstudied in the plant viruses. DIs from the tombusviruses and the carmoviruses, two related groups of small spherical RNA plant viruses, have been recently characterized. In general, these DIs are deletion mutants of the viral genome consisting of 5′- and 3′-terminal sequences derived predominantly from the viral noncoding regions. In addition, small internal regions of variable length derived from viral coding regions are often present in the DIs.

These systems are ideal for study because there is complete structural information for the virions, and infectious RNA can be prepared in vitro from cDNA clones. The presence of DI systems in RNA plant viruses provides valuable research tools available previously only in animal virus systems (Schlesinger, 1988). These include opportunities to elucidate the *cis*-acting regulatory sequences in the DI RNAs important in RNA replication, gene expression, and encapsidation. These DIs will also serve as model systems to explore plant RNA virus replication and recombination mechanisms, and they may prove useful as transient expression vectors. Finally, they offer a good system for exploring the mechanisms of interference that lead to symptom modulation and virus persistence in plant hosts.

ACKNOWLEDGMENTS. This work was supported by grant DE-FG03-88er13908 from the U.S. Department of Energy.

REFERENCES

Adam, G., K. Gaedigk, and K. W. Mundry. 1983. Alterations of a plant Rhabdovirus during successive mechanical transfers. *Z. Pflanzenkr. Pflanzensch.* **90**:28–35.

Burgyan, J., F. Grieco, and M. Russo. 1989. A defective interfering RNA molecule in cymbidium ringspot virus infections. *J. Gen. Virol.* **70**:235–239.

Carrington, J. C., L. A. Heaton, D. Zuidema, B. I. Hillman, and T. J. Morris. 1989. The genome structure of turnip crinkle virus. *Virology* **170**:219–226.

Francki, R. I. B. 1985. Plant virus satellites. *Annu. Rev. Microbiol.* **39**:151–174.

Gallitelli, D., and R. Hull. 1985. Characterization of satellite RNAs associated with tomato bushy stunt virus and five other definitive Tombusviruses. *J. Gen. Virol.* **66:**1533–1543.

Hearne, P., D. A. Knorr, B. I. Hillman, and T. J. Morris. The complete genome structure and infectious RNA synthesized from clones of tomato bushy stunt virus. *Virology,* in press.

Heaton, L. A., J. C. Carrington, and T. J. Morris. 1989. Turnip crinkle virus infection from RNA synthesized in vitro. *Virology* **170:**214–218.

Hillman, B. I. 1986. Genome organization, replication and defective RNAs of tomato bushy stunt virus, Ph.D. thesis, University of California, Berkeley.

Hillman, B. I., J. C. Carrington, and T. J. Morris. 1987. A defective interfering RNA that contains a mosaic of a plant virus genome. *Cell* **51:**427–433.

Hillman, B. I., P. Hearne, D. Rochon, and T. J. Morris. 1989. Organization of the tomato bushy stunt virus genome: characterization of the coat protein gene and 3' terminus. *Virology* **169:**42–52.

Hillman, B. I., D. E. Schlegel, and T. J. Morris. 1985. Effects of low-molecular weight RNA and temperature on tomato bushy stunt virus symptom expression. *Phytopathology* **75:**361–365.

Huang, A. S. 1988. Modulation of viral disease processes by defective interfering particles, p. 195–208. *In* E. Domingo, J. J. Holland, and P. Ahlquist (ed.), *RNA Genetics,* vol. 3. CRC Press, Inc., Boca Raton, Fla.

Ismail, I. D., and J. J. Milner. 1988. Isolation of defective interfering particles of sonchus yellow net virus from chronically infected plants. *J. Gen. Virol.* **69:**999–1006.

Jones, R. K., A. O. Jackson, and T. J. Morris. Interaction of tomato bushy stunt virus and its defective interfering RNA in inoculated protoplasts. *Virology,* in press.

Knorr, D. A., and T. J. Morris. Molecular characterization and infectious transcripts from clones of defective interfering RNAs of tomato bushy stunt virus. Manuscript in preparation.

Li, X. H., L. A. Heaton, T. J. Morris, and A. E. Simon. 1989. Defective interfering RNAs of turnip crinkle virus intensify viral symptoms. *Proc. Natl. Acad. Sci. USA* **86:**9173–9177.

Morris, T. J., and J. C. Carrington. 1988. Carnation mottle virus and viruses with similar properties, p. 73–112. *In* R. Koenig (ed.), *The Plant Viruses,* vol. 3. *Polyhedral Virions with Monopartite RNA Genomes.* Plenum Publishing Corp., New York.

Morris, T. J., and B. I. Hillman. 1989. Defective interfering RNAs of a plant virus. *UCLA Symp. Mol. Cell. Biol. New Ser.* **101:**185–197.

Nuss, D. L. 1988. Deletion mutants of double stranded RNA genetic elements found in plants and fungi, p. 188–210. *In* E. Domingo, J. J. Holland, and P. Ahlquist (ed.), *RNA Genetics,* vol. 2. CRC Press, Inc., Boca Raton, Fla.

Perrault, J. 1981. Origin and replication of defective interfering particles. *Curr. Top. Microbiol. Immunol.* **93:**151–207.

Schlesinger, S. 1988. The generation and amplification of defective interfering RNAs, p. 167–185. *In* E. Domingo, J. J. Holland, and P. Ahlquist (ed.), *RNA Genetics,* vol. 2. CRC Press, Inc., Boca Raton, Fla.

Simon, A. E. 1988. Satellite RNAs of plant viruses. *Plant Mol. Biol. Rep.* **6:**240–252.

Simon, A. E., and S. H. Howell. 1986. The virulent satellite of turnip crinkle virus has a major domain homologous to the 3' end of the helper virus genome. *EMBO J.* **5:**3423–3428.

Verkleij, F. N., and D. Peters. 1983. Characterization of a defective form of tomato spotted wilt virus. *J. Gen. Virol.* **64:**677–686.

Chapter 19

Infectious Theiler's Virus cDNA Clones: Intratypic Recombinants and Viral Polyprotein Processing

R. P. Roos, J. Fu, W. Kong, S. Stein, T. Bodwell, L. Rosenstein,
M. Routbort, and B. L. Semler

Theiler's murine encephalomyelitis viruses (TMEV) are a group of mouse picornaviruses that are most closely related to the cardiovirus genus of picornaviruses on the basis of nucleotide and predicted amino acid sequences (Nitayaphan et al., 1986; Ohara et al., 1988b; Ozden et al., 1986; Pevear et al., 1987). TMEV strains can be divided into two subgroups on the basis of their different biological properties (Lipton, 1980; Lorch et al., 1981) and their reactivity with a panel of neutralizing monoclonal antibodies (Nitayaphan et al., 1985). The GDVII strain and other members of the GDVII subgroup of TMEV are highly neurovirulent; 1 PFU produces a fatal polioencephalomyelitis in weanling mice. In contrast, DA and other members of the TO subgroup of TMEV are not as neurovirulent; 10^6 PFU of the DA strain does not kill a weanling mouse but produces a persistent, demyelinating infection in mice. The reason for persistence of the TO subgroup strain is unclear but presumably relates to a restriction of virus expression in certain neural cells (Cash et al., 1985).

A major goal of TMEV research is to identify the molecular basis for the different subgroup-specific diseases and to elucidate the responsible mechanisms. The TMEV model is an especially attractive one for these studies because of the availability of the mouse as both the natural and experimental host and because of the somewhat unusual properties of the TO subgroup strains (demyelinating activity, virus persistence, and restricted expression). We produced infectious TMEV cDNA clones in order to clarify these issues, since the manipulation of infectious cDNA clones has been extremely valuable in analyzing the functions of alleles of other picornaviruses. This chapter describes the use of these infectious clones to generate intratypic chimeric cDNAs and recombinant viruses and to investigate TMEV polyprotein processing. The recombinant TMEV studies indicated that a segment from the GDVII 1B- to the 2C-coding region is a major determinant of neurovirulence. The GDVII area 5' to this region also contributed to neurovirulence when coupled with the GDVII 1B-2C-coding region but not when present alone. Investigations of polyprotein processing demonstrated that 3C is a major proteinase for TMEV, but a second proteinase also appeared to be active. The results also suggested that a protein is synthesized in reticulocyte lysates from transcripts derived from DA cDNA, but not GDVII cDNA, that is a product of an alternative initiation site in the leader coding region and is out of phase with the polyprotein reading frame.

INTRATYPIC CHIMERIC cDNAs

We constructed a series of chimeric cDNAs in which a segment of an infectious DA cDNA clone (pDAFL3) (Roos et al., 1989b) or a segment of an infectious GDVII cDNA clone (pGDVIIFL2) (J. Fu, S. Stein, L. Rosenstein, T. Bodwell, M. Routbort, B. L. Semler, and R. P. Roos, submitted for publication) was substituted by the corresponding segment of the other strain. These chimeric cDNAs were then transcribed, and the transcripts were transfected

R. P. Roos, J. Fu, W. Kong, S. Stein, T. Bodwell, L. Rosenstein, and M. Routbort, Department of Neurology, University of Chicago School of Medicine, Chicago, Illinois 60637.   *B. L. Semler,* Department of Microbiology and Molecular Genetics, College of Medicine, University of California, Irvine, California 92717.

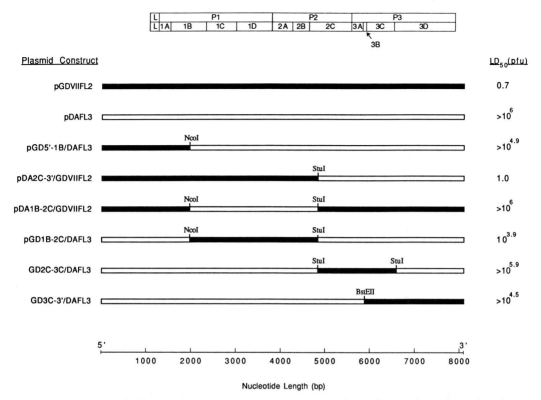

FIGURE 1. Maps of chimeric TMEV cDNAs used to isolate recombinant viruses. The coding region of TMEV is shown at the top; the structures of GDVII, DA, and chimeric cDNAs are shown below. The name of the plasmid with the TMEV insert is given on the left. Neurovirulence is shown at the right as LD_{50}.

into L929 cells to produce recombinant viruses. The viruses were then inoculated into 4-week-old DBA/2 mice to determine the neurovirulence phenotype, i.e., whether the virus caused death of a weanling mouse within 3 weeks.

The results of the mouse inoculations are presented in Fig. 1. As shown, the 50% lethal dose (LD_{50}) of DAFL3 virus was $>10^6$, whereas GDVIIFL2 virus had an LD_{50} of approximately 1. Inoculation of mice with a number of recombinant viruses produced a wide spectrum of disease severity. Insertion into the DA genome of a segment of the GDVII genome that runs from the *Nco*I site in the VP2(1B)-coding area at nucleotide (nt) 1964 to the *Stu*I site in the 2C-coding area at nt 4843 changed the attenuated phenotype of DA, so that the recombinant virus (GD1B-2C/DAFL3) produced death of weanling mice. However, the LD_{50} was only $10^{3.9}$; i.e., this recombinant virus was not as neurovirulent as GDVIIFL2 virus. As expected, insertion of the corresponding segment from the DA genome (from the *Nco*I site in the 1B-coding

area to the *Stu*I site in the 2C-coding area) into the GDVII genome attenuated the GDVII phenotype, so that this recombinant virus (DA1B-2C/GDVIIFL2) no longer killed weanling mice. A virtually complete GDVII neurovirulence phenotype was obtained after inoculation of DA2C-3'/GDVIIFL2 recombinant virus, which substitutes into the DA genome a large segment that includes GDVII nt 1 through 4843, i.e., adds GDVII nt 1 to 1963 to the GDVII 1B-2C segment. However, substitution into the DA genome of a segment of GDVII that includes only nt 1 to 1963 failed to produce death after inoculation of the recombinant virus (GD5'-1B/DAFL3). Interestingly, substitution of nt 1965 to 4843 or nt 1 to 4843 from GDVII into the DA genome changed the plaque size of the recombinant virus from the small size seen with DA virus to the large size seen with GDVII virus (Lipton, 1980). We are presently preparing other chimeric cDNAs to better define the key genomic region(s) that is critical in determining neurovirulence. In addition, it will be of special interest to determine the loci responsible for virus demyelination, persistence, and the restricted growth of DA in specific cells in vivo.

We do not know how the 1B-2C segment affects TMEV neurovirulence. Intertypic poliovirus chimeric cDNA studies prepared from poliovirus type 1 and mouse-adapted poliovirus type 2 have identified a short amino acid segment (which is a major trypsin-sensitive neutralization site located in poliovirus type 2 VP1) that enables the chimeric poliovirus type 1 to kill mice after intracerebral inoculation (Martin et al., 1988; Murray et al., 1988). It is presumed that this segment allows poliovirus to bind to a poliovirus receptor on mouse neurons and thereby permits virus replication in these cells. It may be that the 1B-2C-coding area of GDVII, but not DA, codes for an amino acid segment that is also critical for binding to specific mouse neurons. The presence of this GDVII segment in the GD1B-2C/DAFL3 recombinant virus may permit entry of the virus into mouse neurons and the production of a fatal polioencephalomyelitis. We have identified a major trypsin-sensitive neutralization site at the carboxyl end of VP1 that affects disease phenotype (Roos et al., 1989c); this site may be the critical one in the 1B-2C-coding area that is responsible for neurovirulence. We are in the process of generating mutations in the trypsin-sensitive site to determine the effect on neurovirulence and mouse neuronal binding.

It may be that the GDVII 5' noncoding region (included in the GD5'-2C/DAFL3 recombinant virus) plays a role in neurovirulence after efficient binding of TMEV to the mouse neuron (by means of a region within the GDVII 1B-2C region). Poliovirus studies have identified a single key nucleotide in the 5' noncoding region as critical in determining neurovirulence, although other areas of the poliovirus genome also seemed to exert an effect (Nomoto et al., 1989). The explanation for how the 5' noncoding region determines neurovirulence is still unclear.

TMEV POLYPROTEIN PROCESSING

The preparation of infectious TMEV cDNA clones allowed us to investigate polyprotein processing and the proteinases that carry out these cleavages (Roos et al., 1989a). We were especially interested in these studies with respect to restricted infection of the DA strain, since polyprotein processing is critical in the regulation of picornaviral gene expression. We analyzed in vitro translation programmed by in vitro-derived transcripts of TMEV clones. To

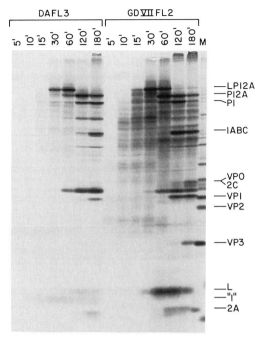

Figure 2. Time course of in vitro translation of in vitro-derived transcripts from pDAFL3 (lanes 1 to 7) and pGDVIIFL2 (lanes 8 to 14) cDNA clones. Transcription and translation reaction conditions were performed as described elsewhere (Roos et al., 1989a). Samples were removed at the indicated time intervals, diluted in Laemmli sample buffer, and subjected to electrophoresis. The marker (M) lane displays [^{35}S]-methionine-labeled proteins from an extract of DA virus-infected BHK-21 cells harvested 14 h after infection. Note that "l" (see text) is present in translations of pDAFL3 but not pGDVIIFL2 transcripts.

identify the active proteinases, we made use of transcription templates that (i) were truncated to various lengths by digestion with different restriction endonucleases and (ii) had linker and/or deletion mutations in putative proteinase-coding areas of pDAFL3.

A time course study of translation programmed by in vitro-derived transcripts from XbaI-linearized pDAFL3 and pGDVIIFL2 is shown in Fig. 2. Samples were removed at various times after the start of translation, and translation was then stopped with Laemmli sample buffer. The identity of the synthesized proteins was ascertained by immunoprecipitation (see Roos et al. [1989a] for details), by consideration of the predicted molecular weight of the proteins, and by examining shifts in the electrophoretic mobility of proteins in mutants. Large

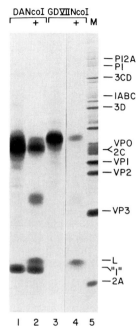

DANcoI GDⅦNcoI
┌─ + ─┐┌─ + ─┐ M

— PI2A
— PI
— 3CD

— IABC
— 3D

VPO
2C
— VPI
— VP2

— VP3

— L
"l"
— 2A

| 1   2   3   4   5

FIGURE 3. In vitro translation of transcripts derived from TMEV cDNA clones linearized by NcoI. Translations, preparation of samples, and the marker (M) lane are described in the legend to Fig. 2. +, Reactions that were incubated with a DA virus-infected BHK-21 cell extract after translation. NcoI digests TMEV in the VP2-coding area. Note that "l" (see text) is present with transcripts derived from NcoI-digested pDAFL3 (lanes 1 and 2) but not pGDVIIFL2 (lanes 3 and 4).

products were synthesized by 30 min that were then processed over the course of translation into structural and nonstructural proteins (Fig. 2).

One interesting feature of the time course experiments was the appearance of a protein as early as 10 min after the start of translation of transcripts derived from pDAFL3 (Fig. 2, lanes 2 to 7) but not pGDVIIFL2 (lanes 8 to 14). A second protein of just slightly slower electrophoretic mobility appeared about 30 to 60 min after the start of translation of transcripts derived from pDAFL3 (lanes 5 to 7) as well as pGDVIIFL2 (lanes 11 to 14). Both proteins were initially identified as the L protein because of their altered mobility or disappearance with translation of transcription templates that had mutations in the L-coding area (Roos et al., 1989a). The early appearance of the lower band and its unchanging size during translation of pDAFL3 transcripts originally raised the possibility that DA L autocatalytically cleaves itself from the polyprotein, as is the case with foot-and-mouth disease virus (Strebel and Beck, 1986). However, linearization (see below) and mutational analysis suggested that the early-appearing protein in pDAFL3 translations was not L but the product of translation from an alternative initiation site at nt 1079 in the L-coding region, downstream from the authentic polyprotein initiation site at nt 1066. The AUG at nt 1079 has a reasonable context for initiation and could be used to synthesize a protein (which we have designated by the letter "l") of about the 17-kilodalton mobility observed with this lower band. GDVII does not have an AUG corresponding to that seen at nt 1079 in DA (Pevear et al., 1988), explaining the absence of "l" in the translations programmed from transcripts of pGDVIIFL2. A protein of slightly slower mobility than "l" that is synthesized later in translation of transcripts derived from both pDAFL3 and pGDVIIFL2 presumably corresponds to authentic L. L appears to run at a somewhat aberrant electrophoretic mobility considering its size, as is also the case with encephalomyocarditis virus L (Campbell and Jackson, 1983).

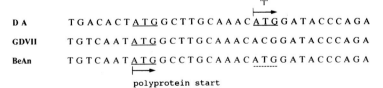

"l"

DA   TGACACTA**TG**GCTTGCAAAC**ATG**GATACCCAGA

GDVII   TGTCAATA**TG**GCTTGCAAACACGGATACCCAGA

BeAn   TGTCAATA**TG**GCCTGCAAACATGGATACCCAGA

polyprotein start

FIGURE 4. Nucleotide sequence from the 5' noncoding region of TMEV showing the presumed authentic initiation site for the polyprotein and a putative initiation site slightly downstream, at DA nt 1079. The latter site could be used to translate a protein (designated "l") of about 17 kilodaltons, out of phase with the polyprotein reading frame. An AUG is present at an equivalent position in BeAn but not GDVII, raising the possibility the "l" is important in the TMEV subgroup-specific neurological diseases.

Linearization studies involved digestion with a variety of restriction endonucleases to generate mRNAs of various lengths which were then used to program an in vitro translation reaction. Figure 3 shows the results of translation reactions programmed by transcripts derived from *Nco*I-digested pDAFL3 and pGDVIIFL2. *Nco*I cuts in the VP2(1B)-coding area, producing a truncated VP0 (since the cleavage of VP4-VP2 [1A-1B] does not occur during in vitro translation but requires viral assembly). Two proteins are seen in translations programmed from transcripts derived from *Nco*I-digested pDAFL3 (Fig. 3, lane 1). The upper band corresponds in mobility to that expected for L-truncated VP0; we have tentatively identified the lower band as "l." When an infected extract is added (lane 2), cleavage of L-truncated VP0 occurs, with the generation of truncated VP0 and authentic L (with a mobility slightly slower than that of "l," which is also detectable). These results suggest that L is not autocatalytically cleaved from the polyprotein, but that the low band seen in lanes 1 and 2 is the product of an alternative initiation site. In contrast to these findings, transcripts derived from *Nco*I-digested pGDVIIFL2 synthesized only L-truncated VP0 (lane 3). The addition of infected extract (lane 4) produced truncated VP0 and L in addition to L-truncated VP0. These findings suggest that GDVII does not have an alternative initiation site that leads to the synthesis of "l."

The absence of an AUG at nt 1079 in GDVII and the presence of an AUG analogous to the one at DA nt 1079 in BeAn (Pevear et al., 1987), the only other member of the TO subgroup that has been sequenced, raise the possibility that "l" plays a role in the subgroup-specific central nervous system diseases (Fig. 4). It is possible that after TO subgroup strain inoculation, "l" is synthesized in certain cells within the central nervous system as it is in reticulocyte lysates and that the production of "l" is important in the resultant restricted persistent infection that is seen. Plans to investigate DA infected demyelinated tissue for the presence of "l" are ongoing.

The results of the other linearization and mutation studies are published elsewhere (Roos et al., 1989a) and will only be briefly reviewed here. Both linearization and mutation studies (Fig. 5) demonstrated that 3C is the major proteinase of TMEV, as is the case with the other picornaviruses that have been studied (Kräusslich and Wimmer, 1988). However, the results suggested that there is a second proteinase besides 3C. This second proteinase may be 2A, since its proteolytic activity has been proven in

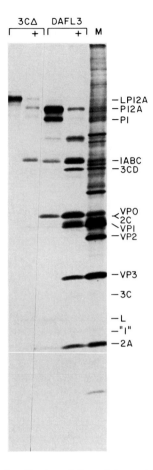

FIGURE 5. In vitro translation of transcripts derived from wild-type pDAFL3 and a DA 3C mutant, pDA3CΔ. Translations, preparation of samples, and the marker (M) lane are described in the legend to Fig. 2. +, Reactions that were incubated with a DA virus-infected BHK-21 cell extract after translation. Translation of transcripts programmed by pDA3CΔ, which has a deletion in the 3C-coding region (Roos et al., 1989a), shows incomplete processing. The addition of an infected extract to the translation reactions produces nearly complete processing, demonstrating the role of 3C as a proteinase and its ability to cleave in *trans*. However, the absence of a full-length polyprotein suggests that another proteinase is also active.

the case of poliovirus (Toyoda et al., 1986) and suggested in the case of other picornaviruses (Clarke and Sangar, 1988). However, a linker insertion mutant in the 2A-coding area of pDAFL3 failed to affect polyprotein processing (Roos et al., 1989a). This latter result suggests that DA 2A may not be a proteinase or that the mutation did not inactivate its proteolytic activity.

SUMMARY

TMEV infectious cDNA clones provided valuable reagents for the construction of chimeric cDNAs and the subsequent production of intratypic recombinant viruses. The recombinant virus studies identified the GVII 1B-2C-coding region as critical for neurovirulence. The GDVII area 5' to this region also contributed to neurovirulence when coupled with the GDVII 1B-2C-coding region but not when present alone. TMEV polyprotein processing was studied with in vitro translation reactions, using the TMEV infectious cDNA clones as transcription templates. Protein 3C is a major TMEV proteinase, although another proteinase also seems active. The results suggested that a protein is synthesized in reticulocyte lysates from an alternative initiation site within the DA leader coding region out of phase with the polyprotein reading frame.

ACKNOWLEDGMENTS. This work was supported by Public Health Service grants 1PO1NS21442 (R.P.R.) and AI22693 (B.L.S.) from the National Institutes of Health. Some of this work was performed while R.P.R. was a recipient of a senior research fellowship from the National Multiple Sclerosis Society and of a National Research Service award from the Public Health Service. B.L.S. is supported by Public Health Service research career development award AI00721 from the National Institutes of Health.

The suggestions of Richard Jackson and the secretarial help of Lee Baksas are gratefully acknowledged.

REFERENCES

Campbell, E. A., and R. J. Jackson. 1983. Processing of the encephalomycarditis virus capsid precursor protein studied in rabbit reticulocyte lysates incubated with N-formyl-[^{35}S]methionine-tRNA$_f^{Met}$. $J.$ $Virol.$ 45:439–441.

Cash, E., M. Chamorro, and M. Brahic. 1985. Theiler's virus RNA and protein synthesis in the central nervous system of demyelinating mice. $Virology$ 144:290–294.

Clarke, B. E., and D. V. Sangar. 1988. Processing and assembly of foot-and-mouth disease virus proteins using subgenomic RNA. $J.$ $Gen.$ $Virol.$ 69:2313–2325.

Kräusslich, H. G., and E. Wimmer. 1988. Viral proteinases. $Annu.$ $Rev.$ $Biochem.$ 57:701–754.

Lipton, H. L. 1980. Theiler's murine encephalomyelitis virus group includes two distinct genetic subgroups that differ physiologically and biologically. $J.$ $Gen.$ $Virol.$ 46:169–177.

Lorch, Y., A. Friedmann, H. L. Lipton, and M. Kotler.

1981. Theiler's murine encephalomyelitis virus group includes two distinct genetic subgroups that differ pathologically and biologically. $J.$ $Virol.$ 40:560–567.

Martin, A., C. Wychowski, T. Couderc, R. Crainic, J. Hogle, and M. Girard. 1988. Engineering a poliovirus type 2 antigenic site on a type 1 capsid results in a chimaeric virus which is neurovirulent for mice. $EMBO$ $J.$ 9:2839–2847.

Murray, M. G., J. Bradley, X.-F. Yang, E. Wimmer, E. G. Moss, and V. R. Racaniello. 1988. Poliovirus host range is determined by a short amino acid sequence in neutralization antigenic site I. $Science$ 241:213–215.

Nitayaphan, S., D. Omilianowski, M. M. Toth, S. Parks, R. R. Rueckert, A. C. Palmenberg, and R. P. Roos. 1986. Relationship of Theiler's murine encephalomyelitis viruses to the cardiovirus genus of picornaviruses. $Intervirology$ 26:140–148.

Nitayaphan, S., M. M. Toth, and R. P. Roos. 1985. Neutralizing monoclonal antibodies to Theiler's murine encephalomyelitis viruses. $J.$ $Virol.$ 53:651–657.

Nomoto, A., N. Kawamura, M. Kohara, and M. Arita. 1989. Expression of the attenuation phenotype of poliovirus type 1, p. 297–306. In B. L. Semler and E. Ehrenfeld (ed.), $Molecular$ $Aspects$ of $Picornavirus$ $Infection$ and $Detection.$ American Society for Microbiology, Washington, D.C.

Ohara, Y., A. Senkowski, J. Fu, L. Klaman, J. Goodall, M. Toth, and R. P. Roos. 1988a. Trypsin-sensitive neutralization site on VP1 of Theiler's murine encephalomyelitis viruses. $J.$ $Virol.$ 62:3527–3529.

Ohara, Y., S. Stein, J. Fu, L. Stillman, L. Klaman, and R. P. Roos. 1988b. Molecular cloning and sequence determination of DA strain of Theiler's murine encephalomyelitis viruses. $Virology$ 164:245–255.

Ozden, S. F., F. Tangy, M. Chamorro, and M. Brahic. 1986. Theiler's virus genome is closely related to that of encephalomyocarditis virus, the prototype cardiovirus. $J.$ $Virol.$ 60:1163–1165.

Pevear, D. C., M. Borkowski, M. Calenoff, C. Oh, B. Ostrowski, and H. L. Lipton. 1988. Insights into Theiler's virus neurovirulence based on a genomic comparison of the neurovirulent GDVII and less virulent BeAn strains. $Virology$ 165:1–12.

Pevear, D. C., M. Calenoff, E. Rozhon, and H. L. Lipton. 1987. Analysis of the complete nucleotide sequence of the picornavirus Theiler's murine encephalomyelitis virus indicates that it is closely related to cardioviruses. $J.$ $Virol.$ 61:1507–1516.

Roos, R. P., W.-P. Kong, and B. L. Semler. 1989a. Polyprotein processing of Theiler's murine encephalomyelitis virus. $J.$ $Virol.$ 63:5344–5353.

Roos, R. P., S. Stein, Y. Ohara, J. Fu, and B. L. Semler. 1989b. Infectious cDNA clones of the DA strain of Theiler's murine encephalomyelitis virus. $J.$ $Virol.$ 63:5492–5496.

Roos, R. P., S. Stein, M. Routbort, A. Senkowski, T. Bodwell, and R. Wollmann. 1989c. Theiler's murine encephalomyelitis virus neutralization escape mutants have a change in disease phenotype. *J. Virol.* **63:**4469–4473.

Strebel, K., and E. Beck. 1986. A second protease of foot-and-mouth disease virus. *J. Virol.* **58:**893–899.

Toyoda, H., M. Nicklin, G. Murray, and E. Wimmer. 1986. A second virus-encoded proteinase involved in proteolytic processing of poliovirus polyprotein. *Cell* **45:**761–770.

IV. PROTEIN TRANSLATION, CLEAVAGE, AND MODIFICATION

Chapter 20

Molecular and Genetic Analysis of Events in Poliovirus Genome Replication

Quentin Reuer, Christopher U. T. Hellen, Sung-Key Jang, Christina Hölscher, Hans-Georg Kräusslich, and Eckard Wimmer

The *Picornaviridae* are a family of small icosahedral viruses that cause a number of important disease syndromes. The family is currently divided into four genera: *Rhinovirus* (the common cold virus), *Enterovirus* (e.g., poliovirus), *Aphthovirus* (foot-and-mouth disease virus), and *Cardiovirus* (e.g., mengovirus and encephalomyocarditis virus [EMCV]). The complete nucleotide sequence of the genomes of some 20 different picornaviruses have been determined, and the three-dimensional structures of members of all four genera have been solved at high resolution. *Picornaviridae* may therefore currently be the best-characterized virus family.

Poliovirus is the prototype picornavirus and has been characterized most extensively, resulting in many discoveries of fundamental biological importance. It was, for example, the first eucaryotic RNA virus for which the complete nucleotide sequence was determined (Kitamura et al., 1981). Poliovirus has a positive-sense single-stranded monopartite genome of 7,443 nucleotides (nt) comprising a genome-linked viral protein (VPg), a long 5' nontranslated region (NTR), a single long open reading frame (ORF), a 3' NTR, and a poly(A) tail. A number of advances have recently been made in understanding the significance of these properties for viral replication, and some of these are discussed below.

INITIATION OF PICORNAVIRUS PROTEIN SYNTHESIS

The 5' NTRs of members of the *Enterovirus* and *Rhinovirus* genera are unusually long, comprising almost 1/10 of the genomic RNA, and are well conserved despite the high error frequency of genome replication of RNA viruses. This suggests constant selection for a wild-type sequence; even point mutations in the 5' NTR can be deleterious. The functions of the 5' NTR are not clear, but minor base changes engineered into the 5' NTR of poliovirus have effects on viral replication (Racaniello and Meriam, 1986; Kuhn et al., 1988a; Kuhn et al., 1988b; Dildine and Semler, 1989), neurovirulence (Evans et al., 1985; for a review, see Nomoto and Wimmer [1987]), and translation (Bienkowska-Szewczyk and Ehrenfeld, 1988; Pelletier et al., 1988; Svitkin et al., 1985; Svitkin et al., 1988; Trono et al., 1988a; Trono et al., 1988b). It is likely that the 5' NTRs of picornaviral RNAs contain a variety of control elements involved in replication, translation, encapsidation, and possibly other viral functions.

Picornaviral mRNAs are uncapped and have 5' NTRs of up to 1,300 nt which contain many noninitiating AUGs, so it is not surprising that they have developed a mechanism for cap-independent translation (Pelletier et al., 1988; Trono et al., 1988b). Internal ribosomal entry for initiation of protein synthesis of EMCV was suggested indirectly by testing the inhibitory effects of cDNAs complementary to certain portions of the 5' NTR of EMCV. Hybridization of a cDNA fragment to the 5' end of the mRNA had little or no effect on translation from the initiation codon for the polyprotein of EMCV, whereas hybridization of a fragment to a downstream sequence within the putative internal

Quentin Reuer, Christopher U. T. Hellen, Sung-Key Jang, Christina Hölscher, and Eckard Wimmer, Department of Microbiology, State University of New York at Stony Brook, Stony Brook, New York 11794. *Hans-Georg Kräusslich*, Deutsches Krebsforschungszentrum, Institut für Virusforschung, Im Neunheimer Feld 280, D-69 Heidelberg 1, Federal Republic of Germany.

ribosomal entry site prevented translation from this site (Shih et al., 1987).

Internal ribosomal entry into picornaviral mRNAs has been demonstrated directly in vitro and in vivo by analyzing the expression of artificial polycistronic mRNAs containing two or more nonoverlapping ORFs (Jang et al., 1988; Jang et al., 1989; Pelletier and Sonenberg, 1988, 1989). Insertion of a portion of either poliovirus or EMCV 5' NTRs into the intercistronic region permitted efficient cap-independent translation of the adjacent downstream ORF even if it was hundreds of nucleotides from the 5' end of the RNA. However, translation of the second or third cistrons of polycistronic mRNAs was very poor if these ORFs were not preceded by picornaviral sequences or if specific deletions were made in the 5' NTRs of picornaviral mRNAs in the intercistronic regions.

cis-Acting elements for internal ribosomal entry have been identified by deletion analysis within the 5' NTRs of poliovirus and EMCV. The sequence nt 320 to 631 is required for cap-independent translation and internal ribosomal entry into poliovirus mRNAs in vivo and in vitro (Pelletier et al., 1988; Pelletier and Sonenberg, 1988; Trono et al., 1988b). Subsequently, Bienkowska-Szewczyk and Ehrenfeld (1988) showed that a poliovirus transcript lacking nt 1 to 567 was translated normally in vitro, but one lacking nt 517 to 627 was not translated. The authors concluded that nt 567 to 627 are necessary and sufficient for internal ribosomal entry. In contrast, a polioviral transcript deleted up to nt 670 was translated more efficiently in vitro than were full-length poliovirus transcripts (Nicklin et al., 1987). This result suggests that translation in vitro of modified monocistronic polioviral transcripts may occur by a different mechanism of ribosomal entry into the mRNA. The sequence between nt 564 and 726 was found not to be essential for viral replication (Kuge and Nomoto, 1987), but it appears that the segment nt 517 to 627 is all-important for translation of poliovirus RNA. This apparent discrepancy could be tested by engineering this sequence into a dicistronic mRNA and testing its effect on translation in vivo and in vitro.

By similar serial deletion experiments, a segment between nt 260 and 484 has been shown to be essential for translation of EMCV mRNA in vitro and in vivo (Jang et al., 1988; Jang et al., 1989). This location is consistent with the position (near nt 450) deduced from hybrid-arrested translation data (Shih et al., 1987). Mapping of the precise internal ribosomal entry sequence element in the EMCV 5' NTR is in progress.

REPLICATION OF POLIOVIRUS RNA: INVESTIGATION OF THE ROLE OF VPg BY CASSETTE MUTAGENESIS

The 5' end of poliovirus genomic RNA does not contain the 7-methylguanosine cap commonly found on eucaryotic RNA but instead is covalently bound to a basic, 22-amino-acid oligopeptide (VPg; Flanegan et al., 1977; Lee et al., 1977; Nomoto et al., 1977a; Pettersson et al., 1978) by a phosphodiester bond between the 5'-terminal residue of the RNA and O^4 of the tyrosine residue in position 3 of VPg (Ambros and Baltimore, 1978; Rothberg et al., 1978). VPg is also found at the 5' end of negative-sense RNA and nascent strands of replicative intermediates but is released from viral mRNA before translation by a cellular unlinking enzyme (Ambros and Baltimore, 1978; Nomoto et al., 1977a; Nomoto et al., 1977b; Pettersson et al., 1977; Rothberg et al., 1978). The process by which VPg or a VPg precursor is linked to RNA has not been identified. Biochemical studies indicate that VPg plays a critical albeit unknown role in replication of poliovirus RNA. Past research suggests that VPg serves as a primer of RNA synthesis by the viral RNA-dependent RNA polymerase (Kuhn and Wimmer, 1987; Semler et al., 1987). Indeed, VPg uridylylated in vitro to VPg-pUpU can be extended to longer RNA molecules. It has also been proposed that VPg serves as a signal for encapsidation of virion RNA into procapsids (Nomoto et al., 1977a); the presence or absence of 5'-terminal VPg may determine whether the RNA serves as virion RNA or becomes associated with the translational machinery of the host cell. A nucleolytic activity has also been assigned to VPg. In this model, VPg cleaves the RNA hairpin that links plus and minus strands of the replicative form (Flanegan et al., 1987).

The predicted importance of VPg in picornavirus replication renders it a suitable target for site-directed mutagenesis experiments. Several viable and nonviable poliovirus VPg mutants have been produced by using poliovirus type 1 (Mahoney) cDNAs containing a VPg mutagenesis cartridge (Kuhn et al., 1988a; Kuhn et al., 1988b). Generation of the unique restriction sites in the cartridge resulted in a lysine-to-arginine substitution at position 10 of VPg. Full-length poliovirus RNA transcribed from pT7-VPg-15, which encodes this substitution, yielded virus that has a multiplication rate similar to that of wild-type poliovirus type 1. We have constructed additional VPg mutants by using the cartridge. Mutant RNAs transcribed from full-length cDNAs containing altered VPg-encoding

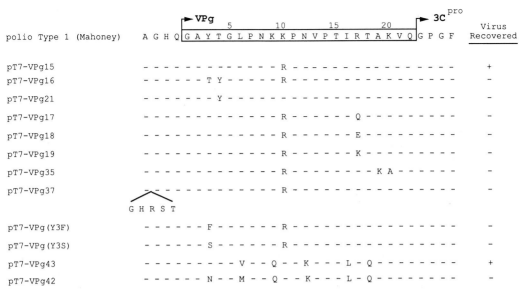

FIGURE 1. Schematic representation of poliovirus type 1 (Mahoney) and mutant VPgs. The amino acid sequence of wild-type VPg is given by the one-letter amino acid code within the boxed region. pT7-VPg15, the product of contruction of the mutagenesis cartridge, contains a lysine-to-arginine substitution at position 10 but displays multiplication properties similar to those of wild-type virus. Mutants are listed below the wild-type sequence, with their amino acid changes replacing the dashed lines. Construct pT7-VPg37 encodes a five-residue insertion, as indicated. Production or absence of infectious virus after transfection of HeLa cells with transcripts is indicated to the right.

regions were considered nonviable if no plaque-forming virus was recovered from transfected HeLa cells by 48 h posttransfection. These studies have allowed us to identify structural properties of VPg and its flanking regions required for maintenance of viability.

Nonviable VPg mutants and two chimeric constructs containing coxsackievirus VPg-encoding regions within the poliovirus genome are listed in Fig. 1. Tyrosine is maintained in position 3 of all picornavirus VPgs characterized to date. Moving the tyrosine to position 4 (VPg16) or addition of a second tyrosine at this position (VPg21) yielded nonviable mutants. Kuhn et al. (1988a) found that VPg16 displayed a pattern of proteolytic processing after in vitro translation that was similar to the wild-type pattern, suggesting that the mutant was nonviable because of a defect in 3AB activity. Cleavage of the VPg21 polypeptide is impaired at the glutamine-glycine cleavage site between protein 3A and VPg.

Replacement of the basic amino acids at positions 17 and 20 yielded nonviable mutants. All VPgs of picornaviruses for which sequences are available contain basic amino acids in approximately the same positions. These positively charged residues may interact with the negatively charged bases of the RNA. Surprisingly, the arginine-to-lysine change of VPg19 resulted in loss of viability. VPg19 did not generate capsid proteins after in vitro translation, but the processing pattern of this mutant was nearly identical to that of VPg20, a viable insertion mutant that contains an additional threonine residue between positions 18 and 19 of the VPg sequence. The protein-processing patterns of VPg17 and VPg18 appeared normal, suggesting that a defect in VPg function rather than impaired processing is responsible for the loss of viability. An alanine residue was moved from position 19 to 20 in VPg35, which was translated to yield a pattern of poliovirus proteins from which the usually prominent 3CD was absent. Glutamine-glycine cleavage sites recognized by viral protease 3C preferentially contain alanine in the +4 position (Nicklin et al., 1986). This may explain the lack of cleavage between 3B and 3C of VPg35, in which alanine has been moved to the +3 position.

We conducted hybridization experiments to determine whether nonviable RNAs were replicated in transfected HeLa cells. Cell lysates were harvested at various times posttransfection and probed with labeled polyribonucleotides specific for the 3' ends of plus- and minus-strand poliovirus RNA. All mutant RNAs that encoded a tyrosine residue in position 3 or 4 of VPg were RNA replication positive. Levels of RNA in transfected cells, as indicated by the amount of hybridized probe, approached or exceeded wild-type levels. VPg(Y3S), VPg(Y3F), and the chimera VPg42 were not replicated after transfection. VPg(Y3S) was of particular interest because the VPg of cowpea mosaic virus (a plant virus that is related to poliovirus) is linked to genomic RNA via a serine residue. The poliovirus RNA replication system does not tolerate serine in this role.

Labeling experiments were done to determine whether VPg was linked to viral RNAs recovered from cells transfected with the nonviable RNAs. Transfected cells were labeled to a high specific activity with ^{32}P, and lysates harvested 20 h posttransfection were phenol/chloroform extracted and treated with RNases A and T_1, which would cleave VPg-linked RNA to VPg-pUp. Immunoprecipitation with VPg-specific antiserum yielded labeled material whose mobility in sodium dodecyl sulfate-polyacrylamide gels was identical to that of VPg-pUp recovered from purified virions. Labeled material was not recovered upon immunoprecipitation with antiserum specific for 3A, indicating that nonviability of the mutants was not due to deficient processing of an RNA-linked VPg precursor.

Insertion of a coxsackievirus-specific VPg-encoding region within the poliovirus genome yielded the viable chimera VPg43. This virus produced smaller plaques than did wild-type poliovirus type 1, and its multiplication in one-step growth experiments was reduced. The defective multiplication may be due to impaired interaction between the poliovirus proteins and the heterologous VPg.

The results presented above lead us to conclude that a linkage-competent VPg, i.e., a VPg with a tyrosine residue in position 3 or 4, is necessary if poliovirus RNA synthesis is to occur. The presence of a tyrosine residue in the N-terminal region may allow uridylylation and concomitant priming of viral RNA synthesis. That VPg mutants can be RNA replication positive and still nonviable indicates that a step in poliovirus replication other than initiation of RNA synthesis can be interrupted by these lethal lesions. This step might be encapsidation, since VPg could serve as a signal labeling the RNA for packaging in procapsids. It should be stressed that none of the phenotypes of mutants generated thus far have allowed us to identify the actual function(s) of VPg.

PROTEOLYTIC PROCESSING OF THE POLIOVIRUS POLYPROTEIN

All picornaviral proteins are generated by proteolytic processing of a single polyprotein that is expressed from the genomic RNA. Amino acid sequence determination of the termini of poliovirus proteins established that they are derived by cleavage between QG, YG, and NS dipeptides. All but one of the processing steps are catalyzed by virus-encoded proteinases (2A^{pro}, 3C^{pro}, and 3CD^{pro}), which are themselves part of the polyprotein (Kräusslich and Wimmer, 1988). Processing of the poliovirus polyprotein occurs as a cascade of cleavage reactions, the order and rate of which are determined initially by the order of appearance of the proteinases and subsequently by their affinity for the different cleavage sites. The efficiency of cleavage at different sites varies considerably even though the residues on either side of the scissile bond are strictly conserved; flanking residues probably provide additional determinants of cleavage site recognition.

The initial cleavage reaction is catalyzed by 2A^{pro} at its own amino terminus and occurs cotranslationally, probably in *cis*. It separates capsid and nonstructural proteins and is a prerequisite for processing of the capsid precursor (P1) to yield capsomer proteins. A second cleavage site that is recognized by 2A^{pro} lies within the viral 3D polymerase sequence; cleavage leads to the appearance of the proteins 3C' and 3D' (Lee and Wimmer, 1988). 2A^{pro} catalyzes cleavage of 2 of the 10 YG pairs within the polyprotein (Toyoda et al., 1986); both sites are preceded by a threonine residue and have a leucine residue at position P4. Site-specific mutations have been made at the VP1-2A cleavage site to investigate the importance of these and other determinants of cleavage site recognition. All but one of the other cleavages within the polyprotein occur between QG dipeptides and are catalyzed by 3C^{pro} and 3CD^{pro}.

The poliovirus capsid is composed of essentially equimolar amounts of four nonidentical proteins (VP1 to VP4) that are synthesized as a structural precursor (P1). Cleavage of P1 to VP0, VP3, and VP1 is catalyzed by 3CD^{pro}; subsequent cleavage of VP0 to VP4 and VP2 during maturation probably occurs by an autocatalytic process involving viral RNA. Polypeptides P1, VP0, and VP4 have common amino acid termini, and their N-terminal glycine residue is blocked by covalently linked myristic acid (Paul et al., 1987). We have investigated the role of myristoylation in the poliovirus life cycle by using site-directed mutagenesis to generate a mutated capsid protein sequence that lacks an essential residue of the myristoylation signal.

For analysis of the VP1-2A site, a vector was constructed that contains poliovirus type 1 (Mahoney) cDNA encoding two amino acids of VP3, all of VP1, 2A and 2B, and half of 2C (nt 2474 to 4600) placed under control of part of the EMCV 5′ NTR (nt 260 to 841). Transcripts produced in vitro from linearized plasmids using T7 polymerase were translated in rabbit reticulocyte lysate to yield a polyprotein that was normally cleaved autocatalytically to yield ΔVP3-VP1 and 2ABΔ2C products. A stop codon was introduced immediately after the last amino acid of 2A^{pro}, and further single mutations were then introduced into the VP1-2A cleavage site by oligonucleotide-directed site-specific mutagenesis by using single-stranded DNA templates that had been enriched in uracil by passage in *Escherichia coli* BW313 (*dut ung*) (Kunkel, 1985). RNA transcripts were translated in rabbit reticulocyte lysate, and products were analyzed by immunoprecipitation and polyacrylamide gel electrophoresis. The wild-type precursor was cleaved to completion, yielding the expected ΔVP3-VP1 and 2A products, as were precursors containing proline and glutamine at the P2 position, isoleucine and arginine at the P1 position, and alanine at the P1′ position (Fig. 2). All other mutations reduced the efficiency of cleavage at the VP1-2A site, resulting in accumulation of the ΔVP3-VP1-2A precursor, although in no instance to more than 20% of the total. However, these mutated precursors were processed in an aberrant manner, yielding reduced levels of ΔVP3-VP1 and 2A and varying amounts of two larger proteins (19.5 and 30 kilodaltons) that were antigenically related to 2A and of two smaller proteins (10 and 21 kilodaltons) that were antigenically related to VP1. This unexpected cleavage at two internal sites in VP1 was most marked in precursor containing proline at the P1 position; less extreme patterns of aberrant cleavage were apparent on translation of precursors containing cysteine, threonine, and leucine substitutions at the P1′ position and isoleucine, leucine, glutamate, and arginine substitutions at the P2 position.

The effects of substitution at the P2, P1, and P1′ positions of the VP1-2A cleavage site indicate that specific residues are not stringent determinants of recognition of this substrate by 2A^{pro}. Residues that line the predicted substrate binding pocket of 2A^{pro} are relatively small (Bazan and Fletterick, 1988), and this may explain the ability of the enzyme both to accommodate a tyrosine residue and to tolerate a

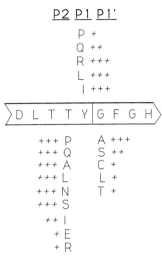

FIGURE 2. Summary of the effects of amino acid substitutions at the VP1-2A cleavage site on the efficiency of autocatalytic processing. Poliovirus VP1-2A precursors containing altered P2, P1, or P1′ residues were tested as substrates in cell-free assays. Substitutions were scored +++ to −−− for their effects on autocatalytic processing. Italic type indicates that the amino acid substitution resulted in aberrant processing of the precursor. The sequence of the natural cleavage is superimposed on the dotted background.

variety of substitutions at the P1 position. The correct cleavage site is specified both by its amino acid sequence and consequent local structure and by its position with respect to that of the active site. Perturbation of the VP1-2A junction decreases the rate of cleavage and consequently increases the half-life of precursors, resulting in an opportunity for cleavage to occur at less favored sites. Possible sites, including a TYG sequence, have been identified in exposed loops of VP1.

A similar strategy was used to mutate the amino-terminal residue of VP4 from Gly to Ala so that the role of myristoylation of this residue in the viral life cycle could be investigated. A fragment containing the mutated amino terminus of the poliovirus polyprotein was cloned into the full-length poliovirus transcription vector pT7XL derived from plasmid pT7 PV1-5 (van der Werf et al., 1986) and into the transcription vector pMN22, which encodes the P1 capsid precursor (Nicklin et al., 1988). Wild-type and mutant plasmids were transcribed in vitro with T7 RNA polymerase, and the synthetic RNA was then used in transfection and translation assays as appropriate. Transfection of wild-type XL RNA into HeLa cells resulted in complete cytopathic effect within 24 to 48 h, whereas no cytopathic effect was observed on transfection with mutant XL RNA or on mock transfection. Hybridization experiments done using probes as described above indicated that replication of both positive- and negative-strand mutant XL RNA did occur but at a level that was considerably lower than in wild-type XL RNA-transfected cells. A defect was apparent in processing of the mutated P1 precursor in vitro. P1 precursor derived from wild-type RNA was processed normally on incubation with extracts from poliovirus-infected cells, whereas cleavage of the P1 precursor containing the Gly-Ala change at the N terminus occurred primarily at the VP3-VP1 site. Cleavage at the VP0-VP3 site was completely resistant to incubation with infected cell extracts. This pattern of processing is reminiscent of cleavage of the P1 precursor by purified 3C^{pro} (Nicklin et al., 1987). The P1 precursor is normally processed by 3CD^{pro}, the immediate precursor of 3C^{pro}, and the viral polymerase 3D^{pol} (Jore et al., 1988; Ypma-Wong et al., 1988). This cleavage is sensitive to nonionic detergent (Nicklin et al., 1988), which led to the suggestion that processing of P1 requires a hydrophobic interaction between the P1 precursor and the 3D domain of 3CD^{pro}. Since myristoylation of the capsid precursor is required for efficient processing by 3CD^{pro}, it is likely that this hydrophobic interaction involves the myristic acid moiety.

ACKNOWLEDGMENTS. This work was supported in part by Public Health Service grants AI 15122 and CA 28146 from the National Institutes of Health to E.W. and by a grant from Boehringer Ingelheim Pharmaceuticals Inc. C. Hölscher was a visiting student from the University of Konstanz (Konstanz, Federal Republic of Germany) supported by Deutscher Akademischer Austauschdienst.

REFERENCES

Ambros, V., and D. Baltimore. 1978. Protein is linked to the 5' end of the poliovirus RNA by a phosphodiester linkage to tyrosine. *J. Biol. Chem.* **253:** 5263–5266.

Bazan, J. F., and R. J. Fletterick. 1988. Viral cysteine proteinases are homologous to the trypsin-like family of serine proteases: structural and functional implications. *Proc. Natl. Acad. Sci. USA* **85:** 7872–7876.

Bienkowska-Szewczyk, K., and E. Ehrenfeld. 1988. An internal 5' noncoding region required for translation of poliovirus RNA in vitro. *J. Virol.* **62:** 3068–3072.

Dildine, S. L., and B. L. Semler. 1989. The deletion of 41 proximal nucleotides reverts a poliovirus mutant containing a temperature-sensitive lesion in the 5' noncoding region of genomic RNA. *J. Virol.* **63:** 847–862.

Evans, D. M. A., G. Dunn, P. D. Minor, G. C. Schild, A. J. Cann, G. Stanway, J. W. Almond, J. W. Currey, and J. V. Maizel, Jr. 1985. Increased neurovirulence associated with a single nucleotide change in a noncoding region of the Sabin type 3 poliovaccine genome. *Nature* (London) **314:** 548–550.

Flanegan, J. B., R. F. Pettersson, V. Ambros, M. J. Hewlett, and D. Baltimore. 1977. Covalent linkage of a protein to a defined nucleotide sequence at the 5'-terminus of virion and replicative intermediate RNAs of poliovirus. *Proc. Natl. Acad. Sci. USA* **74:** 961–965.

Flanegan, J. B., D. C. Young, G. J. Tobin, M. M. Stokes, C. D. Murphy, and S. M. Obertst. 1987. Mechanism of RNA replication by the poliovirus RNA polymerase, HeLa cell host factor, and VPg, p. 273–284. *In* M. A. Brinton and R. R. Rueckert (ed.), *Positive Strand Viruses.* Alan R. Liss, Inc., New York.

Jang, S. K., M. V. Davies, R. J. Kaufman, and E. Wimmer. 1989. Initiation of protein synthesis by internal entry of ribosomes into the 5' nontranslated region of encephalomyocarditis virus RNA in vivo. *J. Virol.* **63:** 1651–1660.

Jang, S. K., H.-G. Kräusslich, M. J. H. Nicklin, G. M. Duke, A. C. Palmenberg, and E. Wimmer. 1988. A segment of the 5' nontranslated region of encephalomyocarditis virus RNA directs internal entry of ribosomes during in vitro translation. *J. Virol.* **62:** 2636–2643.

Jore, J., B. de Geus, R. J. Jackson, P. H. Pouwels, and B. E. Enger-Valk. 1988. Poliovirus protein 3CD is the active protease for processing of the precursor protein P1 *in vitro. J. Gen. Virol.* **149:** 1627–1636.

Kitamura, N., B. L. Semler, P. G. Rothberg, G. R. Larsen, C. J. Adler, A. J. Dorner, E. A. Emini, R. Hanecak, J. J. Lee, S. van der Werf, C. J. Anderson, and E. Wimmer. 1981. Primary structure, gene organization and polypeptide expression of poliovirus RNA. *Nature* (London) **291:** 547–553.

Kräusslich, H.-G., and E. Wimmer. 1988. Viral proteinases. *Annu. Rev. Biochem.* **57:** 701–754.

Kuge, S., and A. Nomoto. 1987. Construction of viable deletion and insertion mutants of the Sabin strain type 1 poliovirus: function of the 5' noncoding sequence in viral replication. *J. Virol.* **61:** 1478–1487.

Kuhn, R. J., H. Tada, M.-F. Ypma-Wong, J. J. Dunn, B. L. Semler, and E. Wimmer. 1988a. Construction of a "mutagenesis cartridge" for poliovirus genome-linked viral protein: isolation and characterization of viable and nonviable mutants. *Proc. Natl. Acad. Sci. USA* **85:** 519–523.

Kuhn, R. J., H. Tada, M.-F. Ypma-Wong, B. L. Semler, and E. Wimmer. 1988b. Mutational analysis of the genome-linked protein VPg of poliovirus. *J. Virol.* **62:** 4207–4215.

Kuhn, R. J., and E. Wimmer. 1987. The replication of picornaviruses, p. 17–51. *In* D. J. Rowlands, B. W. J. Mahy, and M. Mayo (ed.), *The Molecular Biology of Positive Strand RNA Viruses.* Academic Press, Inc., New York.

Kunkel, T. 1985. Rapid and efficient site-specific mutagenesis without phenotypic selection. *Proc. Natl. Acad. Sci. USA* **82**:488–492.

Lee, C.-K., and E. Wimmer. 1988. Proteolytic processing of poliovirus polyprotein: elimination of 2A^pro-mediated alternative cleavage of polypeptide 3CD by *in vitro* mutagenesis. *Virology* **166**:405–414.

Lee, Y. F., A. Nomoto, B. M. Detjen, and E. Wimmer. 1977. The genome-linked protein of picornaviruses. I. A protein covalently linked to poliovirus genome RNA. *Proc. Natl. Acad. Sci. USA* **74**:59–63.

Nicklin, M. J. H., K. S. Harris, P. V. Pallai, and E. Wimmer. 1988. Poliovirus proteinase 3C: large scale expression, purification, and specific cleavage activity on natural and synthetic substrates in vitro. *J. Virol.* **62**:4586–4593.

Nicklin, M. J. H., H.-G. Kräusslich, H. Toyoda, J. J. Dunn, and E. Wimmer. 1987. Poliovirus polypeptide precursors. Expression in vitro and processing by 3C and 2A proteinases. *Proc. Natl. Acad. Sci. USA* **84**:4002–4006.

Nicklin, M. J. H., H. Toyoda, M. G. Murray, and E. Wimmer. 1986. Proteolytic processing in the replication of polio and related viruses. Bio/Technology **4**:33–42.

Nomoto, A., B. M. Detjen, R. Pozzatti, and E. Wimmer. 1977a. The location of the polio genome protein in viral RNA and its implications for RNA synthesis. *Nature* (London) **268**:208–213.

Nomoto, A., N. Kitamura, F. Golini, and E. Wimmer. 1977b. The 5-terminal structures of poliovirion RNA and poliovirus mRNA differ only in the genome-linked protein VPg. *Proc. Natl. Acad. Sci. USA* **74**:5346–5349.

Nomoto, A., and E. Wimmer. 1987. Genetic studies of the antigenicity and the attenuation phenotype of poliovirus. *Symp. Soc. Gen. Microbiol.* **40**:107–134.

Paul, A. V., A. Schultz, S. E. Pincus, S. Oroszlan, and E. Wimmer. 1987. Capsid protein VP4 of poliovirus is N-myristoylated. *Proc. Natl. Acad. Sci. USA* **84**:7827–7831.

Pelletier, J., G. Kaplan, V. R. Racaniello, and N. Sonenberg. 1988. Cap-independent translation of poliovirus mRNA is conferred by sequence elements within the 5' noncoding region. *Mol. Cell. Biol.* **8**:1103–1112.

Pelletier, J., and N. Sonenberg. 1988. Internal binding of ribosomes to the 5' noncoding region of a eukaryotic mRNA: translation of poliovirus. *Nature* (London) **334**:320–325.

Pelletier, J., and N. Sonenberg. 1989. Internal binding of eucaryotic ribosomes on poliovirus RNA: translation in HeLa cell extracts. *J. Virol.* **63**:441–444.

Pettersson, R. F., V. Ambros, and D. Baltimore. 1978. Identification of a protein linked to nascent poliovirus RNA and to the polyuridylic acid of negative-strand RNA. *J. Virol.* **27**:357–365.

Pettersson, R. F., J. B. Flanegan, J. K. Rose, and D. Baltimore. 1977. 5'-Terminal nucleotide sequence of poliovirus polyribosomal RNA and virion RNA are identical. *Nature* (London) **268**:270–272.

Racaniello, V. R., and C. Meriam. 1986. Poliovirus-temperature sensitive mutants containing a single nucleotide deletion in the 5'-noncoding region of the viral RNA. *Virology* **155**:498–507.

Rothberg, P. G., T. J. R. Harris, A. Nomoto, and E. Wimmer. 1978. O^4-(5'-uridyl) tyrosine is the bond between the genome-linked protein and the RNA of poliovirus. *Proc. Natl. Acad. Sci. USA* **75**:4868–4872.

Semler, B. L., R. J. Kuhn, and E. Wimmer. 1987. Replication of the poliovirus genome, p. 23–48. *In* E. Domingo, J. J. Holland, and P. Ahlquist (ed.), *RNA Genetics.* CRC Press, Inc., Boca Raton, Fla.

Shih, D. S., I.-W. Park, C. L. Evans, J. M. Jaynes, and A. C. Palmenberg. 1987. Effects of cDNA hybridization on translation of encephalomyocarditis virus RNA *J. Virol.* **61**:2033–2037.

Svitkin, Y. V., S. V. Maslova, and V. I. Agol. 1985. The genomes of attenuated virulent poliovirus strains differ in their in vitro translation efficiencies. *Virology* **147**:243–252.

Svitkin, Y. V., T. V. Pestova, S. V. Malsova, and V. I. Agol. 1988. Point mutations modify the response of poliovirus RNA to a translation initiation factor: a comparison of neurovirulent and attenuated strains. *Virology* **166**:394–404.

Toyoda, H., M. J. H. Nicklin, M. G. Murray, C. W. Anderson, J. J. Dunn, F. W. Studier, and E. Wimmer. 1986. A second virus-encoded protease involved in proteolytic processing of poliovirus polyprotein. *Cell* **45**:761–770.

Trono, D., R. Andino, and D. Baltimore. 1988a. An RNA sequence of hundreds of nucleotides at the 5' end of poliovirus RNA is involved in allowing viral protein synthesis. *J. Virol.* **62**:2291–2299.

Trono, D., J. Pelletier, N. Sonenberg, and D. Baltimore. 1988b. Translation in mammalian cells of a gene linked to the poliovirus 5' noncoding region. *Science* **241**:445–448.

van der Werf, S., J. Bradley, E. Wimmer, F. W. Studier, and J. J. Dunn. 1986. Synthesis of infectious poliovirus RNA by purified T7 RNA polymerase. *Proc. Natl. Acad. Sci. USA* **83**:2330–2334.

Ypma-Wong, M. F., P. G. Dewalt, V. H. Johnson, J. G. Lamb, and B. L. Semler. 1988. Protein 3CD is the major proteinase responsible for cleavage of the P1 capsid precursor. *Virology* **166**:265–270.

Chapter 21

Unique Features of Initiation of Picornavirus RNA Translation

Michael T. Howell, Ann Kaminski, and Richard J. Jackson

It has long been suspected that initiation on picornavirus RNAs might be atypical. The extremely long 5' untranslated region (UTR) characteristic of these RNAs, with numerous apparently silent AUG codons and extensive secondary structure (Kitamura et al., 1981; Palmenberg et al., 1984; Pilipenko et al., 1989; Rivera et al., 1988; Skinner et al., 1989; Vartapetian et al., 1983), seemed difficult to accommodate in the scanning ribosome model (Kozak, 1989) for initiation site selection. In addition, the in vitro translation characteristics of these RNAs show many unusual features. Encephalomyocarditis virus (EMC) RNA is translated in the rabbit reticulocyte lysate system with exceptionally high efficiency and an unusually high salt optimum for an uncapped RNA (Jackson, 1989). It is also unique in being much more efficient in the presence of KCl rather than potassium acetate and in being activated by potassium thiocyanate (KSCN) (Jackson, 1989). In addition, there is some evidence that initiation on this RNA has an unusually low dependence on ATP (Jackson, 1982, 1989). Poliovirus RNA, by contrast, is translated in the reticulocyte lysate system with low efficiency and low fidelity (Dorner et al., 1984; Phillips and Emmert, 1986). However, in extracts from HeLa or L cells, translation is both efficient and accurate. When factors isolatable from HeLa or L cells are added to the reticulocyte lysate, the synthesis of authentic poliovirus products is stimulated and that of aberrant products, which arise by initiation mainly in the 3' portion of the RNA, is suppressed (Dorner et al., 1984; Phillips and Emmert, 1986; Jackson, 1989).

The definitive experiment showing unique initiation characteristics was the discovery that insertion of the 5' UTR (of either EMC or poliovirus RNA) between the two cistrons of a dicistronic mRNA allowed the downstream cistron to be translated independently of the upstream: the protein product of the downstream cistron appeared in higher yield and earlier than the upstream (Jackson, 1988; Jang et al., 1988; Pelletier and Sonenberg, 1988). This finding argues that the picornavirus 5' UTR must contain an internal entry site for ribosomes, with the result that ribosomes can translate the downstream cistron without having previously translated the upstream. This approach of using dicistronic mRNAs remains the most decisive test for internal initiation. The other criteria, such as showing that initiation is unaffected by whether the RNA is capped or uncapped (Pelletier et al., 1988a, 1988b), are less incisive and open to ambiguities of interpretation.

A surprisingly large length of the picornaviral 5' UTR is necessary for this internal initiation process: the required sequence of EMC RNA extends from about nucleotides (nt) 260 to 834 (Jang et al., 1988), whereas that of poliovirus RNA is from about nt 140 to 630 (Pelletier and Sonenberg, 1988). This finding has led to suggestions that these RNA segments possess an internal entry site to which ribosomes bind in the first stage and then select the authentic initiation codon by a process of scanning from the internal entry site (model 1 in Fig. 1), a model that has the attraction of showing some features in common with the standard scanning ribosome model. However, there is a problem with this model of scanning from an internal entry site in that picornaviral 5' UTRs have many apparently silent AUG codons (Kitamura et al., 1981; Palmenberg et al., 1984), and one may ask why the

---

Michael T. Howell, Ann Kaminski, and Richard J. Jackson, Department of Biochemistry, University of Cambridge, Cambridge CB2 1QW, United Kingdom.

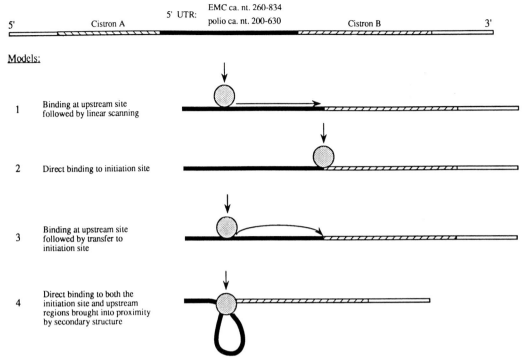

FIGURE 1. Models for internal initiation on picornavirus 5' UTR sequences. The upper line depicts the dicistronic constructs used to test for internal initiation. Symbols: ▨ , coding sequence; ▬ , picornaviral 5' UTR sequence. Below are shown the four models for recognition of the initiation site by 40S ribosomal subunits (see text).

scanning process should overlook these. This chapter describes our experiments which address this question with respect to both EMC and poliovirus RNA.

THE AUTHENTIC INITIATION SITE OF EMC RNA IS NOT SELECTED BY A SCANNING MECHANISM

The disposition of the AUG codons in the 5'-proximal portion of EMC RNA makes this a particularly suitable system to test whether ribosomes approach the initiation site by a scanning mechanism (Palmenberg et al., 1984). The authentic initiation codon, the 11th AUG (AUG-11) from the 5' end of the viral RNA at nt 834, is only a short distance downstream from AUG-10 but in a different reading frame (Fig. 2). As the local (Kozak) nucleotide sequence context of AUG-10 does not seem to be particularly unfavorable, a scanning ribosome should initiate at AUG-10 in preference to AUG-11.

This question was studied by using rabbit reticulocyte lysates to translate RNAs produced by in vitro transcription from constructs in which the EMC 5' UTR is fused to a reporter gene, the poliovirus 2A-coding sequence (Jang et al., 1988). With these transcripts, initiation at the authentic site on the EMC sequence, AUG-11, should yield a protein of 155 amino acid residues, of which the N-terminal five residues are encoded by the EMC sequences and the rest by the poliovirus 2A coding sequence. In contrast, initiation at the upstream out-of-frame AUG-10 would give only a short polypeptide chain (Fig. 2). To detect products of initiation at AUG-10, a frameshift was introduced about 60 nt residues downstream by opening the plasmid at a unique *Bgl*II site, end filling with DNA polymerase Klenow fragment, and religating. The result of this frameshift is that initiation at AUG-10 now results in the synthesis of a protein of 159 amino acid residues, essentially the 2A polypeptide with a few extra residues, whereas initiation at AUG-11 would give only a short product. Deletions extending for various distances from the 5' end of the EMC sequence were made in these constructs as shown in Fig.

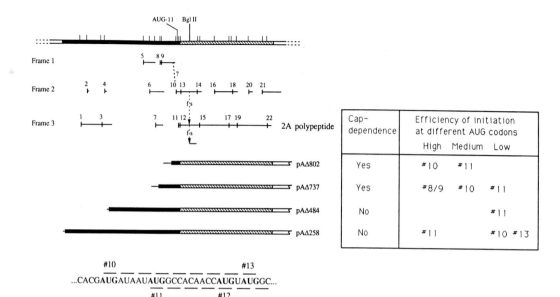

FIGURE 2. Schematic of the plasmid constructs used to test for ribosomal scanning of EMC 5′ UTR sequences. The upper line represents the relevant segment of the parent construct, pAΔ258. Symbols: ■, EMC sequence; ▨, poliovirus 2A-coding sequence; vertical bars, AUG codon. The next three lines show the open reading frames present in each of the three reading frames, with AUG codons denoted by vertical bars and numbered consecutively from the 5′ end. The broken lines designated fs show the switch in reading frames caused by fill-in of the BglII site. Below this are shown the three deletion mutants studied (pAΔ484, pAΔ737, and pAΔ802) in addition to the parent pAΔ258; the EMC 5′ UTR and poliovirus 2A-coding sequences are depicted as above, and transcribed polylinker sequences are symbolized by single lines. The nomenclature of these deletion derivatives is based on the number of residues of the EMC RNA sequence deleted (from the 5′ end of the virion RNA sequences); the authentic initiation site, AUG-11, is at nt 834 of the virion RNA sequence. The table summarizes the results of translation assays similar to those shown in Fig. 3 but using both capped and uncapped transcripts at several different concentrations. Note that although the published sequence places AUG-8 and AUG-9 in a different reading frame from AUG-10 (Palmenberg et al., 1984), sequencing of all of our clones has shown the presence of an additional A residue between AUG-9 and AUG-10, which brings AUG-8 and AUG-9 into the same unbroken reading frame as AUG-10, consistent with the results of translation assays shown in Fig. 3.

2; the nomenclature of the constructs is according to the number of nucleotides deleted from the 5′ end of the EMC viral RNA sequence, and the suffix "fs" denotes the clones with the frameshift. Transcripts of all of these constructs were translated in both capped and uncapped forms, each at four different RNA concentrations. Figure 3 shows the gel electrophoretic analysis of the products synthesized from capped transcripts at one particular (subsaturating) RNA concentration. The 5 to 30% gradient gel resolves the 159-amino-acid residue translation product initiated at AUG-10 of the clones that have the frameshift from the 155-residue protein initiated at AUG-11 of the constructs lacking the frameshift (Fig. 3).

With the shortest length of EMC 5′ UTR

tested, initiation at AUG-10 of the construct with the frameshift was rather more efficient than initiation at AUG-11 of the clone lacking the frameshift (Fig. 3). With a longer length of EMC 5′ UTR that included AUG-8 and AUG-9, efficient initiation occurred at one or both of these AUG codons (as they are only 9 nt apart, we cannot be certain that the products of initiation at these two codons would be resolved) and at AUG-10 but was very inefficient at AUG-11. The translation of all of these RNAs was strongly stimulated (more than sevenfold) by capping the RNA. Thus, the translation of these RNAs would appear to follow the usual scanning mechanism of ribosome binding first to the capped 5′ end, followed by linear scanning up to the 5′-proximal AUG in a favorable sequence

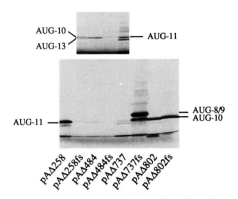

FIGURE 3. Products of translation of the deletion and frameshift derivatives comprising EMC 5' UTR sequences linked to the poliovirus 2A-coding sequence. Capped RNA transcripts were translated at a concentration of 8 μg/ml for 60 min. Samples corresponding to 1 μl of translation assay mix were separated by electrophoresis on a 5 to 30% polyacrylamide gradient gel, which was dried and fluorographed. The top panel shows a threefold-longer exposure than the bottom panel. Products are designated according to the initiating AUG codon as shown in Fig. 2, which also depicts the plasmid constructs. The suffix "fs" denotes constructs in which a frameshift was introduced by fill-in of the *Bgl*II site.

context (Kozak, 1989). Since the contexts of AUG-8 and AUG-9 are not particularly favorable, some of the scanning 40S ribosomes would be expected to bypass these codons and initiate at AUG-10. The important point is that with all of these RNAs that appear to be translated by the conventional scanning mechanism, AUG-10 is used with high efficiency and in preference to AUG-11.

Transcripts of the pAΔ484 clones, which have a longer segment of EMC 5' UTR, were translated very inefficiently indeed, regardless of whether the RNA was capped or whether the frameshift was present (Fig. 3). This particular segment of EMC 5' UTR has been shown to be incapable of promoting independent translation of the downstream cistron when it is inserted in the intercistronic region of a dicistronic mRNA (Jang et al., 1988). Therefore, it does not include the complete EMC 5' UTR sequences necessary to constitute an internal entry site for ribosomes.

When our constructs included the complete EMC 5' UTR segment previously found to be capable of promoting internal entry by ribosomes (Jang et al., 1988), we observed very efficient translation initiated at AUG-11 (Fig. 3), which was uninfluenced by capping the transcripts. In contrast, initiation at AUG-10 of the

derivative that included the frameshift was very inefficient, at about 1% relative to the utilization of AUG-11. As AUG-10 was efficiently utilized in the shorter constructs which were translated with all of the characteristics of the scanning ribosome model, the low frequency of initiation at AUG-10 (contrasted with a high efficiency at AUG-11) in the case of the construct with the longest EMC 5' UTR segment implies that AUG-11 cannot be selected by a process which involves ribosomes scanning over the region between AUG-8 and AUG-10.

These experiments would therefore seem to eliminate model 1 of Fig. 1, in which ribosomes bind first to an upstream entry site and then scan the UTR sequence until they reach the correct AUG codon for initiation. Of the alternative models, the second, involving direct binding of the ribosome to the authentic initiation site, satisfies the criteria of no scanning of the upstream sequences but does not provide an adequate explanation of why such a large segment of the EMC 5' UTR is necessary for the internal entry mechanism, a fact which strongly suggests that ribosomes must contact the far upstream sequences at some stage of the initiation process. The third model attempts to reconcile the notion of a far upstream binding site with the absence of scanning by invoking a nonscanning transfer of the ribosome from the upstream entry site to AUG-11. Although this model is consistent with the available facts, the idea of a transfer step is hard to imagine unless the folding of the EMC 5' UTR brought the upstream entry site close to AUG-11 so that the transfer would be over a very short distance. These considerations lead logically to model 4, in which the ribosome binds directly to a site consisting of far upstream sequences brought into close proximity to AUG-11 by the folding of the EMC 5' UTR. It is this model which we currently favor and are testing by looking for evidence that a ribosome bound at the initiation site simultaneously contacts far upstream sequences.

ROLE OF SCANNING IN INITIATION OF POLIOVIRUS RNA TRANSLATION

In contrast to the very precise mechanism which selects AUG-11 and rejects the nearby AUG-10 for initiation of translation of EMC RNA, initiation on poliovirus RNA seems much more flexible in that it appears to select the first AUG codon downstream of about nt 600 of the viral RNA sequence, regardless of the actual position of this AUG or the sequences (other than the immediate local context) around it.

Although the authentic initiation site of poliovirus type 1 is AUG-9 at nt 743 (Kitamura et al., 1981), a deletion extending from nt 600 to 727 has been shown to have no effect on viral infectivity, and a deletion from nt 564 to 727 gave a small-plaque phenotype (Kuge and Nomoto, 1987). In addition, deletions from nt 631 to 733 of constructs containing the 5' UTR of poliovirus type 2 had no effect on the in vitro translation characteristics (Pelletier et al., 1988a, 1988b) and did not affect the ability of the poliovirus 5' UTR to promote independent initiation of the downstream cistron of a dicistronic mRNA (Pelletier and Sonenberg, 1988). Furthermore, despite the strong similarity between the 5' UTR sequences of the polioviruses and human rhinoviruses, the segment corresponding approximately to nt 610 to 742 of poliovirus RNA is missing from the rhinovirus sequences (Rivera et al., 1988). It was also shown that insertion of 72 nt residues into the poliovirus 5' UTR at nt 702 had no effect on infectivity if the inserted sequence lacked AUG codons but produced a small-plaque phenotype when the insertion included an AUG codon with a poor local (Kozak) context (Kuge et al., 1989).

These facts point to a mechanism of initiation involving ribosome scanning, if not from the 5' end of the RNA then from a site situated upstream of about nt 600. To assess the role of scanning in poliovirus RNA translation, we have inserted a reporter gene, the influenza virus NS cDNA, at various points in the UTR of a construct comprising the entire 5' UTR and P1-coding segment of poliovirus type 1. In vitro transcripts of these clones were translated in both capped and uncapped forms in the rabbit reticulocyte lysate system, and synthesis of the influenza NS and poliovirus P1 polypeptides was assayed by using gel electrophoresis. In addition, we examined the effect of supplementing the lysate with the HeLa cell factors (partially purified), which are known to enhance the otherwise low fidelity and efficiency of poliovirus RNA translation in reticulocyte lysates (Dorner et al., 1984; Phillips and Emmert, 1986; Jackson, 1989). The constructs and the results are summarized in Fig. 4.

When the NS cDNA was inserted, with its own initiation codon, at the 5' end of the poliovirus UTR or near the 5' end at nt 66, it was translated efficiently and in what appeared to be a cap-dependent manner, since capping greatly stimulated NS polypeptide synthesis. The synthesis of poliovirus P1, which is very inefficient in the reticulocyte lysate system, was unaffected by the insertion (in either orientation) of the NS

cDNA at either of these two sites, nor was it influenced by capping the transcripts. On the other hand, addition of the HeLa factors stimulated P1 synthesis while inhibiting, perhaps by nonspecific action, the translation of the NS cistron. These results imply a mechanism of scanning from the 5' end for initiation of NS synthesis but a different mechanism for initiation of P1 synthesis.

The NS cDNA sequence was also inserted at nt 392 without its own initiation codon but in frame with the AUG codons at nt 378 and 390 of the poliovirus 5' UTR. The latter of these is in a favorable local Kozak context (. . . GCC AUGG) and of all the AUG codons in the poliovirus 5' UTR is the one with a context most likely to be recognized efficiently by scanning ribosomes, which in the case of our construct would lead to NS synthesis. In fact, our experiments revealed only a low level of NS synthesis, which was uninfluenced by capping and was not stimulated by addition of HeLa factors. This result suggests either that scanning from the 5' end of the poliovirus UTR does not penetrate effectively as far as nt 390 or that the scanning ribosome bypassed the AUG codon at nt 390 for some unknown reason. We believe that the latter explanation can be eliminated because we did not see any of the smaller products of translation of the NS cistron which our other studies of this gene have shown to be characteristic products formed when the authentic initiation site is missing or is inefficiently recognized by scanning ribosomes.

When the NS cistron was inserted (with its own initiation codon) further downstream, at nt 630 of the poliovirus 5' UTR, it was translated very efficiently, but in contrast to what was seen with insertions at the other sites, this was not influenced by capping but was stimulated by HeLa factors. A further point, unique to insertion at this site, was that the insertion greatly reduced the synthesis of poliovirus P1 but, curiously, did not abolish it completely.

We interpret these observations as showing, first, that ribosomes scanning from the 5' end of the poliovirus UTR cannot penetrate efficiently as far as nt 390, most probably because of secondary-structure constraints. (It should be noted, however, that scanning from the 5' end does seem to reach at least nt 66, which would require scanning through a hairpin loop of quite high stability that is conserved in polioviruses and rhinoviruses [Rivera et al., 1988].) If the process of scanning from the 5' end does not reach as far as nt 390, this implies that the ribosomes which initiate at nt 743 reach this

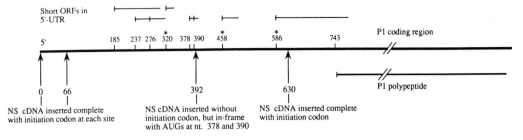

FIGURE 4. Schematic of the constructs used to test for ribosomal scanning on the poliovirus 5′ UTR. The upper part shows the 5′ UTR of poliovirus type 1 RNA, with the short open reading frames (ORFs) and the positions of the AUG codons numbered. Asterisks denote the AUG codons conserved in all three poliovirus serotypes, human rhinoviruses 2 and 14, and coxsackievirus B3 (Rivera et al., 1988; Skinner et al., 1989). The positions at which the influenza NS cDNA reporter gene was inserted are shown. These constructs were transcribed to give either capped or uncapped transcripts, which were translated at several different concentrations in the rabbit reticulocyte lysate in the presence or absence of supplementary HeLa cell factors (Jackson, 1989). The table gives a qualitative summary of the results with respect to NS and P1 synthesis. In all cases, P1 synthesis was uninfluenced by capping. Insertion of the NS cDNA in the reverse orientation at each site had virtually the same effect on P1 synthesis as did insertion in the correct orientation.

site by binding first to an internal entry site and then scanning the RNA starting from a point which our experiments would place as somewhere between nt 400 and 630 but which, according to the results of other groups, can probably be narrowed down to a point between nt 550 and 600.

Does this scanning from an internal entry site follow the normal behavior of the scanning ribosome? One previous observation consistent with this notion is that insertion of a stable hairpin loop at nt 631 of poliovirus type 2 RNA abolished initiation at the downstream initiation site (Pelletier and Sonenberg, 1988). As this same hairpin structure had been shown to inhibit initiation when inserted into the 5′ UTR of a capped mRNA which is believed to be translated by the conventional scanning mechanism, this observation suggests that ribosomes scan past nt

631 of the poliovirus sequence. For our part, we have looked at this question by using a derivative of the influenza virus NS cDNA in which the initiation site has been reiterated, giving two in-frame AUG codons situated 57 nt apart. When this is translated in vitro, products from initiation at both AUG codons are observed, with the degree of preference for the 5′-proximal AUG related to whether the local context of this AUG codon is close to the optimal defined by Kozak (1989). This NS cDNA with the reiterated initiation site has been fused to lengths of the poliovirus type 1 5′ UTR varying between the first (5′-proximal) 630 nt up to the first 747 nt. In all cases where less than 743 nt of poliovirus 5′ UTR were included, two products arising from initiation at both AUG initiation codons of the NS cistron were observed, but the longer product initiated at the 5′-proximal AUG codon

predominated irrespective of the actual length of poliovirus 5' UTR present. This finding strongly implies a selection of the initiation site by the normal scanning rules. This conclusion is reinforced by observations of the constructs with more than 745 nt of poliovirus sequence, therefore including the correct initiation site of the poliovirus RNA, which in some constructs was in frame with the NS cistron. In these cases, three products were observed; the largest, which was also the most abundant, initiated at nt 743 of the poliovirus sequence, whereas the next largest, initiated at the 5'-proximal AUG of the NS cistron, was synthesized only in low yield.

These experiments have so far been conducted only with NS cDNA derivatives in which both AUG codons are in a very good local sequence context, and they need repeating with constructs in which the 5'-proximal AUG has a poor context. However, the results so far are entirely consistent with the idea that ribosomes first bind to an internal entry site in the poliovirus 5'-UTR and then select the AUG codon for initiation by a scanning process that starts from a point situated upstream of nt 630. A critical issue in resolving the question of the starting point for scanning is whether the AUG codon at nt 586 is used for initiation. This AUG is conserved in all known polioviruses and rhinoviruses, and although its context (. . . CUUAU GG . . .) is not ideal, our experiments with the translation of other mRNAs suggest that its utilization by scanning ribosomes should be sufficiently frequent to be detectable. To our knowledge, products arising from initiation at this site have not been detected, which implies that the scanning starts from a point downstream of nt 586.

This leads to the rather uncomfortable conclusion that although poliovirus and EMC RNAs share the common feature of an internal ribosome entry site, they differ in that initiation at AUG-11 of EMC RNA does not involve a scanning approach, whereas initiation at nt 743 of poliovirus RNA does. The two could perhaps be unified by supposing that the basic mechanism in both cases is binding to an internal entry site followed by scanning starting from some point determined by its proximity to that site, but that in the case of EMC RNA the starting point for scanning lies between AUG-10 and AUG-11, with the result that AUG-11 is selected with absolutely minimal scanning. This hypothesis predicts that if AUG-11 (and the nearby AUG-12 and AUG-13) of EMC RNA were to be mutated, the efficiency of initiation would not be seriously compromised but the selected site would be the next downstream AUG codon, whereas AUG-10 would remain silent.

Acknowledgments. We thank our colleagues and collaborators Mary Dasso, Sung-Key Jang, Jan Jore, Bert Semler, and Eckard Wimmer for their gifts of clones used in this work and for helpful advise.

This work was supported by a grant from the Medical Research Council and a grant for overseas collaboration from NATO Scientific Affairs Division.

REFERENCES

Dorner, A. J., B. L. Semler, R. J. Jackson, R. Hanecak, E. Duprey, and E. Wimmer. 1984. In vitro translation of poliovirus RNA: utilization of internal initiation sites in reticulocyte lysate. *J. Virol.* **50:** 507–514.

Jackson, R. J. 1982. The control of initiation of protein synthesis in reticulocyte lysates, p. 362–418. In R. Perez-Bercoff (ed.), *Protein Biosynthesis in Eukaryotes.* Plenum Publishing Corp., New York.

Jackson, R. J. 1988. Picornaviruses break the rules. *Nature* (London) **334:**292–293.

Jackson, R. J. 1989. Comparison of EMC and poliovirus with respect to translation initiation and processing *in vitro*, p. 51–71. In B. L. Semler and E. Ehrenfeld (ed.), *Molecular Aspects of Picornavirus Infection and Detection.* American Society for Microbiology, Washington, D.C.

Jang, S. K., H.-G. Kräusslich, M. J. H. Nicklin, G. M. Duke, A. C. Palmenberg, and E. Wimmer. 1988. A segment of the 5' nontranslated region of encephalomyocarditis virus RNA directs internal entry of ribosomes during in vitro translation. *J. Virol.* **62:** 2636–2643.

Kitamura, N., B. L. Semler, P. G. Rothberg, G. R. Larsen, C. J. Adler, A. J. Dorner, E. A. Emini, R. Hanecak, J. J. Lee, S. van der Werf, C. W. Anderson, and E. Wimmer. 1981. Primary structure, gene organisation and polypeptide expression of poliovirus RNA. *Nature* (London) **291:**547–553.

Kozak, M. 1989. The scanning model for translation: an update. *J. Cell Biol.* **108:**229–241.

Kuge, S., N. Kawamura, and A. Nomoto. 1989. Genetic variation occurring on the genome of an in vitro insertion mutant of poliovirus type 1. *J. Virol.* **63:**1069–1075.

Kuge, S., and A. Nomoto. 1987. Construction of viable deletion and insertion mutants of the Sabin strain of type 1 poliovirus: function of the 5'-noncoding sequences in viral replication. *J. Virol.* **61:**1478–1487.

Palmenberg, A. C., E. M. Kirby, M. R. Janda, N. L. Drake, G. M. Duke, K. F. Potratz, and M. S. Collett. 1984. The nucleotide and deduced amino acid sequences of the encephalomyocarditis viral polyprotein coding region. *Nucleic Acids Res.* **12:**2969–2985.

Pelletier, J., G. Kaplan, V. R. Racaniello, and N. Sonenberg. 1988a. Cap-independent translation of poliovirus mRNA is conferred by sequence elements within the 5′ noncoding region. *Mol. Cell. Biol.* **8**:1103–1112.

Pelletier, J., G. Kaplan, V. R. Racaniello, and N. Sonenberg. 1988b. Translational efficiency of poliovirus mRNA: mapping inhibitory *cis*-acting elements within the 5′ noncoding region. *J. Virol.* **62**:2219–2227.

Pelletier, J., and N. Sonenberg. 1988. Internal initiation of translation of eukaryotic mRNA directed by a sequence derived from poliovirus RNA. *Nature* (London) **334**:320–325.

Phillips, B. A., and A. Emmert. 1986. Modulation of the expression of poliovirus proteins in reticulocyte lysates. *Virology* **148**:255–267.

Pilipenko, E. V., V. M. Blinov, L. I. Romanova, A. N. Sinyakov, S. V. Maslova, and V. I. Agol. 1989. Conserved structural domains in the 5′-untranslated region of picornaviral genomes: an analysis of the segment controlling translation and neurovirulence. *Virology* **168**:201–209.

Rivera, V. M., J. D. Welsh, and J. V. Maizel. 1988. Comparative sequence analysis of the 5′ noncoding region of the enteroviruses and rhinoviruses. *Virology* **165**:42–50.

Skinner, M. A., V. R. Racaniello, G. Dunn, J. Cooper, P. D. Minor, and J. W. Almond. 1989. New model for the secondary structure of the 5′ non-coding RNA of poliovirus is supported by biochemical and genetic data that also show that RNA secondary structure is important in neurovirulence. *J. Mol. Biol.* **207**:379–392.

Vartapetian, A. B., A. S. Mankin, E. A. Skripkin, K. M. Chumakov, V. D. Smirnov, and A. A. Bogdanov. 1983. The primary and secondary structure of the 5′-end region of encephalomyocarditis virus RNA. *Gene* **26**:189–195.

Chapter 22

eIF-4F-Independent Translation of Poliovirus RNA and Cellular mRNA Encoding Glucose-Regulated Protein 78/Immunoglobulin Heavy-Chain Binding Protein

Dennis G. Macejak, Simon J. Hambidge, Lyle Najita, and Peter Sarnow

All eucaryotic cellular mRNA molecules contain a 7-methylguanylate cap at their 5′ termini. This 5′ cap structure is recognized by cap binding proteins (CBPs) which subsequently direct the binding of the 40S ribosomal subunit to the 5′-end complex. The RNA molecule is then scanned by the ribosomal ternary subunit complex until an appropriate initiation codon is recognized for the start of translation (Kozak, 1983).

Poliovirus interferes with this initiation process by proteolytically inactivating p220, a component of the CBP complex eIF-4F (Etchison et al., 1982). The virus translates its mRNA, which is not capped, by an as yet unresolved cap-independent initiation process. Recently, Pelletier and Sonenberg (1988) showed that a 490-nucleotide sequence element within the 743-nucleotide-long 5′ noncoding (NC) region of poliovirus RNA can bind ribosomes internally and that it is sufficient to direct internal initiation of translation to a heterologous reporter mRNA (Pelletier and Sonenberg, 1988). Similarly, Jang and colleagues (1988) have observed that a segment of the 5′ NC region of encephalomyocarditis virus RNA can direct internal entry of ribosomes as well. These internal ribosome-binding sites eliminate the requirement for an intact CBP complex and provide a mechanism for initiating translation at internal AUG codons.

Recently, our laboratory has discovered that one cellular mRNA, encoding the glucose-regulated protein 78/immunoglobulin binding protein GRP78/BiP, is translated at an enhanced rate in poliovirus-infected cells at a time when eIF-4F-dependent translation is inhibited (Sarnow, 1989). This finding provides an example of a cellular mRNA that appears to be translated eIF-4F independently like the viral RNA.

In this chapter, we describe a system to study the mechanism of cap-independent translation of viral mRNA and cellular GRP78/BiP mRNA molecules in more detail.

TRANSFECTION OF IN VITRO-MADE RNA MOLECULES INTO MAMMALIAN CELLS AS A METHOD TO STUDY CAP-INDEPENDENT TRANSLATION

We wanted to examine the role of a 5′-end cap structure in the eIF-4F-independent translation of poliovirus and GRP78/BiP mRNAs in vivo. It has been difficult to express uncapped, nonpolioviral RNA molecules in mammalian cells because most transfected plasmids are transcribed by RNA polymerase II, yielding capped mRNAs. Therefore, we have constructed plasmids from which capped and uncapped RNA molecules can be transcribed in vitro (Fig. 1). These RNA molecules can then be transfected into mammalian cells, and their abilities to act as templates for translation can be determined.

Briefly, 5′ NC regions from poliovirus and GRP78/BiP cDNAs were cloned into plasmid vectors containing the bacteriophage T7 promoter element and the coding region of the firefly luciferase (Luc) gene (De Wet et al., 1987). Upon addition of T7 RNA polymerase

Dennis G. Macejak, Simon J. Hambidge, Lyle Najita, and Peter Sarnow, Department of Biochemistry, Biophysics and Genetics and Department of Microbiology and Immunology, University of Colorado Health Sciences Center, Denver, Colorado 80262.

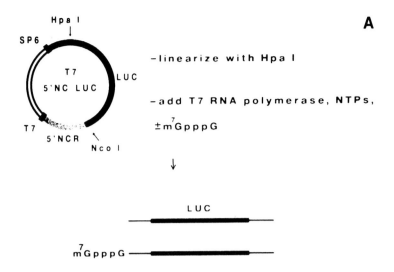

5'NONCODING REGIONS: firefly luciferase (30–54) 24 bp
poliovirus   (1–747) 747 bp
human Bip/GRP78 (3–220) 217 bp

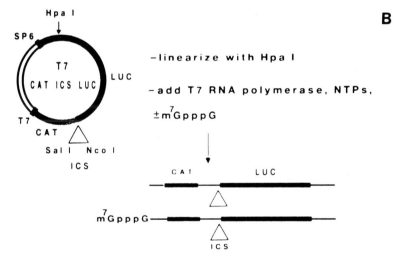

Intracistronic spacers: random polylinker   136 bp
poliovirus 5'NC (50–747)   697 bp
poliovirus P3 (5000–5290) 290 bp
human Bip/GRP 78 (3–220) 217 bp

FIGURE 1. Synthesis of monocistronic (A) and bicistronic (B) RNAs in vitro. Plasmids containing the promoter for T7 RNA polymerase and the cDNAs of the 5' NC region (NCR) of luciferase (LUC), poliovirus, or GRP78/BiP mRNA immediately upstream of the coding sequence for luciferase were used to synthesize monocistronic hybrid RNAs (A). In addition, the 5' NC regions were cloned between the CAT and Luc coding regions as an intracistronic spacer (ICS). After linearization with *Hpa*I and addition of T7 RNA polymerase and 1 mM nucleoside triphosphates (NTPs), uncapped RNAs were generated. Capped RNAs were synthesized by including a cap analog in the reaction mixture. bp, Base pairs.

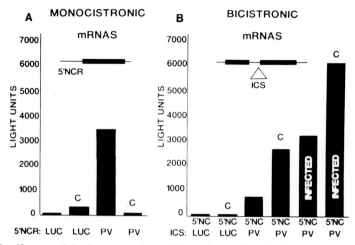

FIGURE 2. Luciferase activity in extracts from HeLa cells transfected with hybrid RNA molecules. Equivalent amounts of capped (C) and uncapped RNAs were transfected into HeLa cells, and cell extracts were made 1 h after transfection. 5′ NCR, 5′ NC regions of luciferase (LUC) or poliovirus (PV) mRNAs. ICS, Intracistronic spacer. Where indicated, cells were infected with poliovirus type 1 (multiplicity of infection, 20) before transfection.

and nucleoside triphosphates, capped and uncapped monocistronic RNA molecules were produced (Fig. 1A). In a separate set of experiments, the 5′ NC regions were engineered as an intracistronic spacer between the bacterial chloramphenicol acetyltransferase (CAT) and Luc genes to obtain capped and uncapped bicistronic chimeric RNA molecules (Fig. 1B). The recombinant RNA molecules were transfected (Sompayrac and Danna, 1981) into HeLa cells, and luciferase activity was measured in transfected cell extracts (De Wet et al., 1987).

Initially, we investigated the translatability of Luc RNA molecules containing their natural 5′ NC regions. We could detect luciferase activity only in extracts transfected with capped and not with uncapped RNA molecules (Fig. 2). Thus, the sensitivity of this system is sufficient to detect luciferase translated from transfected capped RNA molecules but not from similar amounts of uncapped RNAs.

Hybrid Luc RNA molecules containing the 5′ NC region of poliovirus (PV-Luc) were translated very efficiently in vivo but only if they were uncapped (Fig. 2A). Surprisingly, capping of PV-Luc RNA molecules reduced their translatability in vivo to levels below that of capped Luc RNA (Fig. 2A). This finding has not been observed in in vitro translation systems, in which capped and uncapped viral RNA molecules are translated equally well (Pelletier et al., 1988). Thus, it appears that the cap structure interferes with eIF-4F-independent translation

of the viral 5′ NC region in vivo but not in vitro. The addition of a cap structure may change structures in the RNA or direct assembly of CBPs on the 5′ end of the RNA and hence alter the accessibility of the internal ribosome-binding site in vivo.

The poliovirus 5′ NC region can also direct translation of a second cistron when placed as the intracistronic spacer in a bicistronic RNA molecule (Pelletier and Sonenberg, 1988). To study the role of a cap structure in the translation of a bicistronic mRNA molecule containing the 5′ NC region of poliovirus as an intracistronic spacer, we produced capped and uncapped bicistronic RNA molecules in vitro (Fig. 1B). In contrast to monocistronic 5′ NC PV-Luc RNAs, translation of the Luc cistron in capped CAT-5′ NC PV-Luc RNAs was three to four times more efficient than in the uncapped counterparts (Fig. 2B). Thus, the cap does not interfere with internal ribosome binding if the cap is displaced by an upstream cistron, in agreement with the observations of Pelletier and Sonenberg (1988). Transfection of CAT-5′ NC Luc-Luc RNA molecules into HeLa cells did not result in translation of the second cistron (Fig. 2B). Therefore, for a "normal" translation initiation site, there is no reinitiation of translation after the ribosomes have read through the first cistron. It is possible that the presence of a cap increases the stability of the RNA molecules, thereby increasing their apparent concentration.

TRANSLATIONS OF 5' NC PV-Luc and CAT-5' NC PV-Luc RNAs ARE GREATLY ENHANCED EARLY IN POLIOVIRUS-INFECTED CELLS

The complementation of a poliovirus mutant, mapping in 2A, by a persistently infected cell line (Sarnow et al., 1986) under conditions in which cellular translation remained unaltered, presented initial evidence for a specific *trans*-activation of viral translation.

To further substantiate this observation, we tested the translation of transfected RNA molecules containing the 5' NC of poliovirus in infected cells. Transfection of monocistronic 5' NC PV-Luc RNAs into poliovirus-infected cells resulted in a 10- to 13-fold increase in Luc activity as compared to mock-infected cells (data not shown). Similarly, transfection of bicistronic CAT-5' NC PV-Luc RNAs into poliovirus-infected cells resulted in a two- to three-fold-enhanced translation for both capped and uncapped molecules (Fig. 2B). The increased translation was observed early after infection at a time when eIF-4F-dependent translation was not inhibited. Therefore, the stimulation in translation was not simply due to the availability of free ribosomes. Rather, the enhancement of translation early in poliovirus infection appeared to be due to a virally induced factor that *trans*-activates the translation of these hybrid RNA molecules.

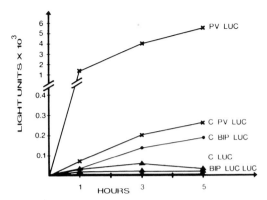

FIGURE 3. Time course of translation of transfected hybrid RNAs in 2B201-infected cells. Equivalent amounts of capped (C) and uncapped RNAs were transfected immediately after virus adsorption (multiplicity of infection, 20). Luciferase (LUC) activity was measured in cell extracts harvested at various times after infection. PV, Poliovirus; BIP, GRP78/BiP.

Luc hybrid RNA was stimulated four- to five-fold, as were both capped and uncapped 5' NC Pv-Luc RNAs (Fig. 3). Thus, our transfection system appears to faithfully mimic the observations we have recently reported for infected cells (Sarnow, 1989). Moreover, the 5' NC region of GRP78/BiP mRNA is sufficient to confer enhanced translatability to a heterologous mRNA in poliovirus-infected cells.

THE 5' NC REGION OF GRP78/BiP IS SUFFICIENT TO ALLOW ENHANCED TRANSLATION OF A HETEROLOGOUS RNA IN POLIOVIRUS-INFECTED CELLS

We have shown that translation of the mRNA encoding GRP78/BiP is specifically enhanced early in poliovirus-infected cells (Sarnow, 1989). To test whether GRP78/BiP mRNA translated eIF-4F independently like the viral RNA, we constructed plasmids for the expression of hybrid RNA molecules containing the entire 5' NC region of GRP78/BiP upstream of the Luc gene. Subsequently, 5' NC GRP78/Bip-Luc, 5' NC Luc-Luc (as a negative control), and 5' NC PV-Luc (as a positive control) RNAs were transfected into HeLa cells that had been infected with 2B201, a poliovirus mutant with a prolonged prereplicative phase (Bernstein et al., 1986). Translations of either capped or uncapped 5' NC Luc-Luc RNAs were not significantly stimulated during the infectious cycle, whereas capped (but not uncapped) 5' NC GRP78/BiP-

IDENTIFICATION OF FACTORS THAT INTERACT WITH THE 5' NC REGION OF POLIOVIRUS

Internal ribosome binding and poliovirus-induced translational *trans*-activation are presumably mediated through the interaction of specific factors with the RNA molecules. To identify potential interactions between the 5' NC region of poliovirus RNA and viral or cellular factors, we have used a modification of the gel retention assay (Fig. 4) utilized by Konarska and Sharp (1986).

Briefly, labeled RNA molecules were incubated with cytoplasmic extracts from either uninfected or poliovirus-infected HeLa cells. A 250-fold molar excess of *Escherichia coli* rRNA or a random copolymer of A, C, and U was added to adsorb nonspecific RNA-binding proteins. After a short incubation at 30°C, the reaction products were analyzed by native polyacrylamide gel electrophoresis. During electrophoresis, free RNA molecules usually migrated

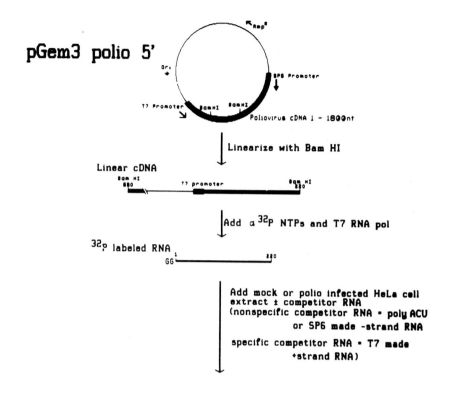

FIGURE 4. RNA gel retention assay. In vitro synthesis of RNA from plasmid pGEM3polio5'. The heavy black line indicates poliovirus cDNA flanked by promoter elements for T7 and SP6 RNA polymerases. Linear cDNA and ^{32}P-labeled RNA are indicated.

more rapidly than RNA molecules to which proteins were bound.

Using several plasmid constructs containing various cDNA segments of the 5' NC region of poliovirus cloned downstream of the T7 promoter element, we synthesized RNA molecules representing various portions of the poliovirus 5' NC region. We found that virtually any RNA molecule of about 250 nucleotides or longer could bind factors even in the presence of an excess of nonspecific competitor RNA molecules. Therefore, we have used RNA molecules of 200 nucleotides or shorter to investigate factor binding in the gel retention assay (Fig. 4).

Figure 5 shows an autoradiograph of a gel retention assay, using a 42-nucleotide viral sequence (nucleotides 178 to 220) as a radiolabeled RNA molecule. Migration of the labeled fragment was not perturbed by the presence of poly(ACU) and a specific competitor (Fig. 5, lane 1). Upon incubation with poly(ACU) and cytoplasmic extracts from uninfected (lane 2) or poliovirus-infected (lane 3) HeLa cells, slower migration of the labeled fragment was observed. However, addition of a 10-fold molar excess of specific competitor RNA (the same RNA fragment but unlabeled) abolished the retention (lanes 4 and 5), indicating that the complex formed was specific for that particular RNA fragment. Furthermore, competition with RNA molecules containing the complement of nucleotide sequences 178 to 220 of the viral RNA did not abolish retention (lanes 6 and 7), suggesting that factor binding is directed by a specificity in sequence or structure and not simply by the length of the RNA molecules. Thus, we have detected a factor, present in both uninfected and poliovirus-infected cell extracts, that binds specifically to a 42-nucleotide sequence element in the 5' NC region of poliovirus RNA (L. Najita and P. Sarnow, submitted for publication). We are currently characterizing this and other factors to assess their roles in eIF-4F-independent translation.

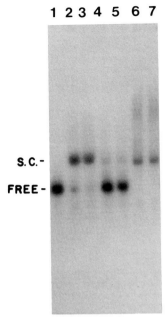

1 2 3 4 5 6 7

S.C.-

FREE-

FIGURE 5. Identification of a cellular factor interacting with poliovirus RNA sequences 178 to 220. Radiolabeled RNA representing poliovirus nucleotides 178 to 220 was incubated with cytoplasmic extracts from uninfected and poliovirus-infected cells. RNA-protein complexes were analyzed by polyacrylamide gel electrophoresis, an autoradiograph of which is shown. The migrations of free, unbound RNA and specific RNA-protein complexes (S.C.) are shown. Lanes: 1, labeled RNA-minus extract; 2 and 3, labeled RNA-plus extracts; 4 to 7, competition with either unlabeled RNA sequences 178 to 220 (lanes 4 and 5) or 220 to 178 (lanes 6 and 7). Lanes 2, 4, and 6 contain extracts from uninfected cells; lanes 3, 5, and 7 contain extracts from infected cells.

CONCLUSIONS

We have developed a method to study the translation of capped and uncapped RNA molecules by transfection of in vitro-synthesized RNA molecules into mammalian cells. This method allows us to study the cap-independent translation of RNA molecules in poliovirus-infected cells. Using this system, we have begun to search for the nucleotide sequences and factors that are mediating eIF-4F-independent translation of poliovirus and GRP78/BiP mRNAs. An RNA gel retention assay is being used to identify factors that activate translation of these RNA molecules in virus-infected cells. Future experiments are being designed to understand this *trans*-activation of translation as well as the internal ribosome binding of poliovirus RNA and certain cellular mRNAs.

ACKNOWLEDGMENTS. This work was supported in part by Public Health Service grants AI-25105 and AG-07347 (P.S.) and NS-07321 (D.G.M. and L.N.), all from the National Institutes of Health, and by a fellowship from the Arthritis Foundation (S.J.H.).

REFERENCES

Bernstein, H. D., P. Sarnow, and D. Baltimore. 1986. Genetic complementation among poliovirus mutants derived from an infectious cDNA clone. *J. Virol.* **60:**1040–1049.

De Wet, J. R., K. V. Wood, M. DeLuca, D. R. Helinski, and S. Subramani. 1987. Firefly luciferase gene: structure and expression in mammalian cells. *Mol. Cell. Biol.* **7:**725–737.

Etchison, D., S. C. Milbourn, I. Edery, N. Sonenberg, and J. W. B. Hershey. 1982. Inhibition of HeLa cell protein synthesis following poliovirus infection correlates with the proteolysis of a 220,000-dalton polypeptide associated with eukaryotic initiation factor 3 and cap binding protein complex. *J. Biol. Chem.* **251:**14806–14810.

Jang, S. K., H.-G. Kräusslich, M. J. H. Nicklin, G. M. Duke, A. C. Palmenberg, and E. Wimmer. 1988. A segment of the 5′ nontranslated region of encephalomyocarditis virus RNA directs internal entry of ribosomes during in vitro translation. *J. Virol.* **62:** 2636–2643.

Konarska, M. M., and P. A. Sharp. 1986. Electrophoretic separation of complexes involved in the splicing of precursors to mRNAs. *Cell* **46:**845–855.

Kozak, M. 1983. Comparison of initiation of protein synthesis in procaryotes, eucaryotes, and organelles. *Microbiol. Rev.* **47:**1–45.

Pelletier, J., G. Kaplan, V. R. Racaniello, and N. Sonenberg. 1988. Cap-independent translation of poliovirus mRNA is conferred by sequence elements within the 5′ noncoding region. *Mol. Cell. Biol.* **8:**1103–1112.

Pelletier, J., and N. Sonenberg. 1988. Internal initiation of translation of eukaryotic mRNA directed by a sequence derived from poliovirus RNA. *Nature* (London) **334:**32–35.

Sarnow, P. 1989. Translation of the glucose-regulated protein 78/immunoglobulin heavy-chain binding protein mRNA is increased in poliovirus-infected cells at a time when cap-dependent translation of cellular mRNAs is inhibited. *Proc. Natl. Acad. Sci. USA* **86:**5795–5799.

Sarnow, P., H. D. Bernstein, and D. Baltimore. 1986. A poliovirus temperature-sensitive RNA synthesis mutant located in a noncoding region of the genome. *Proc. Natl. Acad. Sci. USA* **83:**571–575.

Sompayrac, L. M., and K. J. Danna. 1981. Efficient infection of monkey cells with DNA of simian virus 40. *Proc. Natl. Acad. Sci. USA* **78:**7575–7578.

Chapter 23

Myristylation of Poliovirus VP4 Capsid Proteins

John Simons, Carol Reynolds, Nicola Moscufo, Lisa Curry,
and Marie Chow

Since the initial observation that the $p15^{gag}$ proteins of Rauscher and Moloney murine leukemia viruses were modified with myristic acid, the C_{14} saturated fatty acid (Henderson et al., 1983), N-myristylation has become a well-established phenomenon. Myristate modification is observed on increasing numbers of cellular proteins from different species and cell types (Buss et al., 1987; Carr et al., 1982; Marchildon et al., 1984; Olson et al., 1985; Towler and Glaser, 1986a; Ozols et al., 1984) and viral proteins (Buss et al., 1984; Buss and Sefton, 1985; Chow et al., 1987; Henderson et al., 1983; Jorgensen et al., 1988; Persing et al., 1987; Rhee and Hunter, 1987; Schultz et al., 1985; Schultz and Oroszlan, 1984; Streuli and Griffin, 1987). Myristylation has been identified primarily with proteins that are membrane associated (e.g., retroviral *gag* and *onc* proteins [Buss et al., 1987; Glover et al., 1988; Jorgensen et al., 1988; Rhee and Hunter, 1987; Schultz et al., 1985; Schultz and Oroszlan, 1984; Sefton et al., 1982], hepatitis B preS1 protein [Persing et al., 1987], cyclic AMP-dependent protein kinase [Carr et al., 1982], and guanine nucleotide-binding regulatory proteins [Buss et al., 1987]). However, more recently myristate modification also has been observed on the capsid proteins of picornaviruses and papovaviruses, both nonenveloped viruses (Chow et al., 1987; Paul et al., 1987; Streuli and Griffin, 1987). This has raised questions about the role(s) of this unusual posttranslational modification in the replication of poliovirus and other picornaviruses and about the general functional role of the myristate moiety in viral proteins that are membrane associated.

STRUCTURAL CHARACTERIZATION OF MYRISTATE MODIFICATION IN POLIOVIRUS

Chemical analyses showed that the myristate is attached specifically to the poliovirus VP4 capsid protein via an amide linkage to the alpha amino group of the N-terminal glycine (Chow et al., 1987; Paul et al., 1987). The structure of this linkage appears to be identical to that previously determined for other *N*-myristoyl-modified proteins (Buss et al., 1987; Carr et al., 1982; Henderson et al., 1983; Schultz et al., 1985). Because the poliovirus VP4 protein and its N-terminal tryptic peptide behave as homogeneous species under widely varying chromatographic conditions, all VP4 proteins within the virion appear to be N-terminally modified with myristate (data not shown). Thus, there are 60 copies of myristate per virus particle.

The electron density map of the poliovirus three-dimensional structure clearly identifies the VP4 N-terminal glycine residue and the myristate moiety and is consistent with the chemical description of the myristate-glycine linkage (Chow et al., 1987). Within the virion, five myristate molecules are situated around each of the 12 fivefold symmetry axes. The C_{14} hydrocarbon tails are arranged such that they form a tight cluster at the base of each axis and then splay apart to cradle a twisted tube structure formed by five copies of N-terminal sequences of VP3. Thus, the myristoyl groups interact with each other and with specific amino acid side chains from VP4 and VP3. In the native struc-

John Simons, Carol Reynolds, and Lisa Curry, Department of Applied Biological Sciences, Massachusetts Institute of Technology, 77 Massachusetts Avenue, Cambridge, Massachusetts 02139.   *Nicola Moscufo and Marie Chow,* Department of Biology, Massachusetts Institute of Technology, 77 Massachusetts Avenue, Cambridge, Massachusetts 02139.

ture, the myristate is not exposed on the outside surface of the virus.

MYRISTATE ADDITION WITHIN POLIOVIRUS-INFECTED CELLS

The VP4 sequences are located at the N terminus of the polyprotein precursor, which is proteolytically cleaved within the infected cell to generate the capsid precursor P1. P1 is cleaved subsequently during virion assembly to yield capsid proteins VP1, VP3, and VP0. VP4 and VP2 are formed by cleavage of VP0 after RNA encapsidation. Analysis of poliovirus-infected cell lysates show that both VP4 precursor proteins, P1 and VP0, are also myristate modified (Chow et al., 1987). This indicates that the modification is probably a cotranslational or early posttranslational event and is consistent with the temporal measurements made for $p60^{src}$ and other myristate-modified cellular proteins (Buss et al., 1984; Buss and Sefton, 1985; Olson et al., 1985; Wilcox et al., 1987). In addition, the precursors to P1 are also myristate modified in infected cell lysates in which proteolytic processing of the polyprotein is inhibited by the presence of zinc ions. Thus, myristate modification is not dependent on correct processing and formation of P1. This is consistent with the modification occurring after translational initiation but during synthesis of the 300-kilodalton polyprotein precursor.

Several lines of evidence suggest that the attachment of myristate to VP4 sequences is mediated via a cellular enzyme, the N-myristoyltransferase (NMT). Myristate modification of P1 is observed in in vitro translation systems by using uninfected cell lysates. In addition, the VP4 protein is myristoylated when it is expressed alone in cells by using eucaryotic expression vectors (unpublished results). Therefore, the cellular NMT will recognize and modify poliovirus VP4 sequences. Given that a cellular enzyme will myristate modify poliovirus VP4 sequences, it is unlikely that the virus will use a portion of its limited coding capacity to encode its own NMT. Thus, although it is possible that the responsible NMT is virus encoded, the enzyme is more likely to be of cellular origin.

Although the myristate-VP4 structure within the virus is well characterized, the exact functional role(s) of this modification within the virus replicative cycle is unclear. The potential importance of this modification is suggested by the ubiquity of this modification among all picornavirus VP4 proteins examined (Chow et al., 1987). Several functional roles have been pro-

posed for the myristate moiety (Chow et al., 1987; Paul et al., 1987). The myristate may be important for capsid assembly, i.e., the localization of P1 to the rough membranes or the stabilization of the pentamer assembly intermediates. The myristate may also be important in the early stages of infection, during binding and entry of virus particles into cells. The potential function(s) of the myristate moiety can be identified through the characterization of viral mutants altered in myristate modification.

CONSTRUCTION OF POLIOVIRUS MYRISTATE MUTANTS

Myristylation clearly requires recognition by the NMT of specific amino acids in the substrate protein. VP4 is myristate modified upon expression in CV-1 cells, indicating that the N-terminal VP4 sequences are sufficient to signal myristate addition. Comparison of the N-terminal sequences of many myristate-modified proteins indicates that they are diverse and show little homology other than the N-terminal glycine (residue 2 from the initiating methionine) and possibly a frequently occurring hydroxyamino acid residue at residue 6 (Chow et al., 1987; Towler et al., 1988b). In addition, substrate studies using synthetic oligopeptides and purified yeast NMT indicate that the hydroxyamino acid residue and the amino acid residue adjacent to the glycine (i.e., residues 6 and 3, respectively) can affect the kinetics of myristate addition (Glover et al., 1988; Towler et al., 1987a, 1988a; Towler et al., 1987b; Towler and Glaser, 1986b; Towler et al., 1988b). Thus, poliovirus mutants containing substitutions at any of these positions, i.e., N-terminal glycine (4002G), alanine at VP4 residue 3 (4003A), or serine at VP4 residue 6 (4006S), should affect the myristate modification.

Mutations were constructed within the N-terminal VP4 sequence by using oligonucleotide-directed site-specific mutagenesis methods (Fig. 1). Because the exact substrate specificities of the human NMT are undefined, a variety of amino acid substitutions were desired. Mutagenic oligonucleotides were synthesized with a random mix of all four nucleotides at the codon site to be mutated. Thus, each synthesized oligonucleotide contained a mixture of nucleotides at the desired mutant position that encode virtually all 20 amino acids. *Escherichia coli dut ung* strains and the method of Kunkel (1985) were used to achieve efficient incorporation (greater than 80%) of the mutagenic oligonucleotide sequences. Mutant sequences were initially iden-

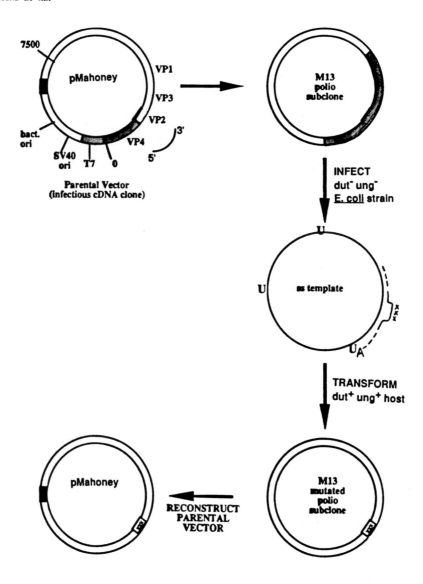

FIGURE 1. In vitro poliovirus mutagenesis scheme. Poliovirus mutations were constructed in mp19 vectors containing a subclone of the poliovirus infectious cDNA sequence that spanned the VP4 capsid sequence. Oligonucleotides were synthesized that contained a random mix of nucleotides at the position of the target codon. The mutated subclone subsequently replaced its cognate region within the infectious cDNA clone (pMahoney) to construct the mutant viral genomes.

tified in an mp19 subclone spanning the VP4 sequences. The mutated poliovirus sequences were subsequently reinserted into the infectious clone to obtain a complete poliovirus cDNA sequence containing the mutant codon. The sequence of the mutated subclone was determined in its entirety by using primer extension dideoxynucleotide sequencing methods to confirm that only the VP4 residue is mutated. The presence of the mutated codon was reconfirmed by sequence analysis after the mutant subclone has been reinserted into the poliovirus cDNA. Mutant clones containing a wide spectrum of amino acid substitutions at residue position 4002G, 4003A, or 4006S were constructed (Fig. 2). (The single-letter amino acid code will be used

Gly (Ala)
Val (Leu)
Leu (Arg)
His Gly
Asp Thr
*   *
N-Gly-Ala-Gln-Val-Ser-Ser--
* *   *
Ala Arg Trp
Val
Met
Tyr
Arg
Glu

FIGURE 2. Amino acid substitutions of poliovirus VP4 N-terminal residues. The N-terminal sequence of VP4 is depicted along with the mutations that have been isolated at the N-terminal Gly, Ala, and Ser residues. Mutations listed below the N-terminal VP4 sequence are nonviable. Mutations listed above the VP4 sequence yield viable virus of various plaque morphologies.

throughout the text. The mutants are designated by the residue number and the nature of the substitution. Thus, 4002G.R indicates that the glycine at VP4 residue 2 has been mutated to an arginine.) The viability and molecular biology of these mutants are characterized upon transfection into HeLa cells.

CHARACTERIZATION OF 4002G MUTANTS

All characterized NMTs show an absolute requirement for N-terminal glycine substrates (Jorgensen et al., 1988; Kamps et al., 1985; Towler et al., 1987a, 1988a; Towler et al., 1987b). Thus, replacement of the 4002G residue should result in a mutant that is incapable of being myristate modified. All mutants with substitutions at the N-terminal glycine residue fail to produce plaques upon transfection with either the cDNA clone or in vitro-synthesized RNA transcripts of the cDNA clone (Fig. 3). Replacement of the mutant residue with the wild-type glycine amino acid restores the infectivity of the cDNA clones. Thus, substitutions at the N-terminal glycine are lethal, and N-myristyl modification is essential for virus viability.

Subsequent characterization of these lethal mutants is hampered by the low transfection efficiencies and the asynchronous infections initiated by these transfections. However, the aggregate data from Northern (RNA) blot and Western blot (immunoblot) analysis indicate that viral RNA and protein synthesis in cells transfected with mutant RNA transcripts occurs at wild-type levels with normal kinetics (data not shown). No myristate modification of P1 is ob-

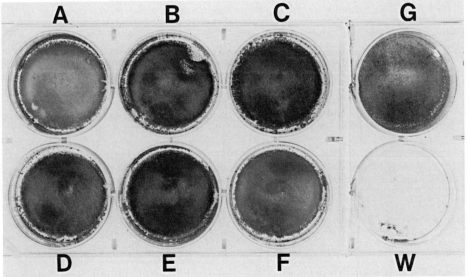

FIGURE 3. RNA transfection of 4002G mutants. RNA was synthesized in vitro from the 4002G mutant cDNA clones, using T7 polymerase (Kaplan and Racaniello, 1988). The RNA transcripts were transfected onto HeLa cell monolayers and overlaid with Dulbecco modified medium–5% fetal bovine serum–1% agarose. Two days later, viable cells were stained with crystal violet. Cells were transfected with RNA transcripts containing 4002 G.V (A), 4002 G.E (B), 4002 G.Y (C), 4002 G.R (D), 4002 G.M (E), 4002 G.A (F), and wild-type (W) sequences. The mock transfection (G) contained the transcription reaction excluding the RNA polymerase.

served in these 4002G mutants. This result is consistent with the implication that the human NMT, like the characterized NMTs, would display an absolute specificity for N-terminal glycine substrates. The normal levels of RNA and protein synthesis indicate that the 4002G mutants appear blocked at a posttranslational step. This phenotype is consistent with that observed in mutants displaying altered levels of myristate modification (see below).

CHARACTERIZATION OF 4003A AND 4006S MUTANTS

Previous characterization of the yeast NMT by using synthetic oligopeptide substrates of various sequences demonstrated that substitutions at these residue positions greatly affected kinetic parameters of the NMT (Towler et al., 1987a, 1988a; Towler et al., 1987b; Towler and Glaser, 1986b; Towler et al., 1988b). Although little is known about the substrate specificities of the human NMT, they are expected to be similar to those of the yeast NMT. Thus, mutants were constructed at these sites with amino acid substitutions of widely varying sizes and chemical properties. These were transfected into CV-1 and HeLa cells to test for viability of the mutant virus at different temperatures and host ranges. Several plaque isolates for each viable mutant were expanded once at low multiplicity of infection (multiplicity of infection, 0.05 to 0.1) in HeLa cells and then passaged once at high multiplicity of infection (i.e., 5). RNA sequence analysis of the resultant virus reconfirmed the identity of the mutant residue (Page et al., 1988). The high-titer virus stock generated from this expansion was used for all subsequent experiments. Substitutions at either 4003A or 4006S generated viral mutants displaying a wide range of growth phenotypes (Fig. 4). Mutants 4003A.R and 4006S.W were nonviable. Minute- to small-plaque phenotypes were observed for 4003A.H, 4003A.D, 4006S.G, 4006S.R, and 4006S.L. Normal wild-type-size plaques were observed for 4003A.G, 4003A.V, 4003A.L, 4006S.T, and 4006S.A. The plaque phenotypes remained constant through several passages, indicating that the mutations are fairly stable genetically and the problem of spontaneous revertants is small. Because similar results have been obtained for all viable mutants, a detailed characterization of only two mutants (4006S.T and 4006S.G) is described below.

Normal plaque morphologies were observed for all 4006S.T mutant isolates. Further analysis showed that this virus is like wild type

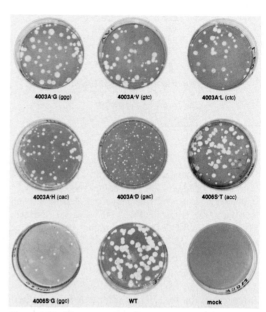

FIGURE 4. Plaque phenotypes of viable 4003A and 4006S viral mutants. HeLa cells were transfected with the wild-type or mutant pMahoney cDNA clones. Individual plaques were picked into phosphate-buffered saline. Plaque phenotypes were assessed in plaque assays on HeLa cell monolayers. Cells were stained with crystal violet after 3 days at 37°C. The mutant codon is given in parentheses.

in every respect (data not shown). Single-step growth curves for the 4006S.T isolates show normal yields of virus with kinetics the same as that of the wild-type virus. RNA synthesis in the presence of actinomycin D proceeds with wild-type kinetics. Protein synthesis appears normal.

In contrast, small plaques are observed for 4006S.G. This small-plaque phenotype is maintained regardless of growth temperature (33, 37, or 39°C) or host cell (HeLa or CV-1). Analysis of the single-step growth curve indicates that infection of HeLa cells by 4006S.G and production of virus occur with the same kinetics as wild-type virus (data not shown). However, the yield of infectious virus per cell is consistently 10-fold lower than that obtained from cells infected with wild-type virus. RNA synthesis is slightly reduced in mutant-infected cells; however, this difference (less than 2-fold) is not sufficient to account for the 10-fold reduction in virus production. Examination of 4006S.G-infected cell lysates labeled with [^{35}S]methionine shows a normal pattern of viral proteins (Fig. 5). Thus, protein synthesis and proteolytic processing of the polyprotein on a gross level appear to be

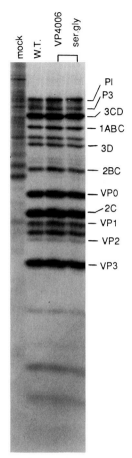

FIGURE 5. Viral protein synthesis. Suspension HeLa cells were infected with wild-type virus (W.T.) or one of two isolates of 4006S.G (VP4006ser.gly) at a multiplicity of infection of 5 or were mock infected. Actinomycin D (5 μg/ml) was added 15 min postinfection. At 3.5 h postinfection, [^{35}S]methionine (10 μCi/ml) was added. Cells were harvested at 4.5 h postinfection, washed with phosphate-buffered saline, and lysed in 1% sodium dodecyl sulfate. The labeled viral proteins were separated in a 10% sodium dodecyl sulfate-polyacrylamide gel, and the gel was autoradiographed.

unaffected. However, labeling of 4006S.G-infected cell lysates with [^{3}H]myristic acid is clearly lower than for wild-type-infected cells. To quantitate the amount of myristate modification, lysates and purified virus double labeled with [^{3}H]myristate and [^{35}S]methionine were isolated from cells infected with wild-type (4006S), 4006S.T, or 4006S.G virus. The labeled viral proteins are separated by polyacrylamide gel electrophoresis, and each viral protein band

is excised and counted (Table 1). The ^{3}H/^{35}S ratio is determined to normalize for any differences in protein synthesis or proteolytic processing. No difference is observed in ratios obtained for VP0 and P1 from wild-type- or 4006S.T-infected cells, indicating that myristate modification of P1 and subsequent processing of P1 to VP0 occur at wild-type levels. However, ^{3}H/^{35}S ratios of P1 and VP0 are reduced in 4006S.G-infected cells, indicating that the levels of myristate modification of P1 are consistently lower than in 4006S.T- or wild-type-infected cells. In addition, assuming that all P1 protein in wild-type-infected cells is myristate modified, the lower ^{3}H/^{35}S ratios observed for P1 in 4006S.G-infected cell lysates suggests that P1 is only partially myristate modified. Thus, there are two populations of P1 protein in 4006S.G-infected cells, one that is myristate modified and one that remains unmodified. This is one of the few instances in which partial myristylation of a protein appears to occur within the cell.

Surprisingly, the ^{3}H/^{35}S ratios obtained for VP4 protein extracted from banded virus were identical for both mutant and wild-type viral strains. Because wild-type VP4 protein is completely modified with myristate, the VP4 protein found in the 4006S.G virus particle is also fully modified. Thus, there must be a selection that occurs during the virion assembly pathway for myristate-modified capsid proteins.

Comparison of the ^{3}H/^{35}S ratios obtained for P1 and VP0 can suggest potentially where the selection process is occurring. Cleavage of P1 to form VP0-VP3-VP1 occurs during pentamer formation, an obligatory intermediate in the assembly pathway. If only myristate-modified P1 protein enters the assembly pathway, then only myristate-modified VP0 should be observed in 4006S.G-infected cell lysates; the VP0 ^{3}H/^{35}S ratio should increase and reflect values similar to that observed in wild-type and 4006S.T-infected cells. Alternatively, if both myristate-modified and unmodified P1 proteins can enter the assembly pathway and the selection of myristate-modified capsid proteins occurs at a later stage, then the VP0 and P1 ^{3}H/^{35}S ratios in the 4006S.G-infected cell should be identical, reflecting the presence of modified and unmodified populations in both proteins. An increase in VP0 ^{3}H/^{35}S ratios is observed in all cell lysates. This reflects the differences in methionine content found in the P1 (24 Met residues per protein) and VP0 (7 Met residues per protein) proteins. Once this difference is normalized, the 4006S.G P1 and VP0 ^{3}H/^{35}S ratios are similar and much lower than observed in wild-type- and 4006S.T-

TABLE 1
Double Labeling of Viral Mutants[a]

Virus	Intracellular lysate				VP4 (virions)	
	P1		VP0			
	cpm	$^3H/^{35}S$	cpm	$^3H/^{35}S$	cpm	$^3H/^{35}S$
WT (Mahoney)	7,764	2.47	22,381	8.8	11,620	57.8
	3,144		2,543		201	
4006S.T 1	7,822	2.45	22,117	8.2	14,375	54.2
	3,197		2,691		265	
2	7,781	2.54	21,113	7.8	12,613	63.0
	3,068		2,679		200	
4006S.G 1	930	0.31	8,471	1.3	14,710	62.1
	3,001		8,168		237	
2	1,143	0.35	12,611	1.25	15,208	51.2
	3,314		10,064		297	

[a] Cells infected wtih wild-type (WT), 4006S.T, or 4006S.G isolates were labeled with [^3H]myristate and [^{35}S]methionine. Cells were harvested and lysed in 1% sodium dodecyl sulfate. The cell lysates and purified virus were separated on 10% sodium dodecyl sulfate-polyacrylamide gels, and the protein bands corresponding to P1, VP0, and VP4 were excised, dissolved in Protosal, and counted in Aquasol. Two isolates of 4006S.T and 4006S.G were quantitated. For each band, the ^3H cpm is the first value and the ^{35}S cpm is the second, corrected for background and spillover.

infected cells. Thus, it appears that the selection for the presence of myristate-modified VP4 capsid proteins occurs after the proteolytic cleavages necessary for pentamer formation. Studies characterizing the assembly intermediates observed in 4006S.G-infected cells will be necessary to identify the exact role of myristate-modified VP4 protein on poliovirus assembly.

ACKNOWLEDGMENTS. This work was supported by Public Health Service grant AI22627 from the National Institutes of Health and grant MV-466 from the American Cancer Society to M.C.

REFERENCES

Buss, J. E., M. P. Kamps, and B. M. Sefton. 1984. Myristic acid is attached to the transforming protein of Rous sarcoma virus during or immediately after synthesis and is present in both soluble and membrane-bound forms of the protein. *Mol. Cell. Biol.* **4**:2697–2704.

Buss, J. E., S. M. Mumby, P. J. Casey, and A. G. Gilman. 1987. Myristoylated α subunits of guanine nucleotide-binding regulatory proteins. *Proc. Natl. Acad. Sci. USA* **84**:7493–7497.

Buss, J. E., and B. M. Sefton. 1985. Myristic acid, a rare fatty acid, is the lipid attached to the transforming protein of Rous sarcoma virus and its cellular homolog. *J. Virol.* **53**:7–12.

Carr, S. A., K. Biemann, S. Shoji, D. C. Parmelee, and K. Titani. 1982. n-Tetradecanoyl is the NH$_2$-terminal blocking group of the catalytic subunit of cyclic AMP-dependent protein kinase from bovine cardiac muscle. *Proc. Natl. Acad. Sci. USA* **79**:6128–6131.

Chow, M., J. F. E. Newman, D. Filman, J. M. Hogle, D. J. Rowlands, and F. Brown. 1987. Myristylation of picornavirus capsid protein VP4 and its structural significance. *Nature* (London) **327**:482–486.

Glover, C. J., C. Goddard, and R. L. Felsted. 1988. N-myristoylation of p60^{src}. Identification of a myristoyl-CoA: glycyl peptide N-myristoyltransferase in rat tissues. *Biochem. J.* **250**:485–491.

Henderson, L. E., H. C. Krutzsch, and S. Oroszlan. 1983. Myristyl amino-terminal acylation of murine retrovirus proteins: an unusual post-translational protein modification. *Proc. Natl. Acad. Sci. USA* **80**:339–343.

Jorgensen, E. C., N. O. Kjeldgaard, F. S. Pedersen, and P. Jorgensen. 1988. A nucleotide substitution in the *gag* N terminus of the endogenous ecotropic DBA/2 virus prevents Pr65^{gag} myristylation and virus replication. *J. Virol.* **62**:3217–3223.

Kamps, M. P., J. E. Buss, and B. M. Sefton. 1985. Mutation of NH$_2$-terminal glycine of p60^{src} prevents both myristoylation and morphological transformation. *Proc. Natl. Acad. Sci. USA* **82**:4625–4628.

Kaplan, G., and V. R. Racaniello. 1988. Construction and characterization of poliovirus subgenomic replicons. *J. Virol.* **62**:1687–1696.

Kunkel, T. A. 1985. Rapid and efficient site-specific mutagenesis without phenotypic selection. *Proc. Natl. Acad. Sci. USA* **82**:488–492.

Marchildon, G. A., J. E. Casnellie, K. A. Walsh, and E. G. Krebs. 1984. Covalently bound myristate in a lymphoma tyrosine protein kinase. *Proc. Natl. Acad. Sci. USA* **81**:7679–7682.

Olson, E. N., D. A. Towler, and L. Glasser. 1985. Specificity of fatty acid acylation of cellular proteins. *J. Biol. Chem.* **260**:3784–3790.

Ozols, J., S. A. Carr, and P. Strittmatter. 1984. Identification of the NH$_2$-terminal blocking group of NADH-cytochrome b$_5$ reductase as myristic acid and the complete amino acid sequence of the membrane-binding domain. *J. Biol. Chem.* **21**:13349–13354.

Page, G. S., A. G. Mosser, J. M. Hogle, D. J. Filman, R. R. Rueckert, and M. Chow. 1988. Three-dimensional structure of poliovirus serotype 1 neutralizing determinants. *J. Virol.* **62**:1781–1794.

Paul, A. V., A. Schultz, S. E. Pincus, S. Oroszlan, and E. Wimmer. 1987. Capsid protein VP4 of poliovirus is N-myristylated. *Proc. Natl. Acad. Sci. USA* **84**:7827–7831.

Persing, D. H., H. E. Varmus, and D. Ganem. 1987. The preS1 protein of hepatitis B virus is acylated at its amino terminus with myristic acid. *J. Virol.* **61**:1672–1677.

Rhee, S. S., and E. Hunter. 1987. Myristylation is required for intracellular transport but not for assembly of D-type retrovirus capsids. *J. Virol.* **61**:1045–1053.

Schultz, A., and S. Oroszlan. 1984. Myristylation of *gag-onc* fusion proteins in mammalian transforming retroviruses. *Virology* **133**:431–437.

Schultz, A. M., L. E. Henderson, S. Oroszlan, E. A. Garber, and H. Hanafusa. 1985. Amino terminal myristylation of the protein kinase p60*src*, a retroviral transforming protein. *Science* **227**:427–429.

Sefton, B. M., I. S. Trowbridge, and J. A. Cooper. 1982. The transforming proteins of Rous sarcoma virus, Harvey sarcoma virus and Abelson virus contain tightly bound lipid. *Cell* **31**:465–474.

Streuli, C. H., and B. E. Griffin. 1987. Myristic acid is coupled to a structural protein of polyoma virus and SV40. *Nature* (London) **326**:619–622.

Towler, D., and L. Glaser. 1986a. Acylation of cellular proteins with endogenously synthesized fatty acids. *Biochemistry* **25**:878–884.

Towler, D., and L. Glaser. 1986b. Protein fatty acid acylation: enzymatic synthesis of an N-myristoylglycyl peptide. *Proc. Natl. Acad. Sci. USA* **83**:2812–2816.

Towler, D., J. Gordon, S. Adams, and L. Glaser. 1988b. The biology and enzymology of eukaryotic protein acylation. *Annu. Rev. Biochem.* **57**:69–99.

Towler, D. A., S. P. Adams, S. R. Eubanks, D. S. Towery, E. Jackson-Machelski, L. Glaser, and J. I. Gordon. 1987a. Purification and characterization of yeast myristoyl CoA:protein N-myristoyltransferase. *Proc. Natl. Acad. Sci. USA* **84**:2708–2712.

Towler, D. A., S. P. Adams, S. R. Eubanks, D. S. Towery, E. Jackson-Machelski, L. Glaser, and J. I. Gordon. 1988a. Myristoyl CoA: protein N-myristoyltransferase activities from rat liver and yeast possess overlapping yet distinct peptide substrate specificities. *J. Biol. Chem.* **263**:1784–1790.

Towler, D. A., S. R. Eubanks, D. S. Towery, S. P. Adams, and L. Glaser. 1987b. Amino-terminal processing of proteins by N-myristoylation. *J. Biol. Chem.* **262**:1030–1036.

Wilcox, C., J.-S. Hu, and E. M. Olson. 1987. Acylation of proteins with myristic acid occurs cotranslationally. *Science* **238**:1275–1278.

Chapter 24

Formation and Function of the Semliki Forest Virus Membrane

Henrik Garoff, Peter Liljeström, Kalervo Metsikkö,
Mario Lobigs, and Johanna Wahlberg

A large number of viruses have acquired an envelope or membrane for the purpose of transmitting their genomes between cells (Dubois-Dalcq et al., 1984; Simons and Fuller, in press). These viruses direct the synthesis of membrane proteins that are destined for either the plasma membrane or different compartments of the exocytic pathway in the infected cells. The membrane proteins have the capacity to recognize encapsidated genome structures in the cell cytoplasm and organize themselves together with the lipid membrane around these structures. This process, which is called virus budding, results in transfer of the viral nucleic acid from the cytoplasm of an infected cell in the form of the virus particle. The viral membrane proteins then direct the targeting of the virus particle to a new host cell and also guide the entry of the viral nucleic acid into the cytoplasm of the same. The latter event occurs through membrane fusion. In this process, the viral membrane proteins induce the fusion of the viral and the host membranes, releasing the encapsidated nucleic acid into the cytoplasmic space.

To study the mechanisms of viral membrane formation and fusion, we have used Semliki Forest virus (SFV), an enveloped virus belonging to the group of alphaviruses (Garoff et al., 1982). SFV has a single-stranded RNA genome of positive polarity. The genome directs the synthesis of a polymerase for replication of genomic (11.5-kilobase or 42S) and subgenomic (4.1-kilobase or 26S) RNA molecules (Fig. 1). The 26S RNA molecule is used for synthesis of the viral structural proteins. These are the capsid (C) protein and the membrane proteins p62

and E1, which are translated from a common coding unit of the 26S RNA in the order C-p62-E1 (see upper part of Fig. 2). The newly synthesized C molecule is used for nucleocapsid with genomic RNA in the cell cytoplasm, whereas the membrane proteins become inserted into the membrane of the endoplasmic reticulum (ER) for further transport to the plasma membrane. Virus budding occurs on the cell surface through association of the nucleocapsids with the viral membrane proteins. SFV entry into a new cell starts with the binding of the particle to some still uncharacterized structure on the cell surface. Next, the particle is internalized via the pathway of receptor-mediated endocytosis. Within the acidic milieu of the endosomes, the membrane proteins of the virus trigger the fusion reaction between the viral and the endosomal membranes (Kielian and Helenius, 1986; Marsh and Helenius, 1989). In recent studies, we have characterized (i) how SFV can direct the synthesis of both cytoplasmically destined C protein and several membrane proteins from a common coding unit, (ii) how membrane protein oligomerization regulates various steps in the assembly processs of the viral membrane, and (iii) how fusion is regulated by the state of the virus membrane-protein association.

NOVEL USE OF SECRETORY SIGNALS DURING BIOSYNTHESIS OF THE VIRAL MEMBRANE PROTEINS

The p62 and E1 proteins represent typical single-spanning membrane proteins which after

Henrik Garoff, Peter Liljeström, Kalervo Metsikkö, Mario Lobigs, and Johanna Wahlberg, Department of Molecular Biology, Center for Biotechnology, The Karolinska Institute, Huddinge University Hospital, S-141 86 Huddinge, Sweden.

ASSEMBLY

DISASSEMBLY

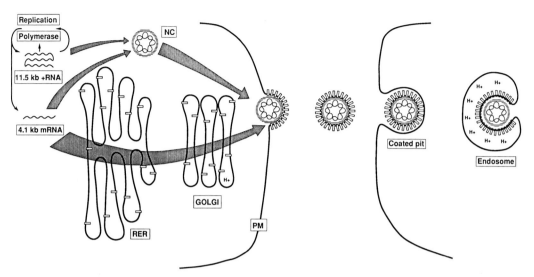

FIGURE 1. Schematic view over the main assembly and disassembly events involved in the life cycle of SFV. kb, Kilobase; NC, nucleocapsid; RER, rough endoplasmic reticulum; PM, plasma membrane.

synthesis have most of their molecular mass, including the N termini, within the ER lumen. In general, such membrane proteins are made with the aid of an N-terminally located cleavable signal peptide (or chain translocation signal) and a C-terminally located membrane anchor (von Heijne, 1985). However, the fact that p62 and E1 of SFV are translated after the C protein from a common coding region implies that internal chain translocation signals must be used in this system.

We have used gene technology and in vitro transcription-translation-translocation assays to localize and characterize the signal peptides of SFV. The results are summarized in Fig. 2. The signal peptide for the p62 protein is contained within the first 16 residues of this protein. Like cellular signal peptides, it requires a free N-terminal end to function, and this is achieved by the autoproteolytic removal of the preceding C chain. Interestingly, the p62 signal peptide is not removed by signal peptidase cleavage. Instead, this region is translocated into the lumen of ER together with the rest of the p62 luminal domain and becomes glycosylated at Asn-14.

The signal peptide for the E1 protein resides in the C-terminal part of the short transmembrane peptide called 6K, which is translated between the p62 and E1 proteins from the 26S RNA (Melancon and Garoff, 1986). Analogously, the insertion of the 6K peptide into the

membrane is ensured by a signal peptide at the C-terminal region of the p62 protein. The E1 and the 6K signal peptides are clearly different from those of the p62 protein and other secretory proteins of the cell in the sense that they function in an internal location. However, it should be mentioned that both the 6K and E1 signal peptides are located very close after the anchor sequences of the p62 and 6K proteins, respectively. Thus, the membrane anchoring of the N-terminally located flanking sequences of these signal peptides could possibly prevent inhibition of signal peptide expression by preceding protein sequences. Moreover, since these flanking transmembrane peptides anchor the nascent chain to the ER membrane, it is possible that translocation of 6K and E1 is a signal recognition peptide-independent process.

The signal peptides of 6K and the E1 protein are each followed by typical cleavage sites for the signal peptidase (von Heijne, 1985). Indeed, by creating specific amino acid substitutions at the signal peptidase cleavage sites, we were able to abolish luminal cleavage of the polyprotein. The cleavages by the signal peptidase have two important consequences for SFV membrane protein biosynthesis. First, the p62-6K-E1 region is processed into individual protein chains. Second, the hydrophobic signal peptides are left at the C-terminal regions of the p62 and 6K chains, respectively. Thus, all signal

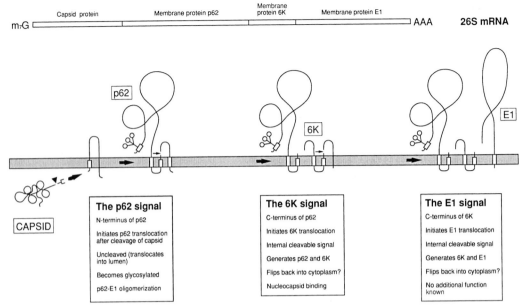

FIGURE 2. Use of translocation signals during synthesis of the structural proteins of SFV. Top, The gene map of the 26S subgenomic RNA. Middle, The process of membrane translocation of the p62, 6K, and E1 proteins. Small arrows on the luminal side denote signal peptidase cleavages. At the bottom, the characteristics of the three signal peptides are given.

peptides of the SFV membrane proteins are retained in the processed proteins. This probably reflects their need for further functions in virus assembly or entry.

OLIGOMERIZATION CONTROL OF MEMBRANE PROTEIN TRANSPORT

The p62 and E1 membrane proteins undergo efficient heterodimerization soon after translocation into the ER (Ziemiecki and Garoff, 1978; Ziemiecki et al., 1980; Rice and Strauss, 1982). The heterodimers are then transported to the cell surface to become incorporated into viral membranes. During a late stage of this transport, the heterodimer undergoes a proteolytic maturation: the p62 chain is cleaved after Arg-66 to yield the transmembrane E2 protein and the small E3 peptide. Viral particles contain essentially only the mature E2E1 heterodimers in their membranes.

The heterodimeric structures have been verified in several ways. These include (i) solubilization with nonionic detergents and subsequent immunoprecipitation with monoclonal antibodies against p62 (E2), E1, or both; (ii) solubilization as described above, followed by sedimentation analysis in sucrose gradients, and

(iii) covalent intermolecular cross-linking and analysis by sodium dodecyl sulfate-polyacrylamide gel electrophoresis.

When the p62 and E1 proteins were expressed from separate coding units of engineered cDNA sequences, it became apparent that the heterodimerization exerts a profound control on the transport process of the membrane proteins; expression of only the E1 protein resulted in its complete retention in the ER compartment. In contrast, the p62 protein was transported to the cell surface even in the absence of E1. When E1 and p62 were coexpressed in the same cell but from different coding units, restoration of the E1 transport defect was observed. Thus, these experiments clearly show that E1 needs the p62 protein for cell surface transport. They also demonstrate most convincingly that p62E1 heterodimerization has to occur in the ER and in no other compartment of the exocytic pathway.

MEMBRANE PROTEIN OLIGOMERIZATION CONTROL OF THE MEMBRANE-NUCLEOCAPSID ASSOCIATION

During virus budding, interactions between the viral membrane proteins and the nucleocap-

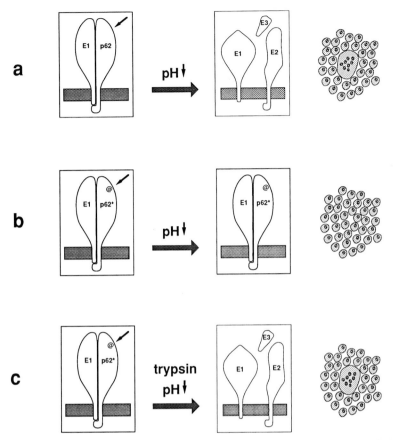

FIGURE 3. Schematic representation of cleavage activation of the fusion function of the SFV heterodimer. (a) Wild-type fusion phenotype. The envelope glycoprotein precursor p62 is proteolytically cleaved to E2 and E3 during transport to the cell surface. Low-pH treatment induces a fusion-competent conformation of the spike heterodimer, triggering cell membrane fusion and polykaryon formation. (b) Fusion phenotype of a p62 cleavage-deficient mutant. Low-pH treatment does not induce polykaryon formation. Fusion can, however, be restored by cleavage with exogenous trypsin of the mutant p62 (c), demonstrating that the p62 cleavage activation is essential for the acid-induced fusion function of the SFV spike heterodimer.

sid are required to curve the membrane around the whole particle. These interactions must primarily be based on interaction between the cytoplasmic or internal portions of the membrane proteins and the nucleocapsid surface. An examination of the cytoplasmic portion of the two membrane proteins of SFV shows that the p62 (E2) protein has 31 residues (including the 6K signal region) and that E1 has only 2 residues (both arginines) on this side of the membrane layer (Garoff et al., 1980). Therefore, it appears that only the p62 (E2) protein has enough molecular mass on the cytoplasmic side of the membrane for nucleocapsid interaction. An important corollary of this is that the E1 protein

needs the p62 (E2) protein also for incorporation into virus particles.

Recently, the interaction between the p62 (E2) cytoplasmic peptide and the nucleocapsid has been verified in two ways. First, Vaux et al. (1988) showed, by using anti-idiotypic antibodies, that the two surfaces display complementarity. Second, we have demonstrated direct binding between a synthetic peptide corresponding to the p62 (E2) cytoplasmic domain and isolated nucleocapsids. Most interestingly, the peptide was able to bind only if presented in an oligomeric form or if bound to a solid matrix. This finding suggests that single p62 (E2) cytoplasmic domains are not sufficient for binding but that

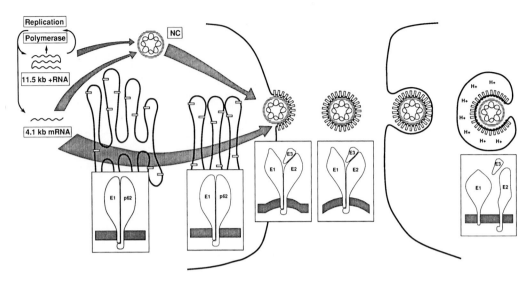

FIGURE 4. Regulation of SFV spike heterodimer association by p62 cleavage and pH. After translocation into the ER, the membrane proteins are transported through mildly acidic compartments to the plasma membrane as acid-resistant p62E1 heterodimers. After cleavage of p62, the virus particles, now containing E2E1 complexes, are released through budding. However, the association between E2 and E1 is sensitive to acidic pH. During entry of the virus through endocytosis, the acidic milieu of the endosome triggers the dissociation of the spike complex, resulting in fusion of the viral and endosomal membranes and release of the nucleocapsid (NC) into the cytoplasmic compartment. kb, Kilobase.

multivalent binding sites must be created. The suggestion finds further support in the recent cryoelectron microscopic analyses of both SFV and Sindbis virus (Vogel et al., 1986; Fuller, 1987). These studies have shown that the spike structures of the alphaviruses are formed from three copies of the E2E1 heterodimer. It could be that productive budding is driven only through the association of these (E2E1)$_3$ structures with the nucleocapsid.

OLIGOMERIZATION CONTROL OF ENTRY FUNCTIONS

Virus budding and fusion represent large opposite processes that normally follow each other during virus replication. An important question is how these processes are regulated so that they do not interfere with each other. One key factor in many other membrane viruses appears to be the synthesis of a fusion protein in an inactive precursor form (see, e.g., Boulay et al. [1987]). Proteolytic conversion of the precursor into a fusion-competent form usually takes place just before virus budding. Could the p62 precursor protein of SFV represent such a reg-

ulatable fusogen? This does not seem to be very likely because, as already discussed, this subunit appears to have a key function in membrane assembly. In contrast, there is good indirect evidence suggesting that the E1 protein carries the fusion activity. First, Omar and Koblet (1988) have created SFV particles essentially free of p62 by proteolysis and showed that these E1 viruses are both fusogenic and infectious. Second, Boggs et al. (1989) have characterized a Sindbis virus variant with a changed pH optimum for fusion and found that the phenotype can be correlated with amino acid substitutions in the E1 polypeptide. Third, Kielian and Helenius (1985) have shown that, in contrast to the E2 protein, the E1 protein gains protease resistance upon exposure to mildly acidic pH. Therefore, it could be that the p62 protein and its cleavage are totally unrelated to the fusion function of SFV.

To test this, we created cleavage-deficient mutants of the p62 protein by changing the Arg-66 residue at the cleavage site to either Leu or Glu. When these mutants were expressed from a 26S cDNA together with the other structural proteins in BHK cells, no cleavage of pulse-labeled p62 could be demonstrated even after 3 h of chase. However, exogenously added trypsin was able to cleave the mutated p62 forms into apparently authentic E2. This cleavage probably occurred at one of the two remaining basic residues at the cleavage site. The trypsin assay enabled us to demonstrate that most of the mutant p62 protein was transported to the cell surface. Using our coimmunoprecipitation assay, we were also able to show that essentially all membrane protein subunits were present as p62E1 heterodimers.

To test the fusion activity of the p62 cleavage site mutants, we injected wild-type or mutant cDNA constructions into adjacent BHK cells on cover slips and then incubated the cells to allow surface expression of the membrane proteins. Fusion activity was assayed by screening for polykaryon formation after activation of potential fusogenic activity of viral membrane proteins by a 1-min incubation at pH 5.5 (Kondor-Koch et al., 1983). The results were clearcut: neither of the mutants induced polykaryon formation (Fig. 3). Fusion could, however, be induced by the mutants if the mutants were treated with exogenous trypsin before the low-pH incubation to allow p62 cleavage. Hence, the results showed that p62 cleavage is essential to activate the fusion function of the SFV spike.

How then can the p62 (E2) protein affect the fusogenic activity if this is a property of the E1 protein? The most reasonable explanation is that p62 cleavage affects subunit association in the heterodimer. One simple possibility is that the mature heterodimer E2E1 has become sensitive to acid-induced dissociation and that the E1 protein must be free from p62 or E2 in order to function in fusion. Support for this model was obtained when we studied the strength of the heterodimeric association of the precursor and the mature form of the heterodimer in buffers of decreasing pH. The coimmunoprecipitation analyses demonstrated a marked resistance to dissociation of the p62E1 complex, whereas the mature E2E1 dissociated already at pH 6.4. Thus, the p62 (E2)E1 oligomerization reaction seems to be a major regulator both for the process of membrane assembly and for the fusion reaction during virus entry. These features are schematically summarized in Fig. 4.

REFERENCES

Boggs, W. M., C. S. Hahn, E. G. Strauss, J. H. Strauss, and D. E. Griffin. 1989. Low pH-dependent Sindbis virus-induced fusion of BHK cells: differences between strains correlate with amino acid changes in the E1 glycoprotein. *Virology* **169**:485–488.

Boulay, F., R. W. Doms, I. Wilson, and A. Helenius. 1987. The influenza hemagglutinin precursor as an acid-sensitive probe of the biosynthetic pathway. *EMBO J.* **6**:2643–2650.

Dubois-Dalcq, M., K. V. Holmes, and B. Rentier. 1984. *Assembly of Enveloped RNA Viruses*. Springer-Verlag, New York.

Fuller, S. D. 1987. The T=4 envelope of Sindbis virus is organized by interactions with a complementary T=3 capsid. *Cell* **48**:932–934.

Garoff, H., A.-M. Frishauf, K. Simons, H. Lehrach, and H. Delius. 1980. Nucleotide sequence of cDNA coding for Semliki Forest virus membrane glycoproteins. *Nature* (London) **288**:236–241.

Garoff, H., C. Kondor-Koch, and H. Riedel. 1982. Structure and assembly of alphaviruses. *Curr. Top. Microbiol. Immunol.* **99**:1–50.

Kielian, M., and A. Helenius. 1985. pH-induced alterations in the fusogenic spike of Semliki forest virus. *J. Cell Biol.* **101**:2284–2291.

Kielian, M., and A. Helenius. 1986. Entry of alphaviruses, p. 91–119. *In* S. S. Schlesinger and M. J. Schlesinger (ed.), *The Togaviridae and Flaviviridae*. Plenum Publishing Corp., New York.

Kondor-Koch, C., B. Burke, and H. Garoff. 1983. Expression of Semliki Forest virus proteins from cloned complementary DNA. I. The fusion activity of the spike glycoprotein. *J. Cell Biol.* **97**:644–651.

Marsh, M., and A. Helenius. 1989. Virus entry into animal cells. *Adv. Virus Res.* **36**:107–151.

Melancon, P., and H. Garoff. 1986. Reinitiation of translocation in the Semliki Forest virus structural polyprotein: identification of the signal for the E1 glycoprotein. *EMBO J.* **5**:1551–1560.

Melancon, P., and H. Garoff. 1987. Processing of the Semliki Forest virus structural polyprotein: role of the capsid protease. *J. Virol.* **61**:1301–1309.

Omar, A., and H. Koblet. 1988. Semliki Forest virus particles containing only the E1 envelope glycoprotein are infectious and can induce cell-cell fusion. *Virology* **166**:17–23.

Rice, C. M., and J. H. Strauss. 1982. Association of Sindbis virion glycoproteins and their precursors. *J. Mol. Biol.* **154**:325–348.

Simons, K., and S. D. Fuller. 1987. The budding of enveloped viruses: a paradigm for membrane sorting, p. 139–150. *In* R. Burnett and H. Vogel (ed.), *Biological Organization: Macromolecular Interactions at High Resolution*. Academic Press, Inc., New York.

Vaux, D. J. T., A. Helenius, and I. Mellman. 1988. Spike-nucleocapsid interaction in Semliki Forest virus reconstructed using network antibodies. *Nature* (London) **336**:36–42.

Vogel, D. J. T., S. W. Provencher, C.-H. von Bonsdorff, M. Adrian, and J. Dubochet. 1986. Envelope structure of Semliki Forest virus reconstructed from cryo-electron micrographs. *Nature* (London) **320:** 533–535.

von Heijne, G. 1985. Structural and thermodynamic aspects of the transfer of proteins into and across membranes. *Curr. Top. Membr. Transp.* **24:**151–179.

Ziemiecki, A., and H. Garoff. 1978. Subunit composition of the membrane glycoprotein complex of Semliki Forest virus. *J. Mol. Biol.* **122:**259–269.

Ziemiecki, A., H. Garoff, and K. Simons. 1980. Formation of the Semliki Forest membrane glycoprotein complexes in the infected cell. *J. Gen. Virol.* **50:**111–123.

Chapter 25

The Tobacco Etch Virus 49-Kilodalton Nuclear Inclusion Protein: a Viral Proteinase with Regulated Cleavage?

T. Dawn Parks, Holly A. Smith, David E. Slade, James C. Carrington,
Robert E. Johnston, and William G. Dougherty

Many RNA-containing viruses use a strategy of genome expression whereby polyprotein precursors are synthesized during translation. These precursors are then cleaved to individual gene products, frequently by virus-encoded proteinases (Wellink and van Kammen, 1988; Kräusslich and Wimmer, 1988). The plant potyviruses, of which tobacco etch virus (TEV) is a member, exhibit this type of gene expression. TEV consists of a single plus-stranded RNA molecule that is polyadenylated at the 3' end and has a small protein (VPg) covalently linked, presumably at the 5' terminus. This RNA is encapsidated by approximately 2,000 capsid protein monomers (for a review, see Dougherty and Carrington [1988]). Other virus groups that also synthesize large polyproteins from RNA genomes with linked VPg molecules are the plant como- and nepoviruses and animal picornaviruses.

The TEV genome, 9,496 nucleotides in length, contains a single open reading frame with the potential to direct synthesis of an ~346-kilodalton (kDa) protein (Allison et al., 1986). This large polyprotein is not detected, however; two virus-encoded proteinases are responsible for cleavage of the polyprotein into smaller polyproteins and individual gene products (Fig. 1). This proteolytic processing results in possibly eight mature products arising from translation of the TEV genome. The 87-kDa protein, thought to be involved in aphid transmission, has been shown to have a proteolytic activity associated with the carboxyl-terminal one-third of the protein. This proteinase cleaves itself from the polyprotein at a specific Gly-Gly dipeptide (Carrington et al., 1989). In addition, the 49-kDa protein has a proteolytic activity localized in its carboxyl-terminal half, as determined by deletion analysis and monoclonal antibody mapping studies. The TEV 49-kDa proteinase is autocatalytically liberated from the polyprotein and also directs cleavage at three other sites (Carrington and Dougherty, 1987a, 1987b; Slade et al., 1989).

The 49-kDa proteinase, as well as the poliovirus 3C protein and other viral proteinases, recently has been proposed to be related to the trypsin family of cellular serine proteases (Bazan and Fletterick, 1988; Gorbalenya et al., 1989). In these viral proteins, however, the active site serine of the catalytic triad has been replaced with cysteine. The other amino acids in this triad are histidine and aspartic acid. In TEV, the His, Asp, and Cys are believed to be located at 49-kDa amino acid positions 234, 269, and 339, respectively (Dougherty et al., 1989b).

Unlike other viral proteinases, the TEV 49-kDa protein recognizes an extended amino acid sequence at each naturally occurring cleavage site. Whereas its poliovirus analog, the 3C proteinase, cleaves at an accessible Gln-Gly or Gln-Ser dipeptide, a specific heptapeptide sequence is required by the TEV 49-kDa proteinase (Ypma-Wong et al., 1988; Carrington and Dougherty, 1988). Nucleotide sequences coding for this amino acid sequence can be moved via recombinant DNA techniques to foreign loca-

T. Dawn Parks, Holly A. Smith, and William G. Dougherty, Department of Microbiology, Oregon State University, Corvallis, Oregon 97331-3804.   *David E. Slade,* Department of Microbiology, North Carolina State University, Raleigh, North Carolina 27695.   *James C. Carrington,* Department of Biology, Texas A&M University, College Station, Texas 77843.   *Robert E. Johnston,* Department of Microbiology and Immunology, University of North Carolina, Chapel Hill, North Carolina 27599.

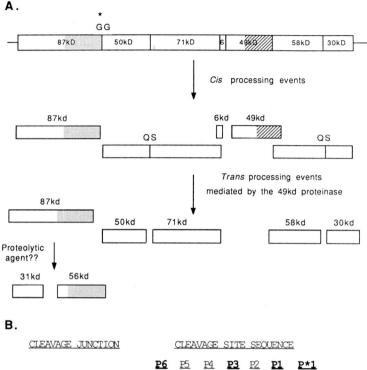

FIGURE 1. (A) Proposed processing scheme of the TEV polyprotein. Translation of the single open reading would result in the synthesis of a 346-kDa polyprotein. The 87-kDa protein cleaves itself from the polyprotein at a Gly-Gly (GG) dipeptide (*; activity mapped to stippled area). Processing events mediated in a *cis* reaction by the 49-kDa proteinase (activity mapped to hashed area) result in the formation of 6- and 49-kDa proteins and two polyproteins. These polyprotein precursors are further cleaved by the 49-kDa proteinase in *trans* reactions, resulting in the formation of 50- and 71-kDa and 58- and 30-kDa protein products. The 87-kDa protein is thought to be further processed to 31- and 56-kDa products by an as yet unidentified proteolytic mechanism. (B) Amino acid sequences found at TEV cleavage junctions. The sequences of the 87-kDa/50-kDa cleavage site (*) and the five 49-kDa cleavage sites are presented. Amino acid positions that are conserved among the 49-kDa cleavage sites are shown in boldface.

tions on the TEV genome and, when expressed as part of a protein, still direct specific cleavage by the TEV 49-kDa proteinase. A synthetic oligonucleotide linker that codes for only the seven-amino-acid cleavage sequence has been inserted into foreign sequences, generating a new functional cleavage site when translated and processed. However, insertion of a sequence coding for only a Gln-Ser dipeptide does not create a functional cleavage site (Carrington and Dougherty, 1988). The consensus seven-amino-acid sequence recognized by the TEV 49-kDa proteinase is as follows:

P6 P5 P4 P3 P2 P1 P'1
Glu - Xaa - Xaa - Tyr - Xaa - Gln - Ser or Gly

This amino acid sequence is found at only five locations on the TEV polyprotein, in each case at a site cleaved by the 49-kDa proteinase.

Cleavage is between the Gln-Ser or Gln-Gly dipeptide.

Four of the seven amino acid positions (Glu, Tyr, Gln, and Ser or Gly at P6, P3, P1, and P'1) are conserved among naturally occurring TEV 49-kDa cleavage sites (Fig. 1B). Extensive site-directed mutagenesis of each of these conserved residues, using the TEV 58-kDa/30-kDa cleavage site as a model, has shown that these amino acids play a pivotal role in defining a TEV cleavage site. Substitution of these residues with other amino acids usually results in the elimination, or at least a substantial reduction, of cleavage (Dougherty et al., 1988; Dougherty et al., 1989a).

In contrast to the four conserved positions in the heptapeptide sequence, three positions, P5, P4, and P2, vary among naturally occurring cleavage sites (Fig. 1B). Site-directed substitutions involving these variable positions show a wide range of cleavage efficiencies. In addition, with two exceptions, mutations of amino acids located outside of the heptapeptide sequence generally have no effect on cleavage (Dougherty et al., 1989a).

On the basis of these observations, we hypothesized that controlled proteolytic processing may be a mechanism of gene regulation during TEV infection and replication. Because TEV expresses its genetic information from a genome-length RNA, producing no detectable subgenomic messages, transcriptional regulation is not likely (Dougherty, 1983). The production (amount and time of appearance) of mature viral proteins might be controlled by the ability of a particular site to be cleaved by the 49-kDa proteinase. Specific amino acids found at the nonconserved (P5, P4, and P2) positions may determine the relative efficiency of cleavage. Initial cell-free studies demonstrate that these residues can affect the rate of proteolytic processing. This finding implicates the 49-kDa proteinase and the cleavage sites it recognizes as a possible means of posttranslational gene regulation during expression of the TEV genome.

THE 49-kDa PROTEINASE IS RELATED TO THE TRYPSIN FAMILY OF SERINE PROTEINASES

The TEV 49-kDa proteinase is similar to other viral proteinases in several respects, including cleavage between certain Gln-Gly or Gln-Ser dipeptides, activity enhancement by dithiothreitol, and a conserved Cys-15 amino acids-His motif. It has been suggested that these

two amino acids may be involved in catalysis, implicating these viral proteinases as a new type of cysteine proteinase (Argos et al., 1984; Carrington and Dougherty, 1987a). However, recent proposals suggest that these virus-encoded proteinases have a tertiary structure similar to that of the trypsin family of serine proteinases (Bazan and Fletterick, 1988; Gorbalenya et al., 1989). Folding of the protein would form an active site with a catalytic triad of His, Asp, and Cys residues. The Cys residue would replace the Ser of cellular trypsin as a nucleophile in the viral enzymes.

Inhibitor and mutational studies were conducted to test this model. In the inhibitor studies, various proteinase inhibitors were incubated with TEV 49-kDa proteinase. The ability of a compound to arrest proteolytic processing was then assayed by the addition of a 34-kDa substrate containing a TEV cleavage site. Successful inhibition resulted in no substrate processing. Inhibitors that had no effect on proteolytic processing resulted in conversion of the 34-kDa substrate to 30- and 4-kDa products. This study included trypsin and papain as control proteinases representative of the serine and cysteine families, respectively. Of the inhibitors tested, three were effective in inhibiting the TEV 49-kDa proteinase. Two of these, N-ethylmaleimide and iodoacetamide, are general alkylating agents and likely modify exposed Cys residues. Zinc ions were also effective in blocking proteolytic processing by the TEV 49-kDa proteinase, as has been demonstrated for other RNA virus-encoded proteinases (Butterworth and Korant, 1974; Pelham, 1979; Peng and Shih, 1984). Zinc inactivation may be accomplished by binding to the active-site His-234, Cys-339, and perhaps the His-355 residue implicated in substrate binding (Bazan and Fletterick, 1988). Other cysteine and serine proteinase inhibitors [phenylmethylsulfonyl fluoride, L-1-chloro-3-(4-tosylamide)-7-amino-2-hepaton (TLCK), aprotinin, L-1-chloro-3-(4-tosylamide)-4-phenyl-2-butanone (TPCK), leupeptin, and E-64] had marginal or no effects on the TEV 49-kDa proteinase. Additional compounds (bestatin, pepstatin A, and EDTA), representing inhibitors of aspartic and metalloproteinases, had no effect on the 49-kDa proteinase. These results revealed that the TEV 49-kDa proteinase did not have characteristics which placed it unambiguously into one of the four classical proteinase groups.

The recent models of Bazan and Fletterick (1988) and Gorbalenya and co-workers (1989) propose that the active site of the TEV 49-kDa proteinase consists of His, Asp, and Cys residues. Site-directed mutagenesis studies have been conducted in an attempt to identify specific putative amino acids involved in catalysis. Mutations have been introduced in various positions in the plasmid pTL37-5473, which contains cDNA encoding a functional TEV proteinase as part of a larger polyprotein precursor (Carrington et al., 1987). Cell-free transcription and translation of these cDNA sequences produces a 75-kDa precursor that is rapidly processed to a 49-kDa proteinase and smaller products. His codons at TEV 49-kDa protein amino acid positions 216, 234, and 249 have been individually converted to Tyr codons in this plasmid. Upon transcription and translation of the mutated sequences, the Tyr replacements at positions 216 and 249 have no apparent effect on the processing ability of the 49-kDa proteinase. The 75-kDa precursor is converted into the expected processed products. However, the His-to-Tyr replacement at position 234 results in a proteinase that does not process, as only the 75-kDa substrate is detected. This suggests, but does not prove, that His-234 is involved in the catalytic triad of the TEV 49-kDa proteinase (Dougherty et al., 1989b).

Site-directed mutations also have been introduced at the Asp residue purportedly involved in catalysis. This amino acid has been changed to Asn, Val, or Glu. Mutant proteinases with Asn or Val at position 269 are no longer able to autoprocess, and only the 75-kDa precursor accumulates. However, Glu replacements at this position are able to process at the N-terminal but not the C-terminal cleavage site of the 49-kDa proteinase.

One predicted major difference between the cellular serine proteinases and their viral counterparts is the substitution of Cys for Ser in the viral enzymes. In the TEV 49-kDa proteinase, Cys-339 has been hypothesized to be the active-site homolog. Ten different amino acids have been substituted for this residue. No processing is observed in the normal assay period (1 h). However, after prolonged incubation (12 h), a polyprotein with a 49-kDa proteinase containing a Cys-to-Ser substitution is able to autoprocess to a small degree. In addition, processing in this mutant is suppressed by the addition of phenylmethylsulfonyl fluoride, a serine proteinase inhibitor.

Together, the inhibitor study and site-directed mutagenesis of proposed active-site amino acids provide initial support for the model that TEV and other viral proteinases are related to the trypsin family of serine proteinases.

POTENTIAL FOR POSTTRANSLATIONAL REGULATION OF GENE EXPRESSION?

The TEV 49-kDa proteinase recognizes an extended heptapeptide sequence (Fig. 1B). In this sequence, four amino acids are conserved, while three vary among the five naturally occurring cleavage sites. In site-directed mutagenesis studies of the TEV 58-kDa/30-kDa cleavage site, amino acid substitutions of the nonconserved amino acids result in substrates which show a wide range of cleavage rates (Dougherty et al., 1988; Dougherty et al., 1989a). We hypothesized that differential proteolytic processing could be a mechanism whereby TEV controlled the production of mature gene products. Earlier observations from in vitro translation studies of TEV RNA suggested this (Dougherty and Hiebert, 1980) in that the 71-kDa cytoplasmic pinwheel inclusion protein always is associated with a higher-molecular-weight precursor, whereas the 30-kDa protein is detected both as a 30-kDa product and as a part of a higher-molecular-weight polyprotein (Fig. 1A).

Our initial experiments (Dougherty and Parks, 1989) to address this question have focused on two naturally occurring TEV cleavage sites, the 50-kDa/71-kDa and 58-kDa/30-kDa cleavage sites. Both of these sites are processed in a bimolecular or trans reaction by the TEV 49-kDa proteinase. Plasmid constructs containing cDNA encoding these two cleavage sites have been developed (Fig. 2A). After cell-free transcription and translation in the presence of [^{35}S]methionine, each radiolabeled substrate was incubated with TEV 49-kDa proteinase (provided in the form of nuclear inclusion bodies). Samples were taken at various time periods, and precursor and products were separated by sodium dodecyl sulfate-polyacrylamide gel electrophoresis. After autoradiography, the radioactivity associated with precursor and product bands was quantitated by using an Ambis Beta scanner. The results from this experiment showed that the 58-kDa/30-kDa cleavage site was processed much faster than the 50-kDa/71-kDa site in our standard assay (Fig. 2B). In ~4.5 min, one-half of the precursor containing the 58-kDa/30-kDa cleavage site sequence had been processed to product (i.e., half-time). In contrast, the half-time of processing of the 50-kDa/71-kDa cleavage site was much slower, ~30 min.

Site-directed mutations that changed the 58-kDa/30-kDa heptapeptide sequence into a 50-kDa/71-kDa cleavage sequence were then introduced into these plasmids. In the reciprocal experiment, the 50-kDa/71-kDa cleavage site

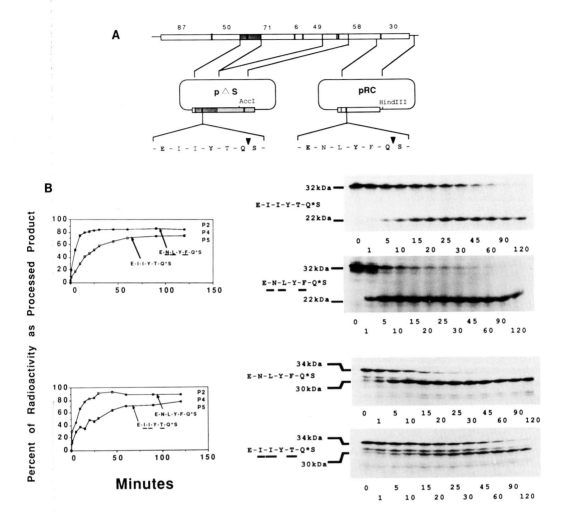

FIGURE 2. Differential processing of two TEV 49-kDa hepapeptide cleavage sequences. (A) Schematic of the TEV genome (top) and plasmids pΔS and pRC. Molecular weights (in thousands) of the gene products are shown. These portions indicated were copied into cDNA and inserted into the transcription vector pTL37 (Carrington et al., 1988). The 50-kDa/71-kDa cleavage site and flanking polyprotein sequences are contained in plasmid pΔS. The 58-kDa/30-kDa cleavage site and flanking polyprotein sequences are contained in plasmid pRC. The cleavage site amino acid sequence for each site is shown. (B) Processing of 50-kDa/71-kDa and 58-kDa/30-kDa cleavage sites and mutant variants. After transcription and translation, substrates were incubated with 49-kDa proteinase, and samples were removed at the time points indicated. After sodium dodecyl sulfate-polyacrylamide gel electrophoresis, the amount of radioactivity in precursor and product bands was determined. The quantitated results are presented in the graphs as the percentage of radioactivity found as processed product. The top graph shows the results of processing at the 50-kDa/71-kDa cleavage site, using substrates containing a naturally occurring 50-kDa/71-kDa cleavage sequence and also a mutated site containing the heptapeptide sequence found at the 58-kDa/30-kDa site. Autoradiographs of these reactions are shown at the right. The lower graph shows the results of processing at the 58-kDa/30-kDa site. Substrates containing the naturally occurring 58-kDa/30-kDa cleavage sequence and a mutated version containing the sequence found at the 50-kDa/71-kDa site are shown. Autoradiographs of these reactions are shown at the right. The amino acid sequences are shown in single-letter code; mutated amino acids are underlined. The radioactive band (31 kDa) between the 34-kDa substrate and the 30-kDa product is believed to be a substrate that arose through internal initiation.

sequence was changed into a 58-kDa/30-kDa sequence. These mutations changed only the three nonconserved amino acids at the P5, P4, and P2 positions into those found at the alternate site. Processing of these substrates revealed that the 58-kDa/30-kDa heptapeptide sequence was processed more rapidly, whether in its natural location or in the context of the 50-kDa/71-kDa site (Fig. 2B). Individual amino acid replacements at each of the nonconserved positions gave intermediate processing profiles (Dougherty and Parks, 1989).

An additional plasmid construct was used to determine whether the same rules govern cleavage when a heptapeptide sequence is moved to a foreign location. This construct (pRC-71N) contained both the TEV 58-kDa/30-kDa cleavage site in its natural polyprotein setting and a 50-kDa/71-kDa cleavage site that had been inserted into the TEV 30-kDa capsid protein gene (Dougherty and Parks, 1989). Cleavage of this substrate again showed that the 58-kDa/30-kDa heptapeptide sequence was cleaved more rapidly, as there was rapid accumulation of a 30-kDa capsid protein-related product, which was subsequently cleaved to a 23-kDa product. This 30-kDa product was the result of cleavage at the naturally occurring 58-kDa/30-kDa cleavage junction. Slower cleavage of the inserted 50-kDa/71-kDa cleavage site processed this 30-kDa protein into a 23-kDa protein. Site-directed mutagenesis of this construct, resulting in two 58-kDa/30-kDa heptapeptide cleavage sequences, gave a different profile. There was very little accumulation of the 30-kDa protein but rapid production of a 23-kDa product. Again, this result suggested that the 58-kDa/30-kDa heptapeptide sequence was cleaved more rapidly than the 50-kDa/71-kDa cleavage sequence.

The results of these mutations support the hypothesis that the nonconserved amino acids could be involved in regulating the rate at which cleavage occurs. We do not yet understand the biochemical basis of these differential rates of cleavage. Possibly a "slow" cleavage site binds to the proteinase at the same rate as a "fast" site but is not cleaved as rapidly. Alternatively, binding to the proteinase may be the determining factor in distinguishing fast and slow cleavage events. The ability to differentially regulate proteolytic processing may be a means by which viruses that express their information as polyproteins can regulate the appearance or disappearance of specific gene products.

ACKNOWLEDGMENTS. W.G.D. acknowledges support from the National Science Foundation, the Department of Energy, and the N. L. Tartar Foundation.

The technical assistance of Susan Cary and David Hustead is greatly appreciated.

REFERENCES

Allison, R., R. E. Johnston, and W. G. Dougherty. 1986. The nucleotide sequence of the coding region of tobacco etch virus genomic RNA: evidence for the synthesis of a single polyprotein. *Virology* **154:** 9–20.

Argos, P., P. Kamer, M. J. H. Nicklin, and E. Wimmer. 1984. Similarity in gene organization and homology between proteins of animal picornaviruses and a plant comovirus suggest a common ancestry of these virus families. *Nucleic Acids Res.* **12:** 7251–7267.

Bazan, J. F., and R. J. Fletterick. 1988. Viral cysteine proteases are homologous to the trypsin-like family of serine proteases: structural and functional implications. *Proc. Natl. Acad. Sci. USA* **85:**7872–7876.

Butterworth, B. E., and B. Korant. 1974. Characterization of the large picornavirus polypeptides produced in the presence of zinc ions. *J. Virol.* **14:** 282–291.

Carrington, J. C., S. M. Cary, and W. G. Dougherty. 1988. Mutational analysis of tobacco etch virus polyprotein processing: *cis* and *trans* proteolytic activities of polyproteins containing the 49-kilodalton proteinase. *J. Virol.* **62:**2313–2320.

Carrington, J. C., S. M. Cary, T. D. Parks, and W. G. Dougherty. 1989. A second proteinase encoded by a plant potyvirus genome. *EMBO J.* **8:**365–370.

Carrington, J. C., and W. G. Dougherty. 1987a. Small nuclear inclusion protein encoded by a plant potyvirus genome is a protease. *J. Virol.* **61:**2540–2548.

Carrington, J. C., and W. G. Dougherty. 1987b. Processing of the tobacco etch virus 49K protease requires autoproteolysis. *Virology* **160:**355–362.

Carrington, J. C., and W. G. Dougherty. 1988. A viral cleavage site cassette: identification of amino acid sequences required for tobacco etch virus polyprotein processing. *Proc. Natl. Acad. Sci. USA* **85:** 3391–3395.

Carrington, J. C., T. D. Parks, S. M. Cary, and W. G. Dougherty. 1987. Vectors for cell-free expression and mutagenesis of protein coding sequences. *Nucleic Acids Res.* **15:**10066.

Dougherty, W. G. 1983. Analysis of viral RNA isolated from tobacco leaf tissue infected with tobacco etch virus. *Virology* **131:**473–481.

Dougherty, W. G., and J. C. Carrington. 1988. Expression and function of potyviral gene products. *Annu. Rev. Phytopathol.* **26:**123–143.

Dougherty, W. G., J. C. Carrington, S. M. Cary, and T. D. Parks. 1988. Biochemical and mutational analysis of a plant virus polyprotein cleavage site. *EMBO J.* **7**:1281–1287.

Dougherty, W. G., S. M. Cary, and T. D. Parks. 1989a. Molecular genetic analysis of a plant virus polyprotein cleavage site: a model. *Virology* **171**:356–364.

Dougherty, W. G., and E. Hiebert. 1980. Translation of potyviral RNA in a rabbit reticulocyte lysate: identification of nuclear inclusion protein as products of tobacco etch virus RNA translation and cytoplasmic inclusion protein as a product of the potyvirus genome. *Virology* **104**:174–182.

Dougherty, W. G., and T. D. Parks. 1989. Molecular genetic and biochemical evidence for the involvement of the heptapeptide cleavage sequence in determining the rate of cleavage at tobacco etch virus cleavage sites in cell-free assays. *Virology* **172**:145–155.

Dougherty, W. G., T. D. Parks, S. M. Cary, J. F. Bazan, and R. J. Fletterick. 1989b. Characterization of the catalytic residues of the tobacco etch virus 49kDa proteinase. *Virology* **172**:302–310.

Gorbalenya, A. E., A. P. Donchenko, V. M. Blinov, and E. V. Koonin. 1989. Cysteine proteases of positive strand RNA viruses and chymotrypsin-like serine proteases: a distinct superfamily with a common structural fold. *FEBS Lett.* **243**:103–114.

Kräusslich, H.-G., and E. Wimmer. 1988. Viral proteinases. *Annu. Rev. Biochem.* **57**:701–754.

Pelham, H. R. B. 1979. Synthesis and proteolytic processing of cowpea mosaic virus proteins in reticulocyte lysates. *Virology* **96**:463–477.

Peng, X. X., and D. S. Shih. 1984. Proteolytic processing of the proteins translated from the bottom component of cowpea mosaic virus. *J. Biol. Chem.* **259**:3197–3201.

Slade, D. E., R. E. Johnston, and W. G. Dougherty. 1989. Generation and characterization of monoclonal antibodies reactive with the 49kDa proteinase of tobacco etch virus. *Virology* **173**:499–508.

Wellink, J., and A. van Kammen. 1988. Proteases involved in the processing of viral polyproteins. *Arch. Virol.* **98**:1–26.

Ypma-Wong, M. F., D. J. Filman, J. M. Hogle, and B. L. Semler. 1988. Structural domains of the poliovirus protein are major determinants for polyprotein cleavage at gln-gly pairs. *J. Biol. Chem.* **263**:17846–17856.

Chapter 26

Evidence for a Second Protease in a Human Rhinovirus

Tim Skern, Wolfgang Sommergruber, Manfred Zorn, Dieter Blaas, Frederike Fessl,
Peter Volkmann, Ingrid Maurer-Fogy, Peter Pallai, Vincent Merluzzi,
and Ernst Kuechler

The central importance of proteolytic processing in the replication of various classes of viruses is now well documented (Kräusslich and Wimmer, 1988). For picornaviruses, and for poliovirus in particular, there is good evidence that the viral genome encodes at least two proteases; of the 11 proteolytic cleavages known to occur on the polyprotein of poliovirus, 9 are carried out by the protease designated 3C and 2 are carried out by the protease designated 2A (Kräusslich and Wimmer, 1988). The first cleavage carried out by 2A lies between the carboxy terminus of VP1 and the amino terminus of 2A itself. This cleavage is the first to occur in poliovirus, taking place before completion of translation of the entire polyprotein and resulting in an immediate separation of the structural from the nonstructural proteins. The second cleavage site of 2A lies in the polymerase protein, 3D; the biological significance of this cleavage is unclear. We wished to examine whether the 2A protein in human rhinovirus 2 (HRV-2) is a protease and, if so, to determine to which class of proteins it belongs.

DEFINITION OF PROTEIN 2A FROM HRV-2 AS A PROTEASE

To show that the 2A protein of HRV-2 is a protease, it was decided to express the VP1-2A-2B region in *Escherichia coli*. Two DNA fragments from this region were assembled (Fig. 1). The first fragment (Fig. 1A) contained the carboxy terminus of VP3, all of VP1, and the amino-terminal third of 2A (nucleotides 2145 to 3320 of HRV-2 [Skern et al., 1985]); the second fragment (Fig. 1B) contained, in addition to these regions, all of 2A and part of 2B (nucleotides 2145 to 3704 of HRV-2). These fragments were then introduced into a bacterial expression vector, pEx34b, so that the rhinovirus proteins would be expressed as a fusion protein joined to the N terminus of the MS2 polymerase (Fig. 1). The expression blocks are under the control of the lambda left promoter; induction can be achieved by transforming the plasmids into a bacterial strain harboring a temperature-sensitive lambda repressor and incubating them at the required temperature (42°C).

Expression of the fusion proteins was carried out as described previously (Skern et al., 1987; Sommergruber et al., 1989); the fusion proteins were found to be insoluble. Therefore, only proteins present in the cell pellet after sonication were examined. Figure 2 shows a typical pattern of expressed proteins when analyzed on a 10% polyacrylamide gel. The construction containing only the first third of the 2A protease gave rise to one protein of 55 kilodaltons (kDa) on induction (Fig. 2A, lane 3). In contrast, the construction possessing an intact 2A gave rise to two proteins (50 and 24 kDa) on induction (Fig. 2A, lane 2). Western blot (immunoblot) analysis of similar gels showed that the 58- and 50-kDa proteins (but not the 24-kDa protein) contained HRV-2 VP1 sequences (Fig. 2B). Two facts pointed to the presence of a

Tim Skern, Manfred Zorn, Dieter Blaas, and *Ernst Kuechler,* Institut für Biochemie der Universität Wien, Währingerstrasse 17, A 1090 Vienna, Austria. *Wolfgang Sommergruber, Frederike Fessl, Peter Volkmann, and Ingrid Maurer-Fogy,* Ernst Boehringer Institut für Arzneimittelforschung, Dr. Boehringergasse 5–11, A 1120 Vienna, Austria. *Peter Pallai and Vincent Merluzzi,* Boehringer Ingelheim Pharmaceuticals, Inc., 90 East Ridge, P.O. Box 368, Ridgefield, Connecticut 06877.

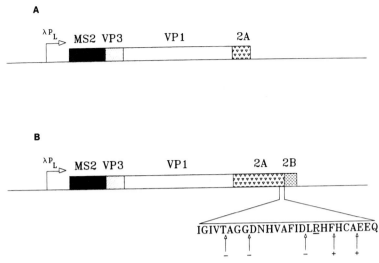

FIGURE 1. Fragments used to express the VP1-2A-2B region of HRV-2 in *E. coli*. (A) VP3′-VP1-2A′; (B) VP3′-VP1-2A-2B′. The expanded sequence in panel B shows the 25 amino acids at the carboxy terminus from 2A; arrows indicate the limits of the BAL 31 nuclease deletions. A plus sign indicates that a deletion was still active; a minus sign indicates that it was inactive. The underlined arginine is referred to in the text.

protease. First, the largest expressed protein was observed with the shortest construction; second, two products were obtained with the construction possessing all of 2A, with combined molecular weights corresponding to that expected for an unprocessed product. It was therefore inferred that 2A was cutting itself off from the fusion protein; the construction shown in Fig. 1A was unable to do this, having only one third of the 2A protein.

To confirm that 2A was indeed the protease responsible for this cleavage, a series of deletion mutants was constructed. The plasmid containing the expression block with 2A and part of 2B (Fig. 1B) was opened by restriction enzyme digestion at the end of the 2B sequence and incubated with BAL 31 nuclease. After religation and transformation, the plasmids were sequenced to determine how much of 2B and 2A had been removed. Protein expression was then induced from plasmids containing a range of deletions across 2A. Removal of 2B sequences had no effect on proteolytic cleavage (data not shown). Deletions removing up to six amino acids were still active. However, all deletions with more than 10 amino acids failed to cleave (Fig. 1). Replacement using site-directed mutagenesis of the arginine residue lying between these limits (underlined in Fig. 1) by glutamine also inhibited proteolysis. This result confirmed that the cleavage was being carried out by the

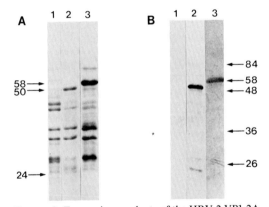

FIGURE 2. Expression products of the HRV-2 VP1-2A region. Positions and molecular sizes (in kilodaltons) of marker proteins are shown on the right. Positions and molecular sizes (in kilodaltons) of proteins expressed from the plasmids are shown on the left. (A) Coomassie blue-stained polyacrylamide gel of proteins expressed from the following plasmids: pEx34c (lane 1); construction from Fig. 1B (lane 2); and construction from Fig. 1A (lane 3). The bands represent the following proteins: 58 kDa, protein with truncated 2A; 50 kDa, cleavage product containing MS2 polymerase and VP1 sequences; 24 kDa, cleavage product containing sequences from 2A and 2B. (B) Western blot obtained from an identical gel probed with HRV-2 antiserum. (From Sommergruber et al. [1989] with permission of Academic Press.)

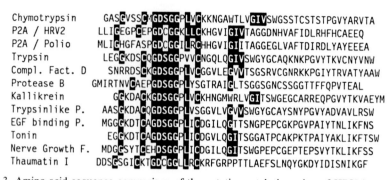

FIGURE 3. Amino acid sequence comparison of the putative catalytic region of HRV-2 protease 2A (amino acids 94 to 142) with those of proteins from the SWISSPROT and Protein Information Resource data banks (see text). Identical amino acids occurring at similar positions are indicated by reverse shading. Partial sequences of the following proteins with scores greater than 43 are shown. The mean score was 23.2 (±6.58). The SWISSPROT identity number and the amino acids shown for each protein are given: bovine chymotrypsin (CTRA$BOVIN), 184 to 234; 2A HRV-2, 94 to 142; 2A poliovirus type 1 (Sabin) (POLG$POLVS), 978 to 1,028; porcine trypsin (TRYP$PIG), 174 to 224; human complement factor D (CFAD$HUMAN), 168 to 218; protease B from *Streptomyces griseus* (PRTB$STRGR), 128 to 178; human kallikrein (KAL$HUMAN), 569 to 619; trypsinlike protein from *Drosophila melanogaster* (TRLI$DROME), 199 to 249; mouse epidermal growth factor binding protein (EGBB$ MOUSE), 202 to 252; rat tonin (TONI$RAT), 177 to 227; mouse nerve growth factor alpha-chain precursor (NGFA$MOUSE), 197 to 247; and thaumatin I from *T. danielli* (THM1$THADA), 49 to 99. (From Sommergruber et al. [1989] with permission of Academic Press.)

2A protein and ruled out any involvement of a bacterial protease. Further experiments were performed with a construction containing two stop codons at the end of the 2A protein, which gave rise to a mature 2A protein of 15 kDa. N-terminal sequencing of this 15-kDa protein showed that 2A cleavage was occurring at the amino acid pair (alanine-glycine) recognized in vivo (Kowalski et al., 1987).

TO WHICH CLASS OF PROTEASES DOES 2A BELONG?

Sequence comparisons of the 2A and 3C proteases from various picornaviruses showed that the amino acid sequence GXCGG (in 2A, X = D) was invariably present. This finding implied that this sequence might be involved in proteolysis, a possibility that was examined by searching the SWISSPROT and Protein Information Resource protein data banks (Cameron, 1988; Sidman et al., 1988) with a 48-amino-acid sequence from HRV-2 2A (including those amino acids mentioned above), using the FAST programs of Lipman and Pearson (1985). The proteins giving the best scores are shown in Fig. 3. It can be seen that the sequence GXCGG is closely related to that surrounding the serine residue of several serine proteases; furthermore,

a glycine and a cysteine are conserved before the putative active site, and a cysteine and a glycine followed by a hydrophobic residue are conserved after this site. Thus, picornaviral 2As would appear to be related to serine proteases; however, the active-site nucleophile is a cysteine residue. Constructions in which site-directed mutagenesis had been used to replace the cysteine residue with a serine were not active, however. Other mutations in this region (such as deleting the glycine residues before or after the active-site nucleophile cysteine) all inactivated the enzyme, except for a mutation involving the proline 3 amino acids before the active-site cysteine (Sommergruber et al., 1989). Introduction of a glycine at this position led to a 30% loss in activity. It seems clear from these results that the 2A protease is a cysteine protease working in the molecular environment of a serine protease. Similar analyses have been performed with the 3C enzymes (Ivanoff et al., 1986; Gorbalenya et al., 1986).

Of the proteins shown in Fig. 3, only thaumatin is not a serine protease; it did, however, have the best fit to the putative active site of 2A. The significance of this homology is unknown, since to our knowledge no proteolytic activity for this protein has been reported. In fact, it has been proposed that this protein is a protease inhibitor, acting as a defense mechanism for the plant species (*Thaumatococcus danielli* Benth) by which it is produced (Richardson et al., 1987). The crystal structure, although determined, does not supply any clues as to the significance of the similarity; it shows the sequence lying on a turn of the molecule (de Vos et al., 1985).

SUMMARY

The work described here confirms that polypeptide 2A of HRV-2 is a protease capable of cleaving at its own N terminus. The active site of the enzyme appears to possess a cysteine nucleophile, although the immediate environment is that of a serine protease.

REFERENCES

Cameron, G. N. 1988. The EMBL data library. *Nucleic Acids Res.* **16**:1865–1867.

De Vos, A., M. Hatuda, H. van der Wel, H. Krubbendam, A. F. Peerdeman, and S.-H. Kim. 1985. Three dimensional structure of thaumatin I, an intensely sweet protein. *Proc. Natl. Acad. Sci. USA* **82**:1406–1409.

Gorbalenya, A. E., V. M. Blinov, and A. M. Donchenko. 1986. Poliovirus-encoded proteinase 3C: a possible evolutionary link between cellular serine and cysteine proteinase families. *FEBS Lett.* **194**:253–257.

Ivanoff, L. A., T. Towatari, J. Ray, B. D. Korant, and S. R. Petteway, Jr. 1986. Expression and site-specific mutagenesis of the poliovirus 3C protease in Escherichia coli. *Proc. Natl. Acad. Sci. USA* **83**:5392–5396.

Kowalski, H., I. Maurer-Fogy, M. Zorn, H. Mischak, E. Kuechler, and D. Blaas. 1987. Cleavage site between VP1 and 2A of human rhinovirus is different in serotypes 2 and 14. *J. Gen. Virol.* **68**:3197–3200.

Kräusslich, H.-G., and E. Wimmer. 1988. Viral proteinases. *Annu. Rev. Biochem.* **57**:701–754.

Lipman, D. J., and W. R. Pearson. 1985. Rapid and sensitive protein searches. *Science* **227**:1435–1441.

Richardson, M., S. Valdes-Rodriguez, and A. Blanco-Labra. 1987. A possible function for thaumatin and a TMV-induced protein suggested by a homology to a maize inhibitor. *Nature* (London) **327**:432–434.

Sidman, K. S., D. G. George, W. C. Barker, and L. Hunt. 1988. The protein identification resource (PIR). *Nucleic Acids Res.* **16**:1869–1871.

Skern, T., C. Neubauer, L. Frasel, P. Gruendler, W. Sommergruber, M. Zorn, E. Kuechler, and D. Blaas. 1987. A neutralizing epitope on human rhinovirus type 2 includes amino acid residues between 153 and 164 of virus capsid protein VP2. *J. Gen. Virol.* **68**:315–323.

Skern, T., W. Sommergruber, D. Blaas, P. Gruendler, F. Frauendorfer, C. Pieler, I. Fogy, and E. Kuechler. 1985. Human rhinovirus 2: complete nucleotide sequence and proteolytic processing signals in the capsid protein region. *Nucleic Acids Res.* **13**:2111–2126.

Sommergruber, W., M. Zorn, D. Blaas, F. Fessl, P. Volkmann, I. Maurer-Fogy, P. Pallai, V. Merluzzi, M. Matteo, T. Skern, and E. Kuechler. 1989. Polypeptide 2A of human rhinovirus 2: identification as a protease and characterisation by mutational analysis. *Virology* **169**:68–77.

Chapter 27

Viral Protease 3C-Mediated Cleavage of Histone H3 in Foot-and-Mouth Disease Virus-Infected Cells

Matthias M. Falk, Pablo R. Grigera, Ingrid E. Bergmann, Andree Zibert, and Ewald Beck

The complete replication cycle of foot-and-mouth disease virus (FMDV), a member of the family *Picornaviridae*, including viral RNA and protein syntheses, processing of the primary translation product (polyprotein), and encapsidation of the single-stranded RNA genome, occurs in the cytoplasm of the infected cell. Virus particles are released by host cell lysis 4 to 6 h after infection. Characteristic of most species of picornaviruses is a specific inhibition of host cell transcription and translation early after infection, an event often referred to as host cell shutoff. The metabolism of the infected cell is completely converted to virus production.

The specific inhibition of host cell protein synthesis is thought to result in part from specific cleavage of the largest subunit (p220) of the eucaryotic translation initiation factor eIF-4F by the viral leader protease (Devaney et al., 1988). Cap-dependent translation of cellular mRNAs is thereby blocked, but translational initiation of the uncapped viral RNA is still possible via cap-independent ribosome entry at an internal ribosome entry site at the 5′ untranslated region of the viral RNA.

In contrast to the mechanism of host cell translation inhibition, little is known about the inhibition of host cell RNA synthesis. The data reported here are possibly related to this transcriptional switch-off.

different cell types with FMDV. Approximately 1.5 h after infection, a new chromatin-associated polypeptide (Pi) with a molecular weight of about 13,000 appears, whereas histone H3 begins to disappear (Fig. 1). By 3 to 5 h later, histone H3 has completely disappeared and protein Pi has accumulated to the same level as the other histones. The other histones are apparently not affected during infection.

Although no functions of the host cell nucleus are involved in the picornaviral life cycle, viral proteins from poliovirus (Bienz et al., 1982; Fernández-Tomás, 1982) and FMDV (Grigera and Sagedahl, 1986) can be isolated from the nuclei of infected cells. Traub and Traub (1978) reported a modification of histone H1 subtypes after mengovirus infection. On the basis of these observations, one could argue that virus-specific gene products might interact with the chromatin of the infected cell, causing the histone modification. Therefore, we postulated that protein Pi represents a proteolytic degradation product of histone H3 catalyzed by a viral protease. To investigate this hypothesis, we isolated protein Pi and determined the amino-terminal amino acid sequence. The derived amino acid sequence fits perfectly to amino acid positions 21 to 40 of histone H3 (Fig. 2). Therefore, we conclude that the first 20 amino acid residues of histone H3 are specifically cleaved off during viral infection.

SPECIFIC HISTONE H3 CLEAVAGE AFTER INFECTION

A modification of the chromatin-associated histone proteins is observed after infection of

VIRUS-INDUCED PROTEOLYSIS OF HISTONE H3

To characterize the virus-induced proteolysis of histone H3 in more detail, we incubated

Matthias M. Falk, Andree Zibert, and Ewald Beck, Zentrum für Molekulare Biologie Heidelberg, D-6900 Heidelberg, Federal Republic of Germany.   *Pablo R. Grigera and Ingrid E. Bergmann*, Centro de Virologia Animal, Serrano 661 (1414), Buenos Aires, Argentina.

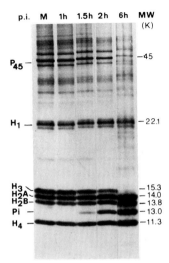

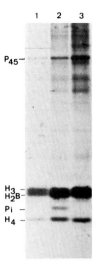

FIGURE 1. Time course of histone H3 cleavage in FMDV-infected cells. Chromatin-associated proteins were isolated from BHK-21 cells at different times after infection, separated by polyacrylamide gel electrophoresis, and stained with Coomassie blue. Lane M, Mock-infected cells; lane MW, molecular weights of proteins. Further details are given in the text.

FIGURE 3. Virus-induced proteolysis of histone H3 in vitro. [^{35}S]methionine-labeled nuclei of uninfected BHK-21 cells were incubated in the presence of cytoplasmic extract derived from infected (lanes 2 and 3) or mock-infected (lane 1) cells. The cysteine protease inhibitor NEM was added in addition in lane 3. (Histones H1 and H2A are not labeled because of the absence of methionine.)

[^{35}S]methionine-labeled isolated nuclei of uninfected cells with cytoplasmic extracts of infected cells in vitro. The partial conversion of histone H3 to protein Pi (Fig. 3, lane 2) indicated the direct participation of FMDV gene products at the histone H3-Pi transition. No cleavage occurred in the presence of a cysteine protease inhibitor, such as ZnCl$_2$, tosylsulfonyl phenylalanyl chloromethyl ketone (TPCK), iodoaceta-

mide, or N-ethylmaleimide (NEM), as shown for NEM in lane 3. This result suggests the participation of the viral 3C protease in this histone H3 cleavage, since this gene product, common to all picornaviruses, is thought to be an ancestral form of a cysteine protease with no homology to papain (Argos et al., 1984). Specific inhibition of the 3C protease of encephalomyocarditis virus

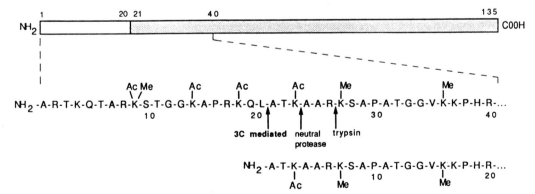

FIGURE 2. Alignment of the N-terminal sequences of histone H3 and protein Pi. Protein Pi was derived from infected BHK-21 cells. The sequence was determined by automated Edman degradation. Ala-1 to Arg-20 of Pi were found to be homologous to Ala-21 to Arg-40 of histone H3. Ac, Acetylation; Me, methylation. Different protease cleavage sites in native histone H3 are indicated by arrows.

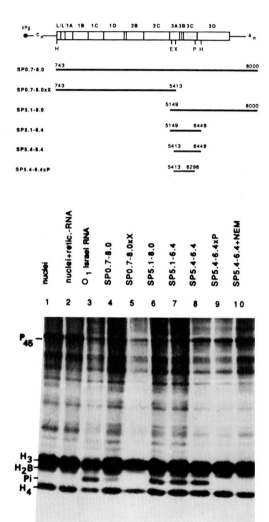

FIGURE 4. FMDV 3C protease-dependent cleavage of histone H3. RNAs representing various parts of the viral genome (indicated at the top) were translated in rabbit reticulocyte lysate and incubated together with [35S]methionine-labeled nuclei isolated from BHK-21 cells. Protein Pi is observed in all reactions with RNAs containing the whole 3C-coding region. Lanes: 1 to 3, controls; 4 to 9, incubation in the presence of RNAs corresponding to different parts of the genome as indicated; 10, incubation in the presence of NEM. Restriction sites: E, EcoRI; H, HindIII; P, PstI; X, XbaI.

by NEM and iodoacetamide has been reported by Pelham (1978) and Gorbalenya and Svitkin (1983).

HISTONE H3 CLEAVAGE IS FMDV 3C PROTEASE DEPENDENT

To investigate which viral gene product is responsible for the specific histone H3 cleavage, we constructed a set of transcription vectors containing cDNAs from various parts of the viral genome (Fig. 4). The RNA transcripts were translated in vitro. Synthesis of active 3C protease was checked by processing of the viral polyprotein in vitro (data not shown). [35S]methionine-labeled native chromatin in the form of isolated nuclei was added after the translation reactions. Chromatin was then isolated from the reaction mixtures and analyzed on a sodium dodecyl sulfate-gel (Fig. 4; Falk et al., 1990). Histone H3 cleavage could always be observed in the presence of intact 3C protease (Fig. 4, lanes 4, 6, 7, and 8). Protein translated from an RNA coding exclusively for the 3C protease and one VPg was also able to cleave histone H3 in this assay (lane 8). C-terminal truncation of the 3C reading frame (lane 9), or addition of a cysteine protease inhibitor as in the assays mentioned above, prevented histone H3 cleavage (lane 10).

From these data, we conclude that (i) histone H3 is degraded specifically in the presence of virus 3C protease during infection, (ii) the 3C protease is the viral gene product responsible for the histone H3-Pi transition, (iii) no other additional viral protein is needed for this specific cleavage, and (iv) the truncated histone H3 remains chromatin associated. From our present data, we cannot determine whether histone H3 is cleaved directly by the viral 3C protease or indirectly by the activation of a nuclear protease.

The observed cleavage site in histone H3 lies in a region also accessible to other proteases (Fig. 2). Cleavage of native histone H3 by the chromatin-associated neutral protease at Lys-23 (Brandt et al., 1975) and by trypsin (Arg-26) has been reported. The trypsin site at Lys-37 is not cleaved, probably because it is inaccessible in the tightly packed native chromatin, as shown by Böhm et al. (1981).

The portion specifically cleaved off by the FMDV 3C protease during infection corresponds to a region of histone H3 postulated to be involved in the regulation of transcriptional activity of chromatin. Many results published in recent years indicate that acetylation and methylation of the hydrophilic amino terminus of histone H3, and also of the other core histones, are prerequisite for transcriptional activation in eucaryotes. Therefore, the observed proteolysis of histone H3 may be related to the switch-off of host cell RNA synthesis observed in picornavirus-infected cells.

Addendum in proof. FMDV protease 3C-dependent cleavage of histone H3 and its relation to cellular transcription inhibition was also found by M. Tesar and O. Marquardt (*Virology* **174**:364–374, 1990).

REFERENCES

Argos, P., G. Kamer, M. J. H. Nicklin, and E. Wimmer. 1984. Similarity in gene organization and homology between proteins of animal picornaviruses and a plant comovirus suggest common ancestry of these virus families. *Nucleic Acids Res.* **12**:7251–7267.

Bienz, K., D. Egger, Y. Rasser, and W. Bossart. 1982. Accumulation of poliovirus proteins in the host cell nucleus. *Intervirology* **18**:189–196.

Böhm, L., G. Briand, P. Sautière, and C. Crane-Robinson. 1981. Proteolytic digestion studies of chromatin core-histone structure. Identification of the limit peptides of histones H3 and H4. *Eur. J. Biochem.* **119**:67–74.

Brandt, W. F., L. Böhm, and C. von Holt. 1975. Proteolytic degradation of histones and site of cleavage in histone F2a1 and F3. *FEBS Lett.* **51**:88–93.

Chong, M. T., W. T. Garrard, and J. Bonner. 1974. Purification and properties of a neutral protease from rat liver chromatin. *Biochemistry* **13**:5128–5134.

Devaney, M. A., V. N. Vakharia, R. E. Lloyd, E. Ehrenfeld, and M. J. Grubman. 1988. Leader protein of foot-and-mouth disease virus is required for cleavage of the p220 component of the cap-binding protein complex. *J. Virol.* **62**:4407–4409.

Falk, M. M., P. R. Grigera, I. E. Bergmann, G. Multhaup, A. Zibert, and E. Beck. 1990. Foot-and-mouth disease virus protease 3C induces specific proteolytic cleavage of host cell histone H3. *J. Virol.* **64**:748–756.

Fernández-Tomás, C. 1982. The presence of viral-induced proteins in nuclei from poliovirus-infected HeLa cells. *Virology* **116**:629–634.

Gorbalenya, A. E., and Y. V. Svitkin. 1983. Protease of encephalomyocarditis virus: purification and role of sulfhydryl groups in processing of the structural proteins precursor. *Biochemistry* (USSR) **48**:385–395.

Grigera, R. G., and A. Sagedahl. 1986. Cytoskeletal association of an aphtovirus-induced polypeptide derived from P3ABC region of the viral polyprotein. *Virology* **154**:369–380.

Pelham, H. R. B. 1978. Translation of encephalomyocarditis virus RNA in vitro yields an active proteolytic processing enzyme. *Eur. J. Biochem.* **85**:457–462.

Traub, U., and P. Traub. 1978. Changes in the microheterogeneity of histone H1 after mengovirus infection of Ehrlich ascites tumor cells. *Hoppe-Seyler's Z. Physiol. Chem.* **359**:581–592.

Chapter 28

Readthrough Suppression in Moloney Murine Leukemia Virus

Y.-X. Feng, J. G. Levin, D. L. Hatfield, T. S. Schaefer, R. J. Gorelick, and A. Rein

Moloney murine leukemia virus (Mo-MuLV) is a member of the mammalian type C family of retroviruses. Retroviruses are not, strictly speaking, positive-strand viruses. However, they do share many fundamental properties with, and must perform many of the same functions as, positive-strand viruses. It is interesting to compare the ways in which the viruses in these groups, which are probably not related to each other, carry out these functions.

The problem considered in this chapter is a mechanism for regulation of gene expression. Retroviruses, like many positive-strand viruses, give rise to RNA molecules that can serve either as mRNA for viral proteins or as genomic RNA. In turn, some virus-encoded proteins are needed at much higher levels than others. One mechanism that has been used to regulate the level of synthesis of different proteins is translational suppression. Thus, the viral genome contains a termination codon at the end of the coding region for the proteins needed in large amounts. Most of the polypeptide chains end at this termination codon, but in a few the termination codon is suppressed, resulting in the synthesis of a fusion protein that includes sequences encoded downstream, as well as upstream, of the termination codon. Such a mechanism has been used by a wide variety of virus groups, including alphaviruses, coronaviruses, hepadnaviruses, and tobamoviruses, as well as retroviruses and their close relatives, retrotransposons. In some cases the suppression is accomplished by ribosomal frameshifting, whereas in other viruses readthrough suppression, i.e., translation of the termination codon as an amino acid, is observed (for reviews, see Hatfield [1985] and Valle and Morch [1988]).

The studies described here are concerned with Mo-MuLV. The Mo-MuLV genome contains an in-frame UAG termination codon between the *gag* and *pol* coding regions (Shinnick et al., 1981). In ~5% of the translation products, the UAG is translated as glutamine (Yoshinaka et al., 1985); this suppression event is required for translation of the *pol* gene. The suppression of this UAG codon is apparently dependent on codon context or secondary structure in the RNA near the termination codon, since suppression is not seen with UAG codons in other mRNAs (Panganiban, 1988; our unpublished work).

The experiments described below bear on two questions: (i) the nature of the tRNA involved in UAG suppression in Mo-MuLV and (ii) whether suppression in Mo-MuLV is specific for UAG.

It has been reported (Kuchino et al., 1987) that Mo-MuLV infection alters the cellular glutamine tRNA pool, raising the level of an isoacceptor that is capable of suppressing UAG. As a test of this possibility, we analyzed suppression in vitro in rabbit reticulocyte lysates. Viral mRNA was synthesized from a subclone of an infectious molecular clone, using SP6 RNA polymerase. This RNA was then translated in reticulocyte lysates that had been depleted of tRNA as described in the legend to Fig. 1; these lysates were supplemented with equal levels of tRNA isolated from either infected or control

Y.-X. Feng and J. G. Levin, Laboratory of Molecular Genetics, National Institute of Child Health and Human Development, Bethesda, Maryland 20892. *D. L. Hatfield*, Laboratory of Experimental Carcinogenesis, National Cancer Institute, Bethesda, Maryland 20892. *T. S. Schaefer*, Department of Agricultural Chemistry, Oregon State University, Corvallis, Oregon 97331. *R. J. Gorelick and A. Rein*, BRI, Basic Research Program, National Cancer Institute-Frederick Cancer Research Facility, Frederick, Maryland 21701.

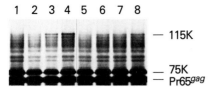

FIGURE 1. Effect of varying the concentration of tRNA from Mo-MuLV-infected and uninfected NIH 3T3 cells on readthrough suppression in reticulocyte lysates. A 5-kilobase fragment of an infectious molecular clone of Mo-MuLV, extending from the *Xba*I site in U3 to the *Hin*dIII site in *pol*, was inserted downstream of an SP6 promoter. Viral mRNA was generated from this clone (which was designated pMXH) after restriction of the DNA with *Sal*I. The product of this transcription includes the entire *gag* gene plus 1.5 kilobases of the *pol* gene. This mRNA was then translated in rabbit reticulocyte lysates (Promega Biotec) that had been treated with micrococcal nuclease for 8 min. The reactions were supplemented with tRNA that had been isolated from uninfected (lanes 1 to 4) or infected (lanes 5 to 8) NIH 3T3 cells. Reactions contained [^{35}S]methionine. Reaction products were immunoprecipitated with antiserum to p30^{gag} and analyzed by sodium dodecyl sulfate-polyacrylamide gel electrophoresis and fluorography. Translation of the *gag* gene alone generates 65- and 75-kilodalton (K) products, whereas readthrough suppression into *pol* generates a 115-kilodalton product. Reactions were supplemented with tRNA as follows: lanes 1 and 5, 2 μg/ml; lanes 2 and 6, 4 μg/ml; lanes 3 and 7, 8 μg/ml; lanes 4 and 8, 20 μg/ml. (Reprinted with permission from Feng et al. [1989].)

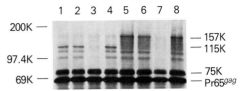

FIGURE 2. Suppression of UAA and UGA in reticulocyte lysates. Viral mRNA was generated from pMXH (see legend to Fig. 1) or pMXH containing UAA or UGA in place of the termination codon UAG, after the respective plasmids were linearized with *Sal*I (lanes 1 to 4) or *Hin*dIII (lanes 5 to 8). These mRNAs were translated in standard rabbit reticulocyte lysates, which had not been treated with micrococcal nuclease but were supplemented with 50 μg of calf liver tRNA per ml, using [^{35}S]methionine to label the translation products. The products were then analyzed as for Fig. 1. Readthrough suppression of mRNA transcribed from *Sal*I-cleaved DNA yields a 115-kilodalton (K) product, whereas suppression of the mRNA from *Hin*dIII-cleaved DNA generates a 157-kilodalton product. Lanes: 1 and 5, pMXH; 2 and 6, pMXH with UGA; 3, 4, 7, and 8, pMXH with UAA. Lanes 4 and 8 were additionally supplemented with 40 μg of NIH 3T3 cell tRNA per ml. (Reprinted with permission from Feng et al. [1989].)

NIH 3T3 cells. tRNAs from either source were equally efficient at supporting the synthesis of a high-molecular-weight protein band, which we have shown elsewhere (Feng et al., 1989) is the product of translational suppression in this system (Fig. 1).

The results presented in Fig. 1 do not support the idea that a virus-induced tRNA is required for suppression in Mo-MuLV. We also found that the total concentration of glutamine acceptor activity was the same in infected and uninfected cells and that the glutamine tRNAs from these cells gave identical chromatographic profiles on reverse-phase chromatographic columns (Feng et al., 1989). Thus, we find no indication that virus infection induces or alters glutamine tRNA; rather, our data suggest that the suppression is mediated by a normal cellular tRNA. We cannot explain the discrepancy between our results and those of Kuchino et al. (1987).

We also investigated the question of whether the suppression in Mo-MuLV is specific for UAG. Mutants with alternate termination codons, i.e., UAA or UGA in place of UAG, were generated by oligonucleotide-directed mutagenesis. When these mutants were tested for suppression in rabbit reticulocyte lysates, we found that UGA (Fig. 2, lanes 2 and 6) was suppressed as efficiently as UAG (lanes 1 and 5). Very little suppression was seen with UAA (lanes 3 and 7), but the level of readthrough product in this case was greatly increased if the lysates were supplemented with additional tRNA (lanes 4 and 8). Supplemental tRNAs from either NIH 3T3 cells or rabbit liver were both effective in this system. We conclude that the signal for suppression is not specific for UAG and that reticulocyte lysates contain tRNAs capable of suppressing UGA as well as UAG. tRNAs that can suppress UAA are apparently present, but in very limited quantities, in normal mammalian cells and in the reticulocyte lysate system.

We also tested for suppression of UAA and UGA in the mutant Mo-MuLVs in vivo. Intact viral genomes containing these alternate termination codons were transfected into CHO cells (a permanent line of hamster cells), and stable transfectants were obtained. These mutants generated fully replication-competent, infectious

TABLE 1

Production of Virus by Cells Containing Mo-UAA and Mo-UGA

Plasmid[a]	Infectivity[b]	RT activity[c] (pmol incorporated/ml)	p30[gag][d] (arbitrary units)
Mo-UAA	3×10^4	15.0	7
Mo-UGA	4×10^4	8.5	4
Mo-MuLV	5×10^4	24.7	12
pGCCos3Neo	$<2 \times 10^0$	0	0

[a] Intact, nonpermuted MuLV genomes were reconstructed in the selectable vector pGCCos3Neo. The presence of the mutations in these clones was confirmed by colony hybridization. The clones were transfected into CHO hamster cells, and stable transfectants were selected with G418 (GIBCO Laboratories). Culture fluids were harvested from these cells and assayed for virus production as indicated.

[b] Assayed by the S^+L^- focus assay (Bassin et al., 1971) and expressed as focus-inducing units per milliliter.

[c] Reverse transcriptase (RT) activity in response to poly(rA)-oligo(dT) was determined as described previously (Levin et al., 1988).

[d] Relative amounts of p30[gag] were determined by protein immunoblotting as described previously (Gorelick et al., 1988).

progeny virus at the same level as did the wild-type control (Table 1). These mutant virions also had specific infectivities equivalent to that of wild-type Mo-MuLV, since the three virus stocks contained equal levels of p30[gag] and reverse transcriptase activity in pelleted virus, as well as equal infectivity titers. We conclude that the level of suppression of UAA and UGA in Mo-MuLV in CHO cells (and in the mouse indicator cells used in the infectivity assays) is sufficient for efficient replication of the virus. Thus, tRNAs capable of suppressing UAA or UGA are evidently present in these cells, and the suppression signals in the viral mRNA are not specific for any one termination codon.

As noted above, modulation of gene expression by readthrough suppression is found in a wide variety of animal and plant RNA viruses. Previous work has shown that the signals for suppression are also effective with any termination codon in the case of the alphavirus Sindbis virus (Li and Rice, 1989) and the tobamovirus tobacco mosaic virus (Ishikawa et al., 1986). Because of this common property, it seemed possible that the suppression signals are similar in the three viruses. We therefore examined the sequences around the suppressible termination codon in the three viruses. Inspection of the

```
Mo-MuLV   CAA AGA CCC CAG ACC UCC CUC CUG ACC CUA GAU GAC |UAG|  2237

Sindbis   CAG AGA CGU AGA CGC AGG AGC AGG AGG ACU GAA UAC |UGA|  5750

TMV       UUG UUA GAU AUG UAU AAG GUC GAU GCA GGA ACA CAA |UAG|  3419


Mo-MuLV   GGA GGU CAG GGU CAG GAG CCC CCC CCU GAA CCC AGG AUA  2276

Sindbis   CUA ACC GGG GUA GGU GGG UAC AUA UUU UCG ACG GAC ACA  5789

TMV       CAA UUA CAG AUU GAC UCG GUG UUC AAA GGU UCC AAU CUU  3458


Mo-MuLV   ACC CUC AAA GUC GGG GGG CAA CCC GUC ACC UUC CUG GUA  2315

Sindbis   GGC CCU GGG CAC UUG CAA AAG AAG UCC GUU CUG CAG AAC  5828

TMV       UUU GUU GCA GCG CCA AAG ACU GGU GAU AUU UCU GAU AUG  3497


Mo-MuLV   GAU ACU GGG GCC CAA CAC UCC GUG CUG ACC CAA AAU CCU  2354

Sindbis   CAG CUU ACA GAA CCG ACC UUG GAG CGC AAU GUC CUG GAA  5867

TMV       CAG UUU UAC UAU GAU AAG UGU CUC CCA GGC AAC AGC ACC  3576


Mo-MuLV   GGA CCC CUA AGU GAU AAG UCU GCC UGG GUC CAA GGG GCU  2393

Sindbis   AGA AUU CAU GCC CCG GUG CUC GAC ACG UCG AAA GAG GAA  5906

TMV       AUG AUG AAU AAU UUU GAU GCU GUU ACC AUG AGG UUG ACU  3615
```

FIGURE 3. mRNA sequences near the readthrough sites in Mo-MuLV (Shinnick et al., 1981), Sindbis virus (Strauss et al., 1984), and tobacco mosaic virus (TMV) (Goelet et al., 1982). The suppressible termination codon is boxed. Positions with identical bases in all three viruses are underlined.

sequences (Fig. 3) reveals no obvious common features. Perhaps the real signal in each case involves secondary or tertiary structure of the mRNA. Indeed, perhaps the three viral mRNAs share such a structure; in the case of ribosomal frameshifting in the coronavirus infectious bronchitis virus, an RNA "pseudoknot" is required for suppression, but a number of different primary sequences can apparently generate the required structure (Brierley et al., 1989).

In summary, readthrough suppression in Mo-MuLV is apparently mediated by a normal cellular tRNA and occurs regardless of the identity of the termination codon. Sindbis and tobacco mosaic viruses appear to use a similar mechanism, but there are no obvious similarities in their primary sequences near the suppressible termination codon. Our current efforts are directed toward identification of the amino acids inserted in response to UAA and UGA and of the sequences required for suppression in Mo-MuLV.

ACKNOWLEDGMENTS. We thank Janet Hanser, Emily Keene, Sandra Ernst, and Hue Nguyen for technical assistance; Ramesh Kumar for critical contributions; Alan Schultz and Tom Wood for help with protein analysis and plasmid constructions, respectively; Marilyn Powers for oligonucleotides; Jack Morris, Rosaura Valle, Marie-Dominique Morch, and Anne-Lise Haenni for informative discussions; and Carol Shawver for typing.

This research was sponsored in part by the National Cancer Institute under Public Health Service contract NO1-CO-74101 with BRI.

REFERENCES

Bassin, R. H., N. Tuttle, and P. J. Fischinger. 1971. Rapid cell culture assay for murine leukaemia virus. *Nature* (London) 229:564–566.

Brierley, I., P. Digard, and S. C. Inglis. 1989. Characterization of an efficient coronavirus ribosomal frameshifting signal: requirement for an RNA pseudoknot. *Cell* 57:537–547.

Feng, Y.-X., D. L. Hatfield, A. Rein, and J. G. Levin. 1989. Translational readthrough of the murine leukemia virus *gag* gene amber codon does not require virus-induced alteration of tRNA. *J. Virol.* 63:2405–2410.

Goelet, P., G. P. Lomonossoff, P. J. G. Butler, M. E.

Akam, M. J. Gait, and J. Karn. 1982. Nucleotide sequence of tobacco mosaic virus RNA. *Proc. Natl. Acad. Sci. USA* 79:5818–5822.

Gorelick, R. J., L. E. Henderson, J. P. Hanser, and A. Rein. 1988. Point mutants of Moloney murine leukemia virus that fail to package viral RNA: evidence for specific RNA recognition by a "zinc-finger-like" protein sequence. *Proc. Natl. Acad. Sci. USA* 85:8420–8424.

Hatfield, D. 1985. Suppression of termination codons in higher eucaryotes. *Trends Biochem. Sci.* 10:201–204.

Ishikawa, M., T. Meshi, F. Motoyoshi, N. Takamatsu, and Y. Okada. 1986. In vitro mutagenesis of the putative replicase genes of tobacco mosaic virus. *Nucleic Acids Res.* 14:8291–8305.

Kuchino, Y., H. Beier, N. Akita, and S. Nishimura. 1987. Natural UAG suppressor glutamine tRNA is elevated in mouse cells infected with Moloney murine leukemia virus. *Proc. Natl. Acad. Sci. USA* 84:2668–2672.

Levin, J. G., R. J. Crouch, K. Post, S. C. Hu, D. McKelvin, M. Zweig, D. L. Court, and B. I. Gerwin. 1988. Functional organization of the murine leukemia virus reverse transcriptase: characterization of a bacterially expressed AKR DNA polymerase deficient in RNase H activity. *J. Virol.* 62:4376–4380.

Li, G., and C. M. Rice. 1989. Mutagenesis of the in-frame opal termination codon preceding nsP4 of Sindbis virus: studies of translational readthrough and its effect on virus replication. *J. Virol.* 63:1326–1337.

Panganiban, A. T. 1988. Retroviral *gag* gene amber codon suppression is caused by an intrinsic *cis*-acting component of the viral mRNA. *J. Virol.* 62:3574–3580.

Shinnick, T. M., R. A. Lerner, and J. G. Sutcliffe. 1981. Nucleotide sequence of Moloney murine leukemia virus. *Nature* (London) 293:543–548.

Strauss, E. G., C. M. Rice, and J. H. Strauss. 1984. Complete nucleotide sequence of the genomic RNA of Sindbis virus. *Virology* 133:92–110.

Valle, R. P. C., and M.-D. Morch. 1988. Stop making sense or regulation at the level of termination in eukaryotic protein synthesis. *FEBS Lett.* 235:1–15.

Yoshinaka, Y., I. Katoh, T. D. Copeland, and S. Oroszlan. 1985. Murine leukemia virus protease is encoded by the *gag-pol* gene and is synthesized through suppression of an amber termination codon. *Proc. Natl. Acad. Sci. USA* 82:1618–1622.

Chapter 29

Synthesis of Dengue Virus Nonstructural Protein NS1 Requires the N-Terminal Signal Sequence and the Downstream Nonstructural Protein NS2A

B. Falgout and C.-J. Lai

Dengue viruses (four serotypes) are the most important members of the flavivirus family in terms of morbidity (Halstead, 1981, 1988). Flaviviruses including dengue virus and the prototype yellow fever virus are serologically related, have the same genome organization and an extensive sequence homology between corresponding genes, and most likely share a common replication strategy (Porterfield, 1980; Rice et al., 1985; Castle et al., 1985; Zhao et al., 1986; Mackow et al., 1987; Deubel et al., 1986; Hahn et al., 1988; Mason et al., 1987). The dengue virus type 4 RNA genome codes for a polyprotein of 3,386 amino acids that has the order N terminus; capsid (C), membrane (M), envelope (E), and nonstructural proteins designated NS1, NS2a, NS2b, NS3, NS4a, NS4b, and NS5; C terminus (Zhao et al., 1986; Mackow et al., 1987). Several dengue virus proteins, such as the membrane protein precursor (preM), E, and NS1, are glycoproteins. Expression of these glycoproteins and other nonglycosylated proteins occurs by a mechanism of proteolytic processing that is also proposed for the processing of other flavivirus proteins (Rice et al., 1985; Castle et al., 1985; Zhao et al., 1986). One type of cleavage that defines the N termini of these glycoproteins is made after a hydrophobic amino acid sequence in the polyprotein by a rough endoplasmic reticulum-associated signalase. Processing also occurs at sites in the polyprotein that feature a common sequence of only a few amino acids, and virally encoded proteinases are presumably responsible for this form of cleavage. Although the complete sequences of several flaviviruses have been obtained, the signal sequences in the polyprotein are still poorly defined, and proteinase activity has not been ascribed to any virus protein. To elucidate the mechanism of flavivirus protein processing, we have studied the expression of dengue virus proteins from cDNA that codes for subsegments of the polyprotein. This chapter describes the analysis of sequence requirements for the synthesis of dengue virus NS1 glycoprotein.

EXPRESSION OF DENGUE VIRUS PROTEINS BY A RECOMBINANT VACCINIA VIRUS

Because of the ease of recombinant virus construction, vaccinia virus was used as a vector for expression of dengue virus polyproteins from cloned cDNA. Earlier we had obtained a recombinant vaccinia virus v(C-M-E-NS1-NS2a) that contained a dengue virus cDNA sequence of approximately 4.0 kilobases near the 5' end coding for the three structural proteins C, M, and E and nonstructural proteins NS1 and NS2a (Zhao et al., 1987). Infection of CV-1 cells with this recombinant virus produced preM, E, and NS1 identical in size to the corresponding proteins produced during dengue virus infection. These dengue virus protein products were glycosylated, as shown by digestion with endoglycosidase (endo) H, which removes sugar moieties from the protein backbone. This finding indicates that processing of these glycoproteins by proteolytic cleavage proceeds normally in the absence of other dengue viral functions distal to NS2a. These results are consistent with the

B. Falgout and C.-J. Lai, Molecular Viral Biology Section, Laboratory of Infectious Diseases, National Institute of Allergy and Infectious Diseases, Bethesda, Maryland 20892.

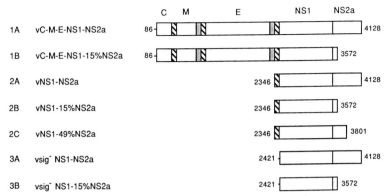

FIGURE 1. Linear map of the dengue virus cDNA sequences expressed by the recombinant vaccinia viruses. The dengue virus DNA fragments and the recombinant vaccinia viruses constructed are indicated. The linear map shows the sequences contained in these DNA fragments. Numbers adjacent to the maps indicate the first and the last dengue virus nucleotides in the fragments. Open boxes represent coding sequences for the dengue virus gene products C, preM (M), E, NS1, and NS2a. The shaded boxes represent the hydrophobic sequences encoded that are putative signals for translocation as well as for stop-transfer.

current view that signal-directed cleavage occurring at the junction between C-preM, preM-E, or E-NS1 is responsible for authentic synthesis of preM and E glycoproteins and presumably C as well. However, signalase-associated processing probably does not account for the synthesis of NS1, since an extended hydrophobic sequence is not present at the NS1-NS2a junction. By alignment with sequences of other flaviviruses, the C terminus of dengue virus type 4 NS1 has been tentatively assigned to alanine at position 1125, immediately upstream of NS2a (Coia et al., 1988; Wright et al., 1989; Mackow et al., 1987). Analysis of flavivirus polyprotein sequences in this region revealed a consensus sequence VXA immediately upstream of the cleavage site. Thus, the cleavage generating the C terminus of NS1 may be mediated by the other proposed mechanism, which involves a virus-specific proteinase. In any case, it is clear that the 4.0-kilobase cDNA segment encodes the sequence information necessary to produce the NS1 glycoprotein.

GLYCOSYLATION AND PROTEOLYTIC PROCESSING OF NS1

The N terminus of NS1 is preceded by two hydrophobic sequences, 14 and 24 amino acids in length, separated by an arginine. This bipartite hydrophobic sequence configuration is also found between the preM-E junction and is apparently a feature shared by all flaviviruses. It has been proposed that the second hydrophobic

sequence which is immediately upstream of E or NS1 functions as a signal for protein processing. To test this hypothesis, and to further elucidate the mechanism of expression of flavivirus proteins, we studied the requirements of an N-terminal signal sequence and the downstream NS2a protein for the processing of NS1 from the viral polyprotein. A series of recombinant vaccinia viruses that contained the coding sequence of NS1 with or without the putative N-terminal signal plus various lengths of downstream NS2a sequences was constructed (Falgout et al., 1989). Figure 1 shows the map of the dengue virus cDNA sequences present in these recombinants. The NS1 products of CV-1 cells infected with these recombinants were analyzed by immunoprecipitation and polyacrylamide gel electrophoresis (Fig. 2). Recombinant v(NS1-NS2a) containing the putative signal sequence of 24 hydrophobic amino acids produced NS1 that was identical in size to that expressed by v(C-M-E-NS1-NS2a) or dengue virus infection. These NS1 products were normally glycosylated, as indicated by digestion with endo H or F. The recombinant v(sig-NS1-NS2a), which lacked the N-terminal signal sequence and contained the methionine codon at the second amino acid of NS1, failed to produce detectable NS1. Presumably, the NS1 polypeptide that was not translocated into the endoplasmic reticulum did not accumulate in a stable form. This analysis indicated that the sequence of 24 hydrophobic amino acids is necessary and sufficient to serve as a signal for directing translocation of

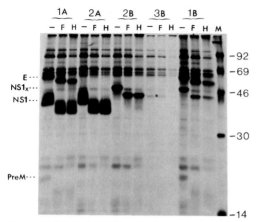

FIGURE 2. Analysis of dengue virus-specific proteins produced by recombinant vaccinia viruses. CV-1 cells were infected with each recombinant vaccinia virus at 2 to 5 PFU per cell. At 16 to 24 h after infection, cells were labeled with [³⁵S]methionine for 2 h, and the lysates were prepared for immunoprecipitation with dengue virus type 4 hyperimmune mouse ascitic fluid. The labeled precipitates were eluted and digested with endo F or H. Samples were analyzed by polyacrylamide gel electrophoresis. Lane M, Protein molecular weight markers (indicated in thousands on the right). The positions of normal dengue virus-specific proteins are indicated on the left. NS1x of recombinant 2B, v(NS1-15%NS2a), is the larger form of NS1. The asterisk shows the NS1 product of v(sig-NS1-15%NS2a).

NS1 to the endoplasmic reticulum lumen, where glycosylation takes place.

Recombinants v(C-M-E-NS1-15%NS2a) and v(NS1-15%NS2a), which encoded polyproteins terminating in the N-terminal region of NS2a, were initially constructed to determine whether NS2a was required for proper processing of NS1. Both recombinants failed to express authentic NS1; instead, they produced a new protein band (termed NS1x) larger than NS1 (Fig. 2). NS1x was glycosylated, since treatment with endo F or H reduced its size to that of the unglycosylated form produced by v(sig-NS1-15%NS2a). These findings indicate that the signal sequence was cleaved from NS1x. Recombinant v(NS1-49%NS2a) also failed to produce NS1 but instead expressed an aberrant NS1 species (NS1y) larger than NS1x. A likely explanation for the size increase is that these aberrant NS1 species are fusion proteins between NS1 and the remaining portions of NS2a resulting from failure of cleavage between NS1 and NS2a. Immunoprecipitation using specific antibodies further confirmed that these larger NS1 products were fusion proteins containing

NS2a sequences. These observations indicate that NS2a plays a role in cleavage at the NS1-NS2a junction.

COMPLEMENTATION TEST FOR NS1-NS2a CLEAVAGE

We wished to determine whether NS2a is a diffusible proteinase acting in *trans* to cleave the bond between NS1 and NS2a. Cells were coinfected with the cleavage-competent v(NS1-NS2a) and the cleavage-deficient v(NS1-15%NS2a) or v(NS1-49%NS2a). The NS1 products were analyzed by immunoprecipitation and polyacrylamide gel electrophoresis. The result indicated that cells coinfected with recombinant viruses synthesized both the expected normal-size NS1 and the larger NS1 species. There was no significant reduction in the level of the aberrant NS1 produced by coinfected cells compared with that of singly infected cells. This result indicates that cleavage of NS1x or NS1y was not effected by NS2a in *trans*. These observations further suggest that NS2a may be a *cis*-acting proteinase cleaving itself from NS1. However, other explanations, such as NS2a providing sequences for recognition by other specific proteases, have not been ruled out.

FUNCTIONAL ANALYSIS OF NS2a

The in vivo expression of authentic NS1 from the NS1-NS2a polyprotein has provided a functional assay for the activity of NS2a. In an effort to map the functional sites of the candidate *cis*-acting viral proteinase, we attempted to identify the NS2a sequence that effects cleavage between NS1 and NS2a. The cDNA coding for NS1-NS2a was digested with BAL 31 to remove the 3′ sequences so that various-length deletions were introduced at the C terminus of NS2a. Recombinant vaccinia viruses containing these cDNA were constructed and used for infection of CV-1 cells. The synthesis of dengue virus NS1 by these recombinants was analyzed as described previously (Fig. 3).

Deletion of 20 amino acids from the 218-amino-acid NS2a had no apparent effect on the synthesis of NS1. Four recombinants containing deletions ranging from 42 to 65 amino acids from the C terminus also produced authentically cleaved NS1. However, the amount of the NS1 product appeared to decrease with increasing length of deletions. Concomitantly, there was increasing accumulation of larger NS1 species, which represented the uncleaved NS1-NS2a.

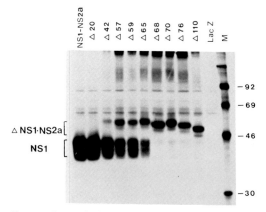

FIGURE 3. Analysis of dengue virus NS1 products expressed by recombinants containing deletions in NS2a. Procedures for infection of cells and radiolabeling were as described in the legend to Fig. 2. The full-length recombinant in lane NS1-NS2a produced authentic NS1, as marker on the left. Recombinants that lack sequences at the C terminus of NS2a are indicated by the length of deletion, ranging from 20 to 110 amino acids. These recombinants produced fusion proteins marked ΔNS1-NS2a with the exception of Δ20. LacZ represents the control virus that did not contain dengue sequences. Lane M, Molecular weight markers (indicated in thousands on the right).

Results of pulse-chase experiments suggest that these aberrant NS1 products were a precursor for NS1. Presumably, these uncleaved NS1-NS2a proteins accumulated as a result of lower cleavage efficiency resulting from NS2a deletion. This finding suggests that the C terminus of NS2a, although not essential for function, may affect the context of the presumed activity. Analysis of other recombinants indicated that deletion of 68 amino acids or more from the C terminus of NS2a essentially abolished the synthesis of NS1 and only the uncleaved NS1-NS2a was produced. It is clear from this analysis that the candidate proteinase activity for cleavage of NS1-NS2a resides in the N-terminal two-thirds (amino acids 1 to 153) of NS2a. Additional experiments such as in vitro processing studies should allow us to further elucidate the mechanism of the proteolytic cleavage and function of NS2a.

REFERENCES

Castle, E., T. Novak, U. Leidner, G. Wengler, and G. Wengler. 1985. Sequence analysis of the viral core protein and the membrane-associated proteins V1 and NV2 of the flavivirus West Nile and of the genome sequence for these proteins. *Virology* **145**: 227–236.

Coia, G., M. D. Parker, G. Speight, M. E. Byrne, and E. G. Westaway. 1988. Nucleotide and complete amino acid sequences of Kunjin virus: definitive gene order and characteristics of the virus-specified proteins. *J. Gen. Virol.* **69**:1–21.

Deubel, V., R. M. Kinney, and D. W. Trent. 1986. Nucleotide sequence and deduced amino acid sequence of the structural proteins of dengue type 2 virus, Jamaica genotype. *Virology* **155**:365–377.

Falgout, B., R. Chanock, and C.-J. Lai. 1989. Proper processing of dengue virus nonstructural glycoprotein NS1 requires the N-terminal hydrophobic signal sequence and the downstream nonstructural protein NS2a. *J. Virol.* **63**:1852–1860.

Hahn, Y. S., R. Galler, T. Hunkapiller, J. M. Dalrymple, J. H. Strauss, and E. G. Strauss. 1988. Nucleotide sequence of dengue 2 RNA and comparison of the encoded proteins with those of other flaviviruses. *Virology* **162**:167–180.

Halstead, S. 1981. The pathogenesis of dengue: molecular epidemiology in infectious diseases. *Am. J. Epidemiol.* **114**:632–648.

Halstead, S. B. 1988. Pathogenesis of dengue: challenge to molecular biology. *Science* **230**:476–481.

Mackow, E., Y. Makino, B. Zhao, Y.-M. Zhang, L. Markoff, A. Buckler-White, M. Guiler, R. Chanock, and C.-J. Lai. 1987. The nucleotide sequence of dengue type 4 virus: analysis of genes coding for nonstructural proteins. *Virology* **159**:217–228.

Mason, P. W., P. C. McAda, T. L. Mason, and M. J. Fournier. 1987. Sequence of the dengue 1 virus genome in the region coding for the three structural proteins and the major nonstructural protein NS1. *Virology* **161**:262–267.

Porterfield, V. S. 1980. Antigenic characteristics and classification of Togaviridae, p. 13–46. In R. W. Schlesinger (ed.), *The Togaviruses*. Academic Press, Inc., New York.

Rice, C. M., E. M. Lenches, S. R. Eddy. S. J. Shin, R. L. Sheets, and J. H. Strauss. 1985. Nucleotide sequence of yellow fever virus: implications for flavivirus gene expression and evolution. *Science* **229**:726–733.

Wright, P. J., M. R. Cauchi, and M. L. Ng. 1989. Definition of the carboxy termini of the three glycoproteins specified by dengue virus type 2. *Virology* **171**:61–67.

Zhao, B., E. Mackow, A. Buckler-White, L. Markoff, R. M. Chanock, C.-J. Lai, and Y. Makino. 1986. Cloning full-length dengue type 4 viral DNA sequences: analysis of genes coding for structural proteins. *Virology* **155**:77–88.

Zhao, B., G. Prince, R. Horswood, K. Eckels, P. Summers, R. Chanock, and C.-J. Lai. 1987. Expression of dengue virus structural proteins and nonstructural protein NS1 by a recombinant vaccinia virus. *J. Virol.* **61**:4019–4022.

V. VIRION STRUCTURE AND ASSEMBLY

Chapter 30

Role of Conformational Transitions in Poliovirus Assembly and Cell Entry

J. M. Hogle, R. Syed, C. E. Fricks, J. P. Icenogle, O. Flore, and D. J. Filman

Over the past several years, X-ray crystallographic studies have provided important insights concerning the architecture, evolution, and immune recognition of picornaviruses (Rossmann et al., 1985; Hogle et al., 1985; Luo et al., 1987; Filman et al., 1989; Acharya et al., 1989). Although these studies have yielded accurate structural models of mature virions, a more complete understanding of the relationships between the virus structure and its biological function also will require information about the changes in virion structure that occur at various stages of the viral life cycle. We have taken particular interest in the conformational changes associated with virus assembly and those occurring during cell attachment and entry. Fortunately, assembly and cell entry both are the subjects of intensive and ongoing biological characterization. Moreover, technological advances in virus crystallography now make it reasonable to undertake a systematic survey of the biological and structural properties of intermediates in the assembly and cell entry pathways and of variants with specific defects in these pathways. This chapter is intended as a review of what the structures of poliovirus and related picornaviruses have taught us about assembly and cell entry and as a brief report of the progress of our research program in these areas.

POLIOVIRUS STRUCTURE

In poliovirus and other members of the picornavirus family, the mature virion consists of a positive-stranded RNA genome (7,000 to 8,000 nucleotides) packaged in an icosahedrally symmetric protein shell. The protein shell is composed of 60 copies of each of four coat proteins, VP1, VP2, VP3, and VP4. VP1, VP2, and VP3 are similar in size and share a common folding pattern. Each of the proteins contains a conserved core that is an eight-stranded antiparallel beta barrel (Fig. 1). The three proteins differ, however, in the size and conformation of the loops that connect the strands of the beta barrels and in the extensions at their amino and carboxyl termini. The small protein VP4 functions in some respects as the detached amino terminus of VP2. In the mature virus particle, the cores pack to form the continuous protein shell, the connecting loops decorate the outer surface of the particle and are the major determinants of the antigenic sites of the virus, and the amino-terminal extensions form an intricate network on the inner surface of the protein shell.

ASSEMBLY

Once poliovirus enters the cell and is uncoated, its RNA is translated in a single large open reading frame to yield a polyprotein. As described in detail elsewhere in this volume, all viral proteins are derived from this polyprotein by successive cleavages catalyzed by virus-encoded proteases. One of the early cleavages, catalyzed by viral protease 2A, releases a 100-kilodalton capsid precursor, P1, from the amino terminus of the polyprotein (Toyoda et al., 1986). P1 contains the capsid protein sequences in the order VP4-VP2-VP3-VP1. The cleavage of

J. M. Hogle, R. Syed, and *D. J. Filman,* Department of Molecular Biology, Research Institute of Scripps Clinic, 10666 North Torrey Pines Road, La Jolla, California 92037.   *C. E. Fricks,* County USC Medical Center, Los Angeles, California 90033.   *J. P. Icenogle,* Centers for Disease Control, Atlanta, Georgia 30333.   *O. Flore,* University of Cagliari, Cagliari, Italy.

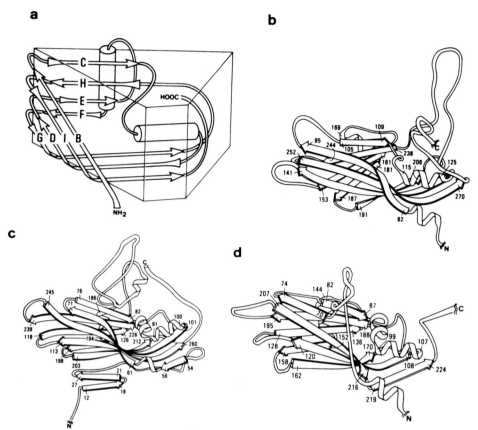

FIGURE 1. Structures of the major capsid proteins of poliovirus. (a) Schematic representation of the conserved wedge-shaped eight-stranded antiparallel beta-barrel core motif shared by VP1, VP2, and VP3. Individual beta strands are shown as arrows and are labeled alphabetically. Flanking helices are indicated by cylinders. (b to d) Ribbon diagrams of VP1 (b), VP2 (c), and VP3 (d). Residue numbers have been included as landmarks. Extensions at the amino and carboxyl termini of VP1 and VP3 have been truncated for clarity.

P1 yields capsid proteins VP1 and VP3 and an immature capsid protein precursor, VP0. These cleavages are catalyzed by 3C protease activity, apparently preferentially by the protease precursor 3CD (Ypma-Wong et al., 1988a). Finally, VP0 is cleaved to yield the capsid proteins VP4 and VP2. This cleavage has not been associated with any viral protease and may be autocatalytic (Hogle et al., 1985; Rossmann et al., 1985).

Several intermediates have been identified in the assembly of infectious virions (see review by Rueckert [1985]). In addition to uncleaved P1, these intermediates include protomers consisting of one copy each of VP0, VP1, and VP3; pentamers consisting of five copies of the protomer; empty capsids consisting of 12 pentamers; and provirions consisting of 60 copies of VP0, VP1, and VP3, enclosing one copy of the viral genomic RNA. Although the assembly pathway has been studied extensively, the kinetic role of these intermediates, particularly the empty capsid and the provirion, are unclear. The assembly process is apparently intimately linked to the proteolytic processing of the capsid proteins. In particular, cleavage of P1 is required for the formation of pentamers (Palmenberg, 1982), and the maturation cleavage of VP0 is thought to be required to make assembly irreversible.

ROLE OF P1 PROCESSING IN ASSEMBLY

Rearrangement of Chain Termini Generated by P1 Cleavage

The virion structure has provided an elegant explanation for the role of P1 processing in

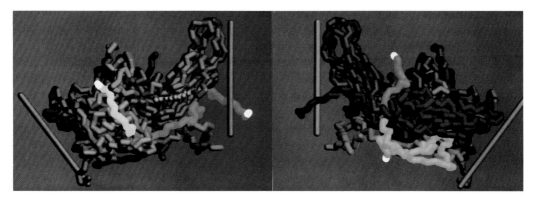

FIGURE 2. Locations of the termini generated by proteolytic processing of the P1 precursor. Alpha-carbon models for two complete protomers are shown. VP1 is blue, VP2 is yellow, VP3 is red, and VP4 is green. In the pentamer and in the virion, the two protomers would share a fivefold axis (shown as the vertical gray bar in each side of the figure), with the outer surface of the particle at the top and the inner surface at the bottom. Particle threefold axes are indicated as slanted gray bars. In pentamers and in virions, the front surfaces of the two protomers would be in contact in the interface between fivefold-related protomers. Residues near the carboxyl terminus of VP2 and the amino terminus of VP3 are highlighted on the left; residues near the carboxyl terminus of VP3 and the amino terminus of VP1 are highlighted on the right.

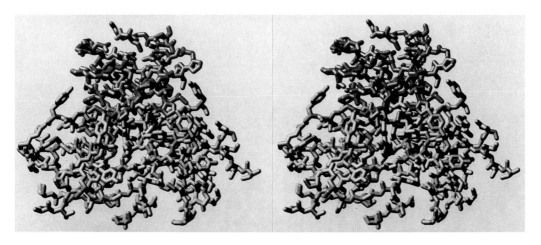

FIGURE 3. Stereo representation of the complex formed by the interaction of the amino termini of VP3, VP4, and VP1 around each fivefold axis. This structure appears to be important for directing the assembly of protomers to form pentamers. The inner surface of the capsid (which is located roughly 90 Å from the center of the particle) is shown at the bottom of the figure. The top of the complex is located at a radius of approximately 130 Å (by comparison, the outer surface of the particle extends to nearly 165 Å near the fivefold axis). At the top of the figure, five intertwined amino termini of VP3 (residues 1 to 11, shown in red) form a twisted tube of parallel beta structure. Five copies of the amino terminus of VP4 (residues 2 to 8 and 23 to 33, shown in green with yellow side chains) form short segments of two-stranded antiparallel beta sheet on the inner surface of the capsid. The amino terminus of VP4 is linked via an amide bond to myristic acid (magenta). Each of the five-residue segments of the polypeptide chain shown in blue is believed to be a portion of the amino terminus of VP1 (Filman et al., 1989).

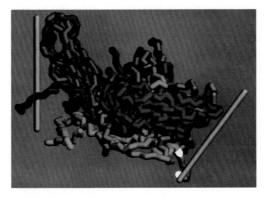

FIGURE 4. Locations of the chain termini generated by the maturation cleavage of VP0. A single protomer is shown. VP1 is blue, VP2 is yellow, VP3 is red, and VP4 is green. The vertical gray bar at the left represents a particle fivefold axis; the slanted gray bar at the right represents a particle threefold axis. The carboxyl terminus of VP4 and the first crystallographically ordered residue of VP2 are highlighted in white. The two residues are located on the inner surface of the mature virion.

regulating pentamer formation. The positions of the amino and carboxyl termini generated by the cleavage of P1 are depicted in Fig. 2. It is apparent that the termini which previously were linked in the uncleaved protomer are distant from one another in the mature virion. Indeed, the amino termini of VP3 and VP1 are located on the inner surface of the capsid, whereas the carboxyl termini of VP2 and VP3 are located on the outer surface. Inspection of Fig. 2 suggests that before cleavage, the carboxyl terminus of VP3 and the amino terminus of VP1 are linked across one surface of the protomer and that the carboxyl terminus of VP0 and the amino terminus of VP3 are linked across the other surface of the protomer. Thus, these two surfaces cannot exhibit the mutual complementarity observed in the mature virion until after cleavage of P1 has permitted the terminal strands to undergo substantial rearrangement. The precise nature of the rearrangement must await structural characterization of the P1 subunit.

Structure of the Amino-Terminal Network

The network that is formed on the inner surface of the protein by VP4 and by the amino-

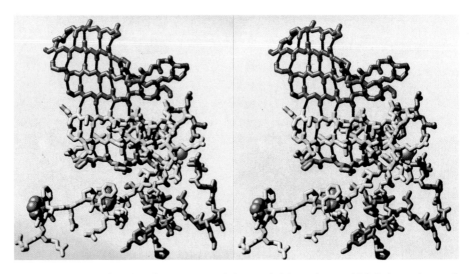

FIGURE 5. Demonstration that the seven-stranded extended beta sheet, which is located near the threefold axis of the mature virion, stabilizes the association of neighboring pentamers. In this stereo view, the outer surface of the capsid is located at the top and the inner surface of the capsid is at the bottom. Black lines indicate the main chain hydrogen bonds of the extended beta sheet. At the top, the C, H, E, and F strands of the VP3 beta barrel (red) form the outermost four strands of the extended sheet. The seventh strand (blue) is formed by residues 36 to 38 of VP1 from the same protomer. In the foreground, residues from a threefold-related protomer are shown. The amino terminal extension of VP2 (yellow) from the threefold-related protomer forms the fifth and sixth strands of the beta sheet, interacts with the carboxyl terminus of VP4 (green), and binds two planar electron density features (magenta) that tentatively have been identified as nucleotides of the viral RNA.

terminal extensions includes a particularly striking structure that is formed by the amino termini of VP3, VP4, and VP1. Five copies of the amino terminus of VP3 intertwine around each of the fivefold axes of the particle, forming a short tube of parallel beta structure (Fig. 3) (Hogle et al., 1985; Rossmann et al., 1985; Luo et al., 1987). This tube is flanked by five copies of the short two-stranded antiparallel beta sheet at the amino terminus of VP4 (Hogle et al., 1985). The amino terminus of VP4 has been shown to be myristoylated (Chow et al., 1987). The myristate serves to mediate the interaction between VP4 and the parallel beta tube formed by the amino termini of VP3 (Chow et al., 1987; Filman et al., 1989). The formation of this structure clearly depends both on the cleavage between VP0 and VP3 to free the amino terminus of VP3 and on the association of protomers to form pentamers. The interaction appears to contribute substantially to the stability of the pentamer, and we propose that it serves to direct pentamer formation during the assembly process.

The changes in the protomer interface induced by the rearrangement of the chain termini and the apparent role of the structure formed by the amino terminus of VP3 both are consistent with a model in which the cleavage of P1 is a prerequisite for pentamer formation (Palmenberg, 1982; Arnold et al., 1987). This model suggests that proteolytic processing of P1 provides both spatial and temporal control of the initiation of assembly. More recent experiments have examined the effect of linker insertions containing 3C protease sites on the processing of P1 in an in vitro translation system (Ypma-Wong et al., 1988b). The results of these experiments suggest that efficient processing of P1 requires the correct and complete folding of the P1 precursor. Thus, P1 processing may play an additional role in the control of assembly by ensuring that only completely and correctly folded precursors can enter the assembly pathway.

THE MECHANISM OF VP0 CLEAVAGE AND ITS ROLE IN ASSEMBLY

Until recently, structural studies of picornaviruses have been unable to provide a satisfactory explanation for the role of VP0 cleavage in virus maturation. The observation of the provirion (which contains a full complement of viral RNA and uncleaved VP0) suggests that the cleavage of VP0 may be the final step in assembly. It should be pointed out, however, that the precise kinetic role of the provirion is uncertain. It remains possible that VP0 cleavage occurs

concurrently with RNA packaging and that the low and apparently variable amounts of provirion that are seen in infections with wild-type virus represent particles that have not been properly processed. Regardless of its timing, the cleavage of VP0 appears to contribute substantially to particle stability, since particles that contain VP0 (such as empty capsids and provirions) are much less stable to heat, detergents, and high ionic strength than are mature virions (Fernandez-Thomas and Baltimore, 1973; Guttman and Baltimore, 1977a). The cleavage of VP0 may thus serve to make the assembly of correctly assembled virions irreversible, providing a final editing mechanism for the assembly process.

Proposed Modes for the Mechanism of VP0 Cleavage

The location in the mature virion of the termini generated by VP0 cleavage is shown in Fig. 4. Although the first few amino acids of VP2 are disordered in most of the known picornavirus structures, it is clear that the carboxyl terminus of VP4 and the amino terminus of VP2 are located close to one another. Moreover, the termini are located on the interior surface of the protein shell and are inaccessible to exogenous proteases. In rhinovirus (Rossmann et al., 1985) and poliovirus (Hogle et al., 1985), the carboxyl terminus of VP4 is hydrogen bonded to the side chain hydroxyl of serine 10. On the basis of the assumption that the structure as seen in the mature virion is relevant to the cleavage, Rossmann and co-workers have proposed that the autocatalytic cleavage of VP0 occurs via a serine protease-like mechanism, with serine 10 serving as the nucleophile and one of the RNA bases serving as the base (Rossmann et al., 1985; Arnold et al., 1987). Although this explanation would account for the apparent dependence of VP0 cleavage on RNA encapsidation, several lines of evidence have been developed recently which suggest that it is incorrect. (i) In the native structure, access of RNA to the cleavage site is blocked by an interaction between the amino termini of VP2 around the particle threefold axes (which is reminiscent of the beta annulus seen in plant virus structures). (ii) There is no nucleophilic side chain in the vicinity of the chain termini in the structure of the related foot-and-mouth disease virus (FMDV) (Acharya et al., 1989). (iii) Most convincingly, virus derived from a construct that lacks a nucleophile at serine 10 cleaves normally (E. Wimmer, personal communication).

The Structure of Empty Capsids and Its Implications for the Mechanism of Cleavage

To investigate the mechanism and role of VP0 cleavage, we have undertaken crystallographic studies of empty capsids (obtained from cells infected in the presence of low levels of guanidine) and of several variants with point mutations that affect VP0 cleavage. One of the cleavage-defective variants appears to accumulate provirions at nonpermissive temperatures (Karla Kirkegaard, personal communication), raising the possibility that we may be able to extend the structural studies to include the provirion. We have recently solved the structure of native-antigenic, reversibly dissociable empty capsids of the Mahoney strain of poliovirus type 1 (P1/Mahoney). Although model building and refinement have not been completed, the preliminary structure provides information that may be relevant to understanding the mechanism and role of VP0 cleavage.

As expected, the structure of the empty capsids is very similar to that of native virions. In particular, the beta-barrel cores and most of the external loops of the two structures are nearly identical. In the empty capsids, however, the network on the inner surface of the protein shell has been altered extensively. Indeed, whereas the amino-terminal extension of VP3 and the interactions between the amino terminus of VP3, the amino terminus of VP0 (VP4), and myristate at the fivefold axes are unperturbed, the remainder of the network is significantly rearranged or entirely disordered.

The most interesting ordered feature of the network in the empty capsids is a 20-residue peptide segment centered at the VP0 cleavage site. The conformation of this peptide is unlike that of the corresponding residues in mature virus, and the scissile bond is located some 21 Å (2.1 nm) from the location observed in virions. There are no potentially catalytic side chains near the location of the scissile bond in the empty capsids, but the chain segment containing this bond is located immediately adjacent to a large hydrophobic depression in the inner surface of the capsid. Provided that the location of the scissile bond in the empty capsids is relevant to the cleavage mechanism, we can suggest three possible mechanisms for cleavage. (i) The encapsidation of RNA induces a disorder-to-order transition in the portion of the amino terminus of VP1 that occupies the hydrophobic pocket in mature virus. This portion of VP1 could then form the catalytic site. (ii) The cleavage of VP0 is catalyzed by a thus far unidentified protease molecule that binds to the hydrophobic pocket. (iii) The cleavage is truly autocatalytic, being catalyzed by the asparagine side chain of the residue that is to become the carboxyl terminus of VP4. Such a mechanism has been proposed in the release of P1 from the polyprotein of FMDV (A. C. Palmenberg, personal communication) and in the autocatalytic maturation cleavage of nodaviruses (Gallagher and Rueckert, 1988).

Implications of an Assembly-Dependent Structure for the Role of VP0 Cleavage

As a consequence of the rearrangement of the amino termini of VP1 and VP0 (VP2), the empty capsids lack an important interaction that spans the boundary between pentamers in mature virions. We first noticed this interaction while attempting to find an explanation for the effect of a point mutation (at VP2 residue 18) in producing a non-temperature-sensitive phenotype in a variant of the Sabin vaccine strain of poliovirus type 3 (P3/Sabin) (Filman et al., 1989). In mature virions, the boundary between threefold-related protomers includes a seven-stranded beta sheet formed by residues from two different pentamers (Fig. 5). The outermost four strands of the sheet are contributed by the C, H, E, and F strands of the beta barrel of VP3 from one protomer; the fifth and sixth strands belong to the two-stranded beta sheet (residues 14 to 25) at the amino terminus of VP2 from a neighboring threefold-related protomer; and the seventh strand is contributed by residues 36 to 38 in the amino-terminal extension of VP1 from the first protomer. A similar structure also is seen in rhinovirus 14 (E. Arnold, personal communication) and in FMDV (Acharya et al., 1989).

The formation of the seven-stranded beta structure obviously depends on the association of pentamers in virions and appears to stabilize that association. The observation that the structure is not formed in empty capsids suggests that its formation is correlated either with cleavage of VP0 or with encapsidation of viral RNA. The mature virus structure is consistent with both possibilities. (i) The residues from VP2 that contribute the fifth and sixth strands to the seven-stranded beta sheet are located close to the amino terminus of VP2, and conformational constraints imposed by uncleaved VP0 may preclude the nearby residues from assuming the appropriate conformation. (ii) The electron density maps of mature virions contain two distinctly planar features in the immediate vicinity of the seven-stranded beta sheet which we have tentatively identified as partially ordered nucleotides (the nucleotides are shown in magenta in

Fig. 5). Future structural studies of provirions may resolve the question of whether RNA binding is a cause or a consequence of the structural differences between empty capsids and mature virions. In any case, the seven-stranded beta sheet clearly contributes to the stability of the mature virion and appears to be relevant to the correlation between encapsidation of RNA, VP0 cleavage, and particle stability.

CELL ENTRY

The recent identification of the cell surface receptors for poliovirus (Mendelsohn et al., 1989) and for the major group rhinoviruses (Tomassini et al., 1989; Staunton et al., 1989; Greve et al., 1989) and the demonstration that picornaviruses enter cells via endosomes (Madshus et al., 1984) represent major advances in our understanding of the early events of picornavirus infection. However, the steps involved in subsequent stages of cell entry, particularly the events associated with the virus crossing the endosomal membrane, remain poorly characterized. When poliovirus is attached to cells at 37°C, a substantial fraction of the attached virus (from 10 to 30% in our hands) elutes from the cell in a conformationally altered form (Fenwick and Cooper, 1962; Joklik and Darnell, 1961). The cell-altered virus (often referred to as eluted virus or 135S particle) has lost the internal capsid protein VP4, is noninfectious, and exhibits changes in sedimentation behavior, antigenicity, and protease sensitivity (Crowell and Philipson, 1971; De Sena and Mandel, 1977). A similar if not identical particle is the dominant form of the virus found inside cells early in infection (Lonberg-Holm et al., 1975; Everaert et al., 1989). The high particle-to-PFU ratio of poliovirus (and of other picornaviruses) makes it difficult to assess the role of intermediates in the early stages of infection. Nonetheless, because two very different classes of antiviral agents appear to exert their antiviral activities by preventing the cell-mediated conformational alteration, it is likely that the cell-altered virus is a necessary intermediate in the cell entry process. Thus, monensin (and other agents that prevent the acidification of endosomes) interferes both with infectivity and with the production of the intracellular form of the altered virus (Madshus et al., 1984). Similarly, a number of agents (including arildone and the so-called WIN compounds from Sterling-Winthrop) that bind to the virus and interfere with viral replication have been shown to block the formation of both the intracellular and the extracellular altered parti-

cles (McSharry et al., 1979; Caliguiri et al., 1980; Fox et al., 1986).

Chemical Characterization of Cell-Altered Particles

We have characterized both the extracellular (eluted) and the intracellular forms of the altered virus by using a variety of sequence-specific probes, including proteases, a panel of antibodies to synthetic peptides corresponding to sequences from capsid protein VP1, and a panel of monoclonal antibodies (Fricks and Hogle, 1990). The probes confirm that the extracellular and intracellular forms of cell-altered virus are similar if not identical and demonstrate that the cell-altered virions are conformationally distinct from native and from heat-inactivated virus. Significantly, the probes also have demonstrated that the loss of VP4 from cell-altered virions is accompanied by the exposure of several (normally internal) residues from the amino terminus of VP1. In particular, the first 30 amino acids of VP1 can be specifically excised from cell-altered virus by digestion with *Staphylococcus* V8 protease (Fricks and Hogle, 1990). Except for the loss of reactivity with antipeptide antibodies that are specific for the amino terminus of VP1, V8-treated eluted virus is indistinguishable from the intact altered particle.

A number of investigators have shown that the cell-altered form of the virus is more hydrophobic than native virions (Lonberg-Holm et al., 1976; Madshus et al., 1984). We have shown that cell-altered virus (but not native or heat-inactivated virus) attaches to liposomes (Fricks and Hogle, 1990). The exposed amino terminus of VP1 appears to be critical for liposome attachment, since removal of the first 30 amino acids of VP1 prevents liposome attachment. Moreover, treatment of preformed complexes of eluted virus and liposomes with V8 protease causes the liposomes to become enriched in a 3-kilodalton peptide that probably corresponds to the detached amino terminus of VP1 (Fricks and Hogle, 1990).

Role of Cell-Altered Particles in Cell Entry

The ability of the amino terminus of VP1 to attach to liposomes suggests that the exposure of these residues may play an important role in the membrane-crossing stage of cell entry. The amino terminus of VP1, perhaps in conjunction with VP4 (which is myristoylated at its amino terminus [Chow et al., 1987]) may facilitate cell entry by forming a pore or by disrupting the endosomal membrane. Curiously, the amino ter-

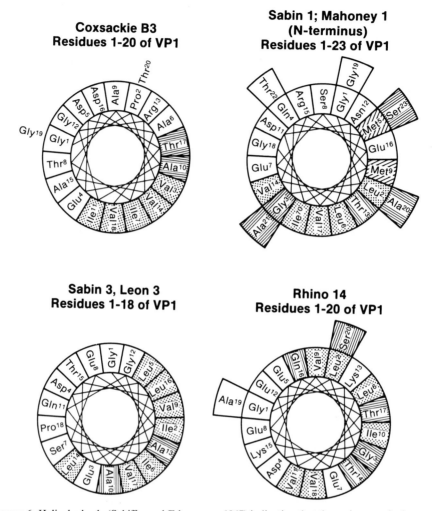

FIGURE 6. Helical wheels (Schiffer and Edmunson, 1967) indicating that the amino-terminal sequence of VP1 in several enteroviruses is consistent with the possible formation of amphipathic helix. Each panel represents an alpha helix, viewed along the helix axis, with side chains corresponding to the amino-terminal sequence of VP1 in coxsackievirus B3, P1/Mahoney and P1/Sabin, P3/Leon and P3/Sabin, and rhinovirus 14. Residue numbers and the identities of side chains are indicated. Hydrophobic side chains are stippled; charged or highly polar side chains are white. Amino acids that could be accommodated either on the polar or on the nonpolar side of an amphipathic helix (Gly, Ala, Ser, and Thr) are striped if they are located on the nonpolar side and white if they are located on the polar side of the helix.

minus of VP1 is among the most poorly conserved sequences in different strains of poliovirus. Thus, 20 of the first 22 amino acid residues of P1/Mahoney differ from those of P3/Sabin and P3/Leon. Nevertheless, in all of the polio-, rhino-, and coxsackievirus strains examined, the first 20 to 30 residues of VP1 exhibit a periodicity of nonpolar residues that is characteristic of an amphipathic helix (Fig. 6).

If the exposure of the amino terminus of VP1 indeed is involved in the entry of the entero- and rhinoviruses into the cell, the mechanism may not be general to other picornaviruses. Structural studies of mengovirus (Luo et al., 1987) and FMDV (Acharya et al., 1989) show that the cardioviruses and aphthoviruses lack residues that are structurally homologous to the first 40 amino acids of poliovirus. Correspondingly, there is also a lack of evidence for the formation of a 135S form of the virion during the entry of cardio- and aphthoviruses. Instead, these viruses, which are acid labile, appear to dissociate directly into pentamers upon the acidification of endosomes (Hall and Rueckert, 1971; Talbot and Brown, 1972).

Role of Cell-Altered Particles in the Immune Response to Poliovirus

There is growing evidence that conformationally altered virus may play an important role in the immune recognition of poliovirus. In a previous study, we showed that synthetic peptides corresponding to two regions of the amino terminus of VP1 (residues 24 to 40 and 61 to 80) were capable of eliciting neutralizing antibodies in mice and in rats (Chow et al., 1985). We have also obtained preliminary evidence that these normally internal sequences are recognized in the human response to poliovirus immunization. Comparisons of immunoprecipitation titers against native virus, 135S eluted virus, and V8-treated 135S eluted virus indicate that sera from two human vaccinees contain high levels of antibodies that are specific for the 135S particles. In one of the two vaccinees, the overwhelming majority of the 135S-specific titer was directed against the first 30 residues of VP1 (Fricks, 1989). A more extensive study by Roivainen et al. (in press) using the peptide-scanning approach has confirmed the presence of high titers to residues 1 to 30 of VP1 in some human vaccinees and has revealed a second epitope, comprising residues 40 to 53 of VP1, that is present in the sera of more than 90% of human vaccinees. Although it has not been demonstrated whether the antibodies that recognize the amino terminus of VP1 neutralize infectivity, the results clearly indicate that receptor-mediated conformational changes may expose novel epitopes in susceptible hosts.

Nature of the Structural Transition

The loss of VP4 and the extrusion of the amino terminus of VP1 pose an interesting question of how normally internal components of the virion become exposed during the conformational transition. An analogy with a well-characterized conformational transition in plant viruses suggests how such externalization might occur. A number of icosahedral plant viruses expand by approximately 10% in radius upon exposure to chelators of divalent cations at slightly alkaline pH in vitro (Incardona and Kaesberg, 1974; Kruse et al., 1982). The expansion has been

particularly well characterized in a group of related T=3 plant viruses (for example, tomato bushy stunt virus and turnip crinkle virus), which bear striking structural similarity to the picornaviruses. In these viruses, expansion is controlled by an interface between capsid proteins which contains a pair of divalent cation-binding sites (Hogle et al., 1983). Upon removal of the divalent cations at basic pH, the interface is destabilized by local concentration of negative charges from the acidic residues that bind the cations (Hogle et al., 1983). Biochemical characterization of tomato bushy stunt and turnip crinkle viruses has shown that the expansion of the plant viruses also results in the externalization of residues near the amino termini of capsid proteins (Harrison et al., 1986). In the structure of the expanded form of tomato bushy stunt virus, which has been solved at low resolution (Robinson and Harrison, 1982), the newly exposed portions of the amino termini of the capsid proteins are extruded through fenestrations caused by the disruption of this interface.

An analysis of the location of residues that contribute to the thermostability of variants of P3/Sabin (Minor et al., 1989) suggests that the analogous interface is also important in controlling stability and conformational rearrangements in poliovirus (Filman et al., 1989). It may be relevant that the point at which the amino-terminal extension of VP1 joins the beta barrel is located immediately below this interface in the native virus particle. The location of these residues and the analogy with plant virus expansion suggest that the cell-mediated conformational change in poliovirus might involve disruption of this interface and that the amino terminus of VP1 might be externalized through the interface.

The same interface might also be the portal for the externalization of VP4, possibly coordinated with externalization of the amino terminus of VP1. During the refinement of the poliovirus models, we observed an electron density feature that we have tentatively identified as five residues from the extreme amino terminus of VP1. These residues form a third beta-strand hydrogen bonded to the two-stranded sheet formed by the amino terminus of VP4 (Filman et al., 1989). The interaction between the amino termini of VP1 and VP4 could provide a mechanism for coordinating their externalization during the formation of the 135S particle.

Additional evidence regarding the nature of the cell-induced conformational change has come from elegant studies of the binding of antiviral drugs to rhinovirus that were conducted by Rossmann and his colleagues in collaboration with Sterling-Winthrop (Smith et al., 1986; Badger et al., 1988). The WIN compounds have been shown to bind virions and to prevent the conformational alterations associated with cell entry and with heat inactivation. In some cases, binding these antiviral drugs also interferes with cell attachment. Rossmann and co-workers have shown that the drugs bind in a hydrophobic pocket in the middle of the beta-barrel core of VP1 and have proposed that the drugs work by preventing VP1 from undergoing structural changes required for rearrangement of the protein shell (Smith et al., 1986; Badger et al., 1988). Interestingly, we have found that the hydrophobic pocket that binds antiviral drugs is occupied in the native poliovirus structures (Filman et al., 1989). Although the nature of the occupant of the pocket has yet to be identified chemically, the electron density feature is consistent with a long-chain hydrocarbon with a polar head group (most probably sphingosine in the P1/Mahoney and P3/Sabin structures). A number of non-temperature-sensitive revertants of P3/Sabin have been shown to contain a mutation in the lipid-binding pocket (Minor et al., 1989; Macadam et al., 1989), suggesting that the antiviral drugs may work by subverting a site normally used to regulate conformational transitions of poliovirus (Filman et al., 1989).

For some time, we have been interested in crystallographic study of cell-altered virions. There are two major obstacles to obtaining material suitable for crystallographic studies: (i) the eluted virus is very hydrophobic and (ii) the number of cells required to produce milligram quantities of eluted virus by using conventional protocols is prohibitively large. The hydrophobic properties of eluted virus may be circumvented through the inclusion of nonionic detergents during purification or by crystallizing V8-treated eluted virus, which has solubility properties similar to those of native virions. At present, we are exploring the possibility of increasing the yield of eluted virus by using cell extracts (De Sena and Mandel, 1976, 1977; Guttman and Baltimore, 1977b) or fixed cells (Madshus et al., 1984) to induce the conformational alteration.

ACKNOWLEDGMENTS. This work was support by Public Health Service grants AI20566 and GM38794 to J.M.H. from the National Institutes of Health.

We thank M. Chow, P. D. Minor, and K. Kirkegaard for helpful discussions, T. Critchlow and M. Kubitz for expert technical assistance, and M. Graber for help in preparing the manuscript. This is publication number 6179-MB of the Research Institute of Scripps Clinic.

REFERENCES

Acharya, R., E. Fry, D. Stuart, G. Fox, D. Rowlands, and F. Brown. 1989. The three-dimensional structure of foot-and-mouth disease virus at 2.9 Å resolution. *Nature* (London) **337:**709–716.

Arnold, E., M. Luo, G. Vriend, M. G. Rossmann, A. C. Palmenberg, G. D. Parks, M. J. H. Nicklin, and E. Wimmer. 1987. Implications of the picornavirus capsid structure for polyprotein processing. *Proc. Natl. Acad. Sci. USA* **84:**21–25.

Badger, J., I. Minor, M. J. Kremer, M. A. Olivera, T. J. Smith, J. P. Griffith, D. M. A. Guerin, S. Krishnaswamy, M. Luo, M. G. Rossmann, M. A. McKinlay, G. D. Diana, F. J. Dutko, M. Fancher, R. R. Rueckert, and B. A. Heinz. 1988. Structural analysis of a series of antiviral agents complexed with human rhinovirus 14. *Proc. Natl. Acad. Sci. USA* **85:** 3304–3308.

Caliguiri, L. A., J. J. McSharry, and G. W. Lawrence. 1980. Effect of aridone on modification of poliovirus *in vitro*. *Virology* **105:**86–93.

Chow, M., J. F. E. Newman, D. Filman, J. M. Hogle, D. J. Rowlands, and F. Brown. 1987. Myristylation of picornavirus capsid protein VP4 and its structural significance. *Nature* (London) **327:**482–486.

Chow, M., R. Yabrov, J. Bittle, J. M. Hogle, and D. Baltimore. 1985. Synthetic peptides from four separate regions of the poliovirus capsid protein VP1 induce neutralizing antibodies. *Proc. Natl. Acad. Sci. USA* **82:**910–914.

Crowell, R. L., and L. Philipson. 1971. Specific alteration of coxsackievirus B3 eluted from HeLa cells. *J. Virol.* **8:**509–515.

De Sena, J., and B. Mandel. 1976. Studies on the *in vitro* uncoating of poliovirus. I. Characterization of the modifying factor and the modifying reaction. *Virology* **70:**470–483.

De Sena, J., and B. Mandel. 1977. Studies on the *in vitro* uncoating of poliovirus. II. Characteristics of the membrane-modified particle. *Virology* **78:**554–566.

Everaert, L., R. Vrijsen, and A. Boeyé. 1989. Eclipse products of poliovirus after cold-synchronized infection of HeLa cells. *Virology* **171:**76–82.

Fenwick, M. L., and P. D. Cooper. 1962. Early interactions between poliovirus and ERK cells. Some observations on the nature and significance of the rejected particles. *Virology* **18:**212–223.

Fernandez-Thomas, C., and D. Baltimore. 1973. Morphogenesis of poliovirus. II. Demonstration of a new intermediate, the provirion. *J. Virol.* **12:**1122–1130.

Filman, D. J., R. Syed, M. Chow, A. J. Macadam, P. D. Minor, and J. M. Hogle.1989. Structural factors that control conformational transitions and serotype specificity in type 3 poliovirus. *EMBO J.* **8:**1567–1579.

Fox, M. P., J. J. Otto, and M. A. McKinlay. 1986. Prevention of rhinovirus uncoating by WIN 51711, a new antiviral drug. *Antimicrob. Agents Chemother.* **30:**110–116.

Fricks, C. E. 1989. Studies of the conformational changes of poliovirus during virus neutralization, cell entry and infection. Ph.D. thesis, University of California, San Diego.

Fricks, C. E., and J. M. Hogle. 1990. Cell-induced conformational change in poliovirus: externalization of the amino terminus of VP1 is responsible for liposome binding. *J. Virol.* **64,** in press.

Gallagher, T. M., and R. R. Rueckert. 1988. Assembly-dependent maturation cleavage in provirions of a small icosahedral insect ribovirus. *J. Virol.* **62:** 3399–3406.

Greve, J. M., G. Davis, A. M. Meyer, C. P. Forte, S. C. Yost, C. W. Marlor, M. E. Kamarck, and A. Mc-Clelland. 1989. The major human rhinovirus receptor is ICAM-1. *Cell* **56:**839–847.

Guttman, N., and D. Baltimore. 1977a. Morphogenesis of poliovirus. IV. Existence of particles sedimenting at 150S and having the properties of provirion. *J. Virol.* **23:**363–367.

Guttman, N., and D. Baltimore. 1977b. A plasma membrane component able to bind and alter virions of poliovirus type 1: studies on cell-free alteration using a simplified assay. *Virology* **82:**25–36.

Hall, L., and R. Rueckert. 1971. Infection of mouse fibroblasts by cardioviruses. Premature uncoating and its prevention by elevated pH and magnesium chloride. *Virology* **43:**152–165.

Harrison, S. C., P. K. Sorger, P. G. Stockley, J. M. Hogle, R. Altman, and R. K. Strong. 1986. Mechanism of RNA virus assembly and disassembly. *UCLA Symp. Mol. Cell. Biol. New Ser.* **54:**379–395.

Hogle, J. M., M. Chow, and D. J. Filman. 1985. Three-dimensional structure of poliovirus at 2.9 Å resolution. *Science* **229:**1358–1365.

Hogle, J. M., T. Kirchhausen, and S. C. Harrison. 1983. The divalent cation binding sites of tomato bushy stunt virus: difference maps at 2.9 Å resolution. *J. Mol. Biol.* **171:**95–100.

Incardona, N. L., and P. Kaesberg. 1974. A pH-induced structural change in bromegrass mosaic virus. *Biophys. J.* **4:**11–21.

Joklik, W. K., and J. E. Darnell. 1961. The adsorption and early fate of purified poliovirus in HeLa cells. *Virology* **13:**439–447.

Kruse, J., K. M. Kruse, J. Witz, C. Chauvin, B. Jacrot, and A. Tardieu. 1982. Divalent ion-dependent reversible swelling of tomato bushy stunt virus and organization of the expanded version. *J. Mol. Biol.* **162:**393–417.

Lonberg-Holm, K., L. B. Gosser, and J. C. Kauer. 1975. Early alteration of poliovirus in infected cells and its specific inhibition. *J. Gen. Virol.* **27:**329–345.

Lonberg-Holm, K., L. B. Gosser, and E. J. Shimshick. 1976. Interaction of liposomes with subviral particles of poliovirus type 2 and rhinovirus type 2. *J. Virol.* **19:**746–749.

Luo, M., G. Vriend, G. Kamer, I. Minor, E. Arnold, M. G. Rossmann, U. Boege, D. G. Scraba, G. M. Duke, and A. C. Palmenberg. 1987. The atomic structure of Mengo virus at 3.0 Å resolution. *Science* **235:**182–191.

Macadam, A. J., C. Arnold, J. Howlett, A. John, S. Marsden, F. Taffs, P. Reeve, N. Hamada, K. Wareham, J. Almond, N. Cammack, and P. D. Minor. 1989. Reversion of the attenuated and temperature-sensitive phenotypes of the Sabin type 3 strain of poliovirus in vaccinees. *Virology* **172:**408–414.

Madshus, I. H., S. Olsnes, and K. Sandvig. 1984. Requirements for entry of poliovirus into cells at low pH. *EMBO J.* **3:**1945–1950.

McSharry, J. J., L. A. Caliguiri, and H. J. Eggers. 1979. Inhibition of uncoating of poliovirus by arildone, a new antiviral drug. *Virology* **97:**307–315.

Mendelsohn, C. L., E. Wimmer, and V. R. Racaniello. 1989. Cellular receptor for poliovirus: molecular cloning, nucleotide sequence, and expression of a new member of the immunoglobulin superfamily. *Cell* **56:**855–856.

Minor, P. D., G. Dunn, D. M. A. Evans, D. I. Magrath, A. John, J. Howlett, A. Phillips, G. Westrop, K. Wareham, J. W. Almond, and J. M. Hogle. 1989. The temperature sensitivity of the Sabin type 3 vaccine strain of poliovirus: molecular and structural effects of a mutation in the capsid protein VP3. *J. Gen. Virol.* **70:**1117–1123.

Palmenberg, A. C. 1982. In vitro synthesis and assembly of picornaviral capsid intermediate structures. *J. Virol.* **44:**900–906.

Robinson, I. K., and S. C. Harrison. 1982. Structure of the expanded state of tomato bushy stunt virus. *Nature* (London) **297:**563–568.

Roivainen, M., A. Narvanen, M. Korkolainen, M.-L. Huhtala, and T. Hovi. *Virology*, in press.

Rossmann, M. G., E. Arnold, J. W. Erickson, E. A. Frankenberger, J. P. Griffith, H.-J. Hecht, J. E. Johnson, G. Kamer, M. Luo, A. G. Mosser, R. R. Rueckert, B. Sherry, and G. Vriend. 1985. Structure of a human common cold virus and functional relationship to other picornaviruses. *Nature* (London) **317:**145–153.

Rueckert, R. R. 1985. Picornaviruses and their replication, p. 705–735. *In* B. Fields (ed.), *Virology.* Raven Press, New York.

Schiffer, M., and A. B. Edmunson. 1967. Use of helical wheels to represent the structure of proteins and to identify segments with helical potential. *Biophys. J.* **7:**121–135.

Smith, T. J., M. J. Kremer, M. Luo, G. Vriend, E. Arnold, G. Kamer, M. G. Rossmann, M. A. McKinlay, G. D. Diana, and M. J. Otto. 1986. The site of attachment in human rhinovirus 14 for antiviral agents that inhibit uncoating. *Science* **233**:1286–1293.

Staunton, D. E., V. J. Merluzzi, R. Rothlein, R. Barton, S. D. Marlin, and T. A. Springer. 1989. A cell adhesion molecule, ICAM-1, is the major surface receptor for rhinoviruses. *Cell* **56**:849–853.

Talbot, P., and F. Brown. 1972. A model for foot-and-mouth disease virus particles. *J. Gen. Virol.* **15**:163–170.

Tomassini, J. E., D. Graham, C. M. DeWitt, D. W. Lineberger, J. A. Rodkey, and R. J. Colonno. 1989. cDNA cloning reveals that the major group rhinovirus receptor on HeLa cells is intercellular adhesion molecule 1. *Proc. Natl. Acad. Sci. USA* **86**:4907–4911.

Toyoda, H., M. J. H. Nicklin, M. G. Murray, C. W. Anderson, J. J. Dunn, F. W. Studier, and E. Wimmer. 1986. A second virus-encoded proteinase involved in proteolytic processing of poliovirus polyprotein. *Cell* **45**:761–770.

Ypma-Wong, M. F., P. G. Dewalt, V. H. Johnson, J. G. Lamb, and B. L. Semler. 1988a. Protein 3CD is the major poliovirus proteinase responsible for cleavage of the P1 capsid precursor. *Virology* **155**:265–270.

Ypma-Wong, M. F., D. J. Filman, J. M. Hogle, and B. L. Semler. 1988b. Structural domains of the poliovirus polyprotein are major determinants for proteolytic cleavage at gln-gly pairs. *J. Biol. Chem.* **263**:17846–17856.

Chapter 31

The Three-Dimensional Structure of Foot-and-Mouth Disease Virus

Ravindra Acharya, Elizabeth Fry, Derek Logan, David Stuart, Fred Brown, Graham Fox, and David Rowlands

Foot-and-mouth disease virus (FMDV) belongs to a large group of small icosahedral animal viruses, the family *Picornaviridae*. Among the characteristics shared by members of this family of viruses is possession of a single-stranded, positive-sense RNA genome of 7,500 to 8,000 nucleotides encapsidated within a T=1 icosahedral capsid of ca. 30-nm diameter, comprising 60 copies of each of four proteins, VP1 to 4. The genomic RNA is linked at the 5' end to a small protein of ca. 20 amino acids (VPg) and is polyadenylated at the 3' end. The protein-coding portion of the RNA occurs as a single continuous open reading frame that is preceded by a long 5' noncoding region (650 to 1,300 nucleotides). The product of translation is thus a polyprotein that is processed by a variety of virus-encoded proteases into the mature protein products. Structural proteins account for approximately one-third of the protein and are encoded toward the 5' end of the open reading frame; the products of the remaining two-thirds are involved in virus replication.

Despite these similarities in structure and genome organization, there is great diversity within the family *Picornaviridae*, and the group has been subdivided into four genera. This classification was based on a number of physicochemical properties such as buoyant density and acid stability of the virus and the overall nucleotide composition of the RNA. Later, more sophisticated criteria of evolutionary relationships based on genome sequence comparisons have substantiated the original classification of the genera. Among the four genera, the enteroviruses (e.g., poliovirus) and rhinoviruses (e.g.,

human rhinovirus) are the most closely related and the aphthoviruses (FMDVs) are the most distantly related. The cardioviruses (e.g., encephalomyocarditis virus) occupy a position between the entero- and rhinoviruses and the aphthoviruses. Another picornavirus, hepatitis A virus, has many unique properties and is not clearly related to any of the four recognized genera.

The genus *Enterovirus* is a large group that includes the coxsackieviruses and echoviruses as well as poliovirus and contains many human pathogens. The best-characterized viruses of the genus *Rhinovirus* are those which are responsible for a large proportion of common colds in humans and include at least 100 serotypes. The cardioviruses, however, are predominantly rodent pathogens, although a wide range of mammalian species can be infected. The FMDVs comprise the genus *Aphthovirus* and infect cloven-hoofed animals. There are seven serotypes of the virus, and there is much antigenic variation within each serotype.

Thus, despite similarities in genome organization and structure, the family *Picornaviridae* includes a wide range of viruses differing in antigenic properties, host range, and tissue specificities. These variations in biological properties must be reflected entirely or to a large extent by differences in the detailed molecular structure of the virus capsids. At present, only X-ray crystallographic analysis can provide such information, and it is therefore crucial to a more thorough understanding of virus function.

The first determination of a picornavirus structure was that of human rhinovirus 14 (Ross-

Ravindra Acharya, Elizabeth Fry, Derek Logan, and David Stuart, Laboratory of Molecular Biophysics, South Parks Road, Oxford OX1 3QU, United Kingdom.   *Fred Brown, Graham Fox, and David Rowlands*, Wellcome Biotech, Langley Court, Beckenham, Kent BR3 3BS, United Kingdom.

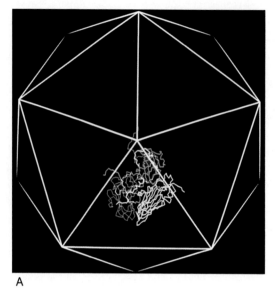

A

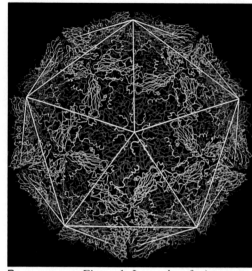

B   *Figure 1. Legend on facing page.*

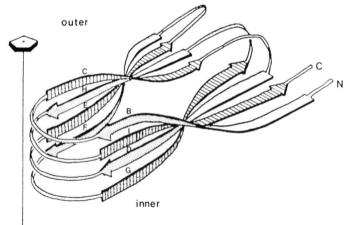

outer

C

N

inner

Figure 2. Legend on facing page.

Figure 3. Legend on facing page.

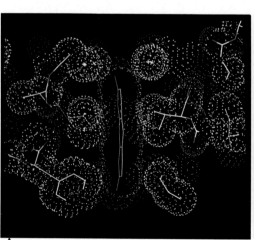

A

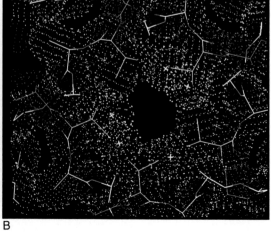

B

mann et al., 1985), which was closely followed by that of poliovirus type 1 (Hogle et al., 1985). These analyses showed that the folding patterns of the major structural proteins, VP1, 2, and 3, were similar and formed a wedge-shaped eight-stranded β barrel, a structure previously seen in the coat proteins of small RNA plant viruses (Harrison et al., 1978). The major differences between VP1, 2, and 3 occur in the loop regions linking the β strands of the core structures. Amino acid substitutions correlating with antigenic variation occur in the surface-oriented loop regions. The VP1 proteins of both viruses are located around the fivefold axes of icosahedral symmetry, and VP2 and 3 alternate around the two- and threefold axes. The small, myristylated protein, VP4, is located entirely at the inner surface of the capsid.

In both poliovirus and rhinovirus, there are deep depressions in the surface surrounding the fivefold axes, and it has been proposed that residues at the bases of these depressions form the points of attachment of the viruses to cell receptor proteins (the canyon hypothesis [Rossmann et al., 1985]). Such a location would shield the necessarily conserved features associated with receptor binding from immune surveillance by steric hindrance, and there is some supportive evidence for this proposal for human rhinovirus (Colonno et al., 1988).

Determination of the structure of a cardiovirus (mengovirus) showed that the main architectural features of the capsid were similar to those seen in poliovirus and rhinovirus (Luo et al., 1987). A major difference is that the canyon, which occurs as a continuous, moatlike depression in the latter viruses, is partially filled in mengovirus, producing a series of pits in the surface. Whether the receptor-binding region of the virus is located in these depressions is unknown.

X-ray crystallographic analysis of a virus is a technically demanding task, and it therefore is worth considering the features of FMDVs that differentiate them from the other picornaviruses and justify determination of the virus structure to understand their physical basis. (i) The major structural proteins, VP1 to 3, of FMDV are smaller than their counterparts in the other picornaviruses, each having a molecular weight of ca. 24,000. This difference is particularly marked in the case of VP1, which is typically ca. 30,000 in the other picornaviruses. (ii) The virus particles have the highest buoyant density in CsCl within the picornaviruses (>1.43 g/ml). The particles are also permeable to organic molecules such as proflavin. (iii) The virus particles are extremely unstable in acid conditions. At pH 6.8 or below, the capsids dissociate into 12S pentameric subunits. (iv) Interchain disulfide bonds link some of the capsid proteins. In nonreducing conditions, VP3 dimers are found; in viruses of serotype O, VP1-VP2 dimers are also seen (D. Rowlands, unpublished observations). (v) The viruses are sensitive to cleavage by proteolytic enzymes such as trypsin. Although proteolysis does not affect particle integrity, some important biological properties are markedly changed. The infectivity of trypsin-treated virus is greatly reduced, and this is correlated with a loss of attachment to cells. With serotype O viruses, there is also a great loss of immunogenicity. These biological effects are associated with cleavage within VP1, the only protein in the virus capsid susceptible to proteolysis. (vi) Synthetic peptides representing those regions of VP1 susceptible to proteolysis in the particles are particularly effective in eliciting virus-neutralizing antibodies. This is especially true of the hypervariable sequence located between residues 141 and 160, although peptides from the C terminus of the protein are also

---

FIGURE 1. (A) The protomer or basic icosahedral subunit comprising one copy of VP1 to VP4 shown against the outline of an icosahedron. Sixty copies of the protomer are arranged around the icosahedral symmetry axes to produce the virus capsid. The standard coloring scheme is used: blue, VP1; green, VP2; red, VP3; and yellow, VP4. (B) The capsid structure shown against the outline of an icosahedron.

FIGURE 2. Schematic of the core of the basic eight-stranded β-barrel structure exhibited by the three external capsid proteins VP1, 2, and 3.

FIGURE 3. (A) Section through the virus capsid in the plane of the fivefold axis showing the hydrophobic pore through which a molecule of proflavin is able to pass. The atoms are depicted with spheres of dots corresponding to their Van der Waals radii with standard coloring: white, carbon; blue, nitrogen; red, oxygen; and yellow, sulfur. A molecule of proflavin is shown in purple. (B) Section through the virus capsid perpendicular to the fivefold axis, showing the fivefold pore constricted by disulfide bonds. This illustrates one of the two possible configurations of disulfide bonds at any fivefold axis. The two possible structures arise because each cysteine can be involved in only one bond at any one time.

active. Antigenic site mapping experiments using monoclonal antibodies have indicated that some antigenic determinants on the virus include residues from both of these regions (Parry et al., 1985). (vii) Both of the highly exposed regions of VP1 that have been shown to be particularly important antigenically are involved in virus binding to cell receptors (Fox et al., 1989). This is in complete contrast to the canyon hypothesis for receptor binding proposed for the other picornaviruses.

We shall not dwell on the details of structure determination, since these have been presented previously in outline (Acharya et al., 1989b). The structure is now rather well refined, using diffraction data extending to spacings corresponding to approximately 2.9 Å (0.29 nm). The broad features of the structure have been described (Acharya et al., 1989a, 1989b); in this chapter, we will reassess the pertinent features of FMDV presented above with reference to the three-dimensional structure.

OVERVIEW OF FMDV STRUCTURE

As expected, the disposition of the four capsid proteins within the FMDV capsid is similar to that observed in other picornaviruses. This arrangement is presented in Fig. 1, where the protomer is shown located on the basic icosahedral lattice of the virus and then built up. Also shown is a very simplified sketch showing the basic organization of the β strands within the so-called jelly roll that comprises the cores of VP1, VP2, and VP3 (Fig. 2). The loops where the antigenic determinants of the picornaviruses reside are identified in terms of the β strands that they link; thus, the FMDV loop is also known as the GH loop.

Structural Consequences of Smaller Capsid Proteins

Shortening is not achieved in general by reduction in length of N- and C-terminal extensions from the β core but rather by truncation of many of the loops. This has a number of effects, the most noticeable being (i) the general thinning of the capsid, whereby the root mean square thickness of the FMDV capsid (excluding VP4) is approximately 33 Å, compared with 39.5 Å for mengovirus and 42.5 Å for human rhinovirus 14, and (ii) the denuding of the region near the icosahedral fivefold axis of loops from VP1. Implications of this are discussed below.

Permeability of the Capsid

A five-stranded β annulus of VP3 surrounds the fivefold icosahedral axis of FMDV as in all of the other picornaviruses. Since the loops of VP1 that normally clothe the capsid external to this structure are absent in FMDV, a pore is exposed which is constricted by a lining of three conserved hydrophobic side chains: Phe-3, Val-5, and Cys-7 of VP3 (going outward from the interior of the capsid). Cys-7 is capable of forming disulfide bridges around the fivefold axis and forms the most severe constriction of the pore. Figure 3A shows a section through the capsid in the plane of an icosahedral fivefold axis. The molecule of proflavin shown will pass through the pore with very little disturbance to the structure of the β annulus. There is no unaccounted-for electron density and thus, as expected from the nature of the side chains, no indication of any ion binding in this channel.

Capsid Dissociation below pH 6.8

Within the protomer, the proteins are tightly associated. This can be illustrated, for example, by comparison of the surface areas of VP1 and VP2 rendered solvent inaccessible by association with VP3, approximately 4,800 Å², with that involved in antibody binding, 600 to 800 Å². The pentameric subunit in all picornaviruses is a stable structure in which the protomers are fastened together by the interweaving of the N terminus of VP3 as it winds into the β annulus around the axis of symmetry. In FMDV, the pentamer assemblies are relatively weakly associated, since this association is primarily mediated by a series of hydrogen bonds between a strand of an N-terminal β-sheet hairpin in VP2 (which flanks the basic core of the β barrel) and a strand of the β barrel of VP3 from a twofold-related pentamer, thus forming an extended β sheet of six strands that straddles the pentamer boundaries. These interactions may be supplemented by ionic components involved in cation ligation at the threefold axes. In rhino- and polioviruses, the β-sheet structure linking pentamers is strengthened by a seventh strand contributed by residues near the N terminus of VP1 also from the twofold-related pentamer. These differences help to explain why the capsids of aphthoviruses (and cardioviruses), in the presence of chloride ions, readily dissociate into pentameric subunits, whereas entero- and cardioviruses retain the icosahedral association of VP1 to 3 to produce modified empty particles lacking VP4 and RNA.

The pH instability of FMDV particles is presumably linked to the ionization of groups closely associated with the pentamer boundaries. This may be explained in terms of the switching of the protonation state of a cluster of histidine residues in this region.

Disulfide Bonds

The disulfide bonds between VP1 and VP2 proteins involve residues 134 of VP1 and 130 of VP2. Their location at the base of the FMDV loop and the absence of cysteine residues at comparable positions in any serotype other than O suggests their importance in determining the antigenic phenotype of this serotype. This view agrees with analyses using monoclonal antibodies and peptides, which indicate that the antigenic sites of serotype O viruses contain complex epitopes whereas, for example, the major antigenic site of serotype A virus has the characteristics of a linear determinant (Parry et al., 1985; Di Marchi et al., 1986; Bolwell et al., 1989).

The observed VP3 homodimers arise through combinations of disulfide bridges between the five copies of Cys-7 that exhibit radial symmetry extending in toward the fivefold axes. At any one of these residues, a disulfide bond can form clockwise or counterclockwise to a neighboring fivefold residue. Only four of the five cysteine residues can be satisfied at any one fivefold axis (Fig. 3B).

Proteolytic Cleavage

Strohmaier et al. (1982) have identified the proteolytic cleavage sites in VP1 of type O_1 Kaufbeuren virus, which is closely related to type O_1 BFS (British Field Strain) virus. These sites are in regions of the virion that are seen to be not only well exposed on the surface but also the most disordered of the surface residues in the three-dimensional structure of type O_1 BFS virus. The points of cleavage for the type O_1 Kaufbeuren virus are after residue 145 for mouse submaxillary gland protease and after residues 138, 154, and 199 for trypsin. After trypsin cleavage, the region from 139 to 154, which corresponds well to the disordered loop in the electron density map (residues 135 to 156), is released from the virus, further supporting the evidence from the structure (see below) that in the intact type O virus there is little interaction between the loop and the rest of the virion. In addition, the exposed C-terminal region (200 to 213) is also released.

Synthetic Peptides

The structure of FMDV type O_1 BFS virus clearly shows that the folding of the polypeptide chain in VP1 brings the two separate segments (residues 141 to 160 of VP1, the FMDV loop, and the C terminus of VP1) that have been shown to elicit virus-neutralizing antibodies into close proximity on the viral surface. This finding substantiates the results of monoclonal antibody mapping, which showed that residues from both regions could contribute to a single complex epitope. Surprisingly, the carboxy-terminal residues that contribute to this epitope come from a symmetry-related VP1 polypeptide. In contrast to the well-defined electron density that we see for the majority of the capsid proteins, that for the C-terminal portion of VP1 fades out at its furthest extent so that the last three residues are hardly visible; similarly, we see no interpretable density for the FMDV loop (to be precise, residues 135 to 156). Whether this disorder in the loop region arises from its ability to exist in a limited number of well-defined orientations or because it forms a self-contained folding unit that is hinged to the rest of the capsid is impossible to tell. The success of peptides from this region in eliciting an immune response may be related to their ability to mimic the flexibility of the loop or maintain the self-contained structure that exists in the intact virus. This implies that the successful design of peptide vaccines in other cases will require consideration of the conformation of the peptide in the natural pathogen.

Cell Attachment

All of the evidence points to the fact that the residues implicated in cell attachment in FMDV are presented on the viral surface and are thus directly accessible to antibodies. Conserved residues near the C terminus of VP1 and the conserved tripeptide Arg-Gly-Asp in the FMDV loop seem particularly important in mediating the cellular interaction (Fox et al., 1989). The Arg-Gly-Asp triplet has been shown in several systems to be critical for attachment to a sizable family of receptors, the integrins. Thus, the receptor for FMDV may also be an integrin. The location of the putative cell attachment residues seems somewhat contradictory, since in order to recognize a necessarily conserved cell receptor, these residues must themselves be conserved, yet we know that viruses constantly mutate exposed regions in order to elude antibody recognition. In rhinoviruses, residues involved in cell attachment are located at the base of surface depressions whose dimensions are thought to render them accessible to cell receptors but not to the relatively obtuse antibody. To attempt an

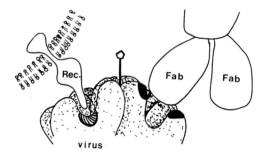

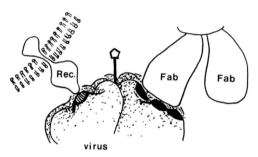

FIGURE 4. Schematic depicting two hypotheses that suggest different means by which members of the picornavirus family may escape immune surveillance. (Top) The canyon hypothesis; (bottom) the FMDV hypothesis.

to understand how uncoating could occur without some form of receptor-mediated capsid destabilization. In the case of FMDV, binding of the cell receptor to regions of the virus that are so loosely associated with the core structure of the particle is unlikely to influence capsid stability. Consequently, it may be that acid sensitivity of the virus is a necessary prerequisite for rapid and efficient uncoating during endocytosis.

SUMMARY

The structure has helped explain various observations peculiar to the genus *Aphthovirus* of the family *Picornaviridae* and in particular to the O_1 serotype. Perhaps these results should temper our assessment of structurally based generalizations within members of one family. FMDV certainly exemplifies the exquisite adaptation of viruses to particular evolutionary niches.

explanation of the FMDV paradox, we need to take into account the large area of antibody interaction (approximately 700 $Å^2$). If the virus were to change any one of a number of the recognized residues surrounding the Arg-Gly-Asp sequence, then it could interfere with antibody binding (Fig. 4). In fact, a great deal of variation is seen in the residues flanking the Arg-Gly-Asp triplet. Furthermore, the peptide Arg-Gly-Asp can compete with monoclonal antibodies for binding to the virus, whereas longer peptides, in which this triplet is flanked by residues different from those in the FMDV loop sequence, do not compete (Fox et al., 1989). Thus, variation around a conserved attachment site may be sufficient to evade immune surveillance.

It is interesting to speculate that two of the unique properties of FMDV, the superficial location of the cell receptor-binding domains and the extreme susceptibility to acid-induced dissociation, may be functionally linked in the mechanism of virus uncoating. There is evidence with other picornaviruses that receptor binding results in structural modification of the particle as a prelude to uncoating. Indeed, with the enteroviruses that are stable at pH 2 to 3, it is difficult

REFERENCES

Acharya, R., E. Fry, D. Stuart, G. Fox, D. Rowlands, and F. Brown. 1989a. Implications of the three dimensional structure of foot and mouth disease virus for its antigenicity and cell attachment, p. 1–7. *In* R. A. Lerner, H. Ginsberg, R. M. Chanock, and F. Brown (ed.), *Vaccines 89: Modern Approaches to New Vaccines including Prevention of AIDS.* Cold Spring Harbor Laboratory, Cold Spring Harbor, N.Y.

Acharya, R., E. Fry, D. I. Stuart, G. Fox, D. Rowlands, and F. Brown. 1989b. The three-dimensional structure of foot and mouth disease virus at 2.9 Å. *Nature* (London) 337:709–716.

Bolwell, C., B. E. Clarke, N. R. Parry, E. J. Ouldridge, F. Brown, and D. J. Rowlands. 1989. Epitope mapping of foot and mouth disease virus with neutralizing monoclonal antibodies. *J. Gen. Virol.* 70:59–68.

Colonno, R. J., J. H. Condra, S. Mizutani, P. L. Callahan, M. Davies, and M. A. Murcko. 1988. Evidence for the direct involvement of the rhinovirus canyon in receptor binding. *Proc. Natl. Acad. Sci. USA* 85:5449–5453.

Di Marchi, R., G. Brooke, C. Gale, V. Cracknell, T. Doel, and N. Mowat. 1986. Protection of cattle against foot and mouth disease by a synthetic peptide. *Science* 232:639–641.

Fox, G., N. V. Parry, P. V. Barnett, B. McGinn, D. J. Rowlands, and F. Brown. 1989. The cell attachment site on foot and mouth disease virus includes the sequence RGD. *J. Gen. Virol.* 70:625–637.

Harrison, S. C., A. J. Olson, C. E. Schutt, F. K. Winkler, and G. Bricogne. 1978. Tomato bushy stunt virus at 2.9 Å resolution. *Nature* (London) 276: 368–373.

Hogle, J. M., M. Chow, and D. J. Filman. 1985. Three-dimensional structure of poliovirus at 2.9 Å resolution. *Science* **229**:1358–1365.

Luo, M., G. Vriend, G. Kamer, I. Minor, E. Arnold, M. G. Rossmann, U. Boege, D. G. Scraba, G. M. Duke, and A. C. Palmenberg. 1987. The atomic structure of Mengovirus at 3.0 Å resolution. *Science* **235**:182–191.

Parry, N. R., E. J. Ouldridge, P. V. Barnett, D. J. Rowlands, and F. Brown. 1985. Identification of neutralizing epitopes of foot and mouth disease virus, p. 211–216. *In* R. A. Lerner, H. Ginsberg, R. M. Chanock, and F. Brown (ed.), *Vaccines 85: Molecular and Chemical Basis of Resistance to Parasitic, Bacterial, and Viral Diseases.* Cold Spring Harbor Laboratory, Cold Spring Harbor, N.Y.

Rossmann, M. G., E. Arnold, J. W. Erickson, E. A. Frankenberger, J. P. Griffith, J. E. Johnson, G. Kamer, M. Luo, A. C. Mosser, R. R. Rueckert, B. Sherry, and G. Vriend. 1985. Structure of a human common cold virus and functional relationship to other picornaviruses. *Nature* (London) **317**:145–153.

Strohmaier, K., R. Franze, and K. H. Adam. 1982. Location and characterization of the antigenic portion of the FMDV immunizing protein. *J. Gen. Virol.* **59**:295–306.

Chapter 32

RNA Packaging in Bean Pod Mottle Virus

Zhongguo Chen, Cynthia Stauffacher, Tim Schmidt, Andrew Fisher, and John E. Johnson

Crystallographic analyses of spherical RNA viruses have revealed coat protein structures at near-atomic resolution. These studies have dramatically advanced understanding of assembly and evolution of simple viruses and have helped to establish structural aspects of interactions between viruses and antibodies and viruses and antiviral agents affecting the capsid (Rossmann and Johnson, 1989). Until recently, these studies have provided no substantial insight for RNA packaging because the RNA structure was disordered in the crystal and invisible in the electron density maps. Figure 1 illustrates the problem of RNA disorder that occurs when an asymmetric nucleic acid molecule is packed inside a symmetric shell. If the RNA were packed identically in each particle and if it affected the surface and therefore the packing of particles in the crystal lattice (Fig. 1A), the RNA would be ordered and the crystallographic analysis would reveal the structure of an entire mRNA. In all but one structure determined to date, the RNA molecules are essentially randomly oriented in the lattice (Fig. 1B) and therefore do not contribute significantly to the high-resolution diffraction pattern. The RNA behaves as a spherically averaged region of density and makes no contribution to the diffraction pattern beyond about 20-Å (2-nm) resolution (Schmidt et al., 1983). The structure of the middle component of bean pod mottle virus (BPMV) has recently been reported (Chen et al., 1989) and shows that nearly 20% of the packaged RNA molecule binds to the capsid interior in a symmetric fashion, making this region of the nucleic acid clearly visible in the 3.0-Å resolution electron density map. RNA order of this type is illustrated in Fig. 1C, where part of the nucleic acid is associated with the symmetric shell and therefore contributes to the diffraction pattern.

In this chapter, we suggest a model for RNA packaging in BPMV middle component that is consistent with the crystallographic analyses, and we present a progress report on other comovirus structures.

Using X-ray crystallography, we have investigated two members of the comovirus group, cowpea mosaic virus (CPMV) (Stauffacher et al., 1987) and BPMV. CPMV is by far the most-studied member of the comovirus group, but its properties have been shown to be general and apply to BPMV (Shanks et al., 1986; Francki et al., 1985). Comoviruses have a bipartite, positive-sense, single-stranded RNA genome that is encapsulated with each RNA molecule in a separate particle (Fig. 2; Goldbach and van Kammen, 1985). Empty capsids are also formed in vivo and constitute approximately 20% of the particles in a typical preparation. Both RNA molecules of CPMV have been sequenced (Van Wezenbeek et al., 1983; Lomonossoff and Shanks, 1983). A portion of the small RNA (RNA2) of BPMV, which contains the coat protein genes, has also been sequenced, and the derived capsid protein sequences have been reported (Chen et al., 1989). Comoviruses display a number of physical and biological properties similar to those of the animal picornaviruses (Goldbach and van Kammen, 1985). Although the genomic information is coded on a single RNA molecule in picornaviruses, the gene orders of the two virus families are similar if the RNAs of CPMV are placed as is shown in Fig. 3 (Franssen et al., 1984). RNA molecules of both comoviruses and picornaviruses are translated as polyproteins that are subsequently processed by two different proteases encoded in the viral genome (Kräusslich and Wimmer, 1988). The capsids of both virus groups are composed of 180 β-barrel domains

Zhongguo Chen, Cynthia Stauffacher, Tim Schmidt, Andrew Fisher, and John E. Johnson, Department of Biological Sciences, Purdue University, West Lafayette, Indiana 47906.

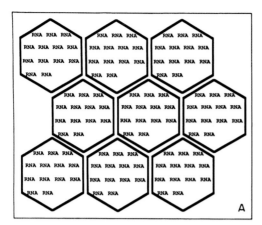
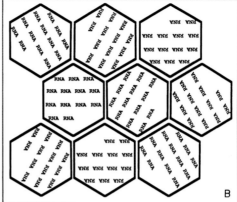
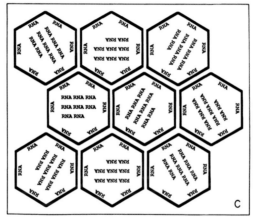

FIGURE 1. Possible relationships between packaged RNA and the symmetric capsid in crystal lattices. RNA is represented by the lettering; the hexagonal outline represents the symmetric capsid. (A) Crystalline arrangement of particles in which the RNA affects the packing of virions. The RNA is ordered and would be visible in the electron density map. Crystallizing virus in a strong magnetic field where the intrinsic dipole of the RNA molecule could affect particle orientation could lead to such an arrangement. (B) The situation observed in most crystalline spherical viruses. The symmetric capsid packs in six structurally equivalent orientations, which orients the RNA in six different ways, leading to disorder. In an icosahedral shell there are 60 equivalent packing arrangements. (C) The situation observed in BPMV, where a portion of the RNA binds to the symmetric capsid and displays the same symmetry as does the virion. The portion of the RNA bound to the protein is ordered in the crystal lattice, while the remaining portion of the RNA is disordered.

(Fig. 3) formed from three different protein types in picornaviruses (all approximately 30 kilodaltons) and two protein types in comoviruses (42 and 24 kilodaltons) (Rossmann et al., 1985; Hogle et al., 1985; Luo et al., 1987; Stauffacher et al., 1987).

BPMV was propagated, purified, and crystallized in an orthorhombic space group (P22$_1$2$_1$, a = 311.2 Å, b = 284.2 Å, c = 350.5 Å) (Sehnke et al., 1988). The structure determination was done with crystals of virus particles containing only the small RNA (RNA2). Photographic film

data were collected from the orthorhombic crystals at the Cornell High Energy Synchrotron Source (λ = 1.566 Å). The data set of 182 A/B film pairs was scaled and postrefined (Rossmann et al., 1979) to produce 698,453 unique reflections where $I > 2\sigma(I)$ with an R_{merge} (R_{merge} = $\sum_h \sum_i | (F_h^2 - F_{hi}^2) | / \sum_h \sum_i F_h^2$) of 9.6%. The final data set included nearly 90% of the complete data between 20- and 3.0-Å resolution. The structure determination of BPMV middle component has been reported (Chen et al., 1989). Currently, the icosahedral asymmetric unit of BPMV contains

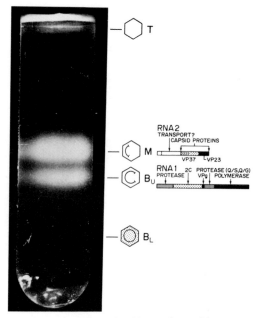

FIGURE 2. A cesium chloride gradient with components of BPMV banded at different densities. The components are identical isometric particles except for their RNA content. T (top component; $\rho = 1.30$ g/ml) contains no RNA. M (middle component; $\rho = 1.40$ g/ml) contains RNA2 (~3,500 nucleotides) that codes for the capsid protein pairs VP37-VP23 and L-S, respectively, in Fig. 3. B_U (bottom upper component; $\rho = 1.42$ g/ml) contains RNA1 (~5,800 nucleotides) that codes for enzymes associated with viral replication and assembly. B_L (bottom lower component [almost invisible in this gradient]; $\rho = 1.47$ g/ml) is identical to B_U in RNA content, but cesium has permeated the capsid and displaced the natural counterions, polyamines, that neutralize the RNA (Virudachalam et al., 1985).

4,332 atoms and the crystallographic agreement factor after refinement is 22.9%, using data in which $I \geq 8\sigma(I)$ (94% of the measured data).

Crystals of pure empty capsid and crystals of pure bottom component (containing RNA1) have been prepared and are isomorphous with crystals of pure middle component (containing RNA2). Partial data sets have been collected by using these crystals (Table 1). Preliminary results from these data sets have been obtained by using difference electron density maps computed with Fourier coefficients $(F_{mid} - F_{top})e^{i\alpha_{mid}}$ and $(F_{mid} - F_{bot})e^{i\alpha_{mid}}$.

BPMV STRUCTURE

The two capsid proteins of BPMV are folded into three antiparallel β-barrel structures; the small protein forms one barrel, and the large protein forms two barrels that are covalently connected to form a single polypeptide (Fig. 4). The 60 copies of each protein type in the virus generate 180 β-barrel domains that are arranged in a manner very similar to a $T = 3$ capsid (Fig. 3). The gene orders for the structural proteins in CPMV and poliovirus (Fig. 3) equate the large protein with VP2 and VP3 and the small protein with VP1, although there is no similarity in the primary sequences of these capsid proteins. The BPMV structure shows that the proteins equated by gene position also have equivalent positions in the capsid. The wedge-shaped β-barrel units are approximately 50 Å in length, 17 Å in width at the narrow end, and 30 Å at the wide end. Their thickness varies from 20 to 35 Å, depending on the size of insertions between the β strands. Figure 4 illustrates the organization of one icosahedral asymmetric unit of the BPMV particle.

PROTEIN-RNA INTERACTIONS

Density comparable in height to that modeled as well-ordered protein, but not connected to any part of the polypeptide chain, was found inside the BPMV capsid. Seven ribonucleotides (designated RNA1 to RNA7) were readily fitted to this density, which showed clear phosphate, sugar, and base densities and thus established the polarity of the chain. The magnitude of this electron density implies that this portion of the RNA exists in virtually all of the 60 equivalent positions in the virus. The overall backbone stereochemistry of the seven-ribonucleotide segment approximates that found in one strand of an A-type RNA helix. The average helical twist angle for the polyribonucleotide is 45° and the mean rise per residue is 3.5 Å, so that the single strand of viral RNA is wound tighter than a strand in an A-type RNA helix, which has a twist angle of ~30° and a rise per residue of 2.6 Å (Arnott et al., 1973). The helical repeat distance for this RNA is roughly 28 Å, which is nearly the same as observed in the A-type helix. The viral RNA consists of eight residues per repeat and completes nearly one full turn in each ordered segment so that first the bases and then the phosphates face the protein.

The quality of the electron density for the bases was best at the 5' end, where the bases face the protein, and weakens toward the 3' end, where they face the solvent in the virus interior. The density for each base represents an average of the different bases that must occur in the 60 asymmetric units of the virus. The bases of the

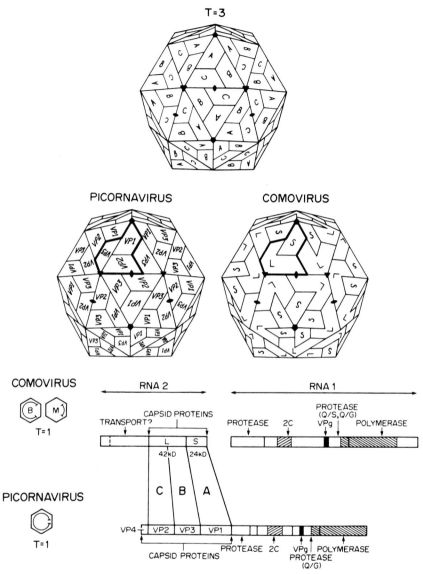

FIGURE 3. Comparison of $T = 3$, picornavirus, and comovirus capsids. In each case, one trapezoid represents a β-barrel. The icosahedral asymmetric unit of the $T = 3$ shell contains three identical subunits labeled A, B, and C. The asymmetric unit of the picornavirus capsid (in the heavy outline) contains three β-barrels, but each has a characteristic amino acid sequence labeled VP1, VP2, and VP3. The comovirus capsid is similar to the picornavirus capsid except that two of the β-barrels (corresponding to VP2 and VP3) are covalently linked to form a single-polypeptide, large-protein subunit (L), while the small-protein subunit (S) corresponds to VP1. Comoviruses and picornaviruses have a similar gene order, and the shaded regions of the nonstructural proteins display significant sequence homology. The relationship between the subunit positions in these viruses and their location in the genes is indicated by the labels A, B, and C in the gene diagram.

TABLE 1

Percentages of Theoretically Possible Reflections in BPMV Bottom and Top Components

Component	Resolution range (Å)	No. of reflections with F > $4\sigma(F)$	Total percentage
Bottom	30.0	0	0.0
	15.0	2,800	67.0
	10.0	7,648	67.0
	7.5	14,578	66.0
	5.0	58,410	64.0
	3.5	156,635	63.0
	3.0	116,089	52.0
	2.9	28,739	16.0
Top	30.0	0	0.0
	15.0	1,630	39.0
	10.0	4,355	38.0
	7.5	8,175	37.0
	5.0	33,167	36.0
	3.5	90,399	36.0
	3.0	72,003	32.0
	2.9	18,213	10.0

entire segment of RNA are stacked. The density for the RNA1 and RNA2 positions is elongated and can accommodate purines, whereas the density for the RNA3 and RNA4 positions is more compact and is only large enough to accommodate pyrimidines. The orientation of the remaining bases is defined, but the density in which they are profiled is progressively smaller and is inadequate to accommodate even a pyrimidine.

The electron density for the seven-ribonucleotide segment is located near the icosahedral threefold axes. A tube of electron density about one-third the height of this well-ordered RNA density connects the 3'-terminal ribonucleotide of one of these segments of the 5'-terminal ribonucleotide of a threefold related segment. Four ribonucleotides (designated RNA1' to RNA4') can be accommodated in this density, but the fit is ambiguous. Including these connecting ribonucleotides, the entire RNA forms a trefoil that consists of roughly 33 ribonucleotides surrounding each threefold icosahedral symmetry axis (Fig. 5). As viewed from outside the particle, the polarity of the chain goes from 5' to 3' in a clockwise direction. This cluster of 33 ribonucleotides may represent loops in the RNA secondary structure that recognize the threefold sites of the assembling virus. The detailed image of the RNA molecule entering and leaving the trefoils and the connections between the trefoils are not observed because

they do not obey icosahedral symmetry. These end effects give rise to the weaker RNA density in the connecting regions because at most, only two of the three regions averaged can actually be present in any given trefoil. The total RNA modeled accounts for roughly 660 ribonucleotides (20 × 33) for the entire particle, which is about 20% of the packaged RNA. Figure 6 shows a schematic representation of the protein RNA interactions in the particle, and Table 2 shows all contact distances that are less than 3.6 Å.

Figure 5 illustrates the trefoil-shaped RNA at each of the icosahedral threefold axes in the particle. The intact RNA molecule must contact the protein at these 20 equivalent positions with virtually identical conformations. A qualitative model for RNA packaging in middle-component BPMV was constructed that is consistent with the crystallographic results by using a template for making a cardboard model of an icosahedron (Williams, 1979; Fig. 7). It is assumed that the RNA connects adjacent icosahedral faces and that the lengths between contacts are approximately equal. Since spectroscopic analyses of BPMV middle component suggest a large amount of base pairing in the RNA, the regions between contacts are likely to be double stranded, whereas the interactions with the protein are clearly single stranded. If the contacts are distributed over the length of the RNA, there would be roughly 140 nucleotides between each contact with the protein. Alternatively, the connections between contacts with the protein could be as short as 20 nucleotides if the RNA connecting trefoils were in an extended single-strand conformation. The region of protein-RNA interactions would then extend over only about one-third of the genome, and the remaining 2,500 nucleotides would be packaged without ordered interactions with the protein.

The model shown in Fig. 7A illustrates a possible configuration for both the ordered and disordered regions of the RNA structure. When this sheet is folded to form an icosahedron (Fig. 7B), it is clear that the connections between trefoils cannot obey the icosahedral symmetry and thus cannot be visible in the electron density. However, except for the break at the point of entering and leaving the trefoils, the regions around the threefold axes repeat perfectly at all 20 triads. Analysis of the BPMV RNA2 sequence shows that the packaging is not dependent on an exactly repeating set of bases. Biochemical experiments are under way to determine which regions of the RNA are interacting with the protein in the middle-component particles.

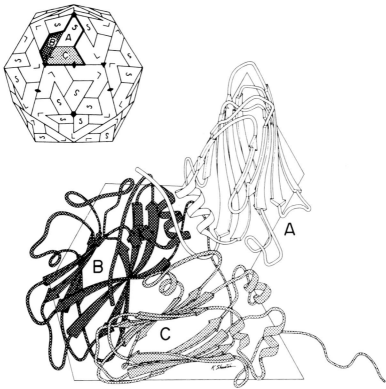

FIGURE 4. Quaternary organization of the small and large subunits of BPMV. The ribbon drawing represents the arrangement of the three β-barrels in the virus quaternary structure. The two domains of the large subunit are shown in different shadings to emphasize that they form one protein. The small subunit is unshaded. The heavy outline in the capsid model encloses the suggested precleavage protomer; the shading matches the β-barrel units found in each trapezoid.

Recently, we have examined electron density maps, using data from crystals of the empty capsid and crystals of the bottom-component particles. The electron density maps show that the different particle structures are largely the same, which is consistent with the formation of isomorphous crystals. As expected, there is no electron density for the RNA in the top-component crystals. There is, however, evidence for substantial movement of the amino terminus of the large protein which interacts with the RNA

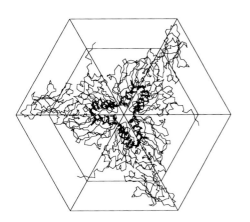

FIGURE 5. Diagram showing that the RNA (heavy lines) from three icosahedral asymmetric units forms a trefoil shape centered about the threefold particle axes. Three large subunits cluster around the trimer axis, forming a pseudo sixfold unit contacting the RNA trefoil. The small subunits are near fivefold axes and do not interact with the RNA. Thirty-three ribonucleotides form this trefoil-shaped cluster, which appears connected in the electron density map but must have at least one point of entry and exit. Connections between these clusters of 33 ribonucleotides are not visible in the electron density map.

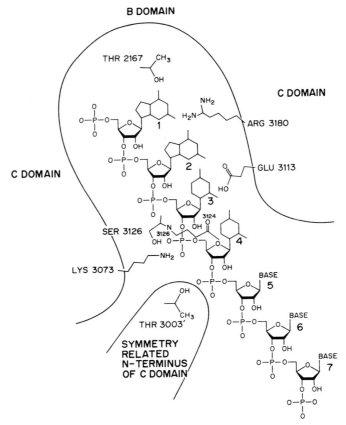

FIGURE 6. Schematic representation of the protein-RNA interactions in BPMV. All side chains less than 3.6 Å from the RNA are shown. Table 2 lists the distances between atoms in the model that are separated by less than 3.6 Å.

in the middle-component structure, suggesting that the N terminus participates in different ways, depending on whether the particle is packaging RNA or assembling it as an empty protein shell.

At the current stage of structure solution, the bottom-component particles do not display electron density for ordered RNA, although the amino-terminal arm is positioned as it is in middle component. The nucleotide sequences of RNA2 (packaged in middle component) and RNA1 (packaged in bottom component) are ex-

TABLE 2
Protein-RNA Interactions

RNA	Contact (distance [Å])		
	Phosphate	Ribose	Base
NA1			N2-Nη2 Arg-3180 (2.70)
NA2			N2-Oε1 Glu-3113 (3.56)
NA3		O2-O Thr-3124 (3.37)	
		O2-N Ser-3126 (3.41)	
NA4	O2P-Oγ Ser-3126 (2.70)	O3-Oγ Ser-3126 (3.05)	
	O2P-Nζ Lys-3073 (3.41)		
NA5	O2P-Oγ1 Thr-4003 (2.70)		

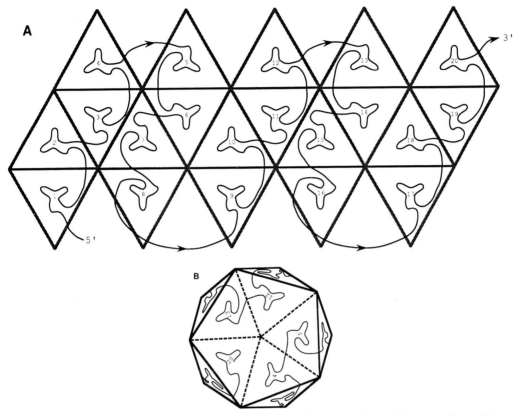

FIGURE 7. Model for RNA packaging that is consistent with the crystallographic observations. (A) Two-dimensional representation of an icosahedron showing the 20 triangular threefold faces. The RNA molecule is threaded through all the faces, maintaining the same polarity about each triad. The actual length of the connections between trefoils is not known. (B) View down a fivefold axis of the icosahedron shown as a two-dimensional sheet in A. The dashed lines indicate the vertex formed by joining the faces labeled 4, 5, 12, 13, 20 in A. The model illustrates the conserved features at the triad and the lack of symmetry in the connecting regions leading to the observed electron density in the crystal structure.

pected to be totally different, since there is no similarity in the two RNA molecules in CPMV. In addition, RNA1 has roughly 5,800 nucleotides and is 34% of the virus particle mass in bottom component. This particle is virtually filled to capacity with RNA. Middle component is only 24% RNA by weight, and it is rather loosely packed (Schmidt et al., 1983). Results at the present stage of analysis suggest that RNA2 and RNA1 interact differently with the coat protein and that they may be packaged in significantly different ways.

The structures of all three components are now being independently refined by using the procedure of Hendrickson and Konnert (1980). The final structures should provide additional insight for understanding the complex process of particle assembly and RNA packaging in viruses.

ACKNOWLEDGMENTS. We thank Sharon Fateley for help in preparing the manuscript.

This work was supported by Public Health Service grant AI18764-08 from the National Institutes of Health and a grant from the Lucille P. Markey Foundation.

REFERENCES

Arnott, S., D. Hukins, D. Dover, W. Fuller, and A. R. J. Hodgson. 1973. Structures of synthetic polynucleotides in the A-RNA and A'-RNA conformations: X-ray diffraction analyses of the molecular conformations of polyadinylic acid · polyuridylic acid and polyinosinic acid · polycytidylic acid. *J. Mol. Biol.* **81:**107–122.

Chen, Z., C. Stauffacher, Y. Li, T. Schmidt, W. Bomu, G. Kamer, M. Shanks, G. Lomonossoff, and J. E. Johnson. 1989. Protein-RNA interactions in an icosahedral virus at 3.0 Å resolution. *Science* **245:** 154–159.

Francki, R. I. B., R. G. Milne, and T. Hatta. 1985. Comovirus group, p. 1–21. *In Atlas of Plant Viruses*, vol. 2. CRC Press, Inc., Boca Raton, Fla.

Franssen, H., J. Leunissen, R. Goldbach, G. Lomonossoff, and D. Zimmern. 1984. Homologous sequences in non-structural proteins from cowpea mosaic virus and picornaviruses. *EMBO J.* **3:**855–861.

Goldbach, R., and A. van Kammen. 1985. Structure, replication and expression of the bipartite genome of cowpea mosaic virus, p. 83–120. *In* J. W. Davies (ed.), *Molecular Plant Virology*, vol. 2. CRC Press Inc., Boca Raton, Fla.

Hendrickson, W. A., and J. H. Konnert. 1980. Incorporation of stereochemical information into crystallographic refinement, p. 13.01–13.26. *In* R. Diamond, S. Ramaseshan, and K. Venkatesan (ed.), *Computing in Crystallography.* Indian Academy of Sciences, Bangalore, India.

Hogle, J. M., M. Chow, and D. J. Filman. 1985. Three-dimensional structure of poliovirus at 2.9 Å resolution. *Science* **229:**1358–1365.

Kräusslich, H.-G., and E. Wimmer. 1988. Viral proteinases. *Annu. Rev. Biochem.* **57:**701–748.

Lomonossoff, G., and M. Shanks. 1983. The nucleotide sequence of cowpea mosaic virus B RNA. *EMBO J.* **2:**2253–2258.

Luo, M., G. Vriend, G. Kamer, I. Minor, E. Arnold, M. G. Rossmann, U. Boege, D. G. Scraba, G. M. Duke, and A. C. Palmenberg. 1987. The atomic structure of Mengo virus at 3.0 Å resolution. *Science* **235:**182–191.

Rossmann, M. G., E. Arnold, J. W. Erickson, E. A. Frankenberger, J. P. Griffith, H. J. Hecht, J. E.

Johnson, G. Kamer, M. Luo, A. G. Mosser, R. R. Rueckert, B. Sherry, and G. Vriend. 1985. Structure of a human common cold virus and functional relationship to other picornaviruses. *Nature* (London) **317:**145–153.

Rossmann, M. G., and J. E. Johnson. 1989. Icosahedral RNA virus structure. *Annu. Rev. Biochem.* **58:**533–573.

Rossmann, M. G., A. G. W. Leslie, S. S. Abdel-Meguid, and T. Tsukihara. 1979. Processing and post-refinement of oscillation camera data. *J. Appl. Crystallogr.* **12:**570–581.

Schmidt, T., J. E. Johnson, and W. E. Phillips. 1983. The spherically averaged structures of cowpea mosaic virus components by X-ray solution scattering. *Virology* **127:**65–73.

Sehnke, P. C., M. Harrington, M. V. Hosur, Y. Li, R. Usha, R. C. Tucker, W. Bomu, C. V. Stauffacher, and J. E. Johnson. 1988. Crystallization of viruses and virus proteins. *J. Crystal Growth* **90:**222–230.

Shanks, M., J. Stanley, and G. Lomonossoff. 1986. The primary structure of red clover mottle virus middle component RNA. *Virology* **155:**697–706.

Stauffacher, C. V., R. Usha, M. Harrington, T. Schmidt, M. V. Hosur, and J. E. Johnson. 1987. The structure of cowpea mosaic virus at 3.5 Å resolution, p. 293–308. *In* D. Moras, J. Drenth, B. Strandberg, D. Suck, and K. Wilson (ed.), *Crystallography in Molecular Biology.* Plenum Publishing Corp., New York.

Van Wezenbeek, P., J. Verver, J. Harmsen, P. Vos, and A. van Kammen. 1983. Primary structure and gene organization of the middle-component RNA of cowpea mosaic virus. *EMBO J.* **2:**941–946.

Virudachalam, R., M. Harrington, J. E. Johnson, and J. Markley. 1985. ^{1}H, ^{13}C and ^{31}P nuclear magnetic resonance studies of cowpea mosaic virus: detection and exchange of polyamines and dynamics of the RNA. *Virology* **141:**43–50.

Williams, R. 1979. *The Geometrical Foundation of Natural Structure*, p. 67. Dover, New York.

Chapter 33

Structure and Synthesis of the Core Protein: Role in Regulation of Assembly and Disassembly of Alphavirus and Flavivirus Cores

Gerd Wengler

In many viral systems, the genome is packed in a nucleoprotein that can be experimentally manipulated as a stable complex. On the other hand, the viral genome must be made accessible to the cellular apparatus of transcription or translation in the initial stages of viral infection in order to initiate the replication cycle. It is not easily seen how a nucleoprotein on the one hand can function as a stable complex and on the other hand can allow liberation of the genome early in infection. Studies of the alphaviruses Sindbis virus and Semliki Forest (SF) virus have indicated that a specific function of the core (C) protein might allow regulation of the assembly and disassembly of cores of these viruses. On the other hand, the problems of nucleoprotein stability and lability may be circumvented by assembling an inherently labile nucleoprotein. A possible mechanism for how such a labile complex can be efficiently assembled follows from analyses of the molecular biology of the West Nile (WN) flavivirus. Both systems are briefly described in this chapter.

STRUCTURE OF THE C PROTEIN OF ALPHAVIRUSES

The primary structures of the C proteins of a number of alphaviruses have been determined (Garoff et al., 1980; Boege et al., 1981; Strauss et al., 1984). The alphavirus C protein contains an amino-terminal segment that is rich in arginine, lysine, and proline and a carboxy-terminal seg-ment that has a more balanced amino acid composition (Fig. 1). These data suggest that the C protein consists of two domains and that the amino-terminal basic domain binds the genome RNA. The coat proteins of a number of plant and animal viruses are composed of an amino-terminal basic domain involved in nucleic acid binding and a carboxy-terminal domain involved in construction of the icosahedral protein shells of these viruses (see Argos and Johnson [1984] for a review). Comparative analyses of the primary structures of the coat proteins of picornaviruses and alphaviruses have recently led to the suggestion that both proteins may be homologous molecules (Fuller and Argos, 1987). Analysis of the crystal structure of the Sindbis virus C protein will clarify this point (Boege et al., 1989).

The C protein of alphaviruses is a multifunctional protein that (i) interacts with single-stranded RNA segments (this function is probably localized in the amino-terminal domain), (ii) assembles into isometric (T = 3) particles (this function is probably localized in the carboxy-terminal domain and can be performed only after binding of C protein to nucleic acid), (iii) interacts with the intracytoplasmic segment of the E2 spike glycoprotein, (iv) releases itself from the growing polyprotein (this proteolytic activity is localized in the carboxy-terminal domain; a specific serine residue is implicated in this function [Fig. 1]); and (v) binds to the large ribosomal subunit both during synthesis and as a mature molecule.

Gerd Wengler, Institut für Virologie, Justus-Liebig-Universität Giessen, D-6300 Giessen, Federal Republic of Germany.

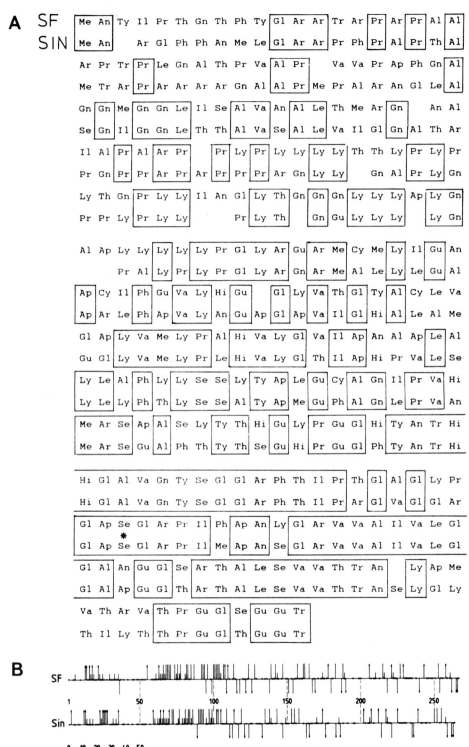

B

SF

Sin

0 10 20 30 40 50

AMINO ACID RESIDUES

IN VITRO ANALYSIS OF THE FUNCTIONS OF THE ALPHAVIRUS C PROTEIN

Analysis of individual functions of the C protein is greatly aided by the availability of in vitro systems that allow study of four of the five functions listed above.

Analyses of the in vitro synthesis of viral structural proteins have shown that the C protein is cleaved rapidly from the growing structural polyprotein in cell-free translation systems in the absence of membranes. These findings indicate that the growing C-protein chain possesses a proteolytic activity which cleaves the C protein from the polyprotein during translation. Elucidation of the primary structures of the C proteins of SF and Sindbis viruses has led to the identification of a conserved Gly–Asp–Ser-219–Gly sequence that is characteristic of the active serine of serine proteases (Boege et al., 1981). It has recently been shown that conversion of this sequence into the sequence Gly-Asp-Arg-Ser-Thr leads to a polyprotein that during in vitro synthesis does not release mature C protein (Melancon and Garoff, 1987).

A system has been developed that allows the reconstitution of alphavirus cores in vitro by using purified C protein and viral genome RNA (Wengler et al., 1984) (Fig. 2). This system permits analysis of the functions of the C protein during interaction with the genome RNA and during assembly of the protein layer present in the cores. Two major conclusions can be drawn from these experiments. (i) The C protein associates into isometric particles only after it has bound to nucleic acids. In accordance with the data of in vivo analyses, corelike particles free of nucleic acids are not formed. (ii) All species of single-stranded nucleic acids analyzed were efficiently assembled into corelike particles in vitro. These data indicate that in a first reaction the C protein binds to single-stranded nucleic acid, and this binding then allows the assembly of a number of similar complexes into a core particle. A number of different nucleic acid segments can be collected into a single core by this process, a finding that may have its in vivo correlate in the assembly of alphavirus cores containing multiple copies of defective interfering RNAs in a single core of normal size.

Furthermore, the availability of purified biologically active C protein allowed us to show that this protein could bind to the large ribosomal subunit in vitro (Wengler et al., 1984). These results led to the conclusion that the ability to bind to the 60S ribosomal subunit not only was a property of the growing C-protein polypeptide chain but was retained in the mature C protein. C protein therefore not only might be transferred from ribosomes into cores but might be transferred from cores back to ribosomes under appropriate conditions. An experiment was performed in which core particles, assembled in vitro from either viral genome RNA or tRNA and radioactively labeled C protein, were added to a rabbit reticulocyte lysate that contained a large excess of ribosomes compared with the number of cores (Fig. 3). It can be seen that the cores dissociate and the C protein is transferred to the large ribosomal subunit. Furthermore, the data indicate that cores containing genome RNA are more stable than tRNA-containing cores.

REGULATION OF THE ASSEMBLY AND DISASSEMBLY OF ALPHAVIRUS CORES BY BINDING OF C PROTEIN TO RIBOSOMES

The data presented above indicate that multiple components compete in vivo for the binding of C protein: (i) the 60S ribosomal subunit, (ii) the viral genome RNA, and (iii) other nucleic acids such as tRNA. These results suggest that

---

Figure 1. Primary structures of the C proteins of SF and Sindbis (SIN) viruses. (A) Alignment of the amino acid sequences of both proteins. Identical amino acid sequences present in both molecules are boxed. Gaps are introduced to maximize homologies. The two-letter code (Al, Ala; Ar, Arg; An, Asn; Ap, Asp; Cy, Cys; Gn, Gln; Gu, Glu; Gl, Gly; Hi, His; Il, Ile; Le, Leu; Ly, Lys; Me, Met; Ph, Phe; Pr, Pro; Se, Ser; Th, Thr; Tr, Trp; Ty, Tyr; Va, Val) is used. The carboxy-terminal regions of both proteins contain the sequence Gl-Ap-Se-Gl, which characteristically contains the active serine residue of serine proteases (∗). (B) Diagrammatic representation of the amino acid sequences of the core proteins of SF and Sindbis (Sin) viruses. The sequences are represented as a horizontal line according to the scale indicated. An upward line shows Lys, an upward arrow represents Arg, a downward line indicates Asp, a downward arrow shows Glu, closed circles represent one of the hydrophobic amino acids Ala, Val, Leu, Ile, Met, Phe, or Trp, and a closed circle with a short upward line indicates Pro.

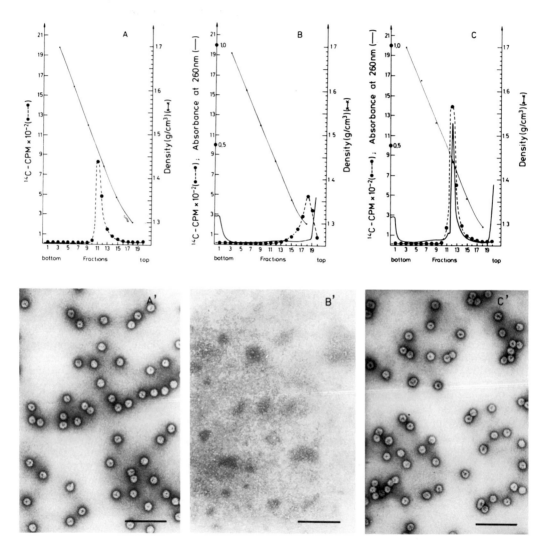

FIGURE 2. Analyses of authentic and in vitro-reconstituted core by equilibrium centrifugation and electron microscopy. Authentic core was isolated from Sindbis virus. Besides the standard in vitro assembly reaction containing unlabeled 42S Sindbis virus genome RNA and ^{14}C-labeled Sindbis virus C protein, two reactions lacking either RNA or protein were set up. All samples were fixed and centrifuged to equilibrium in CsCl density gradients. The distribution of optical density (nucleic acid) and radioactivity (C protein) in the gradients was determined (A to C). The absorbance at the bottom and top of the CsCl density gradient analyses represents schlieren at the interface between the gradient and the buffer used to displace the gradient during collection. Nucleoprotein particles centrifuged to equilibrium were pooled and analyzed by electron microscopy after negative staining with uranyl acetate. Micrographs were taken at ×60,000 magnification in the electron microscope; bars represent 200 nm. The results of analyses of authentic core and of particles generated in vitro in a complete reaction are shown in panels A and A′ and C and C′, respectively. The products generated in vitro in the absence of RNA are analyzed in panels B and B′. The material present in the five top fractions of the gradient presented in panel B is analyzed in the electron micrograph shown in panel B′.

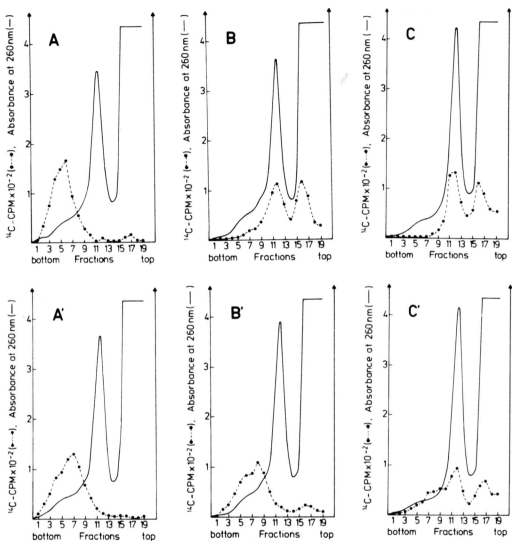

FIGURE 3. Transfer of C protein from core particles to ribosomes in reticulocyte lysates in vitro. In the first step, C particles containing either 42S genome RNA or tRNA were prepared in two reactions containing either 42S RNA or tRNA and ^{14}C-labeled C protein having a specific radioactivity of 1.1×10^6 cpm/mg of protein. The reaction containing 42S RNA was sonicated for 1 min at 30°C after assembly of particles to reduce the number of aggregated particles. Cores were isolated from these reactions by sucrose density gradient centrifugation and then used in the following experiment, shown in the figure. Six vials, each containing 150 μl of reticulocyte lysate, were set up at 0°C, and a 50-μl sample of core particles containing either tRNA or 42S RNA was added to each of three vials. The reactions were then incubated at either 0 or 30°C and subjected to sucrose density gradient centrifugation on 10 to 30% (wt/wt) sucrose density gradients for 50 min at 50,000 rpm and 4°C in a Beckman SW60 rotor. The distribution of optical density and radioactivity in the gradients was determined and is shown as follows: reactions with tRNA-containing particles were incubated at 0°C for 30 min (A), at 30°C for 5 min (B), or at 30°C for 30 min (C); reactions with 42S RNA-containing particles were incubated at 0°C for 30 min (A'), at 30°C for 5 min (B'), or at 30°C for 30 min (C').

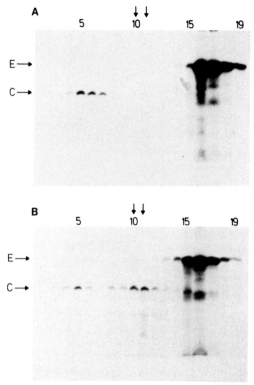

FIGURE 4. Characterization of structural protein transferred to cellular ribosomes early in infection. [^{35}S]methionine-labeled Sindbis virus was allowed to adsorb to chicken embryo fibroblasts in the cold, followed by a 30-min incubation at either 0 or 37°C. Cells were then lysed mechanically, and the lysate was adjusted to contain 0.5% of the nonionic detergent Nonidet P-40 to remove the viral membrane from cell-associated virus. Nuclei were removed by centrifugation, and ribosomes and polyribosomes in the postnuclear fraction were converted into ribosomal subunits by addition of 3 mM EDTA. This fraction was then subjected to sucrose density gradient centrifugation (60 min, 50,000 rpm, 4°C, 10 to 30% [wt/wt] sucrose, SW60 rotor) to display the viral cores (150S) and the ribosomal subunits (60S and 40S). Material from each gradient fraction was subjected to SDS-PAGE (10% polyacrylamide, 0.33% bisacrylamide). Autoradiographs of gels used to analyze the gradients that were used to fractionate lysates of cells incubated at 0°C (A) and 37°C (B) are shown. The fractions analyzed are indicated by numbers above the lanes; fraction 19 corresponds to the top of the gradient. The position of the large ribosomal subunit, as determined from the optical density profile, is indicated by vertical arrows. Positions of migration of viral C and E proteins during SDS-PAGE are indicated at the left.

the regulation of assembly and disassembly of alphavirus cores occurs in the following manner. In stage 1, viral C protein is transferred early in infection from the core of the incoming virus particles to the cellular ribosomes that are present in the cytoplasm in large excess compared with viral cores. In stage 2, synthesis of RNA replicase from the viral genome leads to the accumulation of stable complexes of polymerase and 42S minus-strand template RNA, which synthesize genome RNA and 26S mRNA coding for the viral structural polyprotein. The C-protein molecules synthesized from this mRNA remain mostly associated with the ribosomes. This allows use of 42S genome RNA for translation into RNA replicase. In stage 3, after a certain degree of saturation of ribosomal C-protein-binding sites has been reached, newly synthesized C protein associates with 42S plus-strand RNA into cores. The translation of 42S plus-strand RNA into viral replicase is inhibited by this process. Complexes of RNA polymerase and 42S minus-strand RNA made during stage 2 are stable and continue to synthesize 42S plus-strand and 26S plus-strand RNA. These considerations are consistent with the experimental data on the replication of alphaviruses (see Schlesinger and Schlesinger [1986] for reviews summarizing these data). Analyses of the fate of the structural proteins of incoming virus particles by using [^{35}S]methionine-labeled virus have shown that a transfer of labeled C protein onto cellular ribosomes can be detected during the early stages of infection of cells by Sindbis virus (Fig. 4) (Wengler and Wengler, 1984). Amino acid pulse-labeling experiments involving SF virus-infected cells have shown that in the early stages of viral multiplication, the majority of newly synthesized C protein remains associated with ribosomes for a rather long time and that during viral multiplication, transfer of newly synthesized C protein from ribosomes into cores occurs with increasing speed and yield as the viral multiplication cycle progresses (Ulmanen et al., 1976, 1979). It might be mentioned that this hypothesis makes no implications as to the existence or nonexistence of a genome RNA sequence that leads to a preferential specific encapsidation of alphavirus genome RNA by the viral C protein. Experimental evidence for the existence of such a sequence is presented in the chapter by S. Schlesinger, B. Weiss, and H. Nitschko (this volume). A more complete discussion of the three stages described above and of experimental analyses of the assembly and disassembly of alphavirus cores has been published (Wengler, 1987).

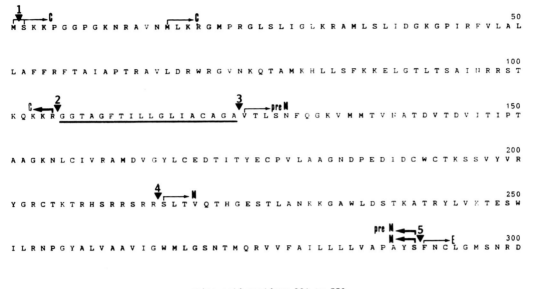

FIGURE 5. Localization of the cleavages involved in generation of the structural proteins of WN virus. The amino acid sequence of the part of the viral polyprotein that gives rise to the structural proteins is presented in a shortened version (the internal sequence of the E protein is not shown), using the one-letter code. The amino-terminal sequences of the structural proteins (Castle et al., 1985; Wengler et al., 1985) are indicated by thin arrows. The carboxy-terminal sequences of the viral structural proteins C, M, and E (Nowak et al., 1989) are indicated by thick arrows. The amino terminus of the WN virus-specific protein NS1 is derived from unpublished data obtained in our laboratory. The six cleavage reactions inferred are indicated by triangles and numbers. The only region of the structural polyprotein that is not found in the structural proteins of purified cell-associated WN virus particles is the underlined sequence between cleavages 2 and 3.

ANALYSIS OF THE SEQUENCE OF THE WN FLAVIVIRUS GENOME AND OF THE TERMINAL SEQUENCES OF ITS STRUCTURAL PROTEINS

We have determined the complete nucleotide sequence of the genome RNA of the WN flavivirus (Castle et al., 1986). These data have shown that a single long open reading frame comprising 10,290 nucleotides is present on the genome. Similar analyses are available for a number of flaviviruses (see Strauss et al. [1987] and Westaway [1987] for recent reviews). Amino-terminal sequence analyses have shown that the structural proteins are localized in the amino-terminal part of the corresponding polypro-

tein in the order C (core protein)–preM (envelope protein of about 22 kilodaltons that is proteolytically cleaved into M protein during virus release)–E (envelope protein of about 50 kilodaltons). The corresponding amino acid sequences are presented in Fig. 5. We have recently characterized the carboxy-terminal sequences of all viral structural proteins (Nowak et al., 1989; Fig. 5). It can be seen that the proteolytic processing is very conservative: cell-associated WN virus particles contain the structural proteins C (amino acid residues 1 to 105), preM (amino acid residues 124 to 290), and E (amino acid residues 291 to 787). Amino acid 788 represents the amino terminus of the ensuing nonstructural protein NS1. The only segment of the polyprotein that is absent from these pro-

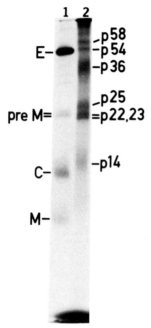

FIGURE 6. SDS-PAGE (15% acrylamide) analyses of proteins translated in vitro. An mRNA coding for all structural proteins of WN virus and for the ensuing 82 amino acids of protein NS1 was translated in vitro in the presence of [^{35}S]methionine in a reticulocyte lysate in the presence of membranes. Shown is a comparison of the [^{35}S]methionine-labeled proteins present in extracellular WN virus (lane 1) with the proteins that were translated from the polyprotein mRNA in vitro in the presence of membranes (lane 2). The apparent molecular weights of the proteins that were synthesized in vitro were estimated from the migration of marker proteins. Positions of the structural proteins E, preM, M, and C are indicated on the left.

teins is the segment comprising amino acid residues 106 to 123.

ANALYSIS OF WN VIRUS STRUCTURAL PROTEIN SYNTHESIS

We have analyzed the synthesis of the WN virus structural proteins by translating an mRNA coding for the corresponding polyprotein in vitro in the presence of membranes (Nowak et al., 1989). Analysis by sodium dodecyl sulfate-polyacrylamide gel electrophoresis (SDS-PAGE) of the proteins that are synthesized in vitro as compared with authentic viral structural proteins is shown in Fig. 6. The proteins p58, p54, p36, p25, p22,23, and p14 synthesized in

vitro were analyzed by fingerprinting, amino-terminal radiosequencing, and other techniques. These data have shown that p54 and p22,23 contain the primary sequences of the viral structural proteins E and preM, respectively. p14 contains the amino-terminal sequence of the viral C protein but migrates slightly slower than authentic C protein during SDS-PAGE (Fig. 6) and can be labeled with [^{35}S]cysteine. Cysteine is absent from the sequence of authentic C protein and is present in the polyprotein for the first time in position 120 (Fig. 5). These data indicate that p14 synthesized in vitro comprises amino acid residues 1 to 123 (Fig. 5). Therefore, in vitro the polyprotein is cleaved between amino acid residues 123 and 124, 290 and 291, and 787 and 788. The properties of the primary sequences cleaved and the fact that these cleavages specifically occur in the presence of membranes in vitro suggest that signalases are responsible for these cleavages.

It seems plausible to assume that a similar cleavage occurs in vivo during synthesis of the viral structural proteins and leads to the synthesis of E and preM and of an elongated version of the viral C protein containing a carboxy-terminal extension comprising amino acid residues 106 to 123. This extension is predicted to be a hydrophobic transmembrane segment functioning as an internal signal sequence that transfers the corresponding region through the rough endoplasmic reticulum membrane and makes the bond between amino acid residues 123 and 124 accessible to cleavage by signalase (Wengler et al., 1985). After signalase cleavage, this extension would anchor protein p14 to the endoplasmic reticulum membrane, and we have therefore called p14 the anchored C protein. The finding that newly synthesized C protein accumulates as a membrane-associated molecule in flavivirus-infected cells (see Westaway [1987] for a review) further supports this interpretation. The analyses described above indicate that a series of proteolytic cleavages is involved in the synthesis of viral structural proteins, the assembly of the proteins into cell-associated virus, and the release of virus from cells (Fig. 7).

ASSEMBLY OF WN VIRUS

Taken together, the data presented above suggest that the following steps are involved in the assembly and disassembly of flaviviruses. The structural polyprotein segment is cleaved by signalase, and the resulting cleavage products, anchored C, preM, and E, are anchored to the endoplasmic reticulum membrane by carboxy-

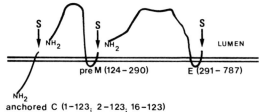

anchored C (1–123, 2–123, 16–123)

FIGURE 7. Schematic representation of the structural proteins of the WN flavivirus after cotranslational cleavage of the viral polyprotein by cellular signalase(s). The amino acid sequence content, the orientation relative to the membrane of the rough endoplasmic reticulum, and the number and localization of transmembrane segments in these proteins are indicated.

terminal membrane anchor segments (Fig. 7). The genome RNA interacts with these modified membrane patches by binding to the C protein. This interaction leads to the formation of cell-associated virus particles that accumulate in intracellular vacuoles. The virus-encoded protein NS1, which is a glycosylated membrane-associated protein, might be present in these modified membranes and play a role in the formation of cell-associated flavivirus particles.

At some stage during this assembly process, the carboxy-terminal anchor segment of the core protein must be removed since it is present neither in the C protein of cell-associated virus nor in virus particles as a separate molecule. Cleavage of anchored C protein between amino acid residues 105 and 106 generates (mature) C protein. This cleavage occurs prior to or during virus assembly. It is probably performed by a cytoplasmic virus-specific protease which cleaves after double basic residues and which is also involved in the synthesis of the viral nonstructural proteins (Rice et al., 1985; Coia et al., 1988). Removal of the carboxy-terminal membrane anchor sequence from the C protein during virus formation converts a membrane-associated, intracellularly assembling genome RNA-protein-membrane complex into an RNA-protein core complex present in virus particles which has lost its strong membrane association via a membrane anchor segment and which therefore can be released from the viral envelope into the cytoplasm early during infection. Preliminary experiments performed in our laboratory have indicated that the viral core complex is a labile structure containing the genome in a rather exposed form. Further experiments are necessary to analyze the regulation of assembly and disassembly of flavivirus nucleoproteins.

Prior to virus release, the viral preM protein is cleaved between amino acid residues 215 and 216. This cleavage is probably performed in Golgi vesicles by a cellular enzyme which is also involved in the processing of cellular polyproteins (Shapiro et al., 1972). The amino-terminal part of the preM protein (residues 124 to 215) is lost from the virus. Residues 216 to 290 remain associated to the virus particles as M protein. A small number of uncleaved preM protein molecules can be found in extracellular WN virus particles.

REFERENCES

Argos, P., and J. E. Johnson. 1984. Chemical stability in simple spherical plant viruses, p. 1–43. *In* F. A. Jurnak and A. McPherson (ed.), *Biological Macromolecules and Assemblies*, vol. 1. *Virus Structures.* John Wiley & Sons, Toronto.

Boege, U., M. Cygler, G. Wengler, P. Dumas, J. Tsao, M. Luo, T. J. Smith, and M. G. Rossmann. 1989. Sindbis virus core protein crystals. *J. Mol. Biol.* **208:**79–82.

Boege, U., G. Wengler, G. Wengler, and B. Wittmann-Liebold. 1981. Primary structures of the core proteins of the alphaviruses Semliki Forest virus and Sindbis virus. *Virology* **113:**293–303.

Castle, E., U. Leidner, T. Nowak, G. Wengler, and G. Wengler. 1986. Primary structure of the West Nile flavivirus genome regions coding for all nonstructural proteins. *Virology* **149:**10–26.

Castle, E., T. Nowak, U. Leidner, G. Wengler, and G. Wengler. 1985. Sequence analysis of the viral core protein and the membrane-associated proteins V1 and NV2 of the flavivirus West Nile virus and of the genome sequence for these proteins. *Virology* **145:** 227–236.

Coia, G., M. D. Parker, G. Speight, M. E. Byrne, and E. G. Westaway. 1988. Nucleotide and complete amino acid sequences of Kunjin virus: definitive gene order and characteristics of the virus-specific proteins. *J. Gen. Virol.* **69:**1–21.

Fuller, S. D., and P. Argos. 1987. Is Sindbis a simple picornavirus with an envelope? *EMBO J.* **6:**1099–1105.

Garoff, H., A. M. Frischauf, K. Simons, H. Lehrach, and H. Delius. 1980. The capsid protein of Semliki Forest virus has clusters of basic amino acids and prolines in its amino-terminal region. *Proc. Natl. Acad. Sci. USA* **77:**6376–6380.

Melancon, P., and H. Garoff. 1987. Processing of the Semliki Forest virus structural polyprotein: role of the capsid protease. *J. Virol.* **61:**1301–1309.

Nowak, T., P. M. Färber, G. Wengler, and G. Wengler. 1989. Analyses of the terminal sequences of West Nile virus structural proteins and of the *in vitro* translation of these proteins allow the proposal of a complete scheme of the proteolytic cleavages involved in their synthesis. *Virology* 169:365–376.

Rice, C. M., E. M. Lenches, R. R. Eddy, S. J. Shin, R. L. Sheets, and J. H. Strauss. 1985. Nucleotide sequence of Yellow Fever virus: implications for flavivirus gene expression and evolution. *Science* 229:726–733.

Schlesinger, S., and M. J. Schlesinger (ed.). 1986. *The Togaviridae and Flaviviridae*. Plenum Publishing Corp., New York.

Shapiro, D., W. E. Brandt, and P. K. Russell. 1972. Change involving a viral membrane glycoprotein during morphogenesis of group B arboviruses. *Virology* 50:906–911.

Strauss, E. G., C. M. Rice, and J. H. Strauss. 1984. Complete nucleotide sequence of the genomic RNA of Sindbis virus. *Virology* 133:92–110.

Strauss, J. H., E. G. Strauss, C. S. Hahn, R. Galler, W. R. Hardy, and C. M. Rice. 1987. Replication of alphaviruses and flaviviruses: proteolytic processing of polyproteins, p. 209–226. *In* M. A. Brinton and R. R. Rueckert (ed.), *Positive Strand RNA Viruses*. Alan R. Liss, Inc., New York.

Ulmanen, I., H. Söderlund, and L. Kääriäinen. 1976. Semliki Forest virus capsid associates with the 60 S ribosomal subunit in infected cells. *J. Virol.* 20:203–210.

Ulmanen, I., H. Söderlund, and L. Kääriäinen. 1979. Role of protein synthesis in the assembly of Semliki Forest virus nucleocapsid. *Virology* 99:265–276.

Wengler, G. 1987. The mode of assembly of alphavirus cores implies a mechanism for the disassembly of the cores in the early stages of infection. *Arch. Virol.* 94:1–14.

Wengler, G., E. Castle, U. Leidner, T. Nowak, and G. Wengler. 1985. Sequence analysis of the membrane protein V3 of the flavivirus West Nile virus and of its gene. *Virology* 147:264–274.

Wengler, G., and G. Wengler. 1984. Identification of a transfer of viral core protein to cellular ribosomes during early stages of alphavirus infection. *Virology* 134:435–442.

Wengler, G., G. Wengler, U. Boege, and K. Wahn. 1984. Establishment and analysis of a system which allows assembly and disassembly of alphavirus core-like particles under physiological conditions *in vitro. Virology* 132:401–410.

Westaway, E. G. 1987. Flavivirus replication strategy. *Adv. Virus Res.* 33:45–90.

Chapter 34

Sindbis Virus RNAs Bind the Viral Capsid Protein Specifically and Are Preferentially Encapsidated

Sondra Schlesinger, Barbara Weiss, and Hans Nitschko

Sindbis virus is a positive-strand-RNA, enveloped virus. The virion RNA (49S RNA) consists of 11,703 nucleotides (Strauss et al., 1984) plus a poly(A) tail. In the virion, the RNA is packaged in an icosahedral nucleocapsid consisting of 180 molecules of the 30-kilodalton viral capsid protein (Fuller, 1987). The T = 3 nucleocapsid is surrounded by a lipid bilayer into which are inserted the viral spike glycoproteins arranged in a T = 4 icosahedral envelope (Fuller, 1987; Harrison, 1986). In cells infected with Sindbis virus, the genomic RNA serves as an mRNA for the nonstructural proteins and as a template for the complementary negative strand (reviewed by Strauss and Strauss [1986]). The negative RNA strand is the template not only for new genomic RNA but also for the subgenomic 26S RNA. The latter codes for the viral structural proteins—the capsid protein and the two glycoproteins.

The Sindbis virus genome (Rice et al., 1987) and the smaller defective interfering (DI) RNAs (Levis et al., 1986) have been cloned as cDNAs directly downstream from the SP6 bacteriophage promoter. These plasmids are providing invaluable tools for determining the nucleotide sequences responsible for the biological properties of this virus. RNAs transcribed from the cDNAs can be tested for their biological activity by transfection into cultured cells. With DI RNA transcripts, the transfections are carried out in the presence of helper virus or virion RNA (Levis et al., 1986; Weiss et al., 1989). In previous studies, we had carried out a deletion analysis of the cDNA of a DI RNA and demonstrated that only deletions at the 5' and 3' termini of the DI genome prevented amplification (Levis et al., 1986). Furthermore, substitu-

tion of 75% of the internal viral sequences with foreign sequences did not destroy the ability of the DI genome to be amplified (Levis et al., 1987). These initial experiments did not distinguish between domains that would be required for replication of the genome and those that might be required for encapsidation. We were particularly interested in determining whether there were any packaging signals in the Sindbis virus genome, since there is specificity in encapsidation. In infected cells, only the 49S RNA is encapsidated; neither the 26S subgenomic RNA nor cellular RNAs are packaged. Wengler et al. (1982, 1984) had shown, however, that in vitro, the viral capsid protein can interact with a variety of RNAs as well as with other negatively charged macromolecules to form nucleocapsid-like structures. Wengler (1987) suggested that the specific incorporation of the viral genomic RNA into nucleocapsids is the result of the high stability of the complex rather than due to a specific packaging signal.

We used three different approaches to investigate the specificity of encapsidation of the Sindbis virus genome. The first was to examine the binding of RNA to purified viral capsid protein that had been immobilized on nitrocellulose. This procedure allowed us to define a region of the Sindbis virus genome that is required for binding. The second was to analyze the assembly of nucleocapsids in vitro. We included in the analysis the effects of competition between RNAs and demonstrated a clear preference for the encapsidation of virion RNA. The third approach was to determine whether the domain associated with binding activity in vitro was essential for packaging in vivo. For this study, we used DI RNAs and were able to

Sondra Schlesinger, Barbara Weiss, and Hans Nitschko, Department of Molecular Microbiology, Washington University School of Medicine, St. Louis, Missouri 63110-1093.

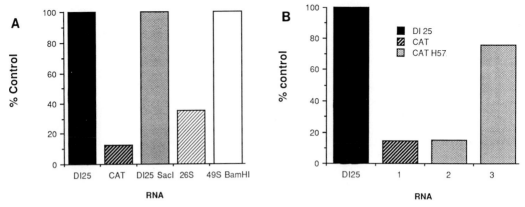

Figure 1. (A) Binding of Sindbis virus-specific RNAs to purified capsid protein. Capsid protein (600 ng per filter) was blotted onto nitrocellulose, followed by incubation with various ^{32}P-labeled RNA transcripts (Weiss et al., 1989). The data are normalized to the binding of the 2,358-nucleotide DI25 RNA. (B) Effect of insertion of 572 nucleotides from 49S virion RNA (the region from nucleotides 683 to 1255 in 49S RNA) into CAT RNAs. Lanes: 1, unmodified CAT RNA transcript; 2, CAT RNA containing the 572-nucleotide insert in the inverse orientation; 3, CAT RNA containing the 572-nucleotide insert in the correct orientation.

dissociate their ability to be replicated in transfected cells from their ability to be packaged. The presence of the region of the genome that confers binding activity to an RNA was also essential for that RNA to be amplified on passaging. Thus, we were able to correlate binding activity in vitro with packaging in vivo.

BINDING OF SINDBIS VIRUS RNAs TO THE VIRAL CAPSID PROTEIN

We examined the binding of RNA to the Sindbis virus capsid protein by immobilizing the capsid protein on nitrocellulose and then incubating the immobilized protein with ^{32}P-labeled RNA transcripts. This procedure permitted us to obtain a quantitative measure of the binding of RNA to the capsid protein. The capsid protein was purified from virions as described by Wengler et al. (1982). The details of this method are described elsewhere (Weiss et al., 1989), but Fig. 1A shows the type of specificity we observed. Several different RNA transcripts were tested for the ability to bind the capsid protein. DI25 RNA has been described previously (Levis et al., 1986). This DI RNA transcript is both replicated and packaged when it is transfected into cultured cells in the presence of helper Sindbis virus. The chloramphenicol acetyltransferase (CAT) RNA was transcribed from a plasmid in which the CAT gene is positioned downstream from the SP6 bacteriophage promoter. It

served as the nonspecific control. Its ability to bind to capsid protein was almost 10-fold lower than that of DI25 RNA. DI25 SacI RNA was transcribed from the same plasmid as DI25, but in this case the plasmid was linearized at a downstream site such that the RNA was 4,241 nucleotides instead of 2,358 nucleotides, which is the size of the biologically active DI RNA. The DI25 SacI RNA is essentially the same size as the 49S BamHI RNA and was used so that comparisons would be made between RNAs of the same size. The 49S BamHI RNA is the 5' one-third of the virion RNA, and this RNA binds to capsid protein to the same extent as does DI25 RNA. In contrast, 26S RNA, the 3' one-third of the genome, binds capsid protein inefficiently. These data suggested that there is a major binding site for the capsid protein in the 5' region of the genomic RNA and that the binding site is also present in the DI RNA. Most of the sequences in DI25 RNA come from the 5' region of the genome (Monroe and Schlesinger, 1984).

By examining (i) 49S RNA transcripts with identical 5' sequences but different amounts of 3' sequences and (ii) deleted forms of the virion RNA, we demonstrated that the region conferring binding specificity to the RNA lies within the region extending from nucleotides 767 to 1407 (Weiss et al., 1989). These nucleotides are in the coding region of the viral nonstructural gene nsP1. The strongest evidence that this is a binding site for capsid protein is the ability of a restriction enzyme fragment covering these nucleotides to convert the CAT RNA from a negative control to an RNA that has significant binding activity (Weiss et al., 1989). We inserted a 572-nucleotide *Hha*I fragment extending from nucleotides 683 to 1255 in the Sindbis virus 49S RNA into the pCAT plasmid immediately downstream from the start site for in vitro transcription. The fragment was inserted into pCAT in two orientations so that it was represented in RNA transcripts in a plus or minus sense. When it was inserted in the plus-sense orientation, binding was enhanced fivefold over the value for unmodified CAT RNA (Fig. 1B). When the insertion was in the opposite orientation, there was no increase in binding over the background value.

IN VITRO ASSEMBLY OF NUCLEOCAPSIDS

Wengler et al. (1982, 1984) and Wengler (1987) were able to assemble nucleocapsids in vitro from purified Sindbis virus capsid protein and a variety of RNAs. The nucleocapsids were characterized according to density and sedimentation profile and by electron microscopy. The conditions that we established for in vitro assembly are described in the legend to Fig. 2. Unless indicated otherwise, the molar ratio of RNA to capsid protein was 1:180. Incubation of either 49S or 26S RNA with capsid protein resulted in the formation of nucleocapsids that sedimented with the same characteristics as authentic viral nucleocapsids (Fig. 2). The addition of a fourfold excess of tRNA to Sindbis virus virion RNA during assembly had almost no discernible effect on the sedimentation profile. In contrast, the addition of a fourfold excess of 49S virion RNA shifted the radioactive peak dramatically. These data indicate that tRNA is unable to compete with virion RNA for assembly. This result was not surprising, since under the conditions of assembly that we used neither tRNA nor rRNA was encapsidated to any significant extent (H. Nitschko and S. Schlesinger, unpublished results).

To analyze competition between 49S RNA and 26S RNA, we added only small amounts of competitor RNA to be able to detect preferential incorporation. In each case, the 49S virion RNA to capsid protein ratio was 1:180. 49S competitor RNA was added at a molar ratio of 0.1, and 26S competitor RNA was added at a molar ratio of 0.3 (These amounts represent the same microgram quantities of each competitor RNA.) The in vitro assembly mimics what is found in infected cells: 49S virion RNA is encapsidated preferentially over 26S RNA. These results can

be explained by two models that are not mutually exclusive. One model is that 49S RNA is preferentially encapsidated as a result of binding specificity affecting the initiation of assembly. A second model is that the specific interactions between 49S RNA and the capsid protein produce a nucleocapsid that is much more stable than that obtained with RNAs such as 26S RNA, which bind weakly or do not bind to the capsid protein. Wengler et al. (1982, 1984) and Wengler (1987) have also reported that the nucleocapsid containing 49S RNA is a more stable structure than the nucleocapsids containing nonspecific RNAs.

THE NUCLEOTIDES REQUIRED FOR SPECIFIC BINDING ARE ALSO REQUIRED FOR PACKAGING OF DI RNAs IN VIVO

DI RNAs provide a useful means for identifying important *cis*-acting sequences because it is possible to alter the sequences of these genomes extensively and still produce biologically active molecules. We compared the binding activities, replication, and packaging of the three DI RNAs described in Fig. 3. The first (CTS14) contains the first 240 nucleotides of DI25, followed by a polylinker and the 572-nucleotide *Hha*I fragment described above (Weiss et al., 1989). There are then foreign sequences from the CAT gene. The remainder of the genome comes from the 3' terminus of the virion RNA. The second (CTS1) is similar to CTS14 except that it lacks the 572-nucleotide *Hha*I fragment. The third (CTS253) contains the first 240 nucleotides of DI25 and then the CAT and simian virus 40 sequences. The 3' terminus is derived from DI25 and contains some sequences from the 5' region of the 49S RNA. It contains 279 nucleotides from the 3' region of the capsid-binding domain (Fig. 3). This DI RNA was previously shown to be amplified on passaging and to be translated to produce enzymatically active CAT (Levis et al., 1987).

The binding of CTS14 to immobilized capsid protein was 89% that of DI25. The binding of CTS1 was essentially the same as that of the nonspecific CAT RNA (18 and 17% of DI25, respectively). The binding of CTS253 was only 29% that of DI25. The inefficient binding of CTS253 RNA may reflect either the lack of the complete binding domain, its improper location at the 3' end of the molecule, or both factors. All three DI RNAs were replicated in cells transfected with the DI RNA transcript and virion RNA as helper (Fig. 4). CTS1 and CTS253 were

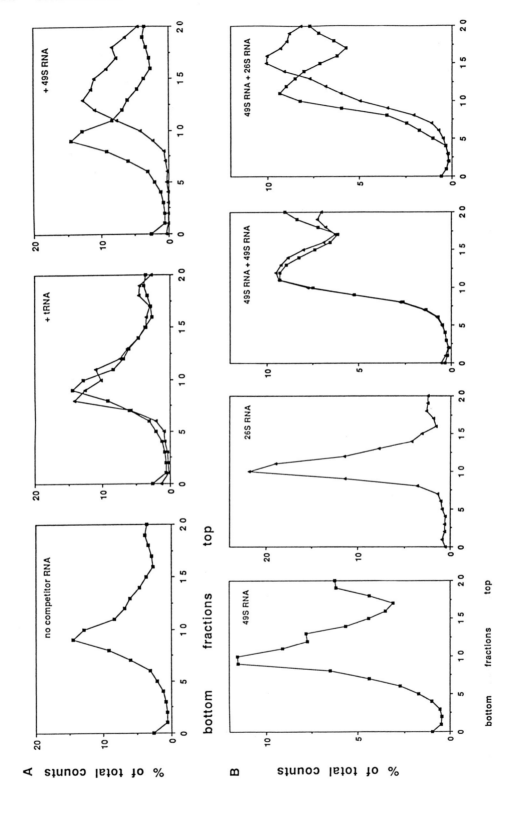

replicated to the same extent in the transfected cells, yet only CTS253 was detected in later passages. Figure 4 also shows the pattern of viral RNAs detected during the formation of passage 3. Both CTS14 and CTS253 RNA were enriched on passaging, but CTS1 was not detectable. These results demonstrated that CTS1 (the DI RNA lacking the 572-nucleotide insert) was lost on passaging and provide strong evidence that the binding domain we identified is required for in vivo assembly (Weiss et al., 1989).

The possibility that internal sequences of the alphavirus genome might be involved in encapsidation was suggested several years ago by Jalanko and Söderlund (1985). They cloned a DI cDNA of Semliki Forest virus into a simian virus 40 vector and expressed the DI RNA in monkey kidney cells. The DI RNA was not replicated by superinfecting helper Semliki Forest virus, nor was it able to interfere with the replication of the virus. The release of virus was inhibited by about 60%. The DI RNA is missing the extreme 5' terminus of the genomic RNA but contains repeated elements from the nonstructural region of the virion genome. The authors suggest that the repeated regions might be involved in encapsidation. The general repeat unit contains approximately 200 nucleotides from nsP2 (starting at nucleotide 2728 of the Semliki Forest sequence), followed by 278 nucleotides (from nucleotides 39 to 317) of the Semliki Forest genome (Lehtovaara et al., 1981, 1982). Thus, that Semliki Forest DI RNA does not have a region from nsP1 comparable to the one we have defined in the Sindbis virus DI RNAs. Semliki Forest DI RNAs are packaged inefficiently (Kääriäinen et al., 1981), and it will be interesting to know whether they contain a capsid-binding domain. There are several domains in the nsP1, nsP2, and nsP4 genes of Sindbis virus RNA that show strong amino acid homology not only with Semliki Forest virus RNA but also with plus-stranded plant RNA viruses (Haseloff et al., 1984; Ahlquist et al., 1985). These homologies are thought to reflect evolutionary

relatedness and conserved function of the proteins. The alphaviruses have also conserved sequence domains at the 5' and 3' termini and at the start of the subgenomic RNA (Strauss and Strauss, 1986). There is increasing evidence that the conserved sequences represent cis-acting regulatory or binding functions (Levis et al., 1986; Levis et al., 1990; R. Kuhn, E. Strauss, and J. Strauss, personal communication). It should now be possible to determine whether there is any conservation in the capsid-binding domain.

INTERACTION OF THE COAT PROTEIN OF OTHER POSITIVE-STRAND VIRUSES WITH VIRAL RNAS

The coat proteins of a number of different positive-strand RNA viruses have been shown to bind to specific regions of their viral RNAs. The most well-defined system of virus assembly is that of tobacco mosaic virus, a helical plant virus particle in which assembly is initiated by a specific interaction between the coat protein disk and an internal sequence in the RNA genome (reviewed by Bloomer and Butler [1986]). Encapsidation of heterologous RNA containing this specific recognition domain occurs both in vitro (Sleat et al., 1986; Gallie et al., 1987) and in vivo (Sacher et al., 1988). Studies of disassembly and reassembly of the plant virus turnip crinkle virus suggest that it is dimers of the coat protein that interact with the viral RNA and that it is most likely this step that leads to the specificity of assembly (Sorger et al., 1986). Domains in the genomes of retroviruses that are required for packaging of the RNA into virions have been identified (Mann et al., 1983; Adam and Miller, 1988).

There are examples of high-affinity binding of proteins to viral RNAs that define functions not necessarily associated with encapsidation. A classic example is the coat protein of the RNA bacteriophages whose interaction with the viral

FIGURE 2. In vitro encapsidation of Sindbis RNAs with purified capsid protein. Sindbis RNA (40 μg/ml) was incubated with purified capsid protein (molar ratio, 1:180) at 37°C for 15 min in a buffer solution containing 40 mM Tris hydrochloride, 200 mM NaCl, and 4 mM EDTA at pH 7. The mixture was then layered on a 10 to 40% sucrose gradient in the same buffer containing 50 μg of gelatin per ml and centrifuged for 65 min at 45,000 rpm in an SW50.1 rotor at 4°C. (A) The first panel shows the pattern obtained when 49S RNA is incubated with capsid protein to form nucleocapsids. The same data (■) are included in the second and third panels for comparison with the data obtained in the presence of a fourfold excess of the indicated competitor. (B) In the two rightmost panels, the 49S RNA at a concentration of 40 μg/ml is represented by the closed squares. The competitor RNAs (▲) were labeled with ^{32}P and were added at a concentration of 4 μg/ml.

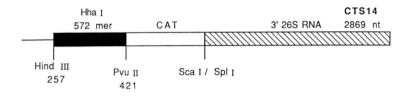

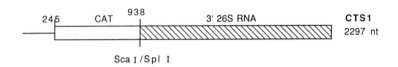

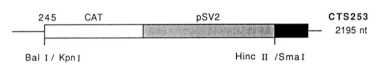

FIGURE 3. Structures of the DI RNAs tested for in vivo packaging. The name and length (in nucleotides [nt]) of each RNA are given on the right. The black bars indicate the region defined as the capsid-binding domain. The 3′ domain fused to CAT in CTS1 and CTS14 was derived from nucleotides 10381 through 11740 in the 49S cDNA. CTS14 has the 572-nucleotide insertion derived from nucleotides 683 to 1255 of 49S virion RNA. It also has a deletion of 164 nucleotides at the 5′ terminus of the CAT gene (Weiss et al., 1989).

RNA serves to repress translation (Weber and Konigsberg, 1975). Alfalfa mosaic virus is another interesting example. The genomic RNAs of this virus contain high-affinity binding sites for the viral coat protein, and binding of the coat protein to each of the genomic RNAs is an essential step in the ability of these RNAs to be infectious. Binding sites were identified both at the 3′ terminus and at internal sites of the largest genomic segment and the subgenomic RNA, but the subgenomic RNA of alfalfa mosaic virus can also be encapsidated. These sites are thought to be involved in replicase recognition (Houwing

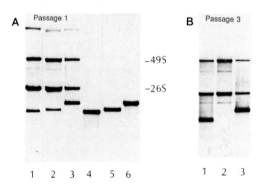

FIGURE 4. Analysis of Sindbis DI RNAs synthesized in cells after transfection or after passaging. (A) Secondary chicken embryo fibroblasts were cotransfected with 49S virion RNA and DI RNA transcripts by using lipofectin (Felgner et al., 1987) and labeled with [3H]uridine in the presence of actinomycin D for 18 h at 30°C. The isolated RNAs were denatured with glyoxal and electrophoresed through a 1% agarose gel (Weiss et al., 1989). Lanes: 1 to 3, RNAs from cells transfected with 49S RNA and CTS253 (lane 1), CTS 1 (lane 2), or CTS 14 (lane 3); 4 to 6, transcripts of CTS253 (lane 4), CTS 1 (lane 5), and CTS14 (lane 6). (B) Electrophoretic analysis of viral RNAs synthesized in passage 3. One-fourth of the medium harvested from passage 2 was used to infect new monolayers. The labeling conditions were as described above for passage 1. Lanes: 1, CTS253; 2, CTS1; 3, CTS14.

and Jaspars, 1982; Zuidema et al., 1983), but one or more may also be important in assembly. Multiple binding sites may also exist on the RNAs of coronaviruses. Northwestern (RNA-protein) blot analysis of the interaction of the coronavirus nucleocapsid protein and viral RNAs has demonstrated that the viral nucleo-capsid protein binds to the 3' end of the viral leader RNA and therefore to all of the viral mRNAs as well as to genomic RNA (Baric et al., 1988; Stohlman et al., 1988). Stohlman et al. (1988) propose that this interaction plays an important role in the transcription process. They also suggest that there must be other sequences affecting encapsidation.

These examples, and that of Sindbis virus, define a domain or domains of the viral genome that impart specificity to the binding of the viral capsid protein. It is not clear whether these domains are defining a linear sequence or a particular structure. Experiments with both to-bacco mosaic virus (Turner et al., 1988) and Qβ bacteriophage (Witherell and Uhlenbeck, 1989) indicate that structure plays a crucial part in determining the specificity. Whether or not this is also true for the Sindbis virus binding domain is now under investigation.

ACKNOWLEDGEMENT. This work was supported by Public Health Service grant AI11377 from the National Institute of Allergy and Infectious Diseases.

REFERENCES

Adam, M. A., and D. Miller. 1988. Identification of a signal in a murine retrovirus that is sufficient for packaging of nonretroviral RNA into virions. *J. Virol.* 62:3802–3806.

Ahlquist, P., E. G. Strauss, C. M. Rice, J. H. Strauss, J. Haseloff, and D. Zimmern. 1985. Sindbis virus proteins nsP1 and nsP2 contain homology to non-structural proteins from several RNA plant viruses. *J. Virol.* 35:536–542.

Baric, R. S., G. W. Nelson, J. O. Fleming, R. J. Deans, J. G. Keck, N. Casteel, and S. A. Stohlman. 1988. Interactions between coronavirus nucleocapsid pro-tein and viral RNAs: implications for viral transcrip-tion. *J. Virol.* 62:4280–4287.

Bloomer, A. C., and P. J. G. Butler. 1986. Tobacco mosaic virus. Structure and self-assembly, p. 19–57. *In* M. H. V. Van Regenmortel and H. Fraenkel-Conrat (ed.), *The Plant Viruses*, vol. 2. Plenum Publishing Corp., New York.

Felgner, P. L., T. R. Gadek, M. Holm, R. Roman, H. W. Chan, M. Wenz, J. P. Northrop, G. M. Ringold, and M. Danielsen. 1987. Lipofection: a highly efficient, lipid-mediated DNA-transfection procedure. *Proc. Natl. Acad. Sci. USA* 84:7413–7417.

Fuller, S. D. 1987. The T = 4 envelope of Sindbis virus is organized by interactions with a complementary T = 3 capsid. *Cell* 48:923–934.

Gallie, D. R., D. E. Sleat, J. W. Watts, P. C. Turner, and T. M. A. Wilson. 1987. In vivo uncoating and efficient expression of foreign mRNAs packaged in TMV-like particles. *Science* 236:1122–1124.

Harrison, S. C. 1986. Alphavirus structure, p. 21–34. *In* S. Schlesinger and M. J. Schlesinger (ed.), *The Togaviridae and Flaviridae*. Plenum Publishing Corp., New York.

Haseloff, J., P. Goelet, D. Zimmern, P. Ahlquist, R. Dasgupta, and P. Kaesberg. 1984. Striking similar-ities in amino acid sequence among nonstructural proteins encoded by RNA viruses that have dissim-ilar genomic organization. *Proc. Natl. Acad. Sci. USA* 81:4358–4362.

Houwing, C. J., and E. M. J. Jaspars. 1982. Protein binding sites in nucleation complexes of alfalfa mosaic virus RNA 4. *Biochemistry* 21:3408–3414.

Jalanko, A., and H. Söderlund. 1985. The repeated regions of Semliki Forest virus defective-interfering RNA interferes with the encapsidation process of the standard virus. *Virology* 141:257–266.

Kääriäinen, L., R. F. Pettersson, S. Keränen, P. Lehtovaara, H. Söderlund, and P. Ukkonen. 1981. Multiple structurally related defective-interfering RNAs formed during undiluted passages of Semliki Forest virus. *Virology* 113:686–697.

Lehtovaara, P., H. Söderlund, S. Keränen, R. F. Pet-tersson, and L. Kääriäinen. 1981. 18S defective interfering RNA of Semliki Forest virus contains a triplicated linear repeat. *Proc. Natl. Acad. Sci. USA* 78:5353–5357.

Lehtovaara, P., H. Söderlund, S. Keränen, R. F. Pet-tersson, and L. Kääriäinen. 1982. Extreme ends of the genome are conserved and rearranged in the defective interfering RNAs of Semliki Forest virus. *J. Mol. Biol.* 156:731–748.

Levis, R., H. Huang, and S. Schlesinger. 1987. Engi-neered defective interfering RNAs of Sindbis virus express bacterial chloramphenicol acetyltransferase in avian cells. *Proc. Natl. Acad. Sci. USA* 84:4811–4815.

Levis, R., S. Schlesinger, and H. Huang. 1990. Pro-moter for Sindbis virus RNA-dependent subge-nomic RNA transcription. *J. Virol.* 64:1726–1733.

Levis, R., B. G. Weiss, M. Tsiang, H. Huang, and S. Schlesinger. 1986. Deletion mapping of Sindbis virus DI RNAs derived from cDNAs defines the se-quences essential for replication and packaging. *Cell* 44:137–145.

Mann, R., R. C. Mulligan, and D. Baltimore. 1983. Construction of a retrovirus packaging mutant and its use to produce helper-free defective retrovirus. *Cell* 33:153–159.

Monroe, S. S., and S. Schlesinger. 1984. Common and

distinct regions of defective-interfering RNAs of Sindbis virus. *J. Virol.* **49**:865–872.

Rice, C. M., R. Levis, J. H. Strauss, and H. V. Huang. 1987. Production of infectious RNA transcripts from Sindbis virus cDNA clones: mapping of lethal mutations, rescue of a temperature-sensitive marker, and in vitro mutagenesis to generate defined mutants. *J. Virol.* **61**:3809–3819.

Sacher, R., R. French, and P. Ahlquist. 1988. Hybrid brome mosaic virus RNAs express and are packaged in tobacco mosaic virus coat protein in vivo. *Virology* **167**:15–24.

Sleat, D. E., P. C. Turner, J. T. Finch, P. J. G. Butler, and T. M. A. Wilson. 1986. Packaging of recombinant RNA molecules into pseudovirus particles directed by the origin-of-assembly sequence from tobacco mosaic virus RNA. *Virology* **155**:299–308.

Sorger, P. K., P. G. Stockley, and S. C. Harrison. 1986. Structure and assembly of turnip crinkle virus. II. Mechanism of reassembly in vitro. *J. Mol. Biol.* **191**:639–658.

Stohlman, S. A., R. S. Baric, G. N. Nelson, L. J. Soe, L. M. Welter, and R. J. Deans. 1988. Specific interaction between coronavirus leader RNA and nucleocapsid protein. *J. Virol.* **62**:4288–4295.

Strauss, E. G., C. M. Rice, and J. H. Strauss. 1984. Complete nucleotide sequence of the genomic RNA of Sindbis virus. *Virology* **133**:92–110.

Strauss, E. G., and J. H. Strauss. 1986. Structure and replication of the alphavirus genome, p. 35–90. *In* S. Schlesinger and M. J. Schlesinger (ed.), *The Togaviridae and Flaviviridae*. Plenum Publishing Corp., New York.

Turner, D. R., L. E. Joyce, and P. J. G. Butler. 1988. The tobacco mosaic virus assembly origin RNA. Functional characteristics defined by directed mutagenesis. *J. Mol. Biol.* **203**:531–547.

Weber, K., and W. Konigsberg. 1975. Proteins of the RNA phages, p. 51–84. *In* N. D. Zinder (ed.), *RNA Phages*. Cold Spring Harbor Laboratory, Cold Spring Harbor, N.Y.

Weiss, B., H. Nitschko, I. Ghattas, R. Wright, and S. Schlesinger. 1989. Evidence for specificity in the encapsidation of Sindbis RNAs. *J. Virol.* **63**:5310–5318.

Wengler, G. 1987. The mode of assembly of alphavirus cores implies a mechanism for the disassembly of the cores in the early stages of infection. *Arch. Virol.* **94**:1–14.

Wengler, G., U. Boege, G. Wengler, H. Bischoff, and K. Wahn. 1982. The core protein of the alphavirus Sindbis virus assembles into core-like nucleoproteins with the viral genome RNA and with other single-stranded nucleic acids in vitro. *Virology* **118**:401–410.

Wengler, G., G. Wengler, U. Boege, and K. Wahn. 1984. Establishment and analysis of a system which allows assembly and disassembly of alphavirus core-like particles under physiological conditions in vitro. *Virology* **132**:401–412.

Witherell, G. W., and O. C. Uhlenbeck. 1989. Specific RNA binding by Qβ coat protein. *Biochemistry* **28**:71–76.

Zuidema, D., M. F. A. Bierhuizen, B. J. C. Cornelissen, J. F. Bol, and E. M. J. Jaspars. 1983. Coat protein binding sites on RNA 1 of alfalfa mosaic virus. *Virology* **125**:361–369.

Chapter 35

Defined Mutations in the Poliovirus Capsid Proteins Cause Specific Defects in RNA Encapsidation, RNA Uncoating, and VP0 Cleavage

Karla Kirkegaard and Susan R. Compton

Our laboratory is interested in the RNA-protein interactions that accomplish the assembly of the poliovirus particle, maintain its structural and functional integrity, and allow its disassembly during cell entry.

The three-dimensional structures of picornaviruses, solved by X-ray crystallography (Rossmann et al., 1985; Hogle et al., 1985; Luo et al., 1987; Acharya et al., 1989), have given us vivid and detailed images of the capsid proteins as they form the icosahedral viral shell. These structures have revealed that the topologies of the capsid proteins, with the carboxy termini on the outside of the particle and the amino termini on the inside, are homologous with each other and with the icosahedral plant viruses. However, little is known of the physical disposition of the viral RNA within the capsid. Recently, planar density that probably results from two ordered and symmetrically placed nucleotides has been found near Trp-38 and Tyr-41 of poliovirus type 1 (Mahoney) VP1 (Filman et al., 1989); these are the only traces of the RNA that have been found thus far. Furthermore, several residues from the amino termini of VP1, VP2, and VP3 are unresolved in the structures; these are the amino acids that are presumably interacting most directly with the viral RNA (Hogle et al., 1985; Rossmann et al., 1985).

During entry into a cell, poliovirions undergo two empirically defined conformational changes, alteration and RNA release (Fig. 1). Detailed physical information about these conformational changes is not yet available, and the cellular location of these events remains a subject of controversy (Crowell and Landau, 1983; Madshus et al., 1984; Rueckert, 1985; Neubauer et al., 1987). Upon initiation of an infection, poliovirions bind reversibly to a specific receptor on the surface of the target cell. Once bound, the virions undergo the first conformational alteration; these altered particles are uninfectious and display a variety of physical changes (Crowell and Landau, 1983; Rueckert, 1985). It is this step that has been reported to be blocked by the antiviral compounds arildone (McSharry et al., 1979; Caliguiri et al., 1980) and Win 51711 (Zeichhardt et al., 1987) and by related compounds (Eggers, 1977; Ninomiya et al., 1984). The second conformational change leads to physical release of the RNA, as measured by the increased nuclease sensitivity of the viral RNA and the decreased light sensitivity of RNA that had been copackaged with chromophores (Crowell and Landau, 1983; Rueckert, 1985). The role of the RNA and the identities of the amino acids that participate in these processes are not yet known. The phenotypes of two of the poliovirus mutants discussed here, however, suggest that the amino-terminal amino acids of VP1 are involved in the release of the RNA during cell entry.

The assembly pathway of poliovirus undoubtedly involves several of the subviral particles that have been identified in infected cells (Fig. 2). The mechanism of RNA encapsidation may involve either the condensation of 12 14S pentamers around an RNA molecule or the threading of an RNA molecule into a 75S procapsid. If the former model is correct, then the 75S procapsids found in infected cells might result from reversible self-assembly of the 14S

Karla Kirkegaard and Susan R. Compton, Department of Molecular, Cellular and Developmental Biology, University of Colorado, Boulder, Colorado 80309.

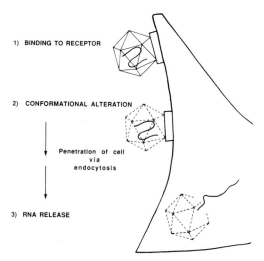

1) BINDING TO RECEPTOR

2) CONFORMATIONAL ALTERATION

Penetration of cell
via
endocytosis

3) RNA RELEASE

FIGURE 1. Conformational changes in poliovirions during receptor-mediated entry into cells.

pentamers in the absence of RNA, supplying a reservoir of 14S particles. The predicted product of either of these assembly pathways is an RNA-containing particle termed the provirion. Such a particle has been observed by some groups and not by others (Fernandez-Tomas and Baltimore, 1973; Guttman and Baltimore, 1977; Rueckert, 1985).

The maturation of the provirion to the final infectious virion requires the cleavage of VP0 to its products, VP2 and VP4. These products have been observed only in RNA-containing parti-

PARTICLE		SEDIMENTATION	COMPONENTS
P1		5S	VP0,VP1,VP3
PENTAMER		14S	5[VP0,VP1,VP3]
PROCAPSID		75S	60[VP0,VP1,VP3]
PROVIRION		125-150S	60[VP0,VP1,VP3] + RNA
VIRION		150S	60[VP4,VP2,VP3,VP1] + RNA

FIGURE 2. Viral and subviral particles observed in poliovirus-infected cells.

cles. Neither VP2 nor VP4 has been seen in infected cells, in 14S pentamers, or in procapsids (Rueckert, 1985). Thus, the maturation cleavage of VP0 appears to be dependent on RNA packaging. One of the surprises from the three-dimensional structures of poliovirus and rhinovirus was that the cleaved VP0 bond was buried within the virion, suggesting an autoproteolytic mechanism for its cleavage (Hogle et al., 1985; Rossmann et al., 1985). The RNA dependence of this autoproteolysis could arise in different ways: by the direct involvement of RNA bases in catalysis (Rossmann et al., 1985; Arnold et al., 1987), for example, or by a conformational change subsequent to RNA packaging that reveals a proteolytic active site composed exclusively of amino acids.

We would like to contribute to the understanding of the functional interactions between the capsid proteins and the viral RNA by the analysis of mutant viruses. Dynamic processes such as viral assembly and cell entry can be arrested at various stages by using mutants defective in individual steps in these processes, and thus the functional steps can be more clearly delineated. Here we describe three defined mutations in the poliovirus capsid proteins, the resulting phenotypes of the mutant viruses, and the associated biochemical defects in virion function.

ISOLATION AND CHARACTERIZATION OF POLIOVIRUS MUTANTS

To combine the advantages of site-directed mutagenesis with those of genetic screening, we identified and characterized mutants with conditional phenotypes after random deletion mutagenesis (Kirkegaard and Nelsen, 1990). The procedure is outlined in Fig. 3. After random small deletions were introduced into a population of infectious poliovirus cDNA-containing plasmids, the mutagenized population was transfected into mammalian cells. Individual plaques were picked and screened for temperature-sensitive, host range, or cold-sensitive phenotypes. Candidate mutants were isolated, and the deletions were mapped and reconstructed into otherwise wild-type poliovirus cDNA molecules by site-directed mutagenesis (Fig. 3).

For two of the mutant viruses discussed here, VP1-101 and VP1-102, the deletion introduced by mutagenesis was found to be responsible for the mutant phenotype. For one of the mutants, VP2-103, a deletion carried by the original isolate was found not to confer a mutant phenotype and to allow the production of phe-

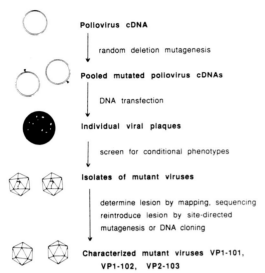

Poliovirus cDNA

↓ random deletion mutagenesis

Pooled mutated poliovirus cDNAs

↓ DNA transfection

Individual viral plaques

↓ screen for conditional phenotypes

Isolates of mutant viruses

↓ determine lesion by mapping, sequencing
reintroduce lesion by site-directed
mutagenesis or DNA cloning

Characterized mutant viruses VP1-101,
VP1-102, VP2-103

FIGURE 3. Strategy for mutagenesis of infectious poliovirus cDNA and selection of conditional poliovirus mutants bearing small deletions in their RNA genomes.

notypically wild-type virus. Instead, the temperature sensitivity of VP2-103 was caused by a single-nucleotide substitution that apparently occurred spontaneously during viral propagation. This finding underscores the importance of reconstructing all suspected lesions in RNA genomes to determine the causal relationship of a candidate mutation and a mutant phenotype, especially if any selection is used in isolation of the mutant.

SMALL DELETIONS IN THE AMINO TERMINUS OF VP1 AFFECT RNA RELEASE DURING CELL ENTRY

Deletion mutants VP1-101 and VP1-102 were identified by virtue of their host range phenotypes; both mutant viruses formed fewer and smaller plaques on CV-1 cells than on HeLa cells, compared with wild-type virus (Table 1). The deletions indicated in Table 1 were shown to be responsible for the host range and small-plaque phenotypes of the mutant viruses.

To discover the nature of the defects causing the mutant phenotypes of VP1-101 and VP1-102, RNA synthesis time courses were performed (Kirkegaard, 1990). For both viruses, viral RNA synthesis in CV-1 cells was normal at very high multiplicities of infection (MOI); at lower MOIs, however, RNA synthesis was markedly reduced on CV-1 cells. Similar exaggerated MOI dependence was observed in viral growth curves. A defect in cell entry in these mutants, lowering the effective MOI under nonpermissive conditions, was consistent with these observations. Cell entry defects caused by the mutated VP1 proteins were further substantiated by the disappearance of the plaque reduction phenotype on CV-1 cells when the mutant virus infections were initiated with RNA molecules, without associated capsid proteins (van der Werf et al., 1986; Sarnow, 1989; Kirkegaard, 1990).

We determined which step in cell entry (Fig. 1) was defective in VP1-101 and VP1-102 viruses by using several different assays. We found that both the initial binding to cellular receptors and the subsequent conformational alteration leading to loss of infectivity were unaffected in the mutants. However, the physical release of RNA from the mutant virions was affected in both mutants. This was determined by assaying wild-type and mutant viruses that had been grown in the presence of neutral red, a chromophore whose incorporation into the assembling virion renders the virus extremely light sensitive (Crowther and Melnick, 1962; Wilson and Cooper, 1963). As the virus enters the cell and the RNA is extruded from the virion, the dye also is released and can no longer inactivate the RNA upon irradiation. Thus, the time course of the loss of light sensitivity of neutral red-containing virus is also a time course of RNA release. For both VP1-101 and VP1-102, a pro-

TABLE 1
Molecular Lesions and Host Range Phenotypes of VP1-101 and VP1-102

Virus	Coding region mutated	Mutation (amino acids deleted)	Relative PFU at 39.5°C (plaque size)	
			HeLa cells	CV-1 cells
Wild type			1,000	150
VP1-101	VP1	8 and 9	1,000	26 (small)
VP1-102	VP1	1 through 4	1,000 (small)	11 (very small)

TABLE 2
Genotype and Temperature-Sensitive Phenotype of VP2-103

Virus	Coding region mutated	Mutation	Relative PFU on HeLa cells (plaque size)	
			32.5°C	39.5°C
Wild type			1,000	980
VP2-103	VP2	CGA to CAA Amino acid 76 Arg to Gln	1,000 (small)	2.5

nounced lag was observed in this process (Kirkegaard, 1990). Thus, both mutations in the amino terminus of VP1, while resulting in viable virus, lead to mutant viruses with delayed rates of RNA release during cell entry. It is possible that these amino acids, thought to be close to the RNA in the three-dimensional structure of the poliovirion, are physically as well as functionally involved in the process of RNA release.

RNA RELEASE AND RNA ENCAPSIDATION ARE GENETICALLY RELATED BUT SEPARABLE FUNCTIONS

In addition to the defect in RNA release during cell entry common to both VP1-101 and VP1-102, another biochemical alteration was detected in VP1-102-infected cells. Analysis of the subviral and viral particles (Fig. 2) synthesized during VP1-102 and wild-type infections revealed that the RNA-containing particles, the virion and provirion peaks, accumulated much more slowly in the mutant-infected cells. This apparently reduced rate of RNA packaging was observed even at extremely high multiplicities of VP1-102 infection that compensated for the defect in cell entry. Normal levels of protein synthesis, host cell shutoff, and 75S and 14S particle accumulation were observed in the mutant infections; only the RNA-containing particles were conspicuously depleted. This effect was more pronounced in CV-1 cells but was observed in HeLa cells as well. VP1-101 infections showed no diminution of RNA-containing particles in either cell type.

The deletion of the first four amino acids of VP1, therefore, affects the encapsidation of RNA into viral particles during assembly as well as its release from viral particles during cell entry. Deletion of amino acids 8 and 9, however, affects only RNA release. Thus, these two related processes are mediated, at least in part, by the same region of polypeptide, the amino terminus of VP1.

Both 75S and 14S particles accumulate in VP1-102-infected cells. As discussed in the introduction, it is not clear which of these particles interacts directly with the RNA in the encapsidation process. Perhaps the RNA-binding properties of the 75S and 14S particles from VP1-102-infected cells, as compared with those of wild-type subviral particles, will help to elucidate this point.

A POINT MUTATION IN VP2 INTERFERES WITH VP0 CLEAVAGE AND CAUSES PROVIRION ACCUMULATION

The mutant virus VP2-103 was isolated on the basis of its extremely temperature-sensitive phenotype (Table 2). The original isolate contained two mutations: a 25-nucleotide deletion in the 5' noncoding region and a single-nucleotide substitution in the VP2-coding region. By site-directed mutagenesis, it was found that the single-nucleotide substitution, which changes amino acid 76 from an arginine to a glutamine, was solely responsible for causing the temperature-sensitive phenotype (Compton et al., submitted).

The primary biochemical defect in VP2-103-infected cells at elevated temperatures appears to be the lack of cleavage of VP0. Cell entry, RNA synthesis, protein synthesis, host cell shutoff, and the synthesis of subviral particles are all comparable to those in wild-type infections. However, 100-fold-fewer infectious particles are made, and no detectable virion peak is observable in sucrose gradients. Instead, a species that sediments at approximately 125S is observed in VP2-103-infected cells. This particle has many properties expected and reported for the elusive provirion (Fig. 2): the particle is quite sensitive to both sodium dodecyl sulfate (SDS) and EDTA, and it contains viral RNA as well as VP0, VP1, and VP3. No VP0 cleavage products have been observed in VP2-103-infected cells at high temperatures, and reduced levels are observed at lower, permissive temperatures (Compton et al., submitted).

Amino acid 76 of VP2 is quite distant spatially from the cleaved VP2-VP4 bond in the three-dimensional structure of poliovirus. Nevertheless, a single-nucleotide substitution at this position specifically affects VP0 cleavage. Whatever the mechanism of VP0 autoproteolysis, we expect it to be quite sensitive to protein folding.

The very existence of provirions implies that VP0 cleavage can be uncoupled from RNA packaging and that RNA packaging is necessary but not sufficient for VP0 cleavage. Perhaps the provirions that can be isolated with increased biochemical purity from VP2-103-infected cells will be amenable to structural and biochemical analysis. These further studies will be directed at understanding the mechanism of VP0 autoproteolysis and the structural or catalytic role of the RNA in this reaction.

REFERENCES

Acharya, R., E. Fry, D. Stuart, G. Fox, D. Rowlands, and F. Brown. 1989. The three-dimensional structure of foot-and-mouth disease virus at 2.9 Å resolution. *Nature* (London) **337:**709–716.

Arnold, E., M. Luo, G. Vriend, M. G. Rossmann, A. C. Palmenberg, G. D. Parks, M. J. H. Nicklin, and E. Wimmer. 1987. Implications of the picornavirus capsid structure for polyprotein processing. *Proc. Natl. Acad. Sci. USA* **84:**21–25.

Caliguiri, L. A., J. J. McSharry, and G. W. Lawrence. 1980. Effect of arildone on modifications of polivirus *in vitro*. *Virology* **105:**86–93.

Compton, S. R., B. Nelsen, and K. Kirkegaard. A temperature-sensitive poliovirus mutant defective in virion maturation. Submitted for publication.

Crowell, R. L., and B. J. Landau. 1983. Receptors in the initiation of picornavirus infections. *Compr. Virol.* **18:**1–42.

Crowther, D., and J. L. Melnick. 1962. The incorporation of neutral red and acridine orange into developing poliovirus particles making them photosensitive. *Virology* **14:**11–21.

Eggers, H. J. 1977. Selective inhibition of uncoating of echovirus 12 by rhodanine. *Virology* **78:**241–252.

Fernandez-Tomas, C. B., and D. Baltimore. 1973. Morphogenesis of poliovirus. II. Demonstration of a new intermediate, the provirion. *J. Virol.* **12:**1181–1183.

Filman, D. J., R. Syed, M. Chow, A. J. Macadam, P. D. Minor, and J. M. Hogle. 1989. Structural factors that control conformational transitions and serotype specificity in type 3 poliovirus. *EMBO J.* **8:**1567–1579.

Guttman, N., and D. Baltimore. 1977. Morphogenesis of poliovirus. IV. Existence of particles sedimenting at 150S having the properties of provirions. *J. Virol.* **23:**363–367.

Hogle, J. M., M. Chow, and D. F. Filman. 1985. Three-dimensional structure of poliovirus at 2.9 Å resolution. *Science* **229:**1358–1365.

Kirkegaard, K. 1990. Mutations in VP1 of poliovirus specifically affect both encapsidation and release of viral RNA. *J. Virol.* **64:**195–206.

Kirkegaard, K., and B. Nelsen. 1990. Conditional poliovirus mutants made by random deletion mutagenesis of infectious cDNA. *J. Virol.* **64:**185–194.

Luo, M., G. Vriend, G. Kamer, I. Minor, E. Arnold, M. G. Rossmann, U. Boege, D. G. Scraba, G. M. Duke, and A. C. Palmenberg. 1987. The atomic structure of Mengo virus at 3.0 Å resolution. *Science* **35:**182–191.

Madshus, I. H., S. Olsnes, and K. Sandvig. 1984. Mechanism of entry into the cytosol of poliovirus type 1: requirement for low pH. *J. Cell Biol.* **98:**1194–1200.

McSharry, J. J., L. A. Caliguiri, and H. J. Eggers. 1979. Inhibition of uncoating of poliovirus by arildone, a new antiviral drug. *Virology* **97:**307–315.

Neubauer, C., L. Frasel, E. Kuechler, and D. Blaas. 1987. Mechanism of entry of human rhinovirus 2 into HeLa cells. *Virology* **158:**255–258.

Ninomiya, Y., C. Ohsawa, M. Aoyama, I. Umeda, Y. Suhara, and H. Ishitsuka. 1984. Antivirus agent, Ro 09-0410, binds to rhinovirus specifically and stabilizes the virus conformation. *Virology* **134:**269–276.

Rossmann, M. G., E. Arnold, J. W. Erickson, E. A. Frankenberger, J. P. Griffith, H. J. Hecht, J. E. Johnson, G. Kamer, M. Luo, A. G. Mosser, R. R. Rueckert, B. Sherry, and G. Vriend. 1985. Structure of a human common cold virus and functional relationship to other picornaviruses. *Nature* (London) **317:**145–153.

Rueckert, R. R. 1985. Picornaviruses and their replication, p. 357–390. *In* B. N. Field, D. M. Knipe, R. M. Chanock, J. L. Melnick, B. Roizman, and R. E. Shope (ed.), *Fundamental Virology*. Raven Press, New York.

Sarnow, P. 1989. Role of 3'-end sequences in infectivity of poliovirus transcripts made in vitro. *J. Virol.* **63:**467–470.

van der Werf, S., J. Bradley, E. Wimmer, F. Studier, and J. Dunn. 1986. Synthesis of infectious poliovirus RNA by purified T7 RNA polymerase. *Proc. Natl. Acad. Sci. USA* **83:**2330–2334.

Wilson, J. N., and P. D. Cooper. 1963. Aspects of the growth of poliovirus as revealed by the photodynamic effects of neutral red and acridine orange. *Virology* **21:**135–145.

Zeichhardt, H., M. J. Otto, M. A. McKinlay, P. Willingmann, and K. O. Habermehl. 1987. Inhibition of poliovirus uncoating by disoxaril (WIN 51711). *Virology* **160:**281–285.

VI. VIRAL RECEPTORS, UPTAKE, AND DISASSEMBLY

Chapter 36

Characterization and Expression of the Receptor Glycoprotein for Mouse Hepatitis Virus

Kathryn V. Holmes, Susan R. Compton, and Richard K. Williams

Mouse hepatitis virus (MHV) is a common murine pathogen. Many serologically distinguishable strains of MHV cause inapparent enteric infection, infantile diarrhea, hepatitis, respiratory infection, and acute and chronic demyelinating neurological diseases (Hirano et al., 1975; Wege et al., 1981b; Talbot and Buchmeier, 1985). Mouse strains differ markedly in their susceptibility to MHV, and the degree of susceptibility of a mouse strain correlates well with the degree of susceptibility to MHV infection of peritoneal macrophages isolated from that strain (Bang and Warwick, 1960; Stohlman et al., 1980; Dupuy et al., 1984; Knobler et al., 1984; Smith et al., 1984; Dindzans et al., 1986). The studies described below were undertaken to investigate the mechanism for the marked species specificity, mouse strain specificity, and tissue specificity of infections with MHV.

IDENTIFICATION OF A 100-KILODALTON (110K) GLYCOPROTEIN RECEPTOR FOR MHV

To study directly the earliest interaction between MHV and its natural target tissues in the mouse, we used a simple and sensitive solid-phase virus-binding assay (Boyle et al., 1987). Membranes purified from the intestinal brush border or hepatocytes of BALB/c mice, which are susceptible to MHV infection, were bound to nitrocellulose sheets in a dot blot apparatus. MHV was allowed to bind, and bound virus was detected with antibody directed against the E2 peplomer glycoprotein of MHV, followed by radioiodinated staphylococcal protein A. The A59

strain of MHV (MHV-A59) bound to both intestinal and liver membrane preparations, indicating that these membranes from BALB/c mice contain a receptor for the virus. In contrast, membranes from SJL/J mice, which are highly resistant to MHV-A59, failed to bind virus. This finding suggested that SJL/J mice may be resistant to MHV because they fail to express a receptor for MHV, express an altered molecule that lacks MHV-binding activity, or express too little receptor to be detectable.

The availability of a receptor-deficient mouse strain, SJL/J, served as an important negative control for subsequent studies on the interaction of the E2 glycoprotein of MHV with receptors on intestine and liver cell membranes from susceptible BALB/c mice. Treatment of membranes with neuraminidase or endoglycosidase F before incubation with virus did not prevent virus attachment, but treatment of membranes with several different proteases destroyed virus-binding activity. These experiments suggested that MHV might be binding to a protein. This hypothesis was confirmed with the use of a virus overlay protein blot assay that detects virus attachment to membrane proteins separated by sodium dodecyl sulfate-polyacrylamide gel electrophoresis and blotted to nitrocellulose. MHV-A59 bound to a single broad band of 100 to 110 kilodaltons (kDa) in BALB/c liver and intestine membranes but did not bind to any SJL/J membrane components (Boyle et al., 1987). The 110K MHV receptor was shown to be a glycoprotein by quantitatively recovering it from solubilized membranes with lectin affinity chromatography (Holmes et al., 1987).

Kathryn V. Holmes and Richard K. Williams, Department of Pathology, Uniformed Services University of the Health Sciences, Bethesda, Maryland 20814. *Susan R. Compton,* Department of Molecular, Cellular and Developmental Biology, University of Colorado, Boulder, Colorado 80309.

BIOLOGICAL EFFECTS OF ANTIRECEPTOR ANTIBODIES

Polyclonal antibodies directed against the MHV receptor were prepared by immunizing receptor-deficient SJL/J mice with intestinal brush border members of receptor-positive BALB/c mice. Surprisingly, this antibody recognized only a single 100K-to-110K protein in immunoblots of BALB/c brush border membranes. Hybridomas prepared from the spleens of these animals were screened for reactivity with the 110K MHV receptor eluted from a sodium dodecyl sulfate-polyacrylamide gel, and more than 15 hybridomas producing monoclonal antibodies (MAbs) directed against the receptor were identified.

One of these receptor-specific MAbs, CC1, was found to block infection of mouse fibroblasts in vitro with MHV-A59 when virus was incubated with cells that had been pretreated for 1 h with MAb. This MAb also blocked infection of mouse fibroblasts with four other strains of MHV (A. Smith and K. V. Holmes, manuscript in preparation). These experiments suggested that MAb CC1 binds to the MHV receptor glycoprotein at a site at or very near the site where MHV E2 attaches. Immunoblots of BALB/c membrane proteins with MAb CC1 showed that CC1 recognized the 100- to 110-kDa band to which virus bound. In addition, MAb CC1 recognized a smaller, approximately 58K protein in membranes from susceptible mice. The 58K protein was only very poorly recognized by MHV virions; therefore, the MAb and virus binding domains on the MHV receptor are not identical.

The relative levels of expression of the MHV receptor glycoprotein on different tissues of MHV-susceptible BALB/c mice were determined by binding of radioiodinated CC1 to membranes prepared from these tissues. Expression of the MHV receptor glycoprotein was greatest in the large and small intestines and liver of BALB/c mice, which are major natural target organs for MHV. Thus, tissue tropism of MHV may be determined at least in part by expression of the 110K receptor glycoprotein.

BIOCHEMICAL CHARACTERISTICS OF THE 110K MHV RECEPTOR GLYCOPROTEIN

The MHV receptor was purified from detergent-solubilized preparations of liver membranes from MHV-susceptible mice by affinity chromatography with MAb CC1, followed by preparative sodium dodecyl sulfate-polyacrylamide gel electrophoresis. The sequence of approximately 15 amino acids from the amino terminus of the receptor molecule was determined by microsequencing and used to prepare both synthetic peptides to this domain and oligonucleotides specific for the deduced nucleotide sequence of this region. These reagents specific for the amino terminus of the receptor protein, together with MAb CC1 and polyclonal antireceptor antibody, are being used to identify, in lambda gt11 libraries of cDNA from susceptible mouse tissues, clones that express the MHV receptor determinant.

The biochemical characteristics of the MHV receptor glycoprotein from BALB/c liver and intestine are as follows. The molecule is a 100K-to-110K glycoprotein with approximately 35% N-linked oligosaccharide. It contains neuraminic acid and binds concanavalin A, wheat germ agglutinin, and ricin. The molecule is highly acidic (pI < 4.5) and membrane bound in intestinal brush border, hepatocytes, mouse fibroblasts, and macrophages, and its virus binding is not affected by reduction or detergents. The biological function(s) of this molecule has not yet been identified.

Because MAb CC1 is not toxic for mouse cell cultures and because receptor-deficient SJL/J mice function normally, it appears likely that the epitope of the MHV receptor recognized by MAb CC1 is not essential for cell growth in vitro or in vivo. When antibody directed against the 15-amino-acid sequence at the amino terminus of the MHV receptor glycoprotein was used to probe immunoblots of SJL/J intestine membranes, a protein approximately 5 kDa smaller than the 110K receptor from BALB/c mice and a protein approximately 5 kDa smaller than the 58-kDa band from BALB/c mice were detected. This observation suggested that SJL/J mice may fail to express the MHV-binding and CC1-binding domains of the glycoprotein which serves as the MHV receptor in BALB/c mice but may express a homologous protein that shares at least some common sequences at the amino end. In part, the presence of such a homologous receptor variant may explain why SLJ/J mice failed to recognize as foreign many other domains of the receptor glycoprotein from BALB/c mice.

COMPARISON OF RECEPTORS FOR CORONAVIRUSES OF DIFFERENT SPECIES

MHV is highly species specific, causing natural infections only in mice, and infecting rats only by unnatural (e.g., intracerebral) routes, at a very young age and with a highly neurotropic virus strain (Hirano et al., 1980; Sorensen et al., 1980; Wege et al., 1981a; Wege et al., 1982; Wege et al., 1984; Sorensen and Dales, 1985; Watanabe et al., 1987). To determine whether this strong species specificity was determined by availability of the MHV-binding domain of the 110K glycoprotein or its homologs, we compared binding of MHV-A59 to intestinal brush border membranes from many species of domestic animals, each of which serves as the natural host for one or two coronaviruses specific for that species (Compton, 1988). MHV bound only to membranes from murine intestine, not to membranes of any other species. This finding strongly supports the role of receptors in determining the species specificity of coronaviruses.

Solid-phase virus-binding assays were developed for coronaviruses from several species, including pig, dog, cat, and human, and used to compare the receptor specificities of these coronaviruses. There is strong cross-reactivity among viruses of all four of these species at the level of coronavirus receptors (Compton, 1988). Thus, species specificity for the coronaviruses of these animals may be determined at least in part by a factor or factors subsequent to the initial virus attachment phase. It appears possible that some coronaviruses share receptor determinants, whereas other coronaviruses recognize unique domains on receptors that are quite species specific. Biochemical characterization of the coronavirus receptors of these species and comparison with the MHV receptor may reveal how members of one virus family have evolved toward specificity for disparate host species.

ACKNOWLEDGMENTS. We are grateful for the excellent technical assistance of Christine Cardellechio, Susan Wetherell, and Patrick Elia.

This work was supported by Public Health Service grants AI 18997, AI 25231 (to K.V.H.), and NRSA F32-AI07810 from the National Institutes of Health to R. K. Williams.

REFERENCES

Bang, F. B., and A. Warwick. 1960. Mouse macrophages as host cells for the mouse hepatitis virus and the genetic basis of their susceptibility. *Proc. Natl. Acad. Sci. USA* **46**:1065–1075.

Boyle, J. F., D. G. Weismiller, and K. V. Holmes. 1987. Genetic resistance to mouse hepatitis virus correlates with absence of virus-binding activity on target tissues. *J. Virol.* **61**:185–189.

Compton, S. R. 1988. Coronavirus attachment and replication. Ph.D. dissertation, The Uniformed Services University of the Health Sciences, Bethesda, Md.

Dindzans, V. J., E. Skamene, and G. A. Levy. 1986. Susceptibility/resistance to mouse hepatitis virus strain 3 and macrophage procoagulant activity are genetically linked and controlled by two non-H2-linked genes. *J. Immunol.* **137**:2355–2360.

Dupuy, J. M., C. Dupuy, and D. Decarie. 1984. Genetically determined resistance to mouse hepatitis virus 3 is expressed in hematopoietic donor cells in radiation chimeras. *J. Immunol.* **133**:1609–1613.

Hirano, N., N. Goto, T. Ogawa, K. Ono, T. Murakami, and K. Fujiwara. 1980. Hydrocephalus in suckling rats infected intracerebrally with mouse hepatitis virus, MHV-A59. *Microbiol. Immunol.* **24**:825–834.

Hirano, N., S. Takenaka, and K. Fujiwara. 1975. Pathogenicity of mouse hepatitis virus for mice depending upon host age and route of infection. *Jpn. J. Exp. Med.* **45**:285–292.

Holmes, K. V., J. F. Boyle, D. G. Weismiller, S. R. Compton, R. K. Williams, C. B. Stephensen, and M. F. Frana. 1987. Identification of a receptor for mouse hepatitis virus. *Adv. Exp. Med. Biol.* **218**:197–202.

Knobler, R. L., L. A. Tunison, and M. B. Oldstone. 1984. Host genetic control of mouse hepatitis virus type 4 JHM strain replication. I. Restriction of virus amplification and spread in macrophages from resistant mice. *J. Gen. Virol.* **65**:1543–1548.

Smith, M. S., R. E. Click, and P. G. Plagemann. 1984. Control of mouse hepatitis virus replication in macrophages by a recessive gene on chromosome 7. *J. Immunol.* **133**:428–432.

Sorensen, O., and S. Dales. 1985. In vivo and in vitro models of demyelinating disease: JHM virus in the rat central nervous system localized by in situ cDNA hybridization and immunofluorescent microscopy. *J. Virol.* **56**:434–438.

Sorensen, O., D. Perry, and S. Dales. 1980. In vivo and in vitro models of demyelinating diseases. III. JHM virus infection of rats. *Arch. Neurol.* **37**:478–484.

Stohlman, S. A., J. A. Frelinger, and L. P. Weiner. 1980. Resistance to fatal central nervous system disease by mouse hepatitis virus, strain JHM. II. Adherent cell-mediated protection. *J. Immunol.* **124**:1733–1739.

Talbot, P. J., and M. J. Buchmeier. 1985. Antigenic variation among murine coronaviruses: evidence for polymorphism on the peplomer glycoprotein, E2. *Virus Res.* **2**:317–328.

Watanabe, R., H. Wege, and V. ter Meulen. 1987. Comparative analysis of coronavirus JHM-induced demyelinating encephalomyelitis in Lewis and Brown Norway rats. *Lab. Invest.* **57**:375–384.

Wege, H., M. Koga, and V. ter Meulen. 1981a. JHM

infections in rats as a model for acute and subacute demyelinating disease. *Adv. Exp. Med. Biol.* **142:** 327–340.

Wege, H., S. Siddell, M. Sturm, and V. ter Meulen. 1981b. Characterisation of viral RNA in cells infected with the murine coronavirus JHM. *Adv. Exp. Med. Biol.* **142:**91–101.

Wege, H., S. Siddell, and V. ter Meulen. 1982. The biology and pathogenesis of coronaviruses. *Curr. Top. Microbiol. Immunol.* **99:**165–200.

Wege, H., R. Watanabe, and V. ter Meulen. 1984. Relapsing subacute demyelinating encephalomyelitis in rats during the course of coronavirus JHM infection. *J. Neuroimmunol.* **6:**325–336.

Chapter 37

The Major-Group Rhinoviruses Utilize the Intercellular Adhesion Molecule 1 Ligand as a Cellular Receptor during Infection

Richard J. Colonno, Robert L. LaFemina, Corrille M. DeWitt, and Joanne E. Tomassini

Human rhinoviruses (HRVs), members of the family *Picornaviridae*, are the major causative agent of one of the more elusive diseases known, the common cold (Gwaltney, 1982). It had been shown in earlier studies that HRVs initiate infection by binding to one of two different receptors on susceptible cells. Of the 101 serotypes tested, 91 competed for a single cellular receptor and are designated the major group; the remaining 10 serotypes competed for a second receptor and constitute the minor group (Abraham and Colonno, 1984; Colonno, 1987; unpublished data). An antireceptor monoclonal antibody (MAb 1A6) was previously isolated and shown to specifically block attachment of the HRV major group (Colonno et al., 1986). The major-group viruses exhibit an absolute requirement for this receptor, since blocking attachment with MAb 1A6 was cytoprotective despite high-titer viral challenge (Colonno et al., 1986). In vivo chimpanzee and human clinical trials involving MAb 1A6 have also demonstrated that this receptor is involved in HRV infection of the nasal cavity (Colonno et al., 1987; Hayden et al., 1988).

Utilizing MAb 1A6 immunoaffinity chromatography, a 90-kilodalton (kDa) glycoprotein was isolated from HeLa cell membranes and subsequently shown to be the receptor protein required for attachment of the HRV major-group viruses to susceptible cells (Tomassini and Colonno, 1986). Recent biochemical characterization of the purified 90-kDa receptor protein determined that it was an acidic glycoprotein having a pI of 4.2 (Tomassini et al., 1989b). Carbohydrates accounted for 30% of the molecular mass of the protein, and seven N-linked glycosylation sites were predicted on the basis of partial digestion with *N*-glycanase (Tomassini et al., 1989b). The studies described below represent additional molecular and biochemical studies that were undertaken to elucidate the normal cellular function of this receptor and to explore the relationship, if any, that exists between the major-group receptor and those of other picornaviruses.

CLONING OF THE MAJOR-GROUP RECEPTOR

In light of the clinical importance of HRV receptors, studies were initiated to identify the major-group receptor by isolating sufficient amounts of receptor protein for direct sequencing and by cloning the gene that encoded the receptor. Attempts to sequence the intact, affinity-purified, 90-kDa protein were unsuccessful and indicated that its N terminus was apparently blocked. Subsequently, chemical and enzymatic digestions of the isolated receptor protein yielded six peptides that gave discernible sequence ranging from 5 to 23 amino acid residues and cumulatively representing 84 residues (Tomassini et al., 1989a). By using degenerate oligonucleotides, representing deduced peptide sequence as primers and probes, several lambda gt11 cDNA libraries were constructed and screened. Four overlapping clones were identified and ligated to generate a single clone (pHRVr1) containing a 3-kilobase (kb) insert (Tomassini et al., 1989a). The full-length clone

Richard J. Colonno, Robert L. LaFemina, Corrille M. DeWitt, and Joanne E. Tomassini, Department of Virus and Cell Biology, Merck Sharp and Dohme Research Laboratories, West Point, Pennsylvania 19486.

had a single large open reading frame initiating at nucleotide 72 that encoded 532 amino acids and contained a 1,333-nucleotide 3' noncoding region. In vitro translation of RNA transcribed from pHRVr1 resulted in a single polypeptide of 55 kDa, a value that is in close agreement to the 54- to 60-kDa size found for deglycosylated receptor protein (Tomassini et al., 1989a; Tomassini et al., 1989b).

DNA sequencing of pHRVr1 revealed a striking homology to the intercellular adhesion molecule 1 (ICAM-1) (Tomassini et al., 1989a). Comparison of the two sequences showed only 6 nucleotide differences from the previously published sequence of ICAM-1 (Simmons et al., 1988; Staunton et al., 1988) and extension of the 5'-terminal sequence by 14 nucleotides. The identity of ICAM-1 as the HRV major-group receptor has also been reported by other laboratories (Greve et al., 1989; Staunton et al., 1989). ICAM-1 is a cell surface ligand for the lymphocyte function-associated antigen 1 adhesion receptor (Makgoba et al., 1988) and exists as a single-chain, 76- to 114-kDa glycoprotein with a polypeptide core of 55 kDa. The ICAM-1 ligand is a member of the immunoglobulin supergene family and is predicted to have five homologous immunoglobulinlike domains defined by amino acids 1 to 88, 89 to 185, 186 to 284, 285 to 385, and 386 to 453 (Staunton et al., 1988). In addition, some sequence homology exists between ICAM-1 and other adhesion proteins of the adult nervous system, namely, neural cell adhesion molecule and myelin-associated glycoprotein (Simmons et al., 1988; Staunton et al., 1988). The interaction of ICAM-1 and lymphocyte function-associated antigen 1 plays an important role in leukocyte adhesion and in the execution of immunological and inflammatory functions mediated by leukocyte adhesion (Dustin et al., 1988). In addition to the virtually identical sequence homology, the HRV major-group receptor and ICAM-1 ligand proteins have equivalent mass, tissue distribution, and carbohydrate moieties (Dustin et al., 1988; Staunton et al., 1988; Tomassini and Colonno, 1986; Tomassini et al., 1989b).

HRV REPLICATION IN VERO CELLS

To confirm that the ICAM-1 receptor was the functional receptor for the major group of HRVs, stable Vero cells were generated by transfection of an ICAM-1 plasmid containing an upstream simian virus 40 early promoter (pSVL-HRVr1), followed by neomycin selection. The

TABLE 1
HRV Infection of Vero ICAM$^+$ Cells^a

HRV serotype	Titer^b (PFU, 10^4)	
	−Antibody	+Antibody
3	990	2
5	190	3
9	34	3
11	35	4
12	92	4
17	350	4
36	32	1
39	140	5
41	25	6
51	4	5

^{a} Confluent cell monolayers in a 48-well cluster plate were infected with the indicated serotypes at a multiplicity of infection of 0.5 as previously described (Abraham and Colonno, 1984). Virus was removed after a 1-h absorption period, and monolayers were washed five times with warm medium before the replacement of 0.1 ml of medium with or without added antibody.

^{b} Plaque assays were performed as previously described (Abraham and Colonno, 1984).

surviving cells were further enriched by fluorescent cell sorting, using MAb 1A6 to yield a population of Vero cells, of which >90% expressed the ICAM-1 ligand. Radiolabeled HRV-36, a major-group virus, showed 24.9% binding to the selected Vero ICAM$^+$ cells, in contrast to only the 3.2% binding observed when untransfected Vero cells were used. The binding to Vero ICAM$^+$ cells was specific, since addition of MAb 1A6 reduced binding of HRV-36 to 2.6%. However, attempts to propagate HRV-36 on Vero ICAM$^+$ cells indicated that HRV-36 grew poorly in Vero ICAM$^+$ cells, suggesting an intracellular block to HRV replication.

To determine whether HRV-36 was unique in its inability to grow in Vero cells, nine additional HRV major-group viruses were selected at random and used to infect Vero ICAM$^+$ cells in the presence or absence of MAb 1A6. Examination of infected monolayers after a 24-h incubation showed that five of the serotypes, HRV-3, -5, -17, and -39, showed extensive cytopathic effects (CPE), whereas HRV-9, -11, -12, -36, and -41 showed some CPE and HRV-51 failed to show any CPE. The supernatants from each infected culture were removed, and progeny virus was titered by plaque assay to quantitate virus growth. Results (Table 1) correlated well with the levels of CPE observed. The HRV serotypes tested showed a wide range of viral yields that increased as much as 495-fold in the case of HRV-3, down to no increase over con-

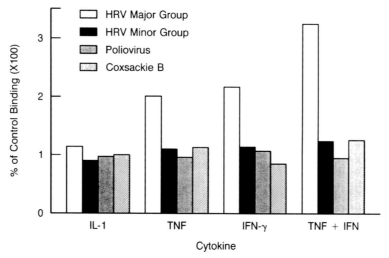

FIGURE 1. Cytokine induction of picornavirus receptors. Confluent HeLa cell monolayers in duplicate 48-well cluster plates were untreated (control) or treated with interleukin-1 (IL-1) at 50 U/ml, TNF at 10 ng/ml, IFN-γ at 2,000 U/ml, or a combination of TNF and IFN-γ in a total volume of 0.2 ml for 24 h. After incubation, cells were washed with fresh medium and used in binding studies with [^{35}S]methionine-labeled HRV-15 (major group), HRV-2 (minor group), poliovirus type 1, or coxsackievirus B1 as previously described (Abraham and Colonno, 1984).

trol levels with HRV-51. The background levels in these experiments were repeatedly high (10^4) and remain unexplained. Infection of Vero ICAM$^-$ cells gave nearly identical background titers in the absence of any observable CPE (data not shown). Taken together, these results suggest that HRVs bound nonspecifically to these cultures despite extensive washing of infected monolayers. This phenomenon correlates well with the 2 to 3% background binding levels observed with radiolabeled HRV-36 in the binding studies described above. Interestingly, the nonspecific binding to cell membranes does not lead to infection of cells.

The variations in progeny virus titers suggest that there are undefined differences among HRV serotypes that allow some of these viruses to replicate in receptor-positive Vero cells more efficiently than other serotypes. A similar intracellular inhibition was observed in the infection of receptor-positive L cells with HRV-2, a minor-group serotype (Yin and Lomax, 1983). These studies demonstrated that HRV-2 required a modification in the P2 nonstructural region before it could replicate in mouse cells. These differences are most likely due to a host cell factor(s), since all 10 of these serotypes grow equally well on HeLa cell monolayers. However, it should be pointed out that virus yields per cell in HeLa cells are about 10 times higher than the best titers observed in the Vero ICAM$^+$ cells (data not shown).

CYTOKINE INDUCTION OF PICORNAVIRUS RECEPTORS

Previous studies have indicated that the ICAM-1 ligand can be induced by several cytokines (Dustin et al., 1988). This induction is variable and dependent on the cells used and cytokines selected. We recently assayed the effects of six cytokines on HeLa cells and demonstrated that treatment with gamma interferon (IFN-γ) and tumor necrosis factor alpha (TNF) resulted in a twofold enhancement in the binding of the major group after a 24-h treatment (Tomassini et al., 1989a). Furthermore, a 3.3-fold enhancement could be demonstrated when IFN-γ and TNF were used together. We extended these studies to include HRV-2 (minor-group serotype), poliovirus, and coxsackievirus B1, since each of these receptors has been mapped to human chromosome 19 (Couillin et al., 1987; Greve et al., 1989; Siddique et al., 1988; D. Blass, personal communication) and because it has been previously postulated that the receptor molecules used by these viruses may share some structural homology (Colonno et al., 1988). HeLa cell monolayers were treated with cytokines for 24 h before virus-binding assays. The degree of binding obtained with each of the other three viruses was little changed, and no enhancement comparable to that observed with the major-group serotype, HRV-15, could be dem-

TABLE 2
Picornavirus Receptors and Attachment Sites

Group	Receptor family	Virion structure	Cellular receptor
Rhinoviruses	HRV major	Canyon	ICAM-1
	HRV minor	Canyon?	120 kDa
Enteroviruses	Poliovirus	Canyon	100 kDa
	Coxsackievirus B	Canyon?	49.5 kDa
Cardioviruses	Mengovirus	Pits	?
Aphthoviruses	FMDV	No crevice	RGD-binding protein

onstrated (Fig. 1). Similar treatment with alpha interferon, interleukin-2, and *Escherichia coli* lipopolysaccharide B also failed to enhance the level of binding of the other viruses treated (data not shown). Although these results clearly show that HeLa cell receptors for these other viruses were not inducible, they do not rule out the possibility that the receptors for poliovirus, group B coxsackieviruses, and minor-group HRVs may be inducible in other cell systems that are more responsive to cytokine treatment.

RECEPTORS AND VIRION ATTACHMENT SITES OF PICORNAVIRUSES

Recently, a cDNA clone encoding the poliovirus receptor has also been reported (Mendelsohn et al., 1989). Similar to ICAM-1, the predicted poliovirus receptor protein appears to be a member of the immunoglobulin supergene family, with three homologous immunoglobulin-like domains and some sequence homology to neural cell adhesion molecule (Mendelsohn et al., 1989). The structural conservation observed between the poliovirus and HRV major-group receptors is consistent with the finding that the deepest regions of the HRV-14 canyon structure are involved in receptor interaction (Colonno et al., 1988). The finding that this region of the canyon is highly conserved among a number of picornaviruses belonging to different receptor families (Blass et al., 1987; Rossmann and Palmenberg, 1988; Hogle et al., 1985) led to the supposition that the receptor molecules that interact with these different viruses have a structural binding domain in common (Colonno et al., 1988). A summary of the existing knowledge regarding all of the picornavirus receptor families and virion attachment sites is shown in Table 2. In contrast to the HRVs and enteroviruses, the cardioviruses and alphaviruses have different virion attachment sites, since a canyon is not visualized in the foot-and-mouth disease virus (FMDV) structure (Acharya et al., 1989) and

is reduced to pits in the mengovirus structure (Luo et al., 1987). Although the atomic structures of the group B coxsackieviruses and encephalomyocarditis viruses have yet to be solved, sequence alignments suggest virion attachment sites composed of a canyon and pit, respectively.

Only the major-group HRV and poliovirus receptors have been cloned to date. However, several putative receptor proteins have been identified for other picornaviruses. The minor-group receptor has been tentatively identified by viral blots as a 120-kDa protein (Mischak et al., 1988; unpublished result). Whereas the poliovirus receptor has been cloned and predicts a core polypeptide of 43 to 45 kDa, the mature receptor protein has yet to be definitively identified. Studies using an antireceptor MAb that blocks poliovirus attachment have enabled the isolation of a putative 100-kDa receptor protein by immunoaffinity chromatography (Shepley et al., 1988). A putative receptor protein for the group B coxsackieviruses has been identified by copurification with virions and has a molecular size of 49 kDa (Mapoles et al., 1985). The receptor(s) utilized by the aphthoviruses remains unidentified. However, the finding that virtually all of the FMDV serotypes have conserved RGD sequences exposed on the surface of virions and the recent demonstration that RGD peptides block virion attachment (Fox et al., 1989) suggest that the FMDV receptor may be an RGD-binding protein. No information is available on the receptors utilized by the cardioviruses.

Clearly, much work remains to be done before we fully understand the functional and structural relationships of the cell surface receptors employed by picornaviruses.

REFERENCES

Abraham, G., and R. J. Colonno. 1984. Many rhinovirus serotypes share the same cellular receptor. *J. Virol.* **51**:340–345.

Acharya, R., E. Fry, D. Stuart, G. Fox, D. Rowlands, and F. Brown. 1989. The three-dimensional structure of foot-and-mouth disease virus at 2.9 Å resolution. *Nature* (London) **337:**709–716.

Blass, D., E. Kuechler, G. Vriend, E. Arnold, M. Luo, and M. G. Rossmann. 1987. Comparison of the three-dimensional structure of two human rhinoviruses (HRV2 and HRV14). *Proteins* **2:**263–272.

Colonno, R. J. 1987. Cell surface receptors for picornaviruses. *BioEssays* **5:**270–274.

Colonno, R. J., P. L. Callahan, and W. J. Long. 1986. Isolation of a monoclonal antibody that blocks attachment of the major group of human rhinoviruses. *J. Virol.* **57:**7–12.

Colonno, R. J., J. H. Condra, S. Mizutani, P. L. Callahan, M. E. Davies, and M. A. Murcko. 1988. Evidence for the direct involvement of the rhinovirus canyon in receptor binding. *Proc. Natl. Acad. Sci. USA* **85:**5449–5453.

Colonno, R. J., J. E. Tomassini, and P. L. Callahan. 1987. Isolation and characterization of a monoclonal antibody which blocks attachment of human rhinoviruses, p. 93–102. *In* M. A. Brinton and R. R. Rueckert (ed.), *Positive Strand RNA Viruses*. Alan R. Liss, Inc., New York.

Couillin, P., F. Huyghe, M. C. Grisard, N. Van Cong, M. F. Louis, H. Hofmann-Radvanyl, and A. Boue. 1987. Echovirus 6, 11, 19; coxsackie B3 sensitivities and poliovirus I, II, III sensitivities are on chromosome 19 respectively on 19pter-q133 and 19q131-q133. *Cytogenet. Cell Genet.* **46:**599.

Dustin, M. L., D. E. Staunton, and T. A. Springer. 1988. Supergene families meet in the immune system. *Immunol. Today* **9:**213–215.

Fox, G., N. R. Parry, P. V. Barnett, D. McGinn, D. J. Rowlands, and F. Brown. 1989. The cell attachment site on foot-and-mouth disease virus includes the amino acid sequence RGD (arginine-glycine-aspartic acid). *J. Gen. Virol.* **70:**625–637.

Greve, J. M., G. Davis, A. M. Meyer, C. P. Forte, S. C. Yost, C. W. Marlor, M. E. Kamarck, and A. McClelland. 1989. The major human rhinovirus receptor is ICAM-1. *Cell* **56:**839–847.

Gwaltney, J. M., Jr. 1982. Rhinoviruses, p. 491–517. *In* E. A. Evans (ed.), *Viral Infection of Man: Epidemiology and Control*, 2nd ed. Plenum Publishing Corp., New York.

Hayden, F. G., J. M. Gwaltney, Jr., and R. J. Colonno. 1988. Modification of experimental rhinovirus colds by receptor blockade. *Antiviral Res.* **9:**233–247.

Hogle, J. M., M. Chow, and D. J. Filman. 1985. Three-dimensional structure of poliovirus at 2.9 Å resolution. *Science* **229:**1358–1367.

Luo, M., G. Vriend, G. Kamer, I. Minor, E. Arnold, M. G. Rossmann, U. Boege, D. G. Scraba, G. M. Duke, and A. C. Palmenberg. 1987. The atomic structure of Mengo virus at 3.0 Å resolution. *Science* **235:**182–191.

Makgoba, M. W., M. E. Sanders, G. E. Ginther Luce, E. A. Gugel, M. L. Dustin, T. A. Springer, and S. Shaw. 1988. Functional evidence that intercellular adhesion molecule-1 (ICAM-1) is a ligand for LFA-1-dependent adhesion in T cell-mediated cytotoxicity. *Eur. J. Immunol.* **18:**637–640.

Mapoles, J. E., D. L. Krah, and R. L. Crowell. 1985. Purification of a HeLa cell receptor protein for group B coxsackieviruses. *J. Virol.* **55:**560–566.

Mendelsohn, C. L., E. Wimmer, and V. Racaniello. 1989. Cellular receptor for poliovirus: molecular cloning, nucleotide sequence, and expression of a new member of the immunoglobulin superfamily. *Cell* **56:**855–865.

Mischak, H., C. Neubauer, B. Berthold, E. Kuechler, and D. Blaas. 1988. Detection of the human rhinovirus minor group receptor on renaturing Western blots. *J. Gen. Virol.* **69:**2653–2656.

Rossmann, M. G., and A. C. Palmenberg. 1988. Conservation of the putative receptor attachment site in picornaviruses. *Virology* **164:**373–382.

Shepley, M. P., B. Sherry, and H. L. Weiner. 1988. Monoclonal antibody identification of a 100-kDa membrane protein in HeLa cells and human spinal cord involved in poliovirus attachment. *Proc. Natl. Acad. Sci. USA* **85:**7743–7747.

Siddique, T., R. McKinney, W. Hung, R. J. Bartlett, G. Bruns, T. K. Mohandas, H. Ropers, C. Wilfert, and A. D. Poses. 1988. The poliovirus sensitivity (PVS) gene is on chromosome 19q12-q13.2. *Genomics* **3:**156–160.

Simmons, D., M. W. Makgoba, and B. Seed. 1988. ICAM, an adhesion ligand of LFA-1, is homologous to the neural cell adhesion molecule NCAM. *Nature* (London) **331:**624–627.

Staunton, D. E., S. D. Marlin, C. Stratowa, M. L. Dustin, and T. A. Springer. 1988. Primary structure of ICAM-1 demonstrates interaction between members of the immunoglobulin and integrin supergene families. *Cell* **52:**925–933.

Staunton, D. E., V. J. Merluzzi, R. Rothlein, R. Barton, S. D. Marlin, and T. A. Springer. 1989. A cell adhesion molecule, ICAM-1, is the major surface receptor for rhinoviruses. *Cell* **56:**849–853.

Tomassini, J. E., and R. J. Colonno. 1986. Isolation of a receptor protein involved in attachment of human rhinoviruses. *J. Virol.* **58:**290–295.

Tomassini, J. E., D. Graham, C. M. DeWitt, D. W. Lineberger, J. A. Rodkey, and R. J. Colonno. 1989a. cDNA cloning reveals that the major group rhinovirus receptor on HeLa cells is intercellular adhesion molecule 1. *Proc. Natl. Acad. Sci. USA* **86:**4907–4911.

Tomassini, J. E., T. R. Maxson, and R. J. Colonno. 1989b. Biochemical characterization of a glycoprotein required for rhinovirus attachment. *J. Biol. Chem.* **264:**1656–1662.

Yin, F. H., and N. B. Lomax. 1983. Host range mutants of human rhinovirus in which nonstructural proteins are altered. *J. Virol.* **48:**410–418.

Chapter 38

Identification of the Major Human Rhinovirus Receptor Reveals Identity with Intercellular Adhesion Molecule 1

Alan McClelland and Jeffrey M. Greve

The rhinoviruses are a group of acid-sensitive picornaviruses consisting of more than 100 distinct serotypes and are the predominant causative agent of the common cold (Hamparian et al., 1987). Structurally, they are perhaps the best-characterized animal viruses. The complete sequence of the plus-strand RNA genome of several serotypes is known (Skern et al., 1985; Stanway et al., 1984; Callahan et al., 1985; Duechler et al., 1987; Hughes et al., 1988), and the three-dimensional structures of human rhinovirus 14 (HRV-14) (Rossmann et al., 1985) and HRV-1A (Kim et al., 1989) have been determined to atomic resolution. This work has revealed an icosahedral protein capsid composed of 60 protomeric units, each containing one copy of the four capsid proteins VP1, VP2, VP3, and VP4. Several exposed regions that protrude from the virus surface and show the greatest variation among serotypes are the dominant antigenic sites on the virus (Sherry et al., 1986). A recessed canyon present in each protomeric subunit of rhinovirus and poliovirus, but absent from foot-and-mouth disease virus (FMDV) (Acharya et al., 1989), has been proposed to be the site of host cell receptor binding (reviewed by Rossmann [1989]).

Attachment to cell surface receptors is the first step in viral infection and is a major although not exclusive determinant of cell tropism and host range restriction. The mechanisms of viral attachment and cell entry are fundamental questions in virology which require the isolation of viral receptors and their genes. Of the more than 100 rhinovirus serotypes, 90% bind to a common receptor site termed the major receptor, whereas the remaining serotypes bind to a second, minor receptor (Abraham and Colonno,

1984; Colonno et al., 1986). We have used parallel genetic and biochemical approaches to identify the major rhinovirus receptor and have demonstrated its identity with intercellular adhesion molecule 1 (ICAM-1), a cell adhesion molecule of the immunoglobulin superfamily (Greve et al., 1989).

TRANSFECTION OF THE RECEPTOR GENE INTO MOUSE CELLS

Virus-binding studies have shown that mouse fibroblasts do not have specific attachment sites for rhinoviruses of the major receptor group (Abraham and Colonno, 1984). In our initial experiments to identify the receptor, we used gene transfer to obtain mouse L cells that express the major human receptor. Genomic DNA from HeLa cells was stably introduced into Ltk$^-$ cells by cotransfection with a selectable marker, and a pool of 130,000 transfectants was screened by fluorescence-activated cell sorting for cells capable of specifically binding UV-inactivated HRV-14. An anti-HRV-14 mouse monoclonal antibody and a fluorescein-conjugated goat anti-mouse antibody were used to detect the virus receptor complex and to select positive cells by fluorescence-activated cell sorting. After several rounds of preparative cell sorting, we isolated a population of transfected cells that were specifically labeled by the virus-binding assay. Southern blotting confirmed the presence of human DNA sequences in the transfectants, supporting the conclusion that they contain the rhinovirus receptor gene.

Alan McClelland and Jeffrey M. Greve, Molecular Therapeutics Inc., Miles Research Center, 400 Morgan Lane, West Haven, Connecticut 06516.

MONOCLONAL ANTIBODIES TO THE RECEPTOR

Axler and Crowell (1968) showed that anticellular serum contained antibodies that inhibited virus attachment, and several monoclonal antibodies that block infection by certain picornaviruses have been obtained (Nobis et al., 1985; Minor et al., 1984; Colonno et al., 1986; Crowell et al., 1986; Hsu et al., 1988). We screened hybridoma supernatants prepared from mice immunized with HeLa cells and identified five hybridomas that prevented infection of HeLa cell monolayers by rhinoviruses of the major receptor group. The antibodies reacted with the surface of HeLa cells and prevented the attachment of radiolabeled virus. When tested against the mouse transfectants described above, we found that the antibodies bound to the transfectants but not to the parent Ltk⁻ cells. This result indicated that the same molecule had been identified by these two independent approaches and allowed us to conclude that this molecule is the major rhinovirus receptor.

During the course of further gene transfer experiments, we identified a population of transfectants that showed a higher level of surface staining with one of the monoclonal antibodies, c78.4A. A cloned cell line, HE1, established from these transfectants, contains 3×10^5 antibody-binding sites per cell, as opposed to 6×10^4 sites on HeLa cells. This cell line is a useful source of receptor protein, yielding five times as much receptor as an equivalent number of HeLa cells.

PURIFICATION OF A BIOLOGICALLY ACTIVE RECEPTOR PROTEIN

Immunoprecipitation of both HeLa and HE1 cell extracts with monoclonal antibody c78.4A identified a single 95-kilodalton polypeptide that contains intrachain disulfide bonds. In the presence of tunicamycin, a 54-kilodalton polypeptide was obtained, indicating that the receptor is extensively N glycosylated. Receptor protein was purified from detergent-solubilized HeLa and HE1 cells by sequential adsorption to wheat germ agglutinin and monoclonal antibody c78.4A-Sepharose, which resulted in a single band on a silver-stained gel.

We assessed the virus-binding activity of the purified receptor by using a pelleting assay. Metabolically labeled receptor was incubated with HRV-3 at 37°C, and the mixture was pelleted through a glycerol cushion in an air-driven ultracentrifuge. Approximately 20% of the labeled receptor pelleted with the virus, indicating association between the two. In the absence of virus or in the presence of an Fab fragment of c78.4A essentially all of the receptor remained in the supernatant, indicating a specific association. Thus, purified receptor in the absence of other cellular components is capable of binding virus.

PROTEIN SEQUENCE OF THE RECEPTOR REVEALS IDENTITY WITH ICAM-1

The receptor appears to have a blocked N terminus, since no sequence was obtained from several attempts to sequence the intact protein. Proteolytic digestion yielded eight peptides, from which a total of 105 residues of the receptor could be confidently assigned, representing approximately 20% of the total protein. A search of all available protein and nucleic acid data bases with these sequences revealed a match with ICAM-1 in the MIPSX (Martinsrieder Institut für Proteinsequenzen) data base (Table 1). All 105 confidently assigned residues and a further 29 of 36 tentative assignments matched exactly the published sequence of ICAM-1 (Simmons et al., 1988; Staunton et al., 1988), providing strong evidence that the major rhinovirus receptor and ICAM-1 are identical.

This conclusion was confirmed by blot hybridization and polymerase chain reaction experiments which showed that the mouse transfectant cell lines contain the human ICAM-1 gene and mRNA. ICAM-1 clones were isolated from a transfectant cDNA library by using oligonucleotide probes, and a full-length ICAM-1 cDNA was obtained by the polymerase chain reaction from the HE1 transfectant. DNA sequencing of this and several other clones showed identity with the published ICAM-1 sequence except for a single A-to-G base substitution, which changes Glu-442 to Lys.

STRUCTURE AND FUNCTION OF ICAM-1

ICAM-1 was identified by a monoclonal antibody that inhibits leukocyte adhesion (Rothlein et al., 1986) and has been shown to mediate cell adhesion by interaction with lymphocyte function-associated antigen 1, a member of the integrin family of adhesion molecules (Marlin and Springer, 1987; Kishimoto et al., 1989). ICAM-1 shows a widespread tissue distribution, and expression is elevated on several cell types by interleukin-1, gamma interferon, and tumor necrosis factor (Dustin et al., 1986). Increased

TABLE 1

Comparison of amino acid sequences of peptides isolated from HRR protein
with the published sequence of ICAM-1[a]

Peptide	Sequence
	** ***** ***** ***** ** * *
RP/LE-19	k dgtFP LPIGE SVTVT RDLEG -YL1R Ar
	K DGTFP LPIGE SVTVT RDLEG TYLCR AR
	406 433
	***** ***** **** * * * *
RP/LE-10	K vYELS NVQED SQPM- YsN-P igQ--a
	K VYELS NVQED SQPMC YSNCP DGQSTA
	50 76
	**** *****
RP/LE-15a	k eLLLP GNNRK
	K ELLLP GNNRK
	40 50
	*** ****
RP/LE-14	k kyRLQ QAQK
	K KYRLQ QAQK
	483 492
	**** *****
RP/T-2	k gTPMK PNTQA
	K GTPMK PNTQA
	492 502
	**** ***** ***** * ***
EB/LE-34Kd	k -LKRE PAVGE PAEVT TrVLV rrd-h ga-f
	k ELKRE PAVGE PAEVT TTVLV RRDHH GANF
	128 157
	**** ***** * *** ***** ** *
EB/T-50Kd	r tLEVD TQGTV V-SLD GLFPV sEA-V -1-1
	R VLEVD TQGTV VCSLD GLFPV SEAQV HLAL
	198 227
	***
RP/LE-15b	k viLPR
	K VILPR
	8 13

[a] The numbering scheme of Staunton et al. (1988) rather than that of Simmons et al. (1988) is used, beginning with Q as residue 1 of mature protein (number 28 of the translated coding sequence of mRNA) because it fits more closely the signal peptidase −1, −3 rule (von Heijne, 1983) and +2 rule (von Heijne, 1984). RP or EB in a peptide code name indicates that the peptide was isolated by reverse-phase chromatography or electroblotting from sodium dodecyl sulfate-gels, respectively. LE or T indicates cleavage with lysyl endopeptidase or trypsin, respectively. In many cases, the given sequences were determined from more than one peptide; for the sake of clarity, only the longest is shown. For each peptide, the sequence is given on the upper line and the corresponding match with ICAM-1 is given on the lower line, with the residue numbers indicated underneath. Uppercase letters are confident assignments, lowercase letters are tentative assignments or cycles at which two residues were obtained, and a dash indicates no assignment at that cycle. The space after the first residue indicates the cleavage point of the enzyme; in all cases, the residue to the left of the space is inferred from the specificity of the enzyme. An asterisk above the peptide sequence indicates a match between a confidently assigned residue and ICAM-1.

expression of ICAM-1 correlates with increased lymphocyte adhesion, suggesting that it may promote lymphocyte influx to inflammatory sites.

ICAM-1 is a transmembrane glycoprotein with an N-terminal extracellular domain of 453 amino acids, a 24-residue membrane-spanning region, and a short cytoplasmic tail of 28 amino acids (Simmons et al., 1988; Staunton et al., 1988). The extracellular portion of the molecule can be divided into five domains that show sequence similarity to each other and to members of the immunoglobulin superfamily, which is strongest for the neural cell adhesion molecules NCAM and MAG. The immunoglobulin repeats of these molecules and of ICAM-1 show the highest similarity with immunoglobulin constant-region domains of the C2 set (Williams and Barclay, 1988). Constant-region domains consist of seven antiparallel beta strands arranged as two beta sheets held together by an intradomain disulfide bond and linked by peptide loops of variable length. In the related variable-domain structure, the loops at the N terminus correspond to the hypervariable regions that make up the antigen-combining site. The cysteine residues that form intrachain disulfide bonds are conserved in the three N-terminal domains of ICAM-1, and we have determined that these bonds exist in the native molecule. Thus, it is likely that at least the first three domains of ICAM-1 have a structure based on the immunoglobulin fold. X-ray crystallographic data will be required to confirm this model of ICAM-1 structure and to provide sufficient resolution of the structure to predict molecular contacts by molecular modeling.

RELATIONSHIP TO OTHER VIRAL RECEPTORS

Relatively few viral receptors have been characterized in detail at the molecular level. It is therefore notable that two other receptors, the CD4 receptor for human immunodeficiency virus (Maddon et al., 1986) and the poliovirus receptor (Mendelsohn et al., 1989), are also members of the immunoglobulin superfamily. The potential significance of this finding should become clear as other viral receptors, particularly picornavirus receptors, are isolated and as the sites of virus binding are characterized. Both the rhinovirus receptor and the poliovirus receptor are now known to be encoded by genes on human chromosome 19 (Greve et al., 1989; Miller et al., 1974). Since several other picornavirus receptor genes have been mapped to this chromosome (Couillin et al., 1987), it is reasonable to speculate that they are members of the same gene family. This argument has also been made on the basis of the structural similarities between rhinovirus and poliovirus. It will be of particular interest to determine the nature of the receptor for FMDV, since this virus does not contain the canyon found in the rhinovirus and poliovirus structures, and it has been proposed that an RGD sequence in the FMDV capsid is part of the cell attachment site, implicating a member of the integrin family as the cellular receptor (Fox et al., 1989).

FUTURE PROSPECTS

The identification of ICAM-1 as the rhinovirus receptor and the availability of the cloned gene and purified product will permit the process of viral attachment and cell penetration to be studied in molecular detail. Manipulation of individual immunoglobulin domains and site-directed mutagenesis of specific residues will aid in the definition of the viral attachment site of the receptor. Crystallization of soluble recombinant forms of the molecule should lead to the resolution of the three-dimensional structure of the extracellular domains and perhaps of a virus-receptor complex. Knowledge of the precise molecular contacts between rhinoviruses and their cellular receptor may aid in the development of novel inhibitors of viral attachment.

The pathogenesis of rhinovirus infections is poorly understood. Indeed, it is not even clear which cells are infected during rhinovirus colds. The absence of widespread cellular damage during infection has led to the hypothesis that rhinovirus infection leads to release of inflammatory mediators that are responsible for symptoms (Winther et al., 1986). Therefore, it is particularly intriguing that the rhinovirus receptor should be a molecule central to the immune response and one that is regulated by cytokines. This raises the possibility that the interaction of rhinoviruses with their cellular receptors has consequences beyond cell attachment, perhaps initiating a signaling process which triggers the events leading to the development of symptoms. Current investigations of ICAM-1 structure, function, and regulation in the context of both the normal immune response and rhinovirus infection hold the promise of significantly advancing our understanding of rhinovirus infections and of viral infections in general.

REFERENCES

Abraham, G., and R. J. Colonno. 1984. Many rhinovirus serotypes share the same cellular receptor. *J. Virol.* **51**:340–345.

Acharya, R., E. Fry, D. Stuart, G. Fox, D. Rowlands, and F. Brown. 1989. The three-dimensional structure of foot-and-mouth disease virus at 2.9 Å resolution. *Nature* (London) **337**:709–716.

Axler, D. A., and R. L. Crowell. 1968. Effect of anticellular serum on the attachment of enteroviruses to HeLa cells. *J. Virol.* **2**:813–821.

Callahan, P. L., S. Mizutani, and R. J. Colonno. 1985. Molecular cloning and complete sequence determination of RNA genome of human rhinovirus type 14. *J. Virol.* **82**:732–736.

Colonno, R. J., P. L. Callahan, and W. J. Long. 1986. Isolation of a monoclonal antibody that blocks attachment of the major group of rhinoviruses. *J. Virol.* **57**:7–12.

Couillin, P., F. Huyghe, M. C. Grisard, N. Van Cong, M. F. Louis, H. Hofmann-Radvanyi, and A. Boue. 1987. Echovirus 6, 11, 19; coxsackie B3 sensitivities and polivorius I, II, III sensitivities are on chromosome 19 respectively on 19pter-q133 and 19q131-q133. *Cytogenet. Cell. Genet.* **46**:599.

Crowell, R. L., A. K. Field, W. A. Schleif, W. L. Long, R. J. Colonno, J. E. Mapoles, and E. A. Emini. 1986. Monoclonal antibody that inhibits infection of HeLa and rhabdomyosarcoma cells by selected enteroviruses through receptor blockade. *J. Virol.* **57**:438–445.

Duechler, M., T. Skern, W. Sommergruber, C. Neubauer, P. Gruendler, I. Fogy, D. Blaas, and E. Kuechler. 1987. Evolutionary relationships within the human rhinovirus genus: comparison of serotypes 89, 2 and 14. *Proc. Natl. Acad. Sci. USA* **84**:2605–2609.

Dustin, M., R. Rothlein, A. K. Bahn, C. A. Dinarello, and T. A. Springer. 1986. Induction of Il 1 and interferon: tissue distribution, biochemistry, and function of a natural adherence molecule (ICAM-1). *J. Immunol.* **137**:245–254.

Fox, G., N. R. Parry, P. V. Barnett, B. McGinn, D. J. Rowlands, and F. Brown. 1989. The cell attachment site on foot-and-mouth disease virus includes the amino acid sequence RGD (arginine-glycine-aspartic acid). *J. Gen. Virol.* **70**:625–637.

Greve, J. G., G. Davis, A. M. Meyer, C. P. Forte, S. C. Yost, C. W. Marlor, M. E. Kamarck, and A. McClelland. 1989. The major human rhinovirus receptor is ICAM-1. *Cell* **56**:839–847.

Hamparian, V. V., R. J. Colonno, M. K. Cooney, E. C. Dick, J. M. Gwaltney, J. H. Hughes, W. S. Jordan, A. Z. Kapikan, W. J. Mogabgab, A. Monto, C. A. Phillips, R. R. Rueckert, J. H. Schieble, E. J. Scott, and D. A. J. Tyrrell. 1987. A collaborative report: rhinoviruses—extension of the numbering system from 89 to 100. *Virology* **159**:191–192.

Hsu, K. L., K. Lonberg-Holm, B. Alstein, and R. L. Crowell. 1988. A monoclonal antibody specific for the cellular receptor for the group B coxsackieviruses. *J. Virol.* **62**:1647–1652.

Hughes, P. J., C. North, C. H. Jellis, P. D. Minor, and G. Stanway. 1988. The nucleotide sequence of human rhinovirus 1B: molecular relationships within the rhinovirus genus. *J. Gen. Virol.* **69**:49–58.

Kim, S., T. J. Smith, M. S. Chapman, M. G. Rossmann, D. C. Pevear, F. J. Dutko, P. J. Felock, G. D. Diana, and M. A. McKinlay. 1989. Crystal structure of human rhinovirus serotype IA (HRVIA). *J. Mol. Biol.* **210**:91–111.

Kishimoto, T. K., R. S. Larson, A. L. Corbi, M. L. Dustin, D. E. Staunton, and T. A. Springer. 1989. The leukocyte integrins. *Adv. Immunol.* **46**:149–182.

Maddon, P. J., A. G. Dalgleish, J. S. McDougal, P. R. Clapham, R. A. Weis, and R. Axel. 1986. The T4 gene encodes the AIDS virus receptor and is expressed in the immune system and the brain. *Cell* **47**:333–348.

Marlin, S. D., and T. A. Springer. 1987. Purified intercellular adhesion molecule-1 (ICAM-1) is a ligand for lymphocyte function-associated antigen-1 (LFA-1). *Cell* **51**:813–819.

Mendelsohn, C. L., E. Wimmer, and V. R. Racaniello. 1989. Cellular receptor for poliovirus: molecular cloning, nucleotide sequence, and expression of a new member of the immunoglobulin superfamily. *Cell* **56**:855–856.

Miller, D. A., O. J. Miller, V. G. Dev, S. Hashmi, R. Tantravahl, L. Medrano, and H. Green. 1974. Human chromosome 19 carries a poliovirus receptor gene. *Cell* **1**:167–170.

Minor, P. D., P. A. Pipkin, D. Hockley, and J. W. Almond. 1984. Monoclonal antibodies which block cellular receptors of poliovirus. *Virus Res.* **1**:203–212.

Nobis, P., R. Zibirre, G. Meyer, J. Kuhne, G. Warnecke, and G. Koch. 1985. Production of a monoclonal antibody against an epitope on HeLa cells that is the functional poliovirus binding site. *J. Gen. Virol.* **66**:2563–2569.

Rossmann, M. 1989. The canyon hypothesis. *J. Biol. Chem.* **264**:14587–14590.

Rossmann, M. G., E. Arnold, J. W. Erickson, E. W. Frankenberger, P. J. Griffith, H. Hecht, J. E. Johnson, G. Kamer, M. Luo, A. G. Mosser, R. R. Rueckert, B. Sherry, and G. Vriend. 1985. Structure of a common cold virus and functional relationship to other picornaviruses. *Nature* (London) **317**:145–153.

Rothlein, R., M. L. Dustin, S. D. Marlin, and T. A. Springer. 1986. A human intercellular adhesion molecule (ICAM-1) distinct from LFA-1. *J. Immunol.* **137**:1270–1274.

Sherry, B., A. G. Mosser, R. J. Colonno, and R. R. Rueckert. 1986. Use of monoclonal antibodies to identify four neutralization immunogens on a common cold picornavirus, human rhinovirus 14. *J. Virol.* **57**:246–257.

Simmons, D., M. W. Makgoba, and B. Seed. 1988. ICAM, an adhesion ligand of LFA-1, is homologous to the neutral cell adhesion molecule NCAM. *Nature* (London) **331**:624–627.

Skern, T., W. Sommergruber, D. Blaas, P. Gruendler, F. Fraundorfer, C. Pieler, I. Fogy, and E. Kuechler. 1985. Human rhinovirus 2: complete nucleotide sequence and proteolytic processing signals in the capsid protein region. *Nucleic Acids Res.* **13**:2111–2126.

Stanway, G., P. J. Hughes, R. J. Mountford, P. D. Minor, and J. W. Almond. 1984. The complete nucleotide sequence of a common cold virus: human rhinovirus 14. *Nucleic Acids Res.* **12**:7859–7875.

Staunton, D. E., S. D. Marlin, C. Stratowa, M. Dustin, and T. A. Springer. 1988. Primary structure of ICAM-1 demonstrates interaction between members of the immunoglobulin and integrin supergene families. *Cell* **52**:925–933.

von Heijne, G. 1983. Patterns of amino acids near signal-sequence cleavage sites. *Eur. J. Biochem.* **133**:17–21.

von Heijne, G. 1984. How signal sequences maintain cleavage specificity. *J. Mol. Biol.* **173**:243–251.

Williams, A. F., and A. N. Barclay. 1988. The immunoglobulin superfamily—domains for cell surface recognition. *Annu. Rev. Immunol.* **6**:381–405.

Winther, B., J. M. Gwaltney, N. Mygind, R. B. Turner, and J. O. Hendley. 1986. Sites of rhinovirus recovery after point inoculation of the upper airway. *J. Am. Med. Assoc.* **256**:1763–1767.

Chapter 39

Antibodies to Intercellular Adhesion Molecule 1: Regulation of Immunological Responses and Inhibition of Rhinovirus-Induced Cytopathogenicity

Vincent J. Merluzzi, Steven D. Marlin, and Robert Rothlein

Leukocyte cellular adhesion is believed to be an integral process in effective immunological responses. A family of molecules that mediate both cell-cell and cell-substrate adhesion has been described. This family has been designated the lymphocyte function-associated antigen 1 (LFA-1) family or the CD18 complex of adhesion molecules (Dustin and Springer, 1988; Dustin et al., 1988; Marlin and Springer, 1987; Rothlein et al., 1986; Rothlein and Springer, 1986). It has been shown that the ligand for LFA-1 is the intercellular adhesion molecule 1 (ICAM-1) (Makgoba et al., 1988a; Makgoba et al., 1988b; Marlin and Springer, 1987; Rothlein et al., 1986). ICAM-1 is necessary for cellular interactions during an immune response mediated via LFA-1-dependent adhesion. In addition, ICAM-1 is an inducible molecule present on fibroblasts and endothelial cells and is responsible, at least in part, for leukocyte adherence and migration through endothelial cells (Dustin and Springer, 1988; Dustin et al., 1988; Rothlein et al., 1988). It has been shown that ICAM-1 can be induced by inflammatory cytokines such as gamma interferon, interleukin-1, and tumor necrosis factor alpha (Rothlein et al., 1988). The induction of ICAM-1 during immunological responses may allow for the effective interaction (adhesion) of leukocytes with each other and with endothelial cells.

Antibodies directed against ICAM-1 have been shown to inhibit in vitro immunological reactions that are dependent on lymphocyte-macrophage adhesion. When human peripheral blood mononuclear cells from two unrelated donors are cultured in vitro in a mixed-lympho-cyte reaction, antibodies to ICAM-1 or LFA-1 inhibit the proliferation of these cells over a 5-day period. When antibodies to ICAM-1 and LFA-1 are added together at suboptimal concentrations, there is an additive, suppressive effect of the mixed-lymphocyte reaction (Merluzzi et al., 1988). Other in vitro assays such as T-cell mitogen proliferation (concanavalin A) and proliferation to specific antigen (tetanus toxoid or keyhole limpet hemocyanin) are also inhibited by the addition of monoclonal antibodies to ICAM-1 to the culture systems. The 50% effective concentration of a monoclonal antibody to ICAM-1 (BIR-R-0001) in the mixed-lymphocyte reaction is 40 ng/ml (V. J. Merluzzi et al., unpublished results).

In vivo studies with antibodies to ICAM-1 have shown that neutrophil influx (induced by phorbol myristic acid) in the rabbit lung is inhibited by the prior injection of anti-ICAM-1 antibodies (Barton et al., 1988). Recent studies have shown that a monoclonal antibody to ICAM-1 given to cynomologous monkeys prolongs kidney allografts (Cosimi et al., 1988).

In addition to studies demonstrating the role that ICAM-1 plays in the immune response as an adhesion molecule, two recent studies have shown and confirmed that ICAM-1 is the receptor for the major subgroup of rhinoviruses, a subgroup of picornaviruses (Greve et al., 1989; Sperber and Hayden, 1988; Staunton et al., 1989). The majority of rhinoviruses and some coxsackieviruses share a common cellular receptor on human cells. Early studies had shown that monoclonal antibodies directed against this receptor on human cells inhibit the attachment

Vincent J. Merluzzi, Steven D. Marlin, and Robert Rothlein, Department of Immunology, Boehringer Ingelheim Pharmaceuticals, Inc., 90 East Ridge, P.O. Box 368, Ridgefield, Connecticut 06877.

of a majority of rhinoviruses (Abraham and Colonno, 1984, 1988; Colonno et al., 1986; Tomassini and Colonno, 1986). Abraham and Colonno (1988) and Colonno et al. (1988) have shown that monoclonal antibodies to this receptor inhibit the majority of rhinoviruses from binding to target cells and also displace some rhinoviruses that have attached to the receptor.

Recently, Greve et al. (1989) have shown that the receptor for the major subgroup of rhinoviruses is identical to ICAM-1. Staunton et al. (1989) have shown that monoclonal antibodies to ICAM-1 inhibit the binding of rhinovirus 14 to ICAM-1-transfected cells as well inhibit the cytopathic effect (CPE) of rhinovirus 54 on HeLa cells. Both of these viruses belong to the major subgroup of rhinoviruses. Antibodies to ICAM-1 did not affect CPE induced by rhinoviruses belonging to the minor subgroup, such as rhinovirus 2 and rhinovirus 49 (Staunton et al., 1989).

In recent studies (V. J. Merluzzi, unpublished results), five major-subgroup picornaviruses (rhinoviruses 4, 5, and 54 and coxsackieviruses A13 and A21) were unable to induce CPE on HeLa cells in the presence of an antibody to ICAM-1. Three minor-subgroup picornaviruses (rhinoviruses 2, 31, and 49), coxsackievirus B1, poliovirus type 1, herpesvirus 1, and influenzavirus A (WSN) (Schulman and Palese, 1977) were unaffected when antibodies to ICAM-1 were present in the CPE assay. Since LFA-1 binds to ICAM-1, it is not inconceivable that anti-LFA-1 antibodies inhibit or neutralize viruses that utilize the ICAM-1 receptor. Antibodies to LFA-1 alpha and LFA-1 beta and mixtures of these antibodies did not inhibit or neutralize virus infectivity of HeLa cells (Merluzzi, unpublished results). In addition, an antibody to an HLA framework antigen (W6/32) present on HeLa cells did not inhibit or affect CPE induced by any virus of the major subgroup. It is therefore unlikely that antibodies to LFA-1 alpha or beta attach to the virus capsid proteins, or if they do, they do not affect attachment or penetration of the virus in HeLa cells. Also, at least one antibody (W6/32) that binds HeLa cells does not nonspecifically inhibit virus attachment to its receptor.

JY lymphoblasts stimulated with phorbol myristic acid form homotypic cell aggregates. Addition of antibodies to ICAM-1 or LFA-1 inhibits this aggregation. Therefore, blocking either LFA-1 or ICAM-1 is sufficient to inhibit aggregation, as would be expected for a ligand-receptor system. Since JY cells express ICAM-1, it is possible that these cells also bind viruses

that belong to the major subgroup of rhinoviruses. Incubation of JY cells with rhinovirus 54 (major subgroup) causes a marked CPE at 48 h. Addition of a monoclonal antibody to ICAM-1, but not LFA-1, inhibits this CPE. These data are consistent with both the expression of ICAM-1 on JY cells and the inhibition of aggregation by anti-ICAM-1 antibody (Merluzzi et al., unpublished data). Additional data showed that when JY cells were incubated with rhinovirus 54, membrane-bound capping of ICAM-1 with time was seen, as detected by immunofluorescence with anti-ICAM-1. The percentage of cells with capped ICAM-1 molecules ranged from 72 to 76% at 33°C. This value is in comparison with JY cells incubated with buffer alone (8 to 11%) and cells incubated with rhinovirus 54 at 4°C (12 to 18%) (Merluzzi et al., unpublished data). The relationship of this capping phenomenon to virus entry into the cell is unknown. Modulation of the ICAM-1 molecule by rhinovirus 54 may be important in limiting other major-group rhinoviruses from attachment to the same cell and may also be important in the regulation of immunological responses in which the interaction of leukocytes with ICAM-1-bearing cells functions in immunophysiology and pathology.

It has been reported that antibodies directed against the major HRV receptor can delay the symptoms related to the cold virus in humans (Hayden et al., 1988). It is presumed, but not proven, that these same antibodies may recognize and bind to ICAM-1. Small molecules, antibodies, or peptides that inhibit LFA-1–ICAM-1 interactions or substances that inhibit the expression of ICAM-1 may have beneficial effects, since they would interfere with the inflammatory process. Substances that block ICAM-1 or inhibit its expression may also inhibit the attachment of certain rhinoviruses and coxsackieviruses to their cell surface receptors. Viral receptors with known physiological functions are rare. A recent paper has reviewed viral receptors of the immunoglobulin supergene family (White and Littman, 1989). One of the more recent and salient analogies of viral receptors with known physiological functions is the human immunodeficiency virus type 1-CD4 virus receptor system (White and Littman, 1989). In this latter system, blocking the CD4 receptor with soluble CD4 or antibodies to CD4 inhibits viral (human immunodeficiency virus type 1) attachment and CPE. It is presumed, but not proven, that blocking this receptor would also cause a suppression of a normal immune response and inflammation caused by helper T cell-leukocyte interactions.

The major rhinovirus receptor is ICAM-1. This receptor may have structural similarities to other picornavirus receptors. Since ICAM-1 is involved in cellular immune functions and leukocyte adherence, this hypothesis suggests that the receptors for other picornaviruses may also be interwoven in cellular immune reactions.

REFERENCES

Abraham, G., and R. J. Colonno. 1984. Many rhinovirus serotypes share the same cellular receptor. *J. Virol.* **51**:409–419.

Abraham, G., and R. J. Colonno. 1988. Characterization of human rhinoviruses displaced by an antireceptor monoclonal antibody. *J. Virol.* **62**:2300–2306.

Barton, R. W., R. Rothlein, J. Ksiazek, and C. Kennedy. 1988. Role of anti-adhesion monoclonal antibodies in rabbit lung inflammation, p. 149–155. *In* T. A. Springer, D. C. Anderson, A. S. Rosenthal, and R. Rothlein (ed.), *Leukocyte Adhesion Molecules*. Springer-Verlag, New York.

Colonno, R. J., P. L. Callahan, and W. J. Long. 1986. Isolation of a monoclonal antibody that blocks attachment of the major group of human rhinoviruses. *J. Virol.* **57**:7–12.

Colonno, R. J., J. H. Condra, S. Mizutani, P. L. Callahan, M.-E. Davies, and M. A. Murcko. 1988. Evidence for the direct involvement of the rhinovirus canyon in receptor binding. *Proc. Natl. Acad. Sci. USA* **85**:5449–5453.

Cosimi, A. B., C. Geoffrion, T. Anderson, D. Conti, R. Rothlein, and R. B. Colvin. 1988. Immunosuppression of cynomologous recipients of renal allografts by Rb.5, a monoclonal antibody to ICAM-1, p. 274–281. *In* T. A. Springer, D. C. Anderson, A. S. Rosenthal, and R. Rothlein (ed.), *Leukocyte Adhesion Molecules*. Springer-Verlag, New York.

Dustin, M. L., and T. A. Springer. 1988. Lymphocyte function associated antigen-1 (LFA-1) interaction with intercellular adhesion molecule-1 (ICAM-1) is one of at least three mechanisms for lymphocyte adhesion to cultured endothelial cells. *J. Cell Biol.* **107**:321–331.

Dustin, M. L., D. E. Staunton, and T. A. Springer. 1988. Supergene families meet in the immune system. *Immunol. Today* **9**:213–215.

Greve, J. M., G. Davis, A. M. Meyer, C. P. Forte, S. C. Yost, C. W. Marlor, E. M. Kamarck, and A. McClelland. 1989. The major human rhinovirus receptor is ICAM-1. *Cell* **56**:839–847.

Hayden, F. G., J. M. Gwaltney, and R. J. Colonno. 1988. Modification of experimental rhinovirus colds by receptor blockade. *Antiviral Res.* **9**:233–247.

Makgoba, M. W., M. E. Sanders, G. E. Ginther Luce, M. L. Dustin, T. A. Springer, E. A. Clarke, P. Mannoni, and S. Shaw. 1988a. ICAM-1: definition by multiple antibodies of a ligand for LFA-1 dependent adhesion of B, T and myeloid cells. *Nature* (London) **331**:86–88.

Makgoba, M. W., M. E. Sanders, G. E. Ginther Luce, E. A. Gugel, M. L. Dustin, T. A. Springer, and S. Shaw. 1988b. Functional evidence that intercellular adhesion molecule-1 (ICAM-1) is a ligand for LFA-1 in cytotoxic T cell recognition. *Eur. J. Immunol.* **18**:637–640.

Marlin, S. D., and T. A. Springer. 1987. Purified intercellular adhesion molecule-1 (ICAM-1) is a ligand for lymphocyte-associated antigen 1 (LFA-1). *Cell* **51**:813–819.

Merluzzi, V. J., R. Rothlein, C. Wood, C. D. Stearns, R. B. Faanes, and K. Last-Barney. 1988. Inhibition of human mixed lymphocyte reactions by monoclonal antibodies to ICAM-1, p. 244–253. *In* T. A. Springer, D. C. Anderson, A. S. Rosenthal, and R. Rothlein (ed.), *Leukocyte Adhesion Molecules*. Springer-Verlag, New York.

Rothlein, R., M. Czajkowski, M. M. O'Neill, S. D. Marlin, E. Mainolfi, and V. J. Merluzzi. 1988. Induction of intercellular adhesion molecule 1 on primary and continuous cell lines by pro-inflammatory cytokines. Regulation by pharmacologic agents and neutralizing antibodies. *J. Immunol.* **141**:1665–1669.

Rothlein, R., M. L. Dustin, S. D. Marlin, and T. A. Springer. 1986. A human intercellular adhesion molecule (ICAM-1) distinct from LFA-1. *J. Immunol.* **137**:1270–1274.

Rothlein, R., and T. A. Springer. 1986. The requirement for lymphocyte function associated antigen 1 in homotypic leukocyte adhesion stimulated by phorbol ester. *J. Exp. Med.* **163**:1132–1149.

Schulman, J. L., and P. Palese. 1977. Virulence factors of influenza A viruses: WSN virus neuraminidase required for plaque production in MDBK cells. *J. Virol.* **24**:170–176.

Sperber, S. J., and F. G. Hayden. 1988. Chemotherapy of rhinovirus colds. *Antimicrob. Agents Chemother.* **32**:409–419.

Staunton, D. E., V. J. Merluzzi, R. Rothlein, R. Barton, S. D. Marlin, and T. A. Springer. 1989. A cell adhesion molecule, ICAM-1, is the major surface receptor for rhinoviruses. *Cell* **56**:849–853.

Tomassini, J. E., and R. J. Colonno. 1986. Isolation of a receptor protein involved in attachment of human rhinoviruses. *J. Virol.* **58**:290–295.

White, J. M., and D. R. Littman. 1989. Viral receptors of the immunoglobulin superfamily. *Cell* **56**:725–728.

Chapter 40

Identification of a Second Cellular Receptor for a Coxsackievirus B3 Variant, CB3-RD

Kuo-Hom Lee Hsu, Severo Paglini, Barbara Alstein, and Richard L. Crowell

There have been a number of recent exciting observations about the cellular receptors for picornaviruses since they were last reviewed (Crowell and Lonberg-Holm, 1986; Colonno et al., 1989). The receptor for the major group of human rhinoviruses (HRVs) has been determined to be intercellular adhesion molecule 1 (ICAM-1) (Greve et al., 1989; Staunton et al., 1989). Thus, the first known function for a picornavirus receptor has been identified. The ICAM-1 molecule presumably serves to bind lymphocytes, carrying lymphocyte function-associated antigen 1 ligands to target cells expressing ICAM-1 (Dustin et al., 1988; Altmann et al., 1989). It will be important to determine how the ICAM-1 molecules facilitate HRV entry into cells (Harrison, 1989) and whether HRV infection of a given cell causes that cell to send out signals (lymphokines, prostacyclin, etc.) (Roberts et al., 1989) that up-regulate the expression of ICAM-1 on neighboring cells to facilitate the successful spread of HRV during infection. The discovery of the gene for the cellular receptor for polioviruses (Mendelsohn et al., 1989) is equally exciting, since the deduced amino acid sequence of the receptor molecule places the gene in the immunoglobulin supergene family with ICAM-1 (White and Littman, 1989). These findings suggest that the gene for the cellular receptor for the prototype members of group B coxsackieviruses belongs to the immunoglobulin supergene family. However, there has been a suggestion that the receptor gene for a strain of foot-and-mouth disease virus belongs to the integrin supergene family (Fox et al., 1989; Hynes, 1987). This suggestion was based on observa-

tions that the virion attachment protein, VAP, contains an RGD domain and that virus binding to receptors is blocked by peptides constructed to resemble this domain. Since the atomic structure of foot-and-mouth disease virus reveals the absence of a canyon (Acharya et al., 1989), in contrast to the structure found for other picornaviruses, e.g., HRV-14 (Rossmann et al., 1985), poliovirus type 1 (PV1) (Hogle et al., 1985), and mengovirus (Luo et al., 1987), perhaps there is no unifying strategy for the binding of all picornaviruses to receptors of similar structures. It will be interesting to learn whether other foot-and-mouth disease virus serotypes have similar RGD domains, since it has been shown that the several serotypes compete for a common receptor (Sekiguchi et al., 1982). Equally important to our understanding of picornavirus receptors will be a definition of this receptor.

It has been recognized for many years that there is a relationship among members of the different species of picornaviruses for utilizing a common but different receptor for binding to susceptible cells (Crowell and Syverton, 1961; Quersin-Thiry and Nihoul, 1961; Crowell, 1966, 1987; Lonberg-Holm et al., 1976; Crowell and Landau, 1983). The grouping of related viruses into families on the basis of receptor competition has been confirmed and extended by use of monoclonal antibodies (MAbs) prepared against the cellular receptors (Campbell and Cords, 1983; Minor et al., 1984; Nobis et al., 1985; Crowell et al., 1986; Colonno et al., 1986; Hsu et al., 1988; Shepley et al., 1988) and by the cloning of receptor genes that express their respective

Kuo-Hom Lee Hsu, Wyeth Ayrest Research, P.O. Box 8299, Philadelphia, Pennsylvania 19101. *Severo Paglini,* Institute of Virology, University of Cordoba, Cordoba, Argentina. *Barbara Alstein and Richard L. Crowell,* Department of Microbiology and Immunology, Hahnemann University School of Medicine, Broad and Vine, Philadelphia, Pennsylvania 19102.

TABLE 1
Receptor Families for Picornaviruses

Picornavirus species and serotypes	Selected reference(s)
Poliovirus 1–3	Crowell, 1966
Coxsackievirus	
B1–B6	Crowell, 1966
A2 and -5	Schultz and Crowell, 1983
A13, -15, -18, and -21	Crowell, 1976
Echovirus 6	Crowell, 1976
Human rhinovirus	Lonberg-Holm et al., 1976
Minor group (1A, 2, 44, 49, etc.)	Abraham and Colonno, 1984
Major group (3, 5, 9, 12, 14, etc.)	
Foot-and-mouth disease virus	Sekiguchi et al., 1982
$A_{12}119$, O_{19}, C_{3RES}	
SAT_{1-3}	
Cardioviruses	Allaway and Burness, 1986
Hepatitis A virus	

receptor proteins (Greve et al., 1989; Staunton et al., 1989; Mendelsohn et al., 1989). These receptor families are summarized in Table 1.

HOST RANGE VIRUS VARIANTS CAN USE TWO RECEPTORS

During the course of our studies of cellular receptors for the group B coxsackieviruses, host range variants were isolated (Reagan et al., 1984). These viruses were selected for growth by human rhabdomyosarcoma cells (RD) in culture, and they acquired the capacity to hemagglutinate human erythrocytes (RBCs). Since these group B coxsackieviruses, designated coxsackievirus B-RD (CB-RD), retained the capacity to grow in HeLa cells, they were considered to have acquired a second receptor-binding site. This conclusion was reached after it was discovered that CB3-RD could saturate HeLa cell receptors to prevent binding of prototype CB3, but the saturation of HeLa cells with prototype CB3 failed to block binding of the CB3-RD variant (Reagan et al., 1984; Crowell et al., 1986). We now report evidence that the CB3-RD host range variant recognizes two distinct proteins that can serve as receptors for facilitating infection. One receptor protein, designated Rp-a (Mapoles et al., 1985), has been purified from HeLa cells and is recognized by an MAb referred to as RmcB (Hsu et al., 1988). This protein serves as the receptor for the prototype strains of the six group B coxsackieviruses, and it has a molecular size of approximately 49 kilodaltons (kDa). The second receptor, found on RD cells, is recognized by RmcA and has a molecular size of approximately 60 kDa. This

receptor also is found in abundance on HeLa cells (Crowell et al., 1986).

Studies of cellular receptors generally have focused on characterizing receptors for prototype viruses. This has been important and helpful to establish base-line characteristics for receptors. However, the relatively high rate of virus mutations among the RNA viruses (Parvin et al., 1986) suggests that virus variants may be selected from any RNA virus population that has been subjected to an altered cell environment during viral replication. Although it is unlikely that host range virus variants will be found as frequently as escape mutants recovered from neutralization by monoclonal antibodies, host range virus mutants are present in significant amounts in virus populations (Reagan et al., 1984; Rozhon et al., 1984; Armstrong, 1939).

The selection of virus host range variants is often mediated by the recovery of viruses that have acquired the capacity to bind to an altered or new cellular receptor on host cells, although other mechanisms may be operative (Lomax and Yin, 1989). The CB3-RD host range variant of CB3-Nancy has some unique phenotypic characteristics that distinguish it from the parental strain of virus (Table 2).

Studies of the receptor on human RBCs were performed to determine whether it was related to the receptor on RD cells. It was found that the MAb RmcA completely blocked attachment of ^{35}S-CB3-RD to RBCs and to HEL cells (a human erythroleukemic cell line) that were pretreated with RmcA (data not shown). Since RmcA also inhibits binding of CB3-RD to RD cells, it was concluded that receptors from both cell types shared an epitope. The number of

TABLE 2
Comparative Characteristics of CB3-Nancy and the Host Range Variant CB3-RD[a]

Strain	Characteristic					
	Host range	Hemagglu-tination	Mouse viru-lence	Cellular receptor (molecular size [kDa])	Receptor mAb probe	Virus com-petition on HeLa cells
CB3-Nancy	HeLa, BGM	−	+	49	RmcB	−
CB3-RD	RD, HeLa, BGM	+	−	49, 60	RmcA	+

[a] See Reagan et al. (1984), Crowell et al. (1986), Crowell et al. (1987), Crowell et al. (1988), and Hsu et al. (1988).

binding sites for CB3-RD and RmcA on each of the cell types (HeLa, RD, RBC, and HEL) was determined by ligand saturation titrations as described previously (Hsu et al., 1988; Hsu and Crowell, 1989) (Table 3). The MAb bound to five- to eightfold more sites on the cells than did virions, which is consistent with the concept of multivalent binding of virions to cells (Lonberg-Holm and Philipson, 1981) and with the relatively large size of virions. Each of the receptors was saturable, and the binding of each ligand was shown to be specific by competition of radiolabeled ligand with unlabeled ligand.

To confirm the hypothesis that the prototype receptor differed from the RD and RBC receptor, the receptors were solubilized from membranes of HeLa cells, Buffalo green monkey kidney (BGM) cells (which share a receptor specificity with HeLa cells), RBCs, and RD cells, respectively, and the extracts were analyzed by sodium dodecyl sulfate (SDS)-polyacrylamide gel electrophoresis under nonreducing conditions. The bands of protein were transferred to nitrocellulose and immunoblotted with ^{125}I-labeled MAbs (RmcA and RmcB) or virus blotted with ^{35}S-CB3-RD or ^{35}S-CB3 (Mischak et al., 1988). Purified CB3 virus-receptor complex (VRC), obtained as described previously (Mapoles et al., 1985), was used as a positive control for the prototype receptor. The

results of the virus blot experiment (Fig. 1) revealed that a major band was identified at approximately 46 to 49 kDa by both ^{35}S-CB3 and ^{35}S-CB3-RD in the lane containing the VRC. This size is comparable to that obtained previously (49.5 kDa [Mapoles et al., 1985]) for the receptor protein, Rp-a. A distinct band at this size also was found from BGM and HeLa cell extracts by both viruses, as expected. A band of lesser intensity, at approximately 40 kDa, may represent a nonglycosylated receptor protein, since the results of separate experiments (not shown) revealed that a similar band was produced after treatment of VRC with endoglycosidase F. Further experiments will be necessary, however, to confirm this interpretation. A significant finding was the identification of a heavy band at approximately 60 kDa blotted with ^{35}S-

TABLE 3
Comparative Number of Binding Sites for CB3-RD and RmcA on Different Human Cells[a]

Cell	No. of binding sites	
	CB3-RD	RmcA
HeLa	6×10^5	3×10^6
RD	4×10^3	3×10^4
RBC	4×10^2	3×10^3
HEL	6×10^3	4×10^4

[a] Also see Hsu et al. (1988).

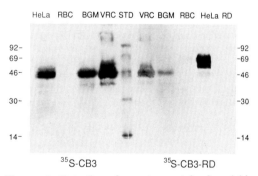

FIGURE 1. Detection of receptor proteins in solubilized membranes from different cells after SDS-polyacrylamide gel electrophoresis (10% polyacrylamide) analysis under nonreducing conditions by the virus blot method. The proteins were electrophoretically transferred to nitrocellulose paper and detected by autoradiography with ^{35}S-CB3 (left side of marker proteins [STD]) and ^{35}S-CB3-RD (right side of marker proteins). The ^{14}C-labeled marker proteins were phosphorylase b (92 kDa), bovine serum albumin (69 kDa), ovalbumin (46 kDa), carbonic anhydrase (30 kDa), and lysozyme (14 kDa).

Table 4
Evidence Suggesting that Some Nonenveloped Viruses Possess More than One Virion Attachment Site for Binding to More than One Molecular Species of Cellular Receptor

Virus	Cell or molecule	Molecule (kDa)	Reference(s)
CB3-RD	RD	60	This chapter
	RBC	60	This chapter
	BGM	49	This chapter
	HeLa	49, 60	This chapter
PV2 and PV1-PV2 chimera	Mouse		Martin et al., 1988
			Murray et al., 1988
	HeLa	43–45	Mendelsohn et al., 1989
Encephalomyocarditis virus	RBC (glycophorin A)	31	Allaway and Burness, 1986
	L		
Reovirus type 3	Thymoma	67	Co et al., 1985
	L	67	Armstrong et al., 1984; Epstein et al., 1984; Pacitti and Gentsch, 1987; Choi and Lee, 1988
	Endothelial	54	Verdin et al., 1989
	RBC (glycophorin A)	31	Paul and Lee, 1987

CB3-RD. The high concentration of the 60-kDa band relative to the band at 49 kDa is consistent with the relatively high number of receptors for RmcA, which identifies the second receptor on HeLa cells and which is approximately 50-fold more than the number of receptors for RmcB (prototype receptor) (Hsu et al., 1988). Unfortunately, either the concentration of receptors on RD cells was too low or the receptor was irreversibly denatured by the SDS detergent to permit detection. We have found a 60-kDa band from RD cells in immunoblots using RmcA in other experiments (not shown). We also have found that the receptor from RBCs is highly sensitive to irreversible denaturation by SDS and is only weakly detected by immunoblots or virus blots (data not shown). Thus, the results of these experiments reveal that there are two separate proteins, of 46 to 49 and 60 kDa, for the prototype and RD cell receptors, respectively. Work is in progress to clone the genes for each of these receptor proteins.

The finding of two receptors for the host range virus variant CB3-RD has suggested that other viruses may have more than one receptor specificity (Table 4). For example, the three immunotypes of reoviruses compete for a common receptor on L cells of approximately 67 kDa (Lee et al., 1981; Epstein et al., 1984; Pacitti and Gentsch, 1987; Choi and Lee, 1988; Armstrong et al., 1984), and types 1 and 3 compete for a different receptor of 54 kDa on endothelial cells (Verdin et al., 1989). Furthermore, type 1 and type 3 reoviruses have distinct histotro-

pisms in mice (Sharpe and Fields, 1985). The type 1 virus infects ependymal cells, whereas type 3 infects neurons. The different diseases in mice have been attributed to differing receptor specificities. A receptor protein of 67 kDa was isolated from mouse thymoma cells by aid of anti-idiotypic antibodies prepared against type 3 reovirus neutralizing antibodies (Co et al., 1985). The receptor resembled the β-adrenergic receptor in size and structure, and it is considered to be a member of the rhodopsin family of receptors (Greene et al., 1987). Evidence that the sigma-1 protein from reovirus type 3 serves as the virion attachment protein is very convincing (Williams et al., 1988). Thus, for some cell types the reoviruses share a common receptor, and for others they have distinct receptor specificities.

Another example of a virus that is likely to have more than one receptor specificity is provided by the mouse-adapted Lansing strain of PV2 (Armstrong, 1939). This virus variant binds to mouse cells while retaining its capacity to bind to HeLa cells (Mendelsohn et al., 1989). Unfortunately, an assay for measuring attachment of this strain of PV2 to mouse cells has not been developed. However, exchanging a nine-amino-acid sequence in VP1 at positions 94 to 102 of PV1 with a corresponding amino acid sequence from PV2 (Lansing) converted PV1 into a mouse-neurovirulent virus (Martin et al., 1988). Furthermore, a chimera of PV1 and PV2 that had six amino acids from PV2 exchanged with those of PV1 (Mahoney) in VP1, in the N-AgI loop, conferred upon PV1 the host range phenotype for mouse infection (Martin et al., 1988; Murray et al., 1988). If these alterations of PV1 (Mahoney), which result in mouse neurovirulence, are a consequence of an acquired capacity of the hybrid virus to bind to mouse cells, then we have another example of a virus that recognizes two cellular receptor specificities. The one virion attachment site for binding to HeLa cells would presumably be located in the canyon (Evans et al., 1989), whereas the second binding site for attachment to mouse cells would be located on the surface of the virion in the N-AgI loop (Murray et al., 1988; Martin et al., 1988). There is both circumstantial (Rossmann et al., 1985; Rossmann and Palmenberg, 1988) and direct (Colonno et al., 1988; Pevear et al., 1989; Heinz et al., 1989) evidence for the canyon as the virion attachment site for binding of prototype picornaviruses to receptors. Further studies will be needed, however, to confirm the identity of two virion attachment sites for the mouse-adapted poliovirus and their corresponding cellular receptors.

In conclusion, we have provided evidence that the host range variant CB3-RD binds to two cellular receptors of different specificities. These receptors are located on different cell types, and they appear to be two distinct proteins of different molecular size, as determined by virus blots and immunoblots with MAbs. We suggest that a number of other viruses can have more than one receptor specificity for initiating infection of different cell types. Further studies of the different cellular receptors for viruses are needed to resolve apparent conflicting results about the identity of receptor specificities when it is possible that viruses have more than one receptor. Finally, the role of multiple receptors as determinants of virus tropisms in the pathogenesis of infections remains to be determined (Crowell et al., 1987; Crowell et al., 1988).

ACKNOWLEDGMENT. Studies from our laboratory were supported by Public Health Service research grant AI-03771 from the National Institute of Allergy and Infectious Diseases.

REFERENCES

Abraham, G., and R. J. Colonno. 1984. Many rhinoviruses share the same receptors. *J. Virol.* **51:**340–345.

Acharya, R., E. Fry, D. Stuart, G. Fox, D. Rowlands, and F. Brown. 1989. The three-dimensional structure of foot-and-mouth disease virus at 2.9 Å resolution. *Nature* (London) **337:**709–716.

Allaway, G. P., and A. T. H. Burness. 1986. Site of attachment of encephalomyocarditis virus on human erythrocytes. *J. Virol.* **59:**768–770.

Altmann, D. M., N. Hogg, J. Trowsdale, and D. Wilkinson. 1989. Cotransfection of ICAM-1 and HLA-DR reconstitutes human antigen-presenting cell function in mouse L cells. *Nature* (London) **338:**512–514.

Armstrong, C. 1939. Successful transfer of the Lansing strain of poliomyelitis virus from the cotton rat to the white mouse. *Public Health Rep.* **54:**2302–2305.

Armstrong, G. D., R. W. Paul, and P. W. K. Lee. 1984. Studies on reovirus receptors of L cells: virus binding characteristics and comparison with reovirus receptors of erythrocytes. *Virology* **138:**37–48.

Campbell, B. A., and C. E. Cords. 1983. Monoclonal antibodies that inhibit attachment of group B coxsackieviruses. *J. Virol.* **48:**561–564.

Choi, A. H. C., and P. W. K. Lee. 1988. Does the β-adrenergic receptor function as a reovirus receptor? *Virology* **163:**191–197.

Co, M. S., G. N. Gaulton, B. N. Fields, and M. I. Greene. 1985. Isolation and biochemical characterization of the mammalian reovirus type 3 cell surface receptor. *Proc. Natl. Acad. Sci. USA* **82:**1494–1498.

Colonno, R. J., G. Abraham, and J. E. Tomassini. 1989. Molecular and biochemical aspects of human rhinovirus attachment to cellular receptors, p. 161–178. *In* B. L. Semler and E. Ehrenfeld (ed.), *Molecular Aspects of Picornavirus Infection and Detection.* American Society for Microbiology, Washington, D.C.

Colonno, R. J., P. L. Callahan, and W. J. Long. 1986. Isolation of a monoclonal antibody that blocks attachment of the major group of human rhinoviruses. *J. Virol.* **57:**7–12.

Colonno, R. J., J. H. Condra, S. Mizutani, P. L. Callahan, M.-E. Davies, and M. A. Murcko. 1988. Evidence for the direct involvement of the rhinovirus canyon in receptor binding. *Proc. Natl. Acad. Sci. USA* **85:**5449–5453.

Crowell, R. L. 1966. Specific cell-surface alteration by enteroviruses as reflected by viral attachment interference. *J. Bacteriol.* **91:**198–204.

Crowell, R. L. 1976. Comparative generic characteristics of picornavirus-receptor interactions, p. 179–202. *In* R. F. Beers, Jr., and E. G. Bassett (ed.), *Cell Membrane Receptors for Viruses, Antigens and Antibodies, Polypeptide Hormones and Small Molecules.* Raven Press, New York.

Crowell, R. L. 1987. Cellular receptors in virus infections. *ASM News* **53:**422–425.

Crowell, R. L., A. K. Field, W. A. Schleif, W. L. Long, R. J. Colonno, J. E. Mapoles, and E. A. Emini. 1986. Monoclonal antibody that inhibits infection of HeLa and rhabdomyosarcoma cells by selected enteroviruses through receptor blockade. *J. Virol.* **57:**438–445.

Crowell, R. L., S. D. Finkelstein, K.-H. L. Hsu, B. J. Landau, P. Stalhandske, and P. W. Whittier. 1988. A murine model for coxsackievirus B3-induced acute myocardial necrosis for study of cellular receptors as determinants of viral tropism, p. 79–92. *In* H. P. Schultheiss (ed.), *New Concepts in Viral Heart Disease.* Springer-Verlag KG, Berlin.

Crowell, R. L., K.-H. L. Hsu, M. Schultz, and B. J. Landau. 1987. Cellular receptors in coxsackievirus infections, p. 453–466. *In* M. A. Brinton and R. R. Rueckert (ed.), *Positive Strand RNA Viruses.* Alan R. Liss, Inc., New York.

Crowell, R. L., and B. J. Landau. 1983. Receptors in the initiation of picornavirus infections. *Compr. Virol.* **18**:1–42.

Crowell, R. L., and K. Lonberg-Holm (ed.). 1986. *Virus Attachment and Entry into Cells.* American Society for Microbiology, Washington, D.C.

Crowell, R. L., and J. T. Syverton. 1961. The mammalian cell-virus relationship. VI. Sustained infection of HeLa cells by coxsackie B3 virus and effect on superinfection. *J. Exp. Med.* **113**:419–435.

Dustin, M. L., K. H. Singer, D. T. Tuck, and T. A. Springer. 1988. Adhesion of T lymphocytes to epidermal keratinocytes is regulated by interferon γ and is mediated by intercellular adhesion molecule 1 (ICAM-1). *J. Exp. Med.* **167**:1323–1340.

Epstein, R. L., M. L. Powers, R. B. Rogart, and H. L. Weiner. 1984. Binding of ^{125}I-labeled reovirus to cell surface receptors. *Virology* **133**:46–55.

Evans, D. J., J. McKeating, J. M. Meredith, K. L. Burke, K. Katrak, A. John, M. Ferguson, P. D. Minor, R. A. Weiss, and J. W. Almond. 1989. An engineered poliovirus chimaera elicits broadly reactive HIV-1 neutralizing antibodies. *Nature* (London) **339**:385–388.

Fox, G., N. R. Parry, P. V. Barnett, B. McGinn, D. J. Rowlands, and F. Brown. 1989. The cell attachment site on foot-and-mouth disease virus includes the amino acid sequence RGD (arginine-glycine-aspartic acid). *J. Gen. Virol.* **70**:625–637.

Greene, M. I., Y. Kokai, G. N. Gaulton, M. B. Powell, H. Geller, and J. A. Cohen. 1987. Receptor systems in tissues of the nervous system. *Immunol. Rev.* **100**:153–184.

Greve, J. M., G. Davis, A. M. Meyer, C. P. Forte, S. C. Yost, C. W. Marlor, M. E. Kamarck, and A. McClelland. 1989. The major human rhinovirus receptor is ICAM-1. *Cell* **56**:839–847.

Harrison, S. C. 1989. Finding the receptors. *Nature* (London) **338**:205–206.

Heinz, B. A., R. R. Rueckert, D. A. Shepard, F. J. Dutko, M. A. McKinlay, M. Fancher, M. G. Rossmann, J. Badger, and T. J. Smith. 1989. Genetic and molecular analyses of spontaneous mutants of human rhinovirus 14 that are resistant to antiviral compounds. *J. Virol.* **63**:2476–2485.

Hogle, J. M., M. Chow, and D. J. Filman. 1985.

Three-dimensional structure of poliovirus at 2.9 Å resolution. *Science* **229**:1358–1365.

Hsu, K.-H. L., and R. L. Crowell. 1989. Characterization of a YAC-1 mouse cell receptor for group B coxsackieviruses. *J. Virol.* **63**:3105–3108.

Hsu, K.-H. L., K. Lonberg-Holm, B. Alstein, and R. L. Crowell. 1988. A monoclonal antibody specific for the cellular receptor for the group B coxsackieviruses. *J. Virol.* **62**:1647–1652.

Hynes, R. O. 1987. Integrins: a family of cell surface receptors. *Cell* **48**:549–554.

Lee, P. W. K., E. C. Hayes, and W. K. Joklik. 1981. Protein sigma 1 is the reovirus cell attachment protein. *Virology* **108**:156–163.

Lomax, N. B., and F. H. Yin. 1989. Evidence for the role of the P2 protein of human rhinovirus in its host range change. *J. Virol.* **63**:2396–2399.

Lonberg-Holm, K., R. L. Crowell, and L. Philipson. 1976. Unrelated animal viruses share receptors. *Nature* (London) **259**:679–681.

Lonberg-Holm, K., and L. Philipson (ed.). 1981. *Receptors and Recognition*, series B, vol. 8. *Virus Receptors*, part 2. Chapman and Hall, Ltd., London.

Luo, M., G. Vriend, G. Kamer, I. Minor, E. Arnold, M. G. Rossmann, U. Boege, D. G. Scraba, G. M. Duke, and A. C. Palmenberg. 1987. The atomic structure of Mengo virus at 3.0 Å resolution. *Science* **235**:182–191.

Mapoles, J. E., D. L. Krah, and R. L. Crowell. 1985. Purification of a HeLa cell receptor protein for group B coxsackieviruses. *J. Virol.* **55**:560–566.

Martin, A., C. Wychowski, T. Coudrec, R. Crainic, J. Hogle, and M. Girard. 1988. Engineering a poliovirus type 2 antigenic site on a type 1 capsid results in a chimaeric virus which is neurovirulent for mice. *EMBO J.* **7**:2839–2847.

Mendelsohn, C. L., E. Wimmer, and V. R. Racaniello. 1989. Cellular receptor for poliovirus: molecular cloning, nucleotide sequence, and expression of a new member of the immunoglobulin superfamily. *Cell* **56**:855–865.

Minor, P. D., P. A. Pipkin, D. Hockley, G. C. Schild, and J. W. Almond. 1984. Monoclonal antibodies which block cellular receptors of poliovirus. *Virus Res.* **1**:203–212.

Mischak, H., C. Neubauer, B. Berger, E. Kuechler, and D. Blaas. 1988. Detection of the human rhinovirus minor group receptor on renaturing Western blots. *J. Gen. Virol.* **69**:2653–2656.

Murray, M. G., J. Bradley, X.-F. Yang, E. Wimmer, E. G. Moss, and V. R. Racaniello. 1988. Poliovirus host range is determined by a short amino acid sequence in neutralization antigenic site I. *Science* **241**:213–215.

Nobis, P., R. Zibirre, G. Meyer, J. Kuhne, G. Warnecke, and G. Koch. 1985. Production of a monoclonal antibody against an epitope on HeLa cells that is the functional poliovirus binding site. *J. Gen. Virol.* **66:**2563–2569.

Pacitti, A. F., and J. R. Gentsch. 1987. Inhibition of reovirus type 3 binding to host cells by sialylated glycoproteins is mediated through the viral attachment protein. *J. Virol.* **61:**1407–1415.

Parvin, J. D., A. Moscona, W. T. Pan, J. M. Leider, and P. Palese. 1986. Measurement of the mutation rates of animal viruses: influenza A virus and poliovirus type 1. *J. Virol.* **59:**377–383.

Paul, R. W., and P. W. K. Lee. 1987. Glycophorin is the reovirus receptor on human erythrocytes. *Virology* **159:**94–101.

Pevear, D. C., M. J. Fancher, P. J. Felock, M. G. Rossmann, M. S. Miller, G. Diana, A. M. Treasurywala, M. A. McKinlay, and F. J. Dutko. 1989. Conformational change in the floor of the human rhinovirus canyon blocks adsorption to HeLa cell receptors. *J. Virol.* **63:**2002–2007.

Quersin-Thiry, L., and E. Nihoul. 1961. Interaction between cellular extracts and animal viruses. II. Evidence for the presence of different inactivators corresponding to different viruses. *Acta Virol.* **5:** 282–293.

Reagan, K. J., B. Goldberg, and R. L. Crowell. 1984. Altered receptor specificity of coxsackievirus B3 after growth in rhabdomyosarcoma cells. *J. Virol.* **49:**635–640.

Roberts, M. K., H. M. Arzipe, C. J. Gauntt, and G. E. Revtyak. 1989. Coxsackievirus B3 attachment stimulates prostacyclin synthesis in cultured endothelial cells. *FASEB J.* **3:**A313.

Rossmann, M. G., E. Arnold, J. W. Erickson, E. A. Frankenberger, J. P. Griffith, H. J. Hecht, J. E. Johnson, G. Kamer, M. Luo, A. G. Mosser, R. R. Rueckert, B. Sherry, and G. Vriend. 1985. Structure of a human common cold virus and functional relationship to other picornaviruses. *Nature* (London) **317:**145–153.

Rossmann, M. G., and A. C. Palmenberg. 1988. Conservation of the putative receptor attachment site in picornaviruses. *Virology* **164:**373–382.

Rozhon, E. J., A. K. Wilson, and B. Jubelt. 1984. Characterization of genetic changes occurring in attenuated poliovirus 2 during persistent infection in mouse central nervous systems. *J. Virol.* **50:**137–144.

Schultz, M., and R. L. Crowell. 1983. Eclipse of coxsackievirus infectivity: the restrictive event for a non-fusing myogenic cell line. *J. Gen. Virol.* **64:** 1725–1734.

Sekiguchi, K., A. J. Franke, and B. Baxt. 1982. Competition for cellular receptor sites among selected aphthoviruses. *Arch. Virol.* **74:**53–64.

Sharpe, A. H., and B. N. Fields. 1985. Pathogenesis of viral infections. Basic concepts derived from the reovirus model. *N. Engl. J. Med.* **312:**486–497.

Shepley, M. P., B. Sherry, and H. L. Weiner. 1988. Monoclonal antibody identification of a 100 kDa membrane protein in HeLa cells and human spinal cord involved in poliovirus attachment. *Proc. Natl. Acad. Sci. USA* **85:**7743–7747.

Staunton, D. E., V. J. Merluzzi, R. Rothlein, R. Barton, S. D. Marlin, and T. A. Springer. 1989. A cell adhesion molecule, ICAM-1, is the major surface receptor for rhinoviruses. *Cell* **56:**849–853.

Verdin, E. M., G. L. King, and E. Maratos-Flier. 1989. Characterization of a common high-affinity receptor for reovirus serotypes 1 and 3 on endothelial cells. *J. Virol.* **63:**1318–1325.

White, J. M., and D. R. Littman. 1989. Viral receptors of the immunoglobulin superfamily. *Cell* **56:**725–728.

Williams, W. V., H. R. Guy, D. H. Rubin, F. Robey, J. N. Myers, T. Kieber-Emmons, D. B. Weiner, and M. I. Greene. 1988. Sequences of the cell-attachment sites of reovirus type 3 and its anti-idiotypic/antireceptor antibody: modeling of their three-dimensional structures. *Proc. Natl. Acad. Sci. USA* **85:**6488–6492.

Chapter 41

Molecular Genetics of Cellular Receptors for Poliovirus

Vincent R. Racaniello, Cathy L. Mendelsohn, Mary Morrison, Marion Freistadt,
Gerardo Kaplan, Eric G. Moss, and Ruibao Ren

Pathogenesis of viral infections depends on both viral and cellular determinants. Work in our laboratory has encompassed the study of both viral and cellular factors that contribute to the development of poliomyelitis, and this chapter summarizes our recent work in these areas. One of our goals is to understand what controls the ability of the Lansing strain of poliovirus type 2 (P2/Lansing) to cause poliomyelitis in mice. It appears that the ability of this strain to infect mice, while other strains cannot, is controlled by an eight-amino-acid sequence on the surface of the virion. This sequence may be involved in the interaction with a host cell receptor on central nervous system (CNS) cells of the mouse. Another goal is to provide a molecular description of the poliovirus-receptor interaction and to elucidate the role of the receptor in pathogenesis. For example, the distribution of the receptor may influence tissue tropism of poliovirus. Therefore, we have embarked on a study of cellular receptors for poliovirus, the first step of which has been isolation of cDNA clones encoding functional receptors.

MOLECULAR BASIS OF POLIOVIRUS HOST RANGE

Although the host range of most strains of poliovirus is limited to primates, several strains have been adapted to other animal species. For the past 7 years we have been studying the P2/Lansing strain of poliovirus, which was adapted to mice in 1939 (Armstrong, 1939). Upon intracerebral inoculation of this virus into mice, the animals develop poliomyelitis that is clinically and histologically similar to the human

disease (Jubelt et al., 1980a; Jubelt et al., 1980b). In contrast, many other strains of poliovirus, including the prototypic Mahoney strain of poliovirus type 1 (P1/Mahoney), do not cause disease after inoculation into mice.

Our initial studies of P2/Lansing were aimed at determining the molecular basis for the ability of this strain to infect mice. This question was addressed by constructing viral recombinants, by manipulation of infectious cDNA, between P2/Lansing and the mouse-avirulent P1/Mahoney. The results of these studies indicated that the mouse-adapted phenotype mapped to the viral capsid proteins (La Monica et al., 1986). This result suggested that functions mediated by the viral capsid—receptor binding, virus entry and uncoating, and encapsidation—were blocked in mice infected with P1/Mahoney. However, inoculation of P1/Mahoney viral RNA into the mouse brain results in one round of replication in the absence of disease (unpublished results). Therefore, the block to P1/Mahoney replication in the mouse CNS is not in encapsidation but rather at an earlier stage of viral replication.

To further localize regions of the capsid required for the mouse-adapted phenotype, neutralization escape mutants of P2/Lansing were isolated that were resistant to neutralization with monoclonal antibodies. In other viral systems, such as reovirus and rabies virus, neutralization escape variants have been useful in elucidating regions of viral proteins important for causing disease. We found that specific mutations within a small region of VP1, amino acids 95 to 102, resulted in reduced neurovirulence in mice (La Monica et al., 1987). Subsequently, it

Vincent R. Racaniello, Cathy L. Mendelsohn, Mary Morrison, Marion Freistadt, Gerardo Kaplan, Eric G. Moss, and Ruibao Ren, Department of Microbiology, Columbia University College of Physicians and Surgeons, 701 West 168th Street, New York, New York 10032.

was shown that a viral recombinant in which VP1 amino acids 95 to 102 of P1/Mahoney were replaced with those of P2/Lansing was neurovirulent in mice (Murray et al., 1988; Martin et al., 1988). Thus, poliovirus host range may be controlled by an eight-amino-acid sequence of VP1, of which only six amino acids differ between P2/Lansing and P1/Mahoney. This region of VP1, known as neutralization antigenic site I or the B-C loop, is a polypeptide loop located on the external surface of the virion, encircling the fivefold axis of symmetry (Hogle et al., 1985).

How much of the B-C loop is required to infect mice? To answer this question, we have constructed derivatives of P2/Lansing in which parts of the B-C loop have been replaced with sequences from P1/Mahoney. The results indicate that full mouse neurovirulence cannot be obtained with loops containing less than the complete P2/Lansing B-C sequence. For example, a virus in which amino acids 95 and 102 of P2/Lansing were replaced with the amino acids from P1/Mahoney is only weakly neurovirulent in mice. Thus, it may be difficult to identify a minimal sequence required for mouse neurovirulence. However, another approach may provide more information on this problem. As expected, substitution of the B-C loop of P2/Lansing with that of P1/Mahoney leads to a virus with little neurovirulence in mice. Typically, when 10^8 PFU of virus is inoculated, 1 in 10 mice becomes paralyzed. However, virus recovered from the brain of the paralyzed mouse has become slightly more neurovirulent. After several passages through the mouse brain, a virus was isolated that has a 50% lethal dose of approximately 10^6 PFU, only 10 times higher than that observed for P2/Lansing. It is likely that this virus has undergone mutations that render it neurovirulent in mice, and work is currently under way to identify these changes. It will be interesting to determine whether these mutations occur within the B-C loop or at a different site that interacts with the B-C loop in the capsid structure.

What is the functional basis for the ability of P2/Lansing to infect mice, while other strains cannot? The simplest explanation is that P2/Lansing is able to recognize a receptor in the mouse CNS, whereas P1/Mahoney cannot. The location of the B-C loop on the virion is compatible with this notion. Five copies of the B-C loop encircle the fivefold axis of symmetry of the icosahedral virion. Just below the B-C loops is a cleft in the virion surface that has been called the canyon, which has been suggested to be the receptor-binding site in rhinovirus (Rossmann

and Palmenberg, 1988; Colonno et al., 1988). There is no evidence to date that the canyon of poliovirus is a receptor attachment site. However, if the canyon is a receptor-binding site, the B-C loop of P2/Lansing could influence receptor binding. Alternatively, the B-C loop of P2/Lansing might bind directly to a mouse cell receptor. Foot-and-mouth disease virus does not contain a canyon but rather appears to bind a cell receptor via a polypeptide loop on the virion surface that is also a major antigenic site (Acharya et al., 1989). The polypeptide loop probably mediates binding to a cell receptor via an RGD sequence, since RGD-containing polypeptides block foot-and-mouth disease virus binding (Fox et al., 1989).

It is certainly possible that the block to P1/Mahoney infection in mice is not at the level of receptor binding but rather at subsequent steps in viral entry. It has not been possible to demonstrate binding of P2/Lansing to mouse brain homogenates, perhaps because of the low abundance or lability of the receptors (Holland, 1961). Therefore, it cannot be determined whether there are receptors for P1/Mahoney in the mouse CNS. This question might be addressed by isolation of cDNA clones encoding the putative murine receptor for P2/Lansing; work toward this goal is discussed below.

IDENTIFICATION OF THE HUMAN CELLULAR RECEPTOR FOR POLIOVIRUS

Although it has been known since the early 1960s that cells susceptible to poliovirus infection contain a specific viral receptor (PVR), isolation of this protein was never achieved. The identification and study of viral receptors is clearly an important goal, since these receptors are required to initiate viral infections and may serve as targets for antiviral chemotherapy. Furthermore, host range and tissue tropism are, for many viruses, influenced by cell receptors. For example, poliovirus replicates only in limited sites in humans (the oropharyngeal and intestinal mucosa, the Peyer's patches of the ileum, and motor neurons within the CNS), yet in its viremic stage poliovirus is exposed to many tissues. It was found that poliovirus-binding sites are present only in susceptible tissues, leading to the suggestion that receptor distribution controls poliovirus tissue tropism (Holland, 1961).

We have used a genetic approach to identifying the PVR, to circumvent problems associated with biochemical purification (Mendelsohn

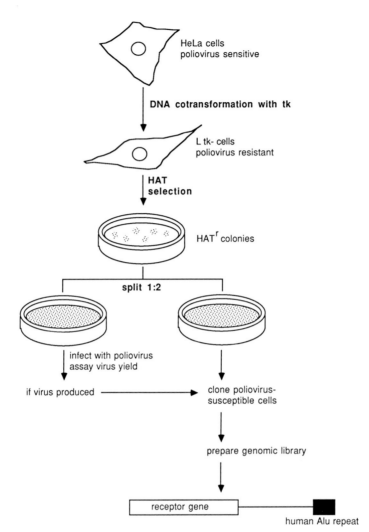

FIGURE 1. Strategy for isolation of a human PVR gene. HeLa cell DNA is cotransformed with a selectable marker into receptor-negative mouse L cells. After hypoxanthine-aminopterin-thymidine (HAT) selection, transformants are split 1:2 and assayed for susceptibility to poliovirus infection. Receptor-expressing cells are cloned from the sibling of plates that contain poliovirus-susceptible cells by rosette assay, using antireceptor monoclonal antibody D171 (Nobis et al., 1985). After isolation of secondary transformants, a genomic library is prepared, and the PVR gene is identified by linkage to a clone human *Alu* repeat sequence. tk, Thymidine kinase. (For further details, see Mendelsohn et al. [1986].)

et al., 1986) (Fig. 1). Briefly, DNA from HeLa cells, which express the PVR, was cotransformed into mouse L cells with a selectable marker. Mouse L cells do not bear PVRs yet can support one round of poliovirus infection initiated with viral RNA (Holland et al., 1959). The cells were placed under the appropriate selection, and the resulting transformants were split 1:2. To assay for receptor expression, transfor-

mants were infected with poliovirus, and the cell medium was assayed for production of infectious virus. The sibling of those plates that contained receptor-positive cells was then subjected to a rosette assay, using an antireceptor monoclonal antibody, D171, to identify receptor-expressing cells. Rosette-positive cells were cloned, DNA from these cells was used to transform L cells, and receptor-positive trans-

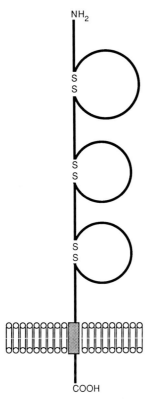

NH₂

COOH

FIGURE 2. Predicted structure of PVR polypeptide. The three immunoglobulinlike domains, formed by disulfide bonding of three cysteine pairs, are shown. The transmembrane domain is depicted by a stippled bar. Not shown are the different lengths (25 or 50 amino acids) of the cytoplasmic tail encoded by two functional cDNA clones. It is not known whether the actual receptor site is a multimer or is associated with other cellular polypeptides.

formants were again identified. In this way, secondary L-cell transformants that express functional poliovirus receptors were isolated. From one such transformant, a genomic library was prepared in bacteriophage lambda, and several *Alu*-reactive clones were isolated. A DNA fragment from one such genomic clone was then used to isolate receptor cDNA clones from HeLa cells.

Two cDNA clones were isolated that, when transformed separately into L cells, led to expression of functional poliovirus receptors (Mendelsohn et al., 1989). The two cDNA clones encode transmembrane glycoproteins with predicted molecular sizes of 43 and 45 kilodaltons, which differ only in the cytoplasmic tails (Fig. 2). The predicted amino acid sequence

reveals that the PVR contains the conserved amino acids and domain structure typical of members of the immunoglobulin superfamily. Many members of the immunoglobulin superfamily have roles in cell-cell adhesion and recognition (Williams and Barclay, 1988). The presence of the PVR in this superfamily suggests that its normal function may involve similar activities.

STRUCTURE OF THE RECEPTOR AND ITS VIRUS-BINDING SITE

The composition and structure of the PVR on the cell surface are not known. For example, the receptor site may be a multimer or may contain proteins other than those encoded by the PVR cDNAs described above. To obtain information on the structure of the PVR site, we have prepared polyclonal antireceptor antisera. Fusion proteins of the PVR and the *trpE* gene product were constructed and expressed in *Escherichia coli*, purified from polyacrylamide gels, and inoculated into rabbits. The resulting antisera react with HeLa cells, as determined by enzyme-linked immunosorbent assay, and block infection with poliovirus. The antisera recognize several polypeptides in HeLa cells, as determined by immunoprecipitation. Work is currently under way to understand how these polypeptides are related to the PVR site.

It will also be important to identify the virus-binding site on the cell receptor. To address this question, deleted versions of the receptor cDNA, that lack one or two domains, have been constructed. The deleted cDNAs will be transformed into mouse L cells to determine whether they can function in virus binding and uptake into cells. Once the virus-binding domain is located, site-directed mutagenesis will be used to more precisely localize the virus-binding site. The receptor may also play a role in virus entry and uncoating, and it may be possible to isolate receptor mutants that are defective in these processes.

Information on the virus attachment site of the receptor will have greater value if the three-dimensional structure of the receptor is known. Knowledge of the three-dimensional structure of the receptor, and especially the virus-binding site of the receptor, may enable rational design of antiviral compounds. We have therefore initiated a collaboration with Jim Hogle, Research Institute of Scripps Clinic, to solve the atomic structure of the PVR. To produce sufficient receptor for crystallization, the receptor has been expressed in insect cells by using baculovirus vectors. By inserting a translational stop codon before the transmembrane domain, a soluble form of the receptor was produced that is secreted into the cell culture medium. This soluble form may be suitable for crystallization studies, and experiments are currently under way to determine whether it can bind poliovirus. Interestingly, when the complete receptor cDNA is expressed in insect cells, functional receptors appear on the cell surface that are able to bind poliovirus. However, binding of poliovirus to insect cells does not lead to productive infection.

TISSUE TROPISM AND THE CELL RECEPTOR FOR POLIOVIRUS

As discussed above, the distribution of the PVR in primates is believed to be a major determinant of poliovirus cell and tissue tropism. This conclusion is based largely on experiments in which virus-binding studies are performed on tissue homogenates (for example, Holland, 1961). To further examine the correlation between tropism and expression of receptors, we plan to examine receptor RNA and protein expression in tissues by using appropriate nucleic acid and antibody probes.

Northern (RNA) blot analysis was used to determine which human tissues express PVR transcripts (Mendelsohn et al., 1989). The results indicate that all of the human tissues examined, including frontal cortex, cerebellar cortex, motor cortex, kidney, and ileum, contain a 3.3-kilobase RNA that is also observed in HeLa cells. Of these tissues, only cerebellar cortex, motor cortex, and ileum are known to be sites of poliovirus replication. Thus, expression of poliovirus-binding sites is not simply controlled by expression of receptor RNA. Perhaps different tissues contain different receptor proteins or proteins that are modified posttranslationally in different ways. Alternatively, receptors may be expressed only in certain parts of different organs in ways that do not physically permit virus infection. Answers to these questions should come from in situ analysis of receptor RNA and protein in tissues, using antibody and cDNA probes.

THE MURINE HOMOLOG OF THE HUMAN PVR

As discussed above, P2/Lansing may be able to infect mice because it can recognize a receptor in the mouse CNS that P1/Mahoney cannot. One way to provide information on this hypothesis would be to isolate the mouse receptor for P2/Lansing and determine whether it can bind P1/Mahoney. A logical candidate for such a P2/Lansing receptor is protein homologous to the human PVR.

Our results indicate that mice indeed contain a gene that is homologous to the human PVR, and genomic and cDNA clones that encode the murine homolog have been isolated. Nucleotide sequence analysis of a partial cDNA clone isolated from mouse liver reveals a polypeptide that contains two domains which appear to be the equivalent of domains 2 and 3 of the human PVR. The human and mouse proteins have 59% identity in domain 2 and 66% identity in domain 3; the cDNA is missing domain 1. The remainder of the protein, from the end of domain 3 through the cytoplasmic tail, is quite different in the two proteins, and the cytoplasmic tail is longer in the murine protein (93 amino acids versus 25 or 50). A cDNA that appears to encode the entire receptor polypeptide has been recently isolated from mouse brain, and nucleotide sequencing of this cDNA is in progress.

Northern blot analysis has revealed that the murine homolog of the human PVR is expressed as two mRNAs in all mouse organs examined, including brain, spinal cord, heart, liver, kidney, and spleen. However, in mice P2/Lansing replicates only in the brain and spinal cord. Therefore, if the receptor homolog indeed serves as the P2/Lansing receptor in mice, then the distribution of its transcript is not sufficient to explain the restriction of P2/Lansing to the mouse CNS.

The crucial question is whether the murine PVR homolog can serve as a receptor for poliovirus, and in particular the P2/Lansing strain. To address this question, the murine PVR cDNA isolated from brain will be transformed into mouse cells, and the transformed cells will be examined for the ability to be infected with P2/Lansing. If the murine PVR homolog can function as a P2/Lansing receptor but not a P1/Mahoney receptor, then the inability of P1/Mahoney to infect mice is due to a block in receptor binding. Should the murine PVR homolog support binding of both serotypes, then the block to P1/Mahoney replication is at a different part of the entry process, possibly uptake or uncoating. It is also possible that the murine PVR homolog cannot serve as a receptor for any poliovirus strain. Binding of P2/Lansing might require posttranslational modification that does not occur in cultured cells transformed with the receptor cDNA. Perhaps the murine PVR homolog is not the receptor used by P2/

Lansing to infect mice. Answers to these questions are required to clarify the functional basis for the mouse-adapted phenotype.

TRANSGENIC MICE EXPRESSING THE HUMAN PVR

A transgenic mouse expressing the human PVR would have several important uses. The factors that govern expression of poliovirus-binding sites and the ability of poliovirus to replicate in tissues could be more easily studied in the mouse than in human tissues. If a transgenic mouse could be isolated that is susceptible to oral inoculation with poliovirus, it would then be possible to study the factors that govern virus spread from the gut to the CNS. A transgenic mouse susceptible to poliovirus infection might also be useful for studying poliovirus attenuation and might be used for testing vaccines.

To derive transgenic mice expressing the human PVR, several cosmid clones that contain the complete human PVR gene have been isolated from a HeLa cell genomic library. After transformation of the cosmid DNAs into L cells, functional virus receptors are expressed on the cell surface. The DNA insert from two different cosmid clones, which contain different amounts of flanking sequences, was excised with the restriction endonuclease *Not*I, purified, and injected into mouse eggs, which were then implanted into pseudopregnant female mice. As of this writing, seven mice have been born that contain intact copies of the human PVR gene integrated into the genome. Offspring of these transgenic animals will be examined for expression of PVR RNA and protein in different organs, and they will also be tested for their susceptibility to poliovirus infection by different routes.

It is clear that poliovirus is an excellent model for studying the molecular basis of viral pathogenesis. The combination of infectious cDNA, the known atomic structure of the viral capsid, a convenient animal model, a molecularly cloned cell receptor, and the possibility of a transgenic mouse model makes possible a multitude of experiments that should provide a great deal of knowledge on the mechanism by which poliovirus causes disease.

REFERENCES

Acharya, R., E. Fry, D. Stuart, G. Fox, D. Rowlands, and F. Brown. 1989. The three-dimensional structure of foot-and-mouth disease virus at 2.9 Å resolution. *Nature* (London) **337**:709–716.

Armstrong, C. 1939. Successful transfer of the Lansing strain of poliomyelitis virus from the cotton rat to the white mouse. *Public Health Rep.* **54**:2302–2305.

Colonno, R., J. Condra, S. Mizutani, P. Callahan, M.-F. Davies, and M. Murcko. 1988. Evidence for the direct involvement of the rhinovirus canyon in receptor binding. *Proc. Natl. Acad. Sci. USA* **85**:5449–5453.

Fox, G., N. R. Parry, P. V. Barnett, B. McGinn, D. J. Rowlands, and F. Brown. 1989. The cell attachment site on foot-and-mouth disease virus includes the amino acid sequence RGD (arginine-glycine-aspartic acid). *J. Gen. Virol.* **70**:625–637.

Hogle, J. M., M. Chow, and D. J. Filman. 1985. Three-dimensional structure of poliovirus at 2.9 Å resolution. *Science* **229**:1358–1365.

Holland, J. J. 1961. Receptor affinities as major determinants of enterovirus tissue tropisms in humans. *Virology* **15**:312–326.

Holland, J. J., J. C. McLaren, and J. T. Syverton. 1959. The mammalian cell virus relationship. IV. Infection of naturally insusceptible cells with enterovirus ribonucleic acid. *J. Exp. Med.* **110**:65–80.

Jubelt, B., B. Gallez-Hawkins, O. Narayan, and R. T. Johnson. 1980a. Pathogenesis of human poliovirus infection in mice. I. Clinical and pathological studies. *J. Neuropathol. Exp. Neurol.* **39**:138–148.

Jubelt, B., O. Narayan, and R. T. Johnson. 1980b. Pathogenesis of human poliovirus infection in mice. II. Age-dependency of paralysis. *J. Neuropathol. Exp. Neurol.* **39**:149–158.

La Monica, N., W. Kupsky, and V. R. Racaniello. 1987. Reduced mouse neurovirulence of poliovirus type 2 Lansing antigenic variants selected with monoclonal antibodies. *Virology* **1**:429–437.

La Monica, N., C. Meriam, and V. R. Racaniello. 1986. Mapping of sequences required for mouse neurovirulence of poliovirus type 2 Lansing. *J. Virol.* **57**:515–525.

Martin, A., C. Wychowski, T. Couderc, R. Crainic, J. Hogle, and M. Girard. 1988. Engineering a poliovirus type 2 antigenic site on a type 1 capsid results in a chimaeric virus which is neurovirulent for mice. *EMBO J.* **7**:2839–2847.

Mendelsohn, C., B. Johnson, K. A. Lionetti, P. Nobis, E. Wimmer, and V. R. Racaniello. 1986. Transformation of a human poliovirus receptor gene into mouse cells. *Proc. Natl. Acad. Sci. USA* **83**:7845–7849.

Mendelsohn, C., E. Wimmer, and V. R. Racaniello. 1989. Cellular receptor for poliovirus: molecular cloning, nucleotide sequence and expression of a new member of the immunoglobulin superfamily. *Cell* **56**:855–865.

Murray, M. G., J. Bradley, X. F. Yang, E. Wimmer, E. G. Moss, and V. R. Racaniello. 1988. Poliovirus host range is determined a short amino acid sequence in neutralization antigenic site I. *Science* **241**:213–215.

Nobis, P., R. Zibirre, G. Meyer, J. Kuhne, G. Warnecke, and G. Koch. 1985. Production of a monoclonal antibody against an epitope on HeLa cells that is the functional poliovirus binding site. *J. Gen. Virol.* **6**:2563–2569.

Rossmann, M. G., and A. C. Palmenberg. 1988. Conservation of the putative receptor attachment site in picornaviruses. *Virology* **164**:373–382.

Williams, A. F., and A. N. Barclay. 1988. The immunoglobulin superfamily—domains for cell surface recognition. *Annu. Rev. Immunol.* **6**:381–405.

VII. ANTIGENIC STRUCTURE AND FUNCTIONS

Chapter 42

Functional Analysis of Surface Glycoproteins of the Coronavirus Murine Hepatitis Virus Strain JHM

Stuart G. Siddell, Roland Stauber, Edward Routledge, and Michael Pfleiderer

The positive-stranded RNA genome of the murine coronavirus murine hepatitis virus strain JHM (MHV-JHM) includes the genes for four virus structural proteins: the nucleocapsid protein (N), the membrane glycoprotein (M), the spike glycoprotein (S), and the hemagglutinin-esterase glycoprotein (HE) (Siddell, 1982). In infected cells, expression of these proteins is mediated by a set of subgenomic mRNAs that are 3' coterminal in relation to the genome and extend to different positions in a 5' direction. The mRNAs are synthesized by a process of leader-primed discontinuous transcription (for details, see Lai [1986]). mRNAs7, -6, -3, and -2-1 have been shown to encode the N, M, S, and HE proteins, respectively (Siddell, 1983; Shieh et al., 1989; Pfleiderer et al., in press). Locations of the major open reading frames (ORFs) in the MHV-JHM genome and their relationships to the subgenomic mRNAs are shown in Fig. 1.

The MHV-JHM virion has two surface glycoproteins. The spike protein is a heterodimer consisting of two noncovalently bound subunits, S_1 and S_2, which are derived by proteolytic processing of an approximately 180,000-molecular-weight precursor, S (Siddell, 1982; Sturman et al., 1985). Multimers of the heterodimer assemble together to produce the characteristic peplomer structures at the surface of the virion. Indirect evidence using S-protein-specific monoclonal antibodies (Collins et al., 1982; Wege et al., 1984) suggests that the MHV S protein can mediate binding of the virus to the cell surface and the fusion of viral and cellular membranes.

The second surface structure of MHV-JHM consists of a disulfide-linked homodimer(s) of the HE protein, the reduced monomer of which has a molecular weight of 65,000 (Siddell, 1982).

The function(s) of the MHV-JHM HE protein is not known. King et al. (1985) have shown that the hemagglutination activity of bovine coronavirus (BCV) is associated with the HE protein, and Vlasak et al. (1988b) have demonstrated that BCV and human coronavirus OC43 can bind to and destroy a receptor on the cell surface that is similar or identical to the influenza C virus receptor. Recently, Vlasak et al. (1988a) showed that in the case of BCV, the receptor-destroying activity of the HE protein had the specificity of an acetylesterase.

In the studies described here, we attempted to directly investigate the biological functions of the MHV-JHM surface glycoproteins by using vaccinia virus recombinants. Moreover, we have isolated a set of S-protein-specific monoclonal antibodies (MAbs) that recognize sequential epitopes and inhibit S-protein functions. Using a bacterial expression system to synthesize a series of carboxy-terminal deleted S-protein polypeptides, we have been able to map these MAbs to specific regions of the S-protein sequence.

FUSION ACTIVITY OF THE MHV-JHM S PROTEIN

To directly demonstrate that the MHV-JHM S protein is responsible for the fusogenic activity of the virus, we cloned the S gene into the unique *Bam*HI site of the vaccinia virus cloning vector pTF7.5 (Fuerst et al., 1987). The construct, pTF7.5/S$^+$, was then used to transfect DBT cells that had been infected with the vaccinia virus recombinant vTF7.3, which expresses the T7 RNA polymerase gene. Figure 2B shows the cytopathic effect 10 h after infection-

---

Stuart G. Siddell, Roland Stauber, Edward Routledge, and Michael Pfleiderer, Institute of Virology, Versbacher Strasse 7, 8700 Würzburg, Federal Republic of Germany.

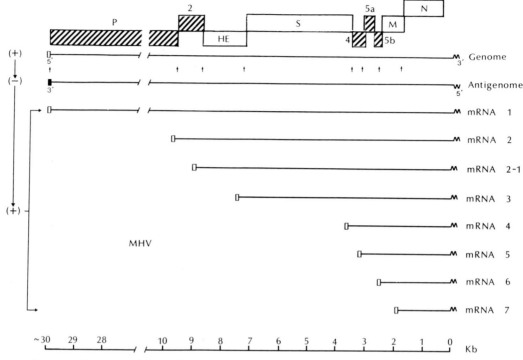

FIGURE 1. Summary of the MHV-JHM replication strategy. The size of the genome is based on the results of Baker et al. (1989). The figure includes data submitted for publication.

transfection. It is evident that the infected-transfected DBT cells display extensive syncytia formation, the characteristic cytopathic result of the MHV-JHM fusogenic activity. This cytopathic effect is quite different from that caused by vaccinia virus infection (Fig. 2A). This result shows that the MHV-JHM S protein alone is sufficient to mediate the membrane fusion of infected cells.

BACTERIAL EXPRESSION OF THE MHV-JHM S PROTEIN AND THE PRODUCTION OF MAbs

To express the MHV-JHM S protein in *Escherichia coli* RRI, we chose the pEV-vrf plasmids developed by Crowl et al. (1985). This system can be used to direct the synthesis of fusion proteins (with short N-terminal extensions) or polypeptides containing the naturally occurring amino terminus. The MHV-JHM S-gene ORF (lacking three N-terminal amino acids) was placed downstream of the λ p_L promoter, in frame with a short, seven-amino-acid, vector-derived ORF. This construct was designated pEV-vrf-1/S. Expression was regulated

by a temperature-sensitive *c*I repressor carried on the pRK248 *c*Its plasmid (Bernard and Helinski, 1979). These features are summarized in Fig. 3. After temperature induction, protein lysates were prepared and analyzed by sodium dodecyl sulfate-polyacrylamide gel electrophoresis (Laemmli, 1970). The S protein expressed after induction was synthesized in appreciable amounts and partitioned to the membrane fraction (Fig. 3). This material was purified and used to immunize 3-month-old BALB/c mice by a variety of routes over a 6-month period. After the final inoculation, spleen cells were fused to NS1 plasmacytoma cells, and hybridomas, selected in hypoxanthine-aminopterin-thymidine medium, were screened for S-protein-specific antibody production, using MHV-JHM-infected or uninfected Sac(−) cell lysates as the capture antigen. Ten hybridomas secreting S-protein-specific antibody were cloned.

CHARACTERIZATION OF THE S-PROTEIN-SPECIFIC MAbs AND LOW-RESOLUTION EPITOPE MAPPING

The ability of the 10 S-protein-specific MAbs to neutralize virus infectivity or to inhibit

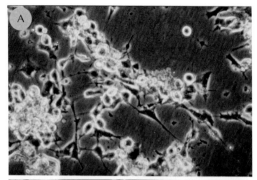

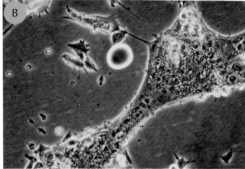

FIGURE 2. Fusion activity of the MHV-JHM S protein. DBT cells were infected with the recombinant vaccinia virus vTF7.3 at a multiplicity of infection of 30. Two hours postinfection, the cells in panel B were transfected with calcium phosphate-coprecipitated pTF7.5/S⁺ DNA (~20 µg). The monolayers were photographed 10 h posttransfection.

the virus-mediated fusion of cell membranes was tested (Table 1). Of particular interest were MAbs 30B, 11F, and 10G. MAb 30B was able to neutralize MHV-JHM but was unable to prevent fusion. Presumably, neutralization in this case was due to the inhibition of virus (i.e., S protein) attachment to the cell surface. MAbs 11F and 10G had lower neutralizing titers than 30B but also inhibited virus-mediated cell-cell fusion (fusion from within), suggesting that these MAbs neutralize virus infectivity by preventing virus-target cell membrane fusion after attachment. It should be noted that MAbs 30B, 10G, and 11F are significantly less effective in virus neutralization than the conformation-dependent MAbs produced in response to immunization with viral S-protein antigen (Wege et al., 1984; E. Routledge, unpublished data). It will be interesting to define the relationships of the epitopes recognized by these two classes of MAbs.

To define the regions of the S protein in which the epitopes recognized by MAbs 30B, 10G, and 11F are located, we used the pEV-vrf-1/S clone to construct a series of plasmids that express carboxy-terminal truncated S-protein polypeptides. First, the *Bam*HI fragment proximal to the 3' end of the MHV-JHM S gene was replaced by a synthetic DNA fragment that contained unique restriction sites appropriate for a unidirectional exonuclease III digestion. These sites were also flanked to the 3' side by termination codons in all three reading frames. After unidirectional deletion, 11 plasmids that

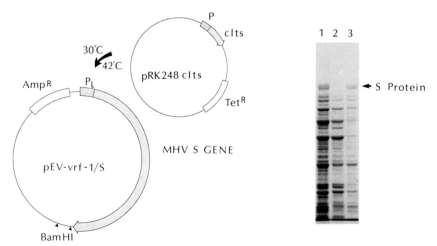

FIGURE 3. Bacterial expression of the MHV-JHM S protein. (Left) Structures of clone pEV-vrf-1/S and plasmid pRK248 *c*Its. (Right) Sodium dodecyl sulfate-polyacrylamide gel electrophoresis analysis of protein lysates from induced bacteria. Lanes: 1, total lysate; 2, soluble fraction; 3, membrane fraction. The position of the MHV-JHM S protein is indicated.

TABLE 1
Analysis of the MHV-JHM S-Protein-Specific MAbs

MAb	Western blot reactivity with given deletion protein												Virus neutrali-zation[a]	Fusion inhibi-tion[b]
	1–1235	1–1224	1–1138	1–861	1–742	1–703	1–602	1–586	1–525	1–520	1–501	1–349		
11F	+	+	+	+	+	+	+	+	+	+	+	+	160–640	640
5E	+	+	+	+	+	+	+	+	+	+	+	+	<40	<40
30B	+	+	+	+	+	+	+	+	+	+	+	+	1,024	<40
17A	+	+	+	−	−	−	−	−	−	−	−	−	ND	ND
10G	+	+	+	−	−	−	−	−	−	−	−	−	40	40–160
15D[c]	+	+	+	+	+	+	−	−	−	−	−	−	ND[d]	ND
18F[e]	+	+	+	+	+	+	−	−	−	−	−	−	ND	ND

[a] Reciprocal of the antibody dilution causing 60% reduction in plaque number.
[b] Reciprocal of the antibody dilution causing 75% reduction in plaque size.
[c] Belongs to the D group described by Wege et al. (1984).
[d] ND, Not done.
[e] Belongs to the F group described by Wege et al. (1984).

represented truncations extending over 70% of the S protein were obtained. The precise extent of each deletion was determined by nucleotide sequence analysis, and the reactivity of the S-protein-specific MAbs with the truncated polypeptides was determined by Western blotting (immunoblotting) (Table 1).

It was found that MAb 30B, which was able to neutralize MHV-JHM, mapped to a region within the amino-terminal (S_1) half of the protein. This result is in agreement with the conclusions reached, for example, by Mockett et al. (1984), who studied infectious bronchitis virus, and Deregt et al. (1989), who have reported the tentative mapping of BCV S-protein-neutralizing epitopes to the amino-terminal half of the molecule. Neutralizing epitopes located on the S_1 moiety of the MHV-JHM S protein have not been previously reported.

Second, the epitopes defined by MAbs 10G and 11F, both of which are able to inhibit S-protein-mediated membrane fusion, mapped to different locations, in the carboxy-terminal (S_2) and amino-terminal (S_1) regions, respectively. The interpretation of this result is not yet clear. Luytjes et al. (1989) have located the epitope recognized by an MHV-A59 fusion-inhibiting MAb (Talbot and Buchmeier, 1985) to a position (amino acids 848 to 856) that lies close to the region that we have defined for MAb 10G. However, there has been no indication that amino-terminal domains on the MHV S protein, as defined by MAb 11F, may be involved in S-protein-mediated membrane fusion. Although the data are clear, this interpretation should be considered tentative because it is based on an oversimplified (linear) view of the S-protein structure.

BIOLOGICAL FUNCTIONS OF THE MHV-JHM HE PROTEIN

Luytjes et al. (1988) have recently cloned and sequenced the unique region of the MHV-A59 mRNA2. They found two large ORFs, of which the downstream ORF lacks an amino-terminal initiation codon. These authors interpreted this ORF as a pseudogene but noted a striking similarity with the HEF1 subunit of the influenza C virus surface glycoprotein (Nakada et al., 1984). Sequence analysis of the MHV-JHM HE protein gene (Shieh et al., 1989; Pfleiderer et al., in press) reveals the same sequence similarity for the MHV-JHM HE protein, which is, however, expressed via an additional mRNA (mRNA2-1) and constitutes a major component of the virion (Siddell, 1982).

Because the influenza C virus HEF protein and the BCV HE protein have been shown to have receptor-binding and -destroying activities specific for O-acetylneuraminic acids, we attempted to directly demonstrate these activities for the MHV-JHM HE protein by using vaccinia virus recombinants. A cDNA containing the HE protein gene was cloned into the unique BamHI site of the pTF7.5 vector with the ATG start codon next to the T7 RNA polymerase promoter. This construct, pTF7.5/HE, was used to transfect DBT cells that had been infected with the recombinant vaccinia virus vTF7.3. Lysates were made from the infected-transfected cells and immunoprecipitated with an MHV-JHM HE MAb (kindly provided by H. Wege). The acetylesterase activities of these immunocomplexes were then tested by using the organic substrate p-nitrophenylacetate. These experiments (Pfleiderer et al., in press) clearly demonstrated that

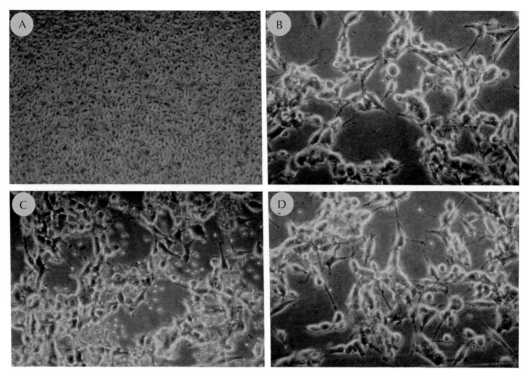

FIGURE 4. Hemabsorption activity of the MHV-JHM HE protein. DBT cells were uninfected (A) or infected with the vaccinia virus recombinant vTF7.3 at a multiplicity of infection of 30 and then transfected with pTF7.5/S⁻ (B) or pTF7.5/HE (C and D) plasmid DNA. Six hours postinfection, monolayers were chilled to 4°C and incubated with rat erythrocytes. Unspecifically bound erythrocytes were washed away at 4°C (A, B, and C) or after elevation of the temperature to 37°C (D).

the HE protein, but not the S protein, has an esterase activity. The activity found in purified MHV-JHM virions can therefore be attributed to this protein.

Since we could demonstrate an HE-specific acetylesterase activity, the next question was whether the protein can also recognize and bind to sialic acid-containing receptors on the cell surface. Therefore, we investigated the ability of the vaccinia virus-expressed HE protein to bind rat erythrocytes, which are known to present the receptor of the influenza C virus in high amounts (G. Herrler, personal communication). DBT cells were infected with the recombinant vaccinia virus vTF7.3, transfected with either pTF7.5/S⁻ or pTF7.5/HE plasmid DNA, and tested for hemadsorption activity. Only DBT cells that express the MHV-JHM HE protein (Fig. 4C) specifically bind erythrocytes. Uninfected DBT cells (Fig. 4A) or DBT cells that express the MHV-JHM S protein (Fig. 4B) do not have this activity. This result shows that MHV-JHM has a second glycoprotein on its

surface which binds to specific receptors on the host cell. Raising the temperature of the cultures that express the HE protein to 37°C results in release of the erythrocytes (Fig. 4D), consistent with acetylesterase activity of the HE protein.

DISCUSSION

Coronavirus MHV-JHM has two surface glycoproteins, the S protein and the HE protein. The experiments reported here and earlier studies (Wege et al., 1984) suggest that both proteins have the ability to bind to cellular receptor(s). In addition, both proteins have a further biological function, the fusion activity of the S protein and the esterase activity of the HE protein.

It has been shown in this chapter that the MHV S protein is the fusion-active component of the virus. This protein does not require acidic pH conditions to develop its activity (Sturman et al., 1985), and it has no amino-terminal hydrophobic sequences in its S_2 subunit (Schmidt et al., 1987). Our studies with MAbs suggest that the MHV-JHM S-protein-mediated fusion activity, like that of the E1 protein of the alphaviruses (Garoff et al., 1980a, 1980b), involves one or more internal domains. Moreover, at least two discontinuous regions of the protein are involved, although it is not yet possible to say whether they form a single conformational domain.

The second surface glycoprotein of MHV-JHM, the HE protein, has been shown in this chapter to have receptor-binding and receptor-destroying activities. Since the closely related MHV-A59 apparently does not express this protein (Luytjes et al., 1988), the importance of these two activities for the biology of MHV-JHM is not yet clear. A detailed analysis of the cell and tissue tropism, as well as the infectivity and pathogenesis, of these viruses is required.

ACKNOWLEDGMENTS. We are indebted to R. Crowl for supplying the pEV-vrf system and to B. Moss for supplying the vaccinia virus system used in these studies. We thank H. Wege for MAbs and Helga Kriesinger for typing the manuscript. The excellent technical assistance of B. Schelle-Prinz is acknowledged.

This work was supported by the Sonderforschungsbereich (grant SFB 165/B1).

REFERENCES

Baker, S. C., C.-K. Shieh, L. H. Soe, M.-F. Chang, D. M. Vannier, and M. M. C. Lai. 1989. Identification of a domain required for autoproteolytic cleavage of murine coronavirus gene A polyprotein. *J. Virol.* **63**:3693–3699.

Bernard, H.-U., and D. R. Helinski. 1979. Use of the λ phage promoter P_L to promote gene expression in hybrid plasmid cloning vehicles. *Methods Enzymol.* **68**:482–492.

Collins, A. R., R. L. Knobler, H. Powell, and M. J. Buchmeier. 1982. Monoclonal antibodies to murine hepatitis virus-4 (strain JHM) define the viral glycoprotein responsible for attachment and cell-cell fusion. *Virology* **119**:358–371.

Crowl, R., C. Seamans, P. Lomedico, and S. McAndrew. 1985. Versatile expression vectors for high-level synthesis of cloned gene products in Escherichia coli. *Gene* **38**:31–38.

Deregt, D., M. D. Parker, G. C. Cox, and L. A. Babiuk. 1989. Mapping of neutralizing epitopes to fragments of the bovine coronavirus E2 protein by proteolysis of antigen-antibody complexes. *J. Gen. Virol.* **70**:647–658.

Fuerst, T. R., P. L. Earl, and B. Moss. 1987. Use of hybrid vaccinia virus-T7 RNA polymerase system for expression of target genes. *Mol. Cell. Biol.* **7**:2538–2544.

Garoff, H., A.-M. Frischauf, K. Simons, H. Lehrach, and H. Delius. 1980a. Nucleotide sequence of cDNA coding for Semliki Forest virus membrane glycoproteins. *Nature* (London) **288**:236–241.

Garoff, H., A.-M. Frischauf, K. Simons, H. Lehrach, and H. Delius. 1980b. The capsid protein of Semliki Forest virus has clusters of basic amino acids and prolines in its amino terminal region. *Proc. Natl. Acad. Sci. USA* **77**:6376–6380.

King, B., B. J. Potts, and D. Brian. 1985. Bovine coronavirus hemagglutinin protein. *Virus Res.* **2**: 53–59.

Laemmli, U. K. 1970. Cleavage of structural proteins during the assembly of the head of bacteriophage T4. *Nature* (London) **227**:680–685.

Lai, M. M. C. 1986. Coronavirus leader-RNA-primed transcription: an alternative mechanism to RNA splicing. *BioEssays* **5**:257–260.

Luytjes, W., P. J. Bredenbeek, A. F. H. Noten, M. C. Horzinek, and W. J. M. Spaan. 1988. Sequence of mouse hepatitis virus A59 mRNA 2: indications for RNA recombination between coronaviruses and influenza C virus. *Virology* **166**:415–422.

Luytjes, W., D. Geerts, W. Posthumus, R. Meloen, and W. Spaan. 1989. Amino acid sequence of a conserved neutralizing epitope of murine coronaviruses. *J. Virol.* **63**:1408–1412.

Mockett, A. P. A., D. Cavanagh, and T. D. K. Brown. 1984. Monoclonal antibodies to the S_1 spike and membrane proteins of avian infectious bronchitis coronavirus strain Massachusetts M41. *J. Gen. Virol.* **65**:2281–2286.

Nakada, S., R. Creager, S. Krystal, R. P. Aaronson, and P. Palese. 1984. Influenza C virus hemagglutinin: comparison with influenza A and B virus hemagglutinins. *J. Virol.* **50**:118–124.

Pfleiderer, M., E. Routledge, and S. G. Siddell. Functional analysis of the coronavirus MHV-JHM surface glycoproteins in vaccinia virus recombinants. *Adv. Exp. Med. Biol.*, in press.

Schmidt, I., M. A. Skinner, and S. G. Siddell. 1987. Nucleotide sequence of the gene encoding the surface projection glycoprotein of the coronavirus MHV-JHM. *J. Gen. Virol.* **68**:47–56.

Shieh, C.-K., H.-J. Lee, K. Yokomori, N. La Monica, S. Makino, and M. M. C. Lai. 1989. Identification of a new transcriptional initiation site and the corresponding functional gene 2b in the murine coronavirus RNA genome. *J. Virol.* **63**:3729–3736.

Siddell, S. G. 1982. Coronavirus JHM: tryptic peptide fingerprinting of virion proteins and intracellular polypeptides. *J. Gen. Virol.* **62**:259–269.

Siddell, S. G. 1983. Coronavirus JHM: coding assignments of subgenomic mRNAs. *J. Gen. Virol.* **64**: 113–125.

Sturman, L. S., C. S. Ricard, and K. V. Holmes. 1985.

Proteolytic cleavage of the E2 glycoprotein of murine coronavirus: activation of cell-fusing activity of virions by trypsin and separation of two different 90K cleavage fragments. *J. Virol.* **56:**904–911.

Talbot, P. J., and M. J. Buchmeier. 1985. Antigenic variation among murine coronaviruses: evidence for polymorphism on the peplomer glycoprotein, E2. *Virus Res.* **2:**317–328.

Vlasak, R., W. Luytjes, J. Leider, W. Spaan, and P. Palese. 1988a. The E3 protein of bovine coronavirus is a receptor-destroying enzyme with acetylesterase activity. *J. Virol.* **62:**4686–4690.

Vlasak, R., W. Luytjes, W. Spaan, and P. Palese. 1988b. Human and bovine coronaviruses recognize sialic acid-containing receptors similar those of Influenza C virus. *Proc. Natl. Acad. Sci. USA* **85:**4526–4529.

Wege, H., R. Dörries, and H. Wege. 1984. Hybridoma antibodies to the murine coronavirus JHM: characterization of epitopes on the peplomer protein (E2). *J. Gen. Virol.* **65:**1931–1942.

Chapter 43

Antigenic Structure and Function of the Flavivirus Envelope Protein E

F. X. Heinz, C. Mandl, H. Holzmann, F. Guirakhoo, W. Tuma, and C. Kunz

The family *Flaviviridae* comprises about 70 serologically related but distinct viruses, most of which are transmitted by infected mosquitoes or ticks. Several flaviviruses represent important human pathogens. The highest worldwide disease incidences are caused by the dengue viruses, yellow fever virus, Japanese encephalitis virus, and tick-borne encephalitis (TBE) virus. Effective vaccines are in widespread use for the prevention of yellow fever virus (live attenuated), Japanese encephalitis virus, and TBE (both purified and inactivated).

The flavivirus genomic RNA contains a single long open reading frame of more than 10,000 nucleotides that codes for all structural and nonstructural proteins (for a review, see Rice et al. [1986]). Figure 1 summarizes our current knowledge of the structural composition of flaviviruses. Three proteins are found in mature virions. These are termed E (envelope), C (capsid), and M (membrane) and have molecular weights of about 50,000 to 60,000, 15,000, and 8,000, respectively. The C protein together with the genomic RNA forms an isometric nucleocapsid. Both the E and M proteins are associated with the lipid envelope. The E protein of most (but not all) flaviviruses is glycosylated and has one or possibly two carbohydrate side chains. The M protein is generated by proteolytic cleavage of its glycosylated precursor protein (preM), presumably by a Golgi protease. The preM protein is a major constituent of immature virus particles, and its processing apparently represents a late event in the maturation of flaviviruses to yield fully infectious mature virions (Wengler and Wengler, 1989).

Biosynthesis of the structural membrane proteins and the first nonstructural protein (NS1) involves internal signal sequences that lead to the transfer of the nascent polypeptide chain into the lumen of the endoplasmic reticulum (Rice et al., 1986; Wengler et al., 1985). As shown by carboxy-terminal sequence analyses, both preM(M) and E (as present in the virion) contain the internal signal sequences for the subsequent protein on the polyprotein precursor (Nowak et al., 1989; Wright et al., 1989). They are therefore believed to contain a double membrane anchor, with the carboxy terminus located outside the membrane.

Being the major virion surface protein, the E protein plays a central role in several important biological functions of flaviviruses: it is the receptor-binding protein and is also probably responsible for membrane fusion activity after acid pH-induced conformational changes. It induces neutralizing and protective antibodies and also represents the basis for the serological classification of flaviviruses into several serocomplexes, each comprising more closely related viruses (De Madrid and Porterfield, 1974; Calisher et al., 1989).

Using TBE virus as a model, we have attempted to characterize the antigenicity, structure, and functions of the E protein at a molecular level.

STRUCTURAL MODEL OF THE E PROTEIN

We have established a working model of the structural organization of the TBE virus E protein (Mandl et al., 1989) that reveals a specific arrangement of the polypeptide chain into several distinct entities (Fig. 2). These can be correlated to different antigenic domains and prob-

F. X. Heinz, C. Mandl, H. Holzmann, F. Guirakhoo, W. Tuma, and C. Kunz, Institute of Virology, University of Vienna, Kinderspitalgasse 15, A-1095 Vienna, Austria.

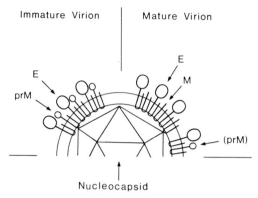

Immature Virion | Mature Virion

E
prM

E
M

(prM)

Nucleocapsid

FIGURE 1. Schematic model of a flavivirus showing immature (intracellular) and mature forms of the virion.

ably mediate different functional activities. It is reasonable to assume a common structural organization of the flavivirus E proteins in general, since there is an absolute conservation of all 12 cysteine residues, and the sequences also reveal virtually superimposable hydrophilicity plots.

As deduced from its reactivity with concanavalin A and experiments using endoglycosidases (Winkler et al., 1987a), the TBE virus E protein is glycosylated. It contains two potential N-glycosylation sites at amino acid positions 154 and 361. Digestion of the native protein with proteases yields an immunoreactive fragment that extends from amino acids 301 to 395 and does not react with concanavalin A (Winkler et al., 1987b). It is therefore concluded that the single carbohydrate side chain is attached to the asparagine at position 154. This N-glycosylation site is conserved among several other flaviviruses, including Japanese encephalitis virus, Murray Valley encephalitis virus, St. Louis encephalitis virus, and the dengue viruses (Mandl et al., 1988).

The structural model shown in Fig. 2 is based on an epitope map (Fig. 3) that reveals the functional activities and topological relationships of 19 distinct epitopes with different serological specificities (Guirakhoo et al., 1989). As deduced from mutual blocking in competitive binding essays using pairs of monoclonal antibodies (MAbs), most of the epitopes cluster to form three major nonoverlapping antigenic domains (A, B, and C). In addition, three MAbs define isolated epitopes, termed i1, i2, and i3. Broadly flavivirus group-reactive epitopes are only found in domain A (A1 and A2), whereas most of the epitopes in domain B are tick-borne complex specific. There is a cluster of subtype-specific epitopes in domain C (Guirakhoo et al., 1989). Each of the domains contains epitopes involved in hemagglutination inhibition, neutralization, or both. In general, however, domain A-specific MAbs seem to have higher specific functional activities than those directed to other sites.

The arrows in Fig. 3 indicate cooperative interactions between antibodies to nonoverlapping sites in the E protein that are due to an enhancement of antibody avidity. Such synergistic effects were found with other flaviviruses as well and also operate in functional assays such as neutralization and passive protection (Heinz et al., 1986; Kimura-Kuroda and Yasui, 1986).

The following approaches were used to deduce the structural model shown in Fig. 2 from the epitope map: characterization of each epitope with respect to conformation dependency, denaturation resistance, and involvement of disulfide bridges; analysis of the antigenic reactivity of protein fragments generated by proteases or cyanogen bromide and determination of their location by NH_2-terminal sequence analysis; selection of MAb escape mutants and comparative sequence analysis to determine amino acid exchanges in the E protein; expression of the E protein and fragments thereof in bacteria; and inclusion of the disulfide bridge assignments determined by Nowak and Wengler (1987) for the E protein of West Nile virus.

Domain A is shown as a discontinuous structural entity composed of a strongly disulfide bridge-stabilized part (amino acids 50 to 125) and a second part separated by about 100 amino acids in the linear sequence (amino acids 200 to 250). This specific arrangement has been derived as follows: epitopes A3, A4, and A5 are defined by quite distant amino acid exchanges in MAb escape mutants (positions 67, 71, 207, and 233, respectively), although the epitope map (Fig. 3) provides evidence for spatial overlap between A3 and A4 and between A4 and A5. Consistent with their putative discontinuous nature, these epitopes are sensitive to denaturation by sodium dodecyl sulfate (SDS) and are also sensitive to conformational changes induced by acidic pH (Guirakhoo et al., 1989). The neutralization site recognized by MAb A3 (defined by the amino acid exchanges at positions 67 and 71 in two independently selected MAb escape mutants) perfectly matches the position of a neutralization site of yellow fever virus determined by Lobigs et al. (1987).

The other two epitopes of antigenic domain

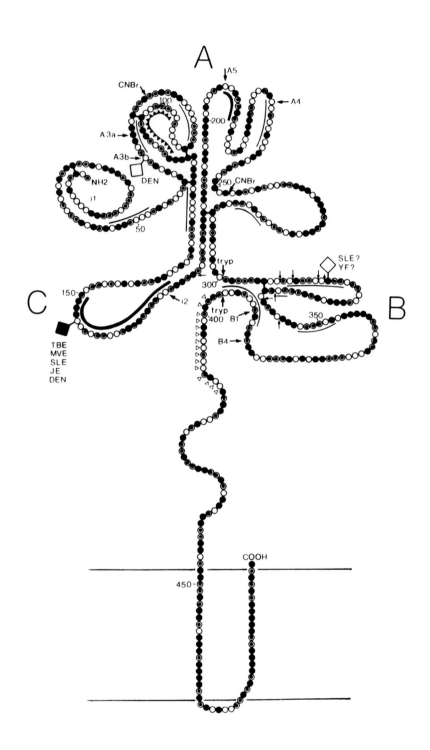

A are defined by nonneutralizing MAbs and can therefore not be mapped by the use of MAb escape mutants. Although the epitope map reveals an overlap between A2 and A3, A1 and A2 are structurally distinct from A3, A4, and A5 since they are resistant to SDS denaturation. They are, however, destroyed by reduction and carboxymethylation, indicating their stabilization by disulfide bridges. In addition, the reactivity of the corresponding MAbs with specific immunoreactive protein E fragments (Mandl et al., 1989) provides evidence that these broadly flavivirus cross-reactive epitopes are located at the disulfide bridge-stabilized part of domain A.

It is interesting to note that epitopes A1 and A2 cannot be detected by Western blotting (immunoblotting) using the E protein expressed in bacteria, although they are resistant to SDS denaturation when the virion-derived E protein is analyzed. Apparently, the disulfide bridges necessary for the stabilization of these epitopes are not properly formed in bacteria (W. Tuma et al., manuscript in preparation). Bacterial expression also reveals that the amino terminus of the E protein is not involved in antigenic domain A. A fusion protein containing amino acids 2 to 34 exclusively reacts with MAb i1, which does not compete with any other MAb used for the epitope map. The reactivity of this epitope does not depend on the disulfide bridge between the cysteines at positions 3 and 30, since it is resistant to reduction and carboxymethylation. There is evidence that these amino-terminal sequences are less accessible to antibody binding in the native protein as compared with denatured and fragmented forms (F. Guirakhoo, unpublished observations).

Domain A also contains the most highly conserved sequence of the E protein (amino acids 98 to 111, located on both sides of a disulfide bridge). There are two points that provide some evidence that this sequence may be involved in membrane fusion activity: (i) low-pH-induced conformational changes affect the epitopes of domain A that surround the conserved sequence, indicating a rearrangement of the polypeptide chain in this area which is probably necessary for the exposure of the fusion-active site, and (ii) the sequence Gly-Leu-Phe-Gly after the disulfide bond corresponds to the amino terminus of influenza virus hemagglutinin 2, which is known to mediate membrane fusion (Wharton et al., 1988). In addition, an inverse sequence (Phe-Leu-Gly) is present at the amino terminus of the paramyxovirus F1 fusion protein and also, as a tandem repeat, within the fusion-active gp41 of human immunodeficiency virus types 1 and 2 and simian immunodeficiency virus (Gallaher, 1987).

All epitopes of domain B are located on a fragment obtained by trypsin digestion of the native protein, which includes amino acids 301 to 395 and is resistant to further proteolysis. The antigenic reactivity of this domain depends on the disulfide bridge between residues 307 and 338, since it is destroyed by reduction and carboxymethylation. However, as revealed by Western blotting and renaturation experiments (Winkler et al., 1987b), the disulfide bridge has a strong tendency for renaturation, and its reformation is associated with the reacquisition of antigenic reactivity. This disulfide bridge is also correctly formed when the protein is expressed in bacteria. Neutralization escape mutants se-

---

FIGURE 2. Model of TBE virus protein E. Open circles represent hydrophilic amino acid residues (Arg, Lys, Asn, Asp, Gln, Glu, and His), dotted circles show intermediate amino acid residues (Pro, Tyr, Ser, Trp, Thr, and Gly), and solid circles show hydrophobic amino acid residues (Ile, Val, Leu, Phe, Cys, Met, and Ala). Position numbers are shown every 50 amino acids. Cysteine residues forming disulfide bridges are connected by solid lines. Arrows depict cleavage sites that liberate immunoreactive fragments by the use of trypsin (tryp) and CNBr. Small arrows indicate potential cleavage sites within these fragments that are not utilized. Two solid lines represent the lipid membrane that is spanned by two transmembrane regions of protein E. The polypeptide chain is folded to indicate the antigenic domains A, B, and C. Arrows together with the names of neutralizing MAbs depict the locations of the mutations identified in the respective antigenic variants of TBE virus by sequence analysis. A line of solid triangles indicates the almost perfectly conserved sequence within domain A. A line of open triangles marks the region of a potential T-cell determinant. A solid diamond represents the carbohydrate side chain of TBE virus. The Murray Valley encephalitis (MVE), St. Louis encephalitis (SLE), Japanese encephalitis (JE), and dengue (DEN) viruses have potential N-glycosylation sites at the homologous position. Yellow fever (YF) and St. Louis encephalitis viruses have such a site within domain B; dengue viruses have an additional site within domain A. The homologous positions of TBE virus are shown by open diamonds. The thin and thick solid lines indicate variable and hypervariable sequences, respectively, as deduced from a sequence comparison of 15 flaviviral E proteins. (Reproduced from Mandl et al. [1989] with permission.)

lected in the presence of MAbs B1 and B4 have single amino acid exchanges at amino acids 384 and 389 near the carboxy terminus of domain B. A homologous structure was identified in the Japanese encephalitis virus E protein by Mason et al. (1989). Using a bacterial expression system, the authors also showed that the antigenic activity of this structural entity depends on the presence of the disulfide bridge as well as the carboxy-terminal sequences. It is therefore very likely that this domain represents a characteristic structural element of flaviviruses in general.

The epitopes of domain C are resistant to reduction and carboxymethylation and seem to be confined to a single loop (Fig. 2) that carries the carbohydrate side chain of the E protein of TBE virus and several other flaviviruses. The high degree of variability of this loop is consistent with most domain C epitopes being type or subtype specific. For TBE virus, the carbohydrate does not seem to be essential for antibody binding but probably plays a role in conferring denaturation resistance to some of these epitopes (Guirakhoo et al., 1989).

After complete deglycosylation with N-glycanase, all domain C antibodies still react with the deglycosylated E protein, although to a somewhat lower extent (Guirakhoo et al., 1989). In the deglycosylated form, however, the epitopes of domain C have lost the resistance to

denaturation with SDS. Apparently, the carbohydrate moiety is essential for shielding these sites from the action of SDS or for allowing renaturation to occur during the immunoassay procedure.

Although there is some degree of one-way blocking between certain domain C antibodies and antibody i3, the corresponding epitope is clearly not part of domain C. It does not lose denaturation resistance upon deglycosylation, and it is also located on different immunoreactive fragments obtained after CNBr cleavage (Guirakhoo et al., 1989). Epitope i3 (which is sequential in nature, as deduced from its resistance to SDS, reduction, and carboxymethylation) may rather be located within the disulfidefree part of domain A, since MAb i3 does not react with a mutant selected in the presence of MAb A4 (Holzmann et al., 1989) that maps to amino acid 233 (Fig. 2).

STUDIES ON THE MOLECULAR BASIS OF ATTENUATION

Each of seven neutralization escape mutants selected in the presence of six different MAbs (compare Fig. 2 and 3) differed from the wild-type virus by only a single amino acid exchange that caused loss of reactivity with the selecting MAb (Holzmann et al., 1989). These mutants allowed us to assess the effect of single amino acid exchanges in different parts of the molecule on the pathogenicity of the virus, using the mouse model. Comparative titrations of identical stocks of each of these mutants and the wild type were performed in primary chicken embryo cells, suckling mice (intracerebral inoculation), and adult mice (subcutaneous inoculation).

By these analyses, mutant B4 differed significantly from all the other viruses. It revealed a strikingly lower titer (by a factor of more than 1 million) in adult mice as compared with suckling mice and chicken embryo cells. Apparently, this variant had lost its ability to kill adult mice upon peripheral inoculation, although it had retained its capacity to replicate to high titers in suckling mice and chicken embryo cells. As revealed by the induction of specific antibodies, the mutant was, however, still capable of replicating in adult mice without causing neurological symptoms. The immune response induced by the attenuated mutant also conferred resistance against challenge with 100 and 1,000 50% lethal doses of a highly virulent strain of TBE virus (Hypr).

We have thus shown that a single amino

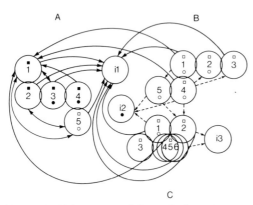

FIGURE 3. Epitope map of the TBE virus glycoprotein E showing three antigenic domains (A, B, and C) and three isolated epitopes (i1, i2, and i3). Overlapping circles indicate mutual blocking, dotted arrows indicate one-way blocking, and solid arrows indicate enhancement of binding as determined in competitive binding assays. Functional activities: ■, strong hemagglutination inhibition activity; □, weak hemagglutination inhibition activity; ●, strong neutralization activity; ○, weak neutralization activity. (Reproduced from Guirakhoo et al. [1989] with permission.)

acid exchange within domain B caused attenuation of neuropathogenicity upon peripheral inoculation of adult mice. The mutant, however, retained its capacity to replicate and to induce a protective immune response.

Antigenic domain B contains several stretches of conserved amino acids, and there is some evidence that amino acid exchanges within this domain may be relevant to the attenuation of flaviviruses in general. Three of the seven nonconservative amino acid exchanges in the E protein of the yellow fever vaccine strain 17D as compared with the wild-type strain Asibi are located within domain B (Hahn et al., 1987). Even more strikingly, three of five nonconservative amino acid exchanges in an attenuated mutant of dengue virus type 2 also map to this domain (Hahn et al., 1988). It is interesting to note that of the seven TBE virus mutants, the amino acid exchange in mutant B4 is the only one that does not affect a variable site but is located within one of the conserved sequence elements. It is therefore tempting to speculate that the conserved sequences in domain B form part of the putative receptor-binding site and that the amino acid exchanges in this area change the receptor specificity, thus leading to loss of neurotropism and attenuation.

REFERENCES

Calisher, C. H., N. Karabatsos, J. M. Dalrymple, R. E. Shape, J. S. Porterfield, E. G. Westaway, and W. E. Brandt. 1989. Antigenic relationships between flaviviruses as determined by cross-neutralization tests with polyclonal antisera. *J. Gen. Virol.* **70:**37–43.

De Madrid, A. T., and J. S. Porterfield. 1974. The flaviviruses (group B arboviruses): a cross-neutralization study. *J. Gen. Virol.* **23:**91–96.

Gallaher, W. R. 1987. Detection of a fusion peptide sequence in the transmembrane protein of human immunodeficiency virus. *Cell* **50:**327–328.

Guirakhoo, F., F. X. Heinz, and C. Kunz. 1989. Epitope model of tick-borne encephalitis virus envelope glycoprotein E: analysis of structural properties, role of carbohydrate side chain, and conformational changes occurring at acidic pH. *Virology* **169:**90–99.

Hahn, C. S., J. H. Dalrymple, J. H. Strauss, and C. M. Rice. 1987. Comparison of the virulent Asibi strain of yellow fever virus with the 17D vaccine strain derived from it. *Proc. Natl. Acad. Sci. USA* **84:** 2019–2023.

Hahn, Y. S., R. Galler, T. Hunkapiller, J. M. Dalrymple, J. H. Strauss, and E. G. Strauss. 1988. Nucleotide sequence of dengue 2 RNA and comparison of the encoded proteins with those of other flaviviruses. *Virology* **162:**167–180.

Heinz, F. X., C. Mandl, G. Winkler, W. Tuma, and C. Kunz. 1986. Cooperative interactions between antibodies to structurally distinct antigenic sites, p. 387–392. *In* F. Brown, R. M. Chanock, and R. A. Lerner (ed.), *Vaccines 86.* Cold Spring Harbor Laboratory, Cold Spring Harbor, N.Y.

Holzmann, H., C. Mandl, F. Guirakhoo, F. X. Heinz, and C. Kunz. 1989. Characterization of antigenic variants of tick-borne encephalitis virus selected with neutralizing monoclonal antibodies. *J. Gen. Virol.* **70:**219–222.

Kimura-Kuroda, J., and K. Yasui. 1986. Antigenic comparison of envelope protein E between Japanese encephalitis virus and some other flaviviruses using monoclonal antibodies. *J. Gen. Virol.* **67:**2663–2672.

Lobigs, M., L. Dalgarno, J. J. Schlesinger, and R. C. Weis. 1987. Location of a neutralization determinant in the E protein of yellow fever virus (17D vaccine strain). *Virology* **161:**474–478.

Mandl, C., F. X. Heinz, and C. Kunz. 1988. Sequence of the structural proteins of tick-borne encephalitis virus (Western subtype) and comparative analysis with other flaviviruses. *Virology* **166:**197–205.

Mandl, C. W., F. Guirakhoo, H. Holzmann, F. X. Heinz, and C. Kunz. 1989. Antigenic structure of the flavivirus envelope protein E at the molecular level, using tick-borne encephalitis virus as a model. *J. Virol.* **63:**564–571.

Mason, P. W., J. M. Dalrymple, M. K. Gentry, J. M. McCown, C. H. Hoke, D. S. Burke, M. J. Fournier, and T. L. Mason. 1989. Molecular characterization of a neutralizing domain of the Japanese encephalitis virus structural glycoprotein. *J. Gen. Virol.* **70:** 2037–2049.

Nowak, T., P. M. Färber, G. Wengler, and G. Wengler. 1989. Analyses of the terminal sequences of West Nile structural proteins and of the in vitro translation of these proteins allow the proposal of a complete scheme of the proteolytic cleavages involved in their synthesis. *Virology* **169:**365–376.

Nowak, I., and G. Wengler. 1987. Analysis of disulfides present in the membrane proteins of the West Nile flavivirus. *Virology* **156:**127–137.

Rice, C. M., E. G. Strauss, and J. H. Strauss. 1986. Structure of the flavivirus genome, p. 279–326. *In* S. Schlesinger and M. J. Schlesinger (ed.), *The Togaviridae and Flaviviridae.* Plenum Publishing Corp., New York.

Wengler, G., E. Castle, U. Leidner, T. Nowak, and G. Wengler. 1985. Sequence analysis of the membrane protein V3 of the flavivirus West Nile and of its gene. *Virology* **147:**264–274.

Wengler, G., and G. Wengler. 1989. Cell-associated West Nile flavivirus is covered with E + pre-M protein heterodimers which are destroyed and reorganized by proteolytic cleavage during virus release. *J. Virol.* **63:**2521–2526.

Wharton, S. A., S. R. Martin, R. W. H. Ruigrok, J. J. Skehel, and D. C. Wiley. 1988. Membrane fusion by peptide analogs of influenza virus hemagglutinin. *J. Gen. Virol.* **69:**1847–1857.

Winkler, G., F. X. Heinz, and C. Kunz. 1987a. Studies on the glycosylation of flavivirus E proteins and the role of carbohydrate in antigenic structure. *Virology* **159:**237–243.

Winkler, G., F. X. Heinz, and C. Kunz. 1987b. Characterization of a disulfide bridge-stabilized antigenic domain of tick-borne encephalitis virus structural glycoprotein. *J. Gen. Virol.* **68:**2239–2244.

Wright, P. J., M. R. Cauchi, and M. L. Ng. 1989. Definition of the carboxytermini of three glycoproteins specified by dengue type 2. *Virology* **171:**61–67.

Chapter 44

T-Lymphocyte Responses to Dengue Viruses

Ichiro Kurane, Alan L. Rothman, Jack F. Bukowski, Udo Kontny, Jurand Janus, Bruce L. Innis, Ananda Nisalak, Suchitra Nimmannitya, Anthony Meager, and Francis A. Ennis

Dengue virus infections are a significant cause of morbidity and mortality in many areas of the world, including southeast Asia and Central and South America (Halstead, 1981). Dengue viruses are transmitted to humans by mosquitoes, and they cause two forms of clinical syndromes: dengue fever and dengue hemorrhagic fever/ dengue shock syndrome (DHF/DSS) (Halstead, 1980). Dengue fever is a self-limited disease and is the most common type of dengue illness. In some situations, patients infected with dengue virus develop life-threatening complications such as hemorrhagic manifestations and shock. The pathogenesis of DHF/DSS has not been elucidated; however, epidemiological studies in Thailand have shown that DHF/DSS is much more commonly observed in secondary dengue virus infections caused by a serotype of virus different from that which caused the primary infection (Halstead, 1980). A higher incidence of DHF/DSS during secondary infections was also observed in recent dengue virus epidemic in Cuba (Kouri et al., 1986). A small percentage of DHF/DSS cases is also observed during primary infections, and most of these patients are infants between 6 and 12 months of age born to dengue virus antibody-positive mothers (Halstead, 1980). On the basis of these epidemiological studies and the in vitro observation that antibodies to dengue viruses at subneutralizing concentrations augment dengue virus infection of Fcγ receptor-positive cells (Halstead and O'Rourke, 1977), it has been hypothesized that anti-dengue virus antibodies increase the number of dengue virus-infected cells, and the lysis of dengue virus-infected cells may lead to DHF/DSS (Halstead, 1980).

It has been reported that virus-specific cytotoxic T lymphocytes (CTL) play an important role in recovery from certain virus infections (Byrne and Oldstone, 1984; Kuwano et al., 1988; Wells et al., 1981). Dengue virus-specific CTL may contribute to recovery from dengue virus infections by lysing dengue virus-infected cells; however, it is also possible that the lysis of dengue virus-infected monocytes by dengue virus-specific CTL may lead to DHF/DSS. Therefore, we have begun to analyze T-cell responses to dengue viruses to understand the role of CTL in the pathogenesis of and in recovery from dengue virus infections, using human (Kurane et al., 1989a; Kurane et al., 1989c) and murine (Rothman et al., 1989) lymphocytes.

T-CELL PROLIFERATION RESPONSES TO DENGUE VIRUS ANTIGENS

Dengue virus-specific human T-cell proliferation was examined by using peripheral blood lymphocytes (PBL) from Thai donors who had antibodies to dengue viruses. Most of the PBL from antibody-positive donors showed serotype-cross-reactive proliferative responses to dengue virus antigens prepared by using Vero cells (Kurane et al., 1989a). However, because the dengue virus infection histories of these Thai donors were not known, serotype specificity of the responses could not be precisely determined. To further examine the serotype speci-

---

Ichiro Kurane, Alan L. Rothman, Jack F. Bukowski, Udo Kontny, Jurand Janus, and Francis A. Ennis, Division of Infectious Diseases, Department of Medicine, University of Massachusetts Medical Center, Worcester, Massachusetts 01655.   *Bruce L. Innis and Ananda Nisalak,* Department of Virology, Armed Forces Research Institute of Medical Sciences, Bangkok, Thailand.   *Suchitra Nimmannitya,* Children's Hospital, Bangkok, Thailand.   *Anthony Meager,* Division of Immunobiology, National Institute for Biological Standards and Control, Hertfordshire EN6 3QG, United Kingdom.

ficity of the responses, PBL from a donor who was known to have been infected with dengue virus type 3 (dengue-3) were used. PBL of this donor responded best to dengue-3 antigen and also responded to dengue-1, -2, and -4 antigens to lower levels. PBL from a donor who was known to have been infected with dengue-1 responded best to dengue-1 antigen and also responded to dengue-2, -3, and -4 antigens to lower levels. Therefore, human T-cell responses in bulk cultures after primary dengue virus infections are predominantly serotype specific but also contain serotype-cross-reactive responses. The responding lymphocytes were characterized as T cells with a $CD3^+$ $CD4^+$ $CD8^-$ phenotype (Kurane et al., 1989c).

We have also examined murine T-cell proliferation responses to dengue virus antigens after single immunization with one serotype of dengue virus. The proliferative responses of the spleen cells were predominantly serotype specific; however, lower levels of cross-reactivity were also observed. Most of the proliferating cells have a $Thy1^+$ $L3T4^+$ $Lyt2^-$ phenotype. These results are consistent with those obtained by using human PBL (Rothman et al., 1989).

DENGUE VIRUS-SPECIFIC T-CELL CLONES

To analyze T-cell specificity, dengue virus-specific T-cell clones were established. Thirteen human T-cell clones were established from a dengue-3-immune donor. All of the clones had a $CD3^+$ $CD4^+$ $CD8^-$ phenotype. The serotype specificities of these T-cell clones were examined in proliferation assays (Table 1). Nine clones responded to dengue-1, -2, and -4 antigens to similar levels as to dengue-3 antigen; therefore, they are serotype cross-reactive. Four clones predominantly responded to dengue-3 antigen; therefore, they are serotype specific. These results indicate that serotype-cross-reactive proliferation in bulk culture reflects the serotype-cross-reactive responses at a clonal level.

Four murine T-cell clones were established from spleen cells of a dengue-2-immune mouse. These clones had a $Thy1^+$ $L3T4^+$ $Lyt2^-$ phenotype. In contrast to the human clones, three clones showed serotype-specific proliferation, and one showed serotype-cross-reactive proliferation. Although serotype-specific clones may be common in mice, these results are consistent with those obtained by using the human T-cell clones, because $CD4^+$ $CD8^-$ T-cell clones are generated and they contain both serotype-cross-reactive and serotype-specific clones.

TABLE 1
Proliferation Responses of T-Cell Clones to Dengue Virus Antigens[a]

Antigen	[³H]thymidine incorporation (cpm)		
	Serotype cross-reactive		Specific (JK31)
	JK3b	JK32	
Dengue-1	2,322	1,920	357
Dengue-2	1,960	1,965	675
Dengue-3	4,816	2,756	3,451
Dengue-4	1,968	1,644	669
Control	398	330	587
None	200	318	81

[a] A total of 10^4 cells were cultured with 2×10^5 gamma-irradiated autologous peripheral blood mononuclear cells in 0.2 ml of RPMI 1640 containing dengue virus and control antigens diluted 1:30 for 3 days. Cells were pulsed with 1.25 µCi of [³H]thymidine for 8 h before harvest.

LYMPHOKINE PRODUCTION BY DENGUE VIRUS-SPECIFIC T LYMPHOCYTES

Production of gamma interferon (IFN-γ) and interleukin-2 (IL-2) by dengue virus-specific human T cells was examined. High titers of IFN-γ were detected in the culture fluids of PBL from dengue virus-immune donors stimulated with dengue virus antigens (Table 2). At the clonal level, all clones produced IFN-γ after stimulation with dengue-3 antigen, and all but one produced IFN-γ after stimulation with dengue antigens of other serotypes. We also examined four clones for IL-2 production. These clones produced IL-2 after stimulation with den-

TABLE 2
IFN-γ Production by Peripheral Blood Mononuclear Cells from Dengue Virus Antibody-Positive Donors after Stimulation with Dengue Virus Antigens[a]

Antigen	IFN-γ (U/ml)		
	Donor 1	Donor 2	Donor 3
Dengue-1	24	7	ND
Dengue-2	50	32	38
Dengue-3	18	ND	43
Dengue-4	29	ND	35
Control	2	2	2
None	3	1	<1

[a] A total of 4×10^5 peripheral blood mononuclear cells were cultured for 6 days with dengue virus and control antigens diluted 1:30. Culture fluids were examined for IFN-γ by radioimmunoassay. Donors 1 and 2 were healthy Thai donors who had antibodies to dengue viruses. The dengue virus infection histories of these donors were not known. Donor 3 was known to be infected with dengue-3.

gue-3 antigen, as determined by CTLL assays. The proliferation responses of T-cell clones were inhibited by anti-human IL-2 and anti-IL-2 receptor antibodies. These results indicate that CD4$^+$ CD8$^-$ human T-cell clones produce IL-2 and respond to the produced IL-2.

CYTOTOXIC ACTIVITY OF DENGUE VIRUS-SPECIFIC T-CELL CLONES

Human T-cell clones were examined for cytotoxic activities against dengue virus antigen-pulsed and dengue virus-infected autologous target cells. Serotype-cross-reactive T-cell clones lysed autologous lymphoblastoid cell lines pulsed with dengue-1, -2, -3, and -4 antigens. They also lysed dengue-2-infected lymphoblastoid cell lines. Serotype cross-reactivity in cytotoxicity was consistent with that observed in proliferation assays. The lysis by these clones was human leukocyte antigen (HLA) class II restricted, and HLA DP and DQ antigens were the restricting antigens for the clones examined (Kurane et al., 1989c).

POSSIBLE ROLE OF DENGUE VIRUS-SPECIFIC T LYMPHOCYTES IN THE PATHOGENESIS OF DENGUE VIRUS INFECTIONS

Dengue virus-specific, CD4$^+$ CD8$^-$ T lymphocytes are generated after primary dengue virus infection. They have two kinds of serotype specificities at the clonal level: serotype cross-reactive and serotype specific. Serotype cross-reactive clones seem to be dominant in humans. These T lymphocytes produce IFN-γ and IL-2 after stimulation with dengue virus antigens and have cytotoxic activity to dengue virus-infected and dengue virus antigen-pulsed target cells in an HLA class II-restricted fashion. We have reported that IFN-γ augments dengue virus infection of human monocytic cells in the presence of antibodies to dengue viruses by increasing the number of Fcγ receptors (Kontny et al., 1988). On the basis of these observations, we hypothesize the role of T lymphocytes in the pathogenesis of DHF/DSS (Fig. 1). During secondary infection with a serotype of dengue virus different from that causing the primary infection, serotype-cross-reactive CD4$^+$ CD8$^-$ T lymphocytes produce IFN-γ. IFN-γ increases the number of dengue virus-infected monocytes and activates these monocytes to produce various mediators. These monocytes are lysed by CTL, and the mediators released by these monocytes

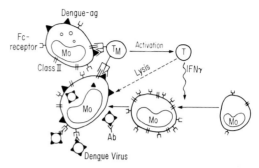

FIGURE 1. Augmentation by IFN-γ of dengue virus infection of human monocytes and lysis by dengue virus-specific CTL: possible immunopathological role of T lymphocytes during secondary dengue virus infections. Dengue-ag, Dengue virus antigen; Mo, monocyte. (Reprinted, with permission, from Kurane et al. [1989b].)

may induce plasma leakage and hemorrhagic manifestations. It has also been reported that IL-2 at high titers induces plasma leakage (Rosenstein et al., 1986). Production of high titers of IL-2 by dengue virus-specific T lymphocytes may be one of the immunopathological mechanisms.

It is likely that these T lymphocytes also contribute to the recovery from dengue virus infections. Even during secondary dengue virus infections, most patients recover from infection without developing DHF/DSS (Halstead, 1980). If dengue-specific CTL contribute to both immunopathology and recovery, it will be very important to determine the mechanisms that determine the outcome of dengue virus-specific T-lymphocyte responses during infections.

The epitopes recognized by these T lymphocytes also need to be determined. Dengue virus proteins prepared by recombinant techniques, purified dengue virus proteins, and synthetic peptides will be used for mapping of T-cell epitopes. Mapping of dominant T-cell epitopes will contribute to the future development of effective subunit vaccines against dengue virus infections.

ACKNOWLEDGMENTS. This work was supported by grant DAMD 17-86-C-6208 from the U.S. Army Medical Research and Development Command and by Public Health Service grant NIH-T32-AI07272 from the National Institutes of Health.

REFERENCES

Byrne, J. A., and M. B. A. Oldstone. 1984. Biology of cloned cytotoxic T lymphocytes specific for lymphocytic choriomeningitis virus: clearance of virus in vivo. *J. Virol.* **51**:682–686.

Halstead, S. B. 1980. Immunological parameters of togavirus disease syndromes, p. 107–173. *In* R. W. Schlesinger (ed.), *The Togaviruses: Biology, Structure, Replication.* Academic Press, Inc., New York.

Halstead, S. B. 1981. Dengue haemorrhagic fever—a public health problem and a field for research. *Bull. WHO* **58**:1–21.

Halstead, S. B., and E. J. O'Rourke. 1977. Dengue viruses and mononuclear phagocytes. I. Infection enhancement by non-neutralizing antibody. *J. Exp. Med.* **146**:201–217.

Kontny, U., I. Kurane, and F. A. Ennis. 1988. Interferon gamma augments Fcγ receptor-mediated dengue virus infection of human monocytic cells. *J. Virol.* **62**:3928–3933.

Kouri, G., M. G. Guzman, and J. Bravo. 1986. Hemorrhagic dengue in Cuba: history of an epidemic. *Pan Am. Health Organ. Bull.* **20**:24–30.

Kurane, I., B. L. Innis, A. Nisalak, C. Hoke, S. Nimmannitya, A. Meager, and F. A. Ennis. 1989a. Human T cell responses to dengue virus antigens. Proliferative responses and interferon gamma production. *J. Clin. Invest.* **83**:506–513.

Kurane, I., U. Kontny, A. L. Rothman, and F. A. Ennis. 1989b. Interferon-γ production by dengue antigen-specific T lymphocytes: possible immuno-

pathological role in secondary dengue virus infections, p. 367–370. *In* R. A. Lerner, H. Ginsberg, R. M. Chanock, and F. Brown (ed.), *Vaccines 89: Modern Approaches to New Vaccines including Prevention of AIDS.* Cold Spring Harbor Laboratory, Cold Spring Harbor, N.Y.

Kurane, I., A. Meager, and F. A. Ennis. 1989c. Dengue virus-specific human T cell clones: serotype cross-reactive proliferation, interferon gamma production, and cytotoxic activity. *J. Exp. Med.* **170**:763–775.

Kuwano, K., M. Scott, J. F. Young, and F. A. Ennis. 1988. HA2 subunit of influenza A H1 and H2 subtype viruses induces a protective cross-reactive cytotoxic T lymphocyte responses. *J. Immunol.* **140**:1264–1268.

Rosenstein, M., S. E. Effinghausen, and S. A. Rosenberg. 1986. Extravasation of intravascular fluid mediated by the systemic administration of recombinant interleukin 2. *J. Immunol.* **137**:1735–1742.

Rothman, A. L., I. Kurane, Y.-M. Zhang, C.-J. Lai, and F. A. Ennis. 1989. Dengue virus-specific murine T lymphocyte proliferation: serotype specificity and response to recombinant viral proteins. *J. Virol.* **63**:2486–2491.

Wells, M. A., F. A. Ennis, and P. Albrecht. 1981. Recovery from a viral respiratory infection. II. Passive transfer of immune spleen cells to mice with influenza pneumonia. *J. Immunol.* **126**:1042–1046.

Chapter 45

Mapping of Sindbis Virus Neutralization Epitopes

Ellen G. Strauss, Alan L. Schmaljohn, David S. Stec, and James H. Strauss

For a number of years, we have been interested in characterizing the antigenic epitopes on glycoproteins E2 and E1 of Sindbis virus that are involved in neutralization of the virus. Depending on the particular mechanism of neutralization, these epitopes could reside close to the domains responsible for specific interaction of the virus with its host receptors to initiate infection and could even overlap with these domains. In any event, we want to identify the nature and location of virus neutralization epitopes in order to understand the processes by which the immune system operates to prevent alphavirus disease.

A panel of monoclonal antibodies (MAbs) to Sindbis virus has been isolated (Schmaljohn et al., 1982; Schmaljohn et al., 1983), which includes seven neutralizing MAbs that react with E2 and a single neutralizing MAb that reacts with E1. By enzyme-linked immunosorbent assay, the MAbs were grouped into three nonoverlapping epitopes: A and B in E2 and the anti-E1 epitope (Schmaljohn et al., 1983). Our approach to further characterization of these epitopes has been to determine the location of the amino acid alterations in a number of antigenic escape variants, i.e., variants that have been selected for the ability to grow in the presence of particular neutralizing MAbs. Variants have been isolated (Stec et al., 1986) that are resistant to a single MAb (v23, v50, v30, etc., resistant to MAbs 23, 50, and 30, respectively) or that have been selected sequentially to be resistant to more than one MAb (for example, v23/50). One such set of variants is the set v33 (resistant to the anti-E1 MAb 33), v33/50 (v33 selected to be resistant to anti-E2 MAb 50), and v33/50/23 (v33/50 selected to be resistant to MAb

23). In addition, there are a number of revertants, i.e., variants that were selected by enrichment techniques to be once more neutralizable by a MAb.

SEQUENCING OF VARIANTS

Preliminary results of this study have been published (Strauss et al., 1987), and we have now completed the localization of the changes in 12 of these variants (resistant to six of the MAbs reacting with E2 and the one anti-E1 neutralizing MAb) as well as four revertants by sequencing their genomic RNAs throughout the region encoding E2 and in some cases throughout the entire structural protein domain. We chose to do this by chain termination sequencing (Sanger et al., 1977; Zimmern and Kaesberg, 1978), using reverse transcriptase with synthetic oligonucleotides as primers (Hahn et al., 1989). We found that for sequencing this portion of the genome, it was most efficient to use intracellular RNA from infected cells treated with actinomycin D at the time of infection. The virus-specific RNA present is predominantly the 26S subgenomic mRNA that encodes the structural proteins. Cells were harvested at 7 h postinfection, nuclei were removed, and total cytoplasmic RNA was prepared by phenol extraction. Although the use of unpurified material increased the background in all of the sequencing lanes, the yield was much improved, and the presence of rRNA seemed to stabilize the preparations against nuclease degradation. Although dideoxy sequencing can have unavoidable sequence-dependent ambiguities, we found that an altered nucleotide gave a discernibly different pattern. To make

Ellen G. Strauss and James H. Strauss, Division of Biology, California Institute of Technology, Pasadena, California 91125.   *Alan L. Schmaljohn,* Virology Division, U.S. Army Medical Research Institute for Infectious Diseases, Fort Detrick, Frederick, Maryland 21701.   *David S. Stec,* Department of Microbiology and Immunology, University of Maryland School of Medicine, Baltimore, Maryland 21201.

such comparisons easier, we routinely ran reactions using the same primer with RNAs from different variants at the same time and on the same gel.

DEDUCED CHARACTERISTICS OF THE NEUTRALIZING EPITOPES

We have sequenced variants resistant to seven different MAbs and revertants to four of these. From all of these mapping data, a number of general principles have become clear. First, the antigenic epitopes represented in this panel of MAbs are dependent on charged residues. All of the E2 alterations are in a very hydrophilic domain (25% charged amino acids), and all but one of the variants are altered in charge. These changes involve loss of charge (a basic amino acid replaced with a neutral residue), change of charge (a basic residue replaced with an acidic residue), or gain of charge (a neutral residue replaced with a basic or acidic residue). In the case of E1 variants resistant to MAb 33, the parental Gly residue is changed in one case to Arg and in the second to Glu, indicating that gain of either a negative or a positive charge will confer resistance to neutralization. Since it is generally believed that E2 is the moiety responsible for receptor binding and attachment, these changes may exert their effects by affecting the overall configuration of the E1-E2 heterodimer in the glycoprotein spikes. It is not yet clear whether these alterations affect virulence in an animal, but it is known that the E1 variants in particular grow poorly, and therefore this region at amino acid 132 may be sensitive to conformational change.

The epitope defined by each MAb seems to be either very small or dependent on a dominant residue. In three instances, multiple independent isolates resistant to the same MAb have been obtained. In one case (variants resistant to MAb 23), all of the isolates affect the same residue, the Lys at E2 position 216. In two variants this residue was changed to Glu, and another Sindbis strain (HRSP) with Glu at this position, which was selected in a different way, is also resistant to MAb 23. A third variant resistant to MAb 23 has Asn replacing the Lys, creating a new potential glycosylation site of the type Asn-Ile-Thr, and the presence of the oligosaccharide is thought to prevent interaction with the antibody. In the second case, of three variants resistant to MAb 50, two change the Lys at position 190 to Met and Asn, respectively, while the third changes the Glu at position 181 to Val. The third example of multiple

independent variants resistant to the same MAb consists of the E1 variants resistant to MAb 33 discussed above.

The variants that have been isolated affect a surprisingly small number of positions. Variants resistant to MAbs 23, 18, and 51, which belong to the same epitope group as defined by competition experiments, all have changes in the same residue, Lys 216 of E2. However, these MAbs appear to have different interactions with the epitope. In v18, the change to Ile-216 leads to complete loss of reactivity with MAb 18, but v18 can still react partially with MAb 23. On the other hand, v23, with Glu-216, is completely resistant to both MAb 23 and MAb 18. MAb 30 also mapped to this epitope by competition, but v30 has a change in residue 184. Interestingly, v30 is resistant to MAbs 30 and 51 but is sensitive to MAb 23 and 18. Thus, it appears that although a single changed residue can confer resistance, other residues within E2 contribute to the specificity of each MAb and can alter the pattern of reactivity. It is possible that changes at these other residues are lethal and therefore that variants affecting these positions cannot be obtained with this type of selection.

It is obviously more difficult to isolate revertants to sensitivity than to isolate resistant variants, and for this reason we have only a few revertants available. In most cases, but not all, the revertants had reacquired the original amino acid.

The location of the neutralization domain of E2, which contains all of the residues changed in the variants resistant to anti-E2 MAbs, as well as the location of the residue changed in the two independent v33 isolates are mapped in Fig. 1 on a variability plot of the glycoproteins of alphaviruses. The aligned amino acid sequences of Sindbis, Semliki Forest, O'Nyong-nyong, Ross River, Venezuelan equine encephalitis, Eastern equine encephalitis, and Western equine encephalitis viruses are plotted; the height of any peak reflects the number of different amino acids present at a given residue, i.e., the greater the variability, the higher the peak. Base-line residues are completely conserved. The locations of transmembrane domains and glycosylation sites are indicated. The shaded domain, which encompasses all of the E2 antigenic variants, is of average variability for E2. A complex polysaccharide chain is attached in the middle of this domain, suggesting that the region is on the outer surface of the virion, interacting with the solvent, and accessible to antibodies. The location of the v33 variants in E1, which is a much more highly conserved protein overall, is in a

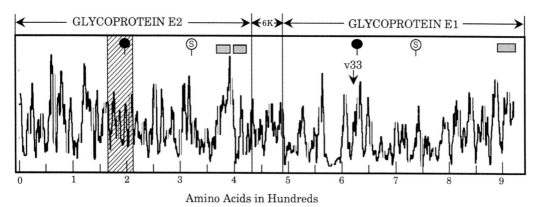

FIGURE 1. Variability plot of the aligned protein files of the glycoproteins of four Old World alphaviruses and three New World alphaviruses. The number of different residues at each position is plotted, using a smoothing function. The boundaries of the proteins, the transmembrane hydrophobic regions (stippled boxes), the locations of the simple (open symbols) and complex (solid symbols) carbohydrate chains in Sindbis virus glycoproteins, and the neutralization domain in E2 (shaded) are shown. A labeled arrow in E1 indicates the residue changed in variants resistant to MAb 33. Viral sequences used to construct this plot are found in the following references: Sindbis (Strauss et al., 1984); Semliki Forest (Takkinen, 1986); O'Nyong-nyong (Levinson et al., in press); Ross River (Faragher et al., 1988); Venezuelan equine encephalitis (Kinney et al., 1986); Eastern equine encephalitis (Chang and Trent, 1987); and Western equine encephalitis (Hahn et al., 1988).

region of above-average variability and is also near a glycosylation site, suggesting that it too is located on the outer surface. Both amino acid 132 of E1 (v33) and amino acid 216 of E2 (v23, v18, v51, and HRSP) are residues that are not conserved among the alphaviruses examined, and indeed the residues found in the variants occur naturally in other members of the group. This implies that both of these positions can accommodate a number of different amino acids without losing viability.

CONCLUDING REMARKS

We are extending this study with two types of complementary approaches. We have constructed λgt11 expression libraries containing small inserts of Sindbis virus coding sequence and are screening these for reactivity to the same panel of MAbs. If these clones encoding small peptides are reactive, this finding would imply that the epitopes have largely a linear rather than conformational character and that it is the altered amino acid per se, rather than some long-range effect on the overall shape of the glycoprotein, that is important for neutralization. Second, we are using anti-idiotypic antibodies made to this panel of MAbs to look for those which can interfere with receptor-mediated virus adsorption.

REFERENCES

Chang, G.-J. J., and D. W. Trent. 1987. Nucleotide sequence of the genome region encoding the 26S mRNA of Eastern equine encephalitis virus and the deduced amino acid sequence of the viral structural proteins. *J. Gen. Virol.* **68:**2129–2142.

Faragher, S. G., A. D. J. Meek, C. M. Rice, and L. Dalgarno. 1988. Genome sequences of a mouse-avirulent and a mouse-virulent strain of Ross River virus. *Virology* **163:**509–526.

Hahn, C. S., S. Lustig, E. G. Strauss, and J. H. Strauss. 1988. Western equine encephalitis virus is a recombinant virus. *Proc. Natl. Acad. Sci. USA* **85:**5997–6001.

Hahn, C. S., E. G. Strauss, and J. H. Strauss. 1989. Dideoxy sequencing of RNA using reverse transcriptase. *Methods Enzymol.* **180:**121–130.

Kinney, R. M., R. J. B. Johnson, V. L. Brown, and D. W. Trent. 1986. Nucleotide sequence of the 26S mRNA of the virulent Trinidad donkey strain of Venezuelan equine encephalitis virus and deduced sequence of the encoded structural proteins. *Virology* **152:**400–413.

Levinson, R., J. H. Strauss, and E. G. Strauss. Complete nucleotide sequence of the genomic RNA of O'Nyong-nyong virus and its use in the construction of phylogenetic trees. *Virology*, in press.

Sanger, F., S. Nicklen, and A. R. Coulson. 1977. DNA sequencing with chain-terminating inhibitors. *Proc. Natl. Acad. Sci. USA* **74:**5463–5467.

Schmaljohn, A. L., E. D. Johnson, J. M. Dalrymple,

and G. A. Cole. 1982. Nonneutralizing monoclonal antibodies can prevent lethal alphavirus encephalitis. *Nature* (London) **297**:70–72.

Schmaljohn, A. L., K. M. Kokubun, and G. A. Cole. 1983. Protective monoclonal antibodies define maturational and pH-dependent antigenic changes in Sindbis virus E1 glycoprotein. *Virology* **130**:144–154.

Stec, D. S., A. Waddell, C. S. Schmaljohn, G. A. Cole, and A. L. Schmaljohn. 1986. Antibody-selected variation and reversion in Sindbis virus neutralization epitopes. *J. Virol.* **57**:715–720.

Strauss, E. G., C. M. Rice, and J. H. Strauss. 1984. Complete nucleotide sequence of the genomic RNA of Sindbis virus. *Virology* **133**:92–110.

Strauss, E. G., A. L. Schmaljohn, D. E. Griffin, and J. H. Strauss. 1987. Structure-function relationships in the glycoproteins of alphaviruses, p. 365–378. *In* M. A. Brinton and R. R. Rueckert (ed.), *Positive Strand RNA Viruses.* Alan R. Liss, Inc., New York.

Takkinen, K. 1986. Complete nucleotide sequence of the nonstructural protein genes of Semliki Forest virus. *Nucleic Acids Res.* **14**:5667–5682.

Zimmern, D., and P. Kaesberg. 1978. 3' Terminal sequence of encephalomyocarditis virus RNA determined by reverse transcriptase and chain-terminating inhibitors. *Proc. Natl. Acad. Sci. USA* **75**: 4257–4261.

VIII. MOLECULAR ASPECTS OF PATHOGENESIS AND VIRULENCE

Chapter 46

Current Approaches to the Problem of Poliovirus Attenuation

Vadim I. Agol

More than 30 years ago, Albert Sabin selected strains belonging to the three existing poliovirus serotypes, and since then these strains have been used as an efficient and safe live poliovirus vaccine. Why are these strains attenuated? This is a simple biological question, and it requires an answer that is likewise simple yet satisfies our current way of thinking in molecular biological terms. To do so, we should pose and answer numerous, very difficult corollary questions.

WHAT ARE THE GENETIC DIFFERENCES BETWEEN ATTENUATED AND NEUROVIRULENT STRAINS?

Numerous attempts to answer this question were not rewarded adequately until the early 1980s, when the primary structures of the RNA genomes of the attenuated poliovirus strains and their neurovirulent ancestors were established (reviewed by Nomoto and Wimmer [1987], Almond [1987], and Racaniello [1988]). In the framework of this discussion, it was important that the empirical multistep selection of the attenuated strains was found to be accompanied by many mutations spread over the entire RNA molecule; for the poliovirus type 1 genome, there were more than 50 nucleotide differences between the neurovirulent parental strain and its attenuated derivative, and for the poliovirus type 3 RNA, there were 10 such differences.

WHICH OF THE GENETIC DIFFERENCES ARE RELEVANT TO ATTENUATION?

The first approximate mapping of the attenuating mutations was accomplished by deriva-
tion of recombinants between attenuated and neurovirulent parents (Agol et al., 1983; Agol et al., 1984). By crossing heterotypic strains, one of which possessed a guanidine-resistant mutation, we were able to select recombinants having the crossover point in the middle of the viral genome (Tolskaya et al., 1983). Four recombinants with every possible arrangement of genomic halves derived from attenuated and virulent heterotypic strains (virulent-virulent, virulent-attenuated, attenuated-virulent, and attenuated-attenuated) were studied (Agol et al., 1983; Agol et al., 1984). The conclusion was that the major determinant(s) of attenuation should be located within the 5'-end-adjacent half of the Sabin type 1 and type 3 genomes. Further experiments (Agol et al., 1985a; Agol et al., 1985b) confirmed this conclusion and also revealed a less strong (modulating) attenuating effect of the 3' half of the Sabin type 1 RNA (reviewed by Agol [1988]).

Because of the scarcity of selectable markers, only relatively rough mapping could be achieved with recombinants obtained in vivo. The resolving power of the recombinant analysis using constructs engineered by means of cDNA technology is obviously much higher. Appropriate experiments performed by several groups indicated that there are only a few attenuating mutations in each of the three Sabin strains, and some of these mutations were mapped at the nucleotide level (reviewed by Nomoto and Wimmer [1987], Almond [1987], and Racaniello [1988]). In particular, strong attenuating mutations were found to be located in the middle of the 5' untranslated region (5' UTR); specifically, they clustered at positions 472 to 481. Independent evidence for strong attenuating mutations

Vadim I. Agol, Institute of Poliomyelitis and Viral Encephalitides, USSR Academy of Medical Sciences, Moscow Region 142782, and Moscow State University, Moscow 119898, USSR.

in this genomic segment was also provided by studies on neurovirulent revertants of the vaccine strains (Evans et al., 1985). In the following presentation, I shall concentrate only on these 5′-UTR mutations, although the involvement of specific amino acid residues of the capsid proteins in attenuation was also demonstrated, and less clear-cut evidence implicates in this phenomenon the noncapsid proteins as well.

WHAT STEP OF VIRUS MULTIPLICATION IS AFFECTED BY THE ATTENUATING 5′-UTR MUTATIONS?

Several years ago, we attempted to determine whether a specific step in virus-cell interaction, the synthesis of virus-specific proteins, could be affected by attenuating mutations (Svitkin et al., 1985). It was discovered that RNAs isolated from the Sabin type 1 and type 3 strains did exhibit a diminished ability to direct protein synthesis in an appropriate cell-free system (extracts from Krebs-2 cells). Importantly, RNA from a neurovirulent revertant (strain WHO-119) of the Sabin type 3 vaccine was as good a template as was the genome of the wild-type predecessor of the vaccine, strain Leon/37. There was no evidence that the rate of polypeptide chain elongation was at all affected, and we concluded that the difference in template activity was due to the impaired initiation. The primary structures of all these RNA species being known, it was easy to conclude that it was the attenuating mutation, a C→U transition at position 472 for the type 3 strain, that was most likely responsible for the translation initiation deficiency.

On the basis of these experiments, we suggested that (i) there is a *cis*-acting translational control element in the middle of the poliovirus 5′ UTR (around position 472) that is involved in the template interaction with ribosomes, translation initiation factors, or both and (ii) the 5′-UTR attenuating mutations affect the function of this control element by decreasing the efficiency of viral polyprotein synthesis. In other words, the attenuated phenotype of the Sabin strains should be due partly to a translation defect (Svitkin et al., 1985).

The existence of a *cis*-acting translation control element within the middle of the picornavirus 5′ UTR was later unequivocally confirmed by numerous laboratories by using a variety of engineered templates. The most popular view implicates this element in internal ribosome binding.

WHAT ARE THE STRUCTURAL CHANGES CAUSED BY THE ATTENUATING MUTATIONS IN THE *cis*-ACTING CONTROL ELEMENT?

A model of the secondary structure of the appropriate region of the poliovirus 5′ UTR was suggested on the basis of evolutionary considerations and experimental testing (Blinov et al., 1988; Pilipenko et al., 1989) (Fig. 1). In fact, this is a consensus model for all enteroviruses and human rhinoviruses sequenced so far (Fig. 2). Interestingly, the attenuating mutations in the 5′ UTR of the type 1 and type 3 Sabin genomes involve the base pairs that are, according to the model, fully conserved in all entero- and rhinovirus RNAs thus far sequenced except those from the Sabin strains. In serotypes of both types of viruses, the attenuating mutations are expected to destabilize secondary-structure elements.

Thus, a highly conserved base pair, A480 · U525 (S1 in Fig. 2), in the virulent type 1 virus (note that homologous nucleotides in the different entero- and rhinovirus genomes may be numbered slightly differently) appears to be a terminal pair within a short secondary-structure element (compare Fig. 2 and 3). As a result of the A480→G transition in the Sabin type 1 strain, the potential to form a base pair between nucleotides 480 and 525 should be decreased, if not entirely lost. It seems that just this secondary-structure destabilization diminishes the template efficiency of the Sabin type 1 virus RNA and confers the attenuated phenotype to the virus. In any case, more suitable variants could readily be selected from the Sabin 1 genome or its derivatives both in vitro (Kuge and Nomoto, 1987) and in vivo (Muzychenko et al., submitted). These variants have either a true reversion to A at position 480 or a second site mutation, U→C, at position 525; in either case, the potential to form a base pair is restored (Fig. 3).

Similarly, the 5′-UTR attenuating mutation in the Sabin 3 RNA affects the conserved base pair C472 · G537 (S3 in Fig. 2), located centrally within a 3-base-pair-long secondary-structure element. As a result of the C472→U transition, the secondary structure should be destabilized. In turn, this destabilizing mutation results in a decrease of both translation efficiency and neurovirulence.

With the Sabin type 2 virus, the situation appears to be somewhat different. The predecessor of the Sabin type 2 vaccine, strain P712, possesses a G residue at position 481. Sabin strain P712 ch 2ab, which is more attenuated, has an A residue at this position. The reversion

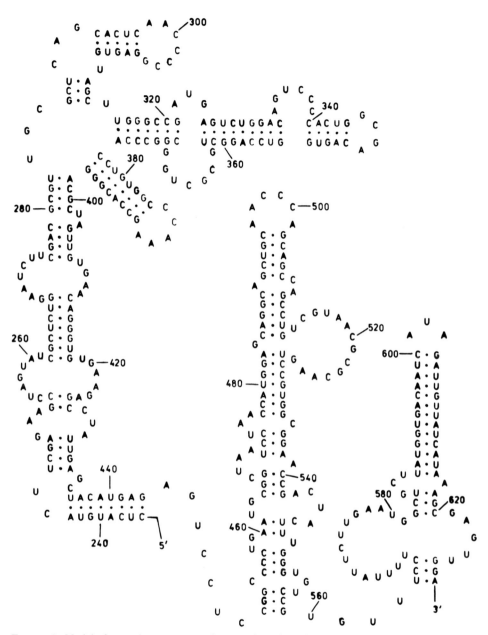

FIGURE 1. Model of secondary structure of a central portion of the 5' UTR of poliovirus type 3.

to G481 was observed in all type 2 strains isolated from vaccine-associated paralytic polio cases (Table 1), and this reversion was accompanied by an increase in neurovirulence and in the relatively low translational template activity of the Sabin strain viral RNA (Muzychenko et al., submitted). Interestingly, the nucleotide change at position 481 is very often covaried with a change at position 398; the Sabin type 2 strain has a U residue here, whereas most other strains have a C residue (Table 1). Such a covariation may hint at an interaction between nucleotides 481 and 398. However, our secondary-structure model suggests that these two residues are located in separate domains. It should be admitted in this regard that the flat second-

P1	P2L	P2P	P3	CB1	CB3	CB4	R1	R2	R14	R89	con
6	6	6	6	6	6	6	6	6	6	6	
C	C	C	C	C	C	C	U	U	C	U	c
A	G	G*	G	U*	U	G	A	G*	A	A	
C	1 1	C	U	C	C	C	C	C	C	C	c
U	U	U	C	C	C	U	U	A	C	C	
G	G	G	G	C	C		C	C		C	
3 4	1 2	1 2	1 2	2 3	1 2	2 3	6	1 2	2 3	3 4	
C	C	C	C			C	C	A	A		
G	G	G	G	A	A	A	U	G	U		
G	G	G*	G	C	C	C	A	A	U		G
A	A	A	A	A	A	A	C	U	C	C	
C	C	C	C	C	C	C	C	C	C	C	C
1 14	1 14	1 14	1 14	13	13	13	1 14	1 14	1 14	1 14	
				G	G	G					
S1 A	A	A	A	A	A	A	A	A	A	A	A
G	G*	G	G	G	G	G	C	C	G	C	
G	G*	G*	G	G*	G*	G*	G	G	G	G	G
C	C	C	U*	C	C	C	U*	U*	U*	U*	
1 1	A	A	A	G	G	G	C	C	C	C	
C	C	C	C	U	U	U	C	C	C	C	
C	C	C	C	C	C	C					
3 1	3 1	3 1	3 1	3 1	3 1	3 1	5 3	5 3	5 3	5 3	
S3 C	C	C	C	C	C	C	C	C	C	C	c
U	U	U	U	U	U	U	C	C	C	C	
4 1	4 1	4 1	4 1	4 1	4 1	4 1	4 1	4 1	4 1	4 1	
G	G	G	G	G	G	G	G	G	G	G	G
G	G	G	G	G	G	G	G	G	G	G	G
C	C	C	C	C	C	C	C	U	C	U	c
2 5	2 5	2 5	2 5	2 5	2 5	2 5	2 5	2 5	2 5	2 5	
A	A	A	A	A	A	A	A	A	A	A	A
A	A	A	A	A	A	A	A	A	A	A	A
2 1	2 1	2 1	2 1	2 1	2 1	2 1	2 1	2 1	2 1	2 1	
C	C	C	C	C	C	C	C	C	C	C	C
C	C	C	C	C	C	C	C	C	C	C	C
C	C	C	C	C	C	C	C	C	C	C	C
1 3	1 3	1 3	1 3	1 3	1 3	1 3	1 3	1 3	1 3	1 3	
G	G	G	G	G	G	G	G	G	G	G	G
G	G	G	G	G	G	G	G	G	G	G	G
C	C	C	C	C	C	C	C	C	C	C	C
7	7	7	7	8	8	8	7	7	8	7	

FIGURE 2. Comparison of the structures of a domain (see Fig. 1) in the genomes of entero- and rhinoviruses: poliovirus type 1 (strain Mahoney; P1), type 2 (strains Lansing [P2L] and P712 ch 2ab [P2S]), and type 3 (strain Leon/37; P3), coxsackieviruses B1 (CB1), B3 (CB3), and B4 (CB4), and human rhinoviruses 1B (R1), 2 (R2), 14 (R14), and 89 (R89). The single-letter code is used; A · U, U · A, C · G, G · C, G · U, and U · G pairs are denoted by A, U, C, G, G*, and U*, respectively. Numbers at the top indicate the number of nucleotides in the top loop; the other numbers correspond to the number of unpaired nucleotides in the ascending and descending branches of the domain (see Fig. 1). The consensus structures are given in the last column. Positions S1 and S3 correspond to the positions that are changed in the genomes of the Sabin type 1 (A480→G) and type 3 (C472→U) strains, respectively.

ary-structure model shown in Fig. 1 is only a rough approximation of the real spatial organization of the relevant segment of the poliovirus genome. Nucleotides 398 and 481, together with neighboring nucleotides, form two short complementary stretches that could be imagined to interact with each other in a kind of pseudoknot-like structure (Fig. 4). (The potential for tertiary interactions of the appropriate genome segments appears to be conserved among entero- and rhinoviruses [E. V. Pilipenko, unpublished observations].) Within this hypothetical structure, a base pair between nucleotides 398 and 481 may exist. This could be a U · A pair in the Sabin type 2 virus, a C · G pair in most of its more neurovirulent relatives, and a U · G pair in some exceptional cases.

Thus, the 5'-UTR attenuating mutations ap-

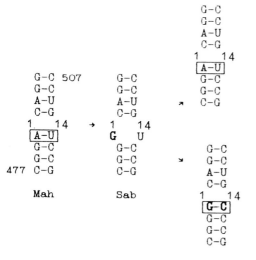

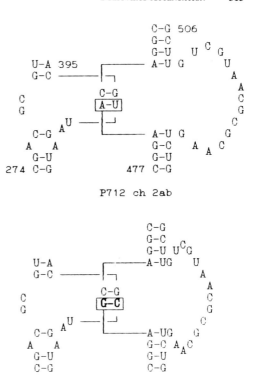

FIGURE 3. Proposed structural interpretation of coupled mutations in the 5′ UTR of poliovirus type 1. Numerals (1 and 14) represent the numbers of looped-out (single-stranded) nucleotides.

FIGURE 4. Proposed structural interpretation of coupled mutations in the 5′ UTR of poliovirus type 2.

pear to alter not only the primary but also the secondary or tertiary structure of a critical region of the viral genome, the *cis*-acting translation control element that is involved also in viral attenuation.

TABLE 1

Coupled Nucleotide Changes in the 5′ UTR of Poliovirus Strains Isolated from Vaccine-Associated Cases of Paralytic Poliomyelitis

Strain	Nucleotide at position:	
	398	481
P712	C	G
P712 ch 2ab	U	A
Zh-a	C	G
Ch-v	C	G
R-te	C	G
S-va	C	G
T-7	C	G
T-8	C	G
II-863[a]	C	G
II-867[a]	C	G
WHO-117	U	G
WHO-151	C	G
WHO-155	U	G
WHO-185	C	G

[a] Isolates from stool (II-863) and brain (II-867) of the same agammaglobulinemic patient, taken at 2 and 3 months, respectively, after the onset of symptoms.

WHAT IS THE NATURE OF THE *trans*-ACTING FACTOR THAT SENSES THE ATTENUATING MUTATIONS?

The structural features of a *cis*-acting control element are important not so much per se but rather to the extent that they ensure the interaction of this element with relevant factors, in our case with *trans*-acting translation initiation factors, ribosomes, or both. We attempted to find a cellular factor having a combination of two properties: (i) it senses the attenuating point mutations within the middle of the 5′ UTR and (ii) the specific interaction of this factor with the *cis* element has meaningful physiological consequences. The rationale of our search was as follows. Poliovirus RNA is known to be translated in rabbit reticulocyte lysates mostly from aberrant sites located inside the polyprotein reading frame (Dorner et al., 1984), although correct initiation at the beginning of this frame also takes place to a lesser extent. As shown originally by Brown and Ehrenfeld (1979) and confirmed in many laboratories, the addition of crude preparations of translation initiation factors from nucleated cells (e.g., HeLa cells) to reticulocyte lysates results in a shift to a more normal pattern of the products of poliovirus RNA translation. When the effect of a similar preparation (the ribosomal wash fraction from Krebs-2 cells) was tested on the translation of RNAs from attenuated and neurovirulent poliovirus strains, the former templates appeared to be much less responsive than the latter (Svitkin et al., 1988). The data suggested that the attenuating mutations in the middle of the 5' UTR were indeed responsible for the difference in sensitivity to an unidentified factor. We inferred that an initiation-correcting factor (ICF) should be present in the active fraction from nucleated cells and that ICF interacts with a region of the viral 5' UTR that encompasses attenuating mutations.

When ICF was partially purified, it continued to discriminate between RNA templates from attenuated and neurovirulent viruses, being much more active in the latter case. Although the exact nature of ICF has not yet been established, our data are consistent with the assumption that ICF corresponds to a complex of translation initiation factors eIF-2 and eIF-2B (Svitkin et al., 1988).

IS THE EXPRESSION OF ATTENUATING MUTATIONS TISSUE SPECIFIC?

Most of the foregoing discussion was based on the assumption that RNA molecules from the Sabin strains are poor translational templates because of the 5'-UTR mutations. Why, then, can these viruses replicate so well in HeLa or monkey kidney cells? The solution to this problem was suggested several years ago (Svitkin et al., 1985) as follows. The ability to sense attenuating mutations may vary among cells, depending on their differentiation status; for example, this ability may be especially high in human neural cells (in fact, this suggestion was based on our finding that the efficiency of poliovirus polyprotein initiation varied among different cell-free protein-synthesizing systems).

Further development of this idea became possible after the tentative identification of ICF as a complex of translation initiation factors eIF-2 and eIF-2B. Let us assume that factor eIF-2 is indeed involved in the recognition of attenuating signals in the poliovirus genome. It is well known that the activity of this factor is subject to a variety of controls involving phosphorylations and dephosphorylations mediated by different enzymes, as well as interactions with diverse auxiliary protein factors. There-

fore, it seems quite likely that the eIF-2-dependent capacity to accept poliovirus RNA as an efficient template and to discriminate between the RNAs of attenuated and neurovirulent strains is governed by the status of cell differentiation; in particular, such a discriminatory ability might be higher in specific neural cells than in commonly used host cells such as HeLa or monkey kidney cells. Therefore, we investigated the result of infection of human neuroblastoma cells with the Sabin strains and their neurovirulent relatives.

We were lucky enough to encounter a line, SK-N-MC, that proved to be amazingly refractory to the Sabin type 1 and type 2 viruses (Tolskaya et al., 1989; Agol et al., 1989). In this virus-cell system, the harvest generally was as low as <1 PFU per cell. Most remarkably, the attenuated viruses did not induce any appreciable cytopathic effect in the neuroblastoma monolayers. On the other hand, the predecessor of the Sabin type 1 vaccine, strain Mahoney, usually produced >100 PFU per cell, and it brought about nearly complete destruction of the infected cell. Intratypic (type 1) recombinants between attenuated and neurovirulent parents (having a crossover point in the central region of viral RNA) were assayed for growth potential in neuroblastoma cells. The results suggested that the major determinants of the growth deficiency lie in the 5' half of the Sabin type 1 genome, but an additional, weaker effect appeared to be contributed by the 3' half; qualitatively, these conclusions are identical to those drawn when the same pair of recombinants were tested for monkey neurovirulence (Agol et al., 1985a).

These data suggest that the inability to grow in neuroblastoma cells has a complex nature and involves different parts of the viral genome. However, the original supposition that the attenuating mutations in the 5' UTR involved in translational control may play an important part in growth restriction in neuroblastoma cells was strongly supported by independent observations made by La Monica and Racaniello (1989). These investigators showed that poliovirus strains differing only at position 472 of their poliovirus type 3-derived 5' UTR produced 10-fold-different harvests of infectious particles in another line of human neuroblastoma cells (but not in HeLa cells), the attenuated strain being less efficient (Muzychenko et al., submitted). In addition, the authors reported that virus-specific protein synthesis in the neuroblastoma cells infected with the attenuated virus seemed to be primarily impaired.

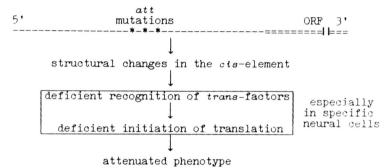

FIGURE 5. Proposed mechanism of the expression of attenuating mutations located in the middle of the poliovirus 5' UTR.

Thus, the notion that the translation machinery of neural cells is especially sensitive to the attenuating mutations in the poliovirus 5' UTR appears to be correct.

CONCLUDING REMARKS

The existing data, though incomplete, can be rationalized in the following simple scheme (Fig. 5). Attenuating mutations located in the middle of the 5' UTR of the genomes of three Sabin strains alter the structure of a key *cis*-acting translational control element in such a way that it now fails to interact efficiently with an important *trans*-acting factor. As a result, the translational template activity of the appropriate RNAs is diminished. This modulating effect is especially pronounced in specific neural cells. That is why the Sabin strains produce relatively mild, if any, damage to the primate central nervous system even if they can reach it.

But we may go even further and suggest that the physiological or pathological changes in the activity of translation initiation factors (e.g., the eIF-2 system) within the target cells affect the fate of poliovirus infection. For example, one may wonder whether such changes have something to do with two well-known but entirely unexplained phenomena: could individual modulations of the translation apparatus in certain cells of the central nervous system explain why the incidence of paralytic illness among wild-type poliovirus-infected nonimmune human beings is so surprisingly low and why, on the other hand, paralytic cases are encountered, though extremely rarely, among recipients of the Sabin vaccine or their contacts?

The discovery that a control element in the middle of the viral 5' UTR is involved in both translation initiation and virus attenuation, in

conjunction with the knowledge of the structural organization of this element, paves the way for modifications of the poliovirus 5' UTR that will allow the development of new attenuated strains that exhibit less residual virulence and more genetic stability than do the Sabin strains.

REFERENCES

Agol, V. I. 1988. Genetic determinants of neurovirulence and attenuation of poliovirus. *Mol. Genet. Mikrobiol. Virusol.* 1:3–9.

Agol, V. I., S. G. Drozdov, M. P. Frolova, V. P. Grachev, M. S. Kolesnikova, V. G. Kozlov, N. M. Ralph, L. I. Romanova, E. A. Tolskaya, and E. G. Viktorova. 1985a. Neurovirulence of the intertypic poliovirus recombinant v3/a1-25: characterization of strains isolated from the spinal cord of diseased monkeys and evaluation of the contribution of the 3' half of the genome. *J. Gen. Virol.* 65:309–316.

Agol, V. I., S. G. Drozdov, V. P. Grachev, M. S. Kolesnikova, V. G. Kozlov, N. M. Ralph, L. I. Romanova, E. A. Tolskaya, A. V. Tyufanov, and E. G. Viktorova. 1985b. Recombinants between attenuated and virulent strains of poliovirus type 1: derivation and characterization of recombinants with centrally located crossover points. *Virology* 143:467–477.

Agol, V. I., S. G. Drozdov, T. A. Ivannikova, M. S. Kolesnikova, M. B. Korolev, and E. A. Tolskaya. 1989. Restricted growth of attenuated poliovirus strains in cultured cells of a human neuroblastoma. *J. Virol.* 63:4034–4038.

Agol, V. I., V. P. Grachev, S. G. Drozdov, N. M. Ralph, L. I. Romanova, E. A. Tolskaya, A. V. Tyufanov, and E. G. Viktorova. 1984. Construction and properties of intertypic poliovirus recombinants: first approximation mapping of the major determinants of neurovirulence. *Virology* 136:41–55.

Agol, V. I., E. G. Viktorova, V. P. Grachev, S. G. Drozdov, M. S. Kolesnikova, V. G. Kozlov, N. M.

Ralph, L. I. Romanova, E. A. Tolskaya, and A. V. Tyufanov. 1983. Study on the nature of the poliovirus pathogenicity using the recombinants between neurovirulent and attenuated strains. *Mol. Genet. Mikrobiol. Virusol.* **11**:34–37.

Almond, J. W. 1987. The attenuation of poliovirus neurovirulence. *Annu. Rev. Microbiol.* **41**:153–180.

Blinov, V. M., E. V. Pilipenko, L. I. Romanova, A. N. Sinyakov, S. V. Maslova, and V. I. Agol. 1988. A comparison of the secondary structures of the 5'-untranslated segment of neurovirulent and attenuated poliovirus strains. *Dokl. Akad. Nauk SSSR* **298**:1004–1006.

Brown, B. A., and E. Ehrenfeld. 1979. Translation of poliovirus RNA in vitro: changes in cleavage pattern and initiation sites by ribosomal salt wash. *Virology* **97**:396–405.

Dorner, A. J., B. L. Semler, R. J. Jackson, R. Hanecak, E. Duprey, and E. Wimmer. 1984. In vitro translation of poliovirus RNA: utilization of internal initiation sites in reticulocyte lysate. *J. Virol.* **50**:507–514.

Evans, D. M. A., G. Dunn, P. D. Minor, G. C. Schild, A. J. Cann, G. Stanway, J. W. Almond, K. Currey, and J. W. Maizel, Jr. 1985. Increased neurovirulence associated with a single nucleotide change in a noncoding region of the Sabin type 3 poliovaccine genome. *Nature* (London) **314**:548–550.

Kuge, S., and A. Nomoto. 1987. Construction of viable deletion and insertion mutants of the Sabin strain of type 1 poliovirus: function of the 5' noncoding sequence in viral replication. *J. Virol.* **61**:1478–1487.

La Monica, N., and V. R. Racaniello. 1989. Differences in replication of attenuated and neurovirulent polioviruses in human neuroblastoma cell line SH-SY5Y. *J. Virol.* **63**:2357–2360.

Muzychenko, A. R., G. Y. Lipskaya, S. V. Maslova, Y. V. Svitkin, E. V. Pilipenko, B. K. Nottay, O. M. Kew, and V. I. Agol. Coupled mutations in the 5'-untranslated region of the Sabin poliovirus strains during in vivo passages: structural and functional implications. Submitted for publication.

Nomoto, A., and E. Wimmer. 1987. Genetic studies of the antigenicity and the attenuation phenotype of poliovirus. *Symp. Soc. Gen. Microbiol.* **40**:107–134.

Pilipenko, E. V., V. M. Blinov, L. I. Romanova, A. N. Sinyakov, S. V. Maslova, and V. I. Agol. 1989. Conserved structural domains in the 5'-untranslated region of picornaviral genomes: an analysis of the segment controlling translation and neurovirulence. *Virology* **168**:201–209.

Racaniello, V. R. 1988. Poliovirus neurovirulence. *Adv. Virus Res.* **34**:217–246.

Svitkin, Y. V., S. V. Maslova, and V. I. Agol. 1985. The genomes of attenuated and virulent poliovirus strains differ in their in vitro translation efficiencies. *Virology* **147**:243–252.

Svitkin, Y. V., T. V. Pestova, S. V. Maslova, and V. I. Agol. 1988. Point mutations modify the response of poliovirus RNA to a translation initiation factor: a comparison of neurovirulent and attenuated strains. *Virology* **166**:394–404.

Tolskaya, E. A., M. S. Kolesnikova, S. G. Drozdov, and V. I. Agol. 1989. Selective resistance of human neuroblastoma cells to a vaccine strain of poliomyelitis virus. *Dokl. Akad. Nauk SSSR* **305**:987–989.

Tolskaya, E. A., L. I. Romanova, M. S. Kolesnikova, and V. I. Agol. 1983. Intertypic recombination in poliovirus. Genetic and biochemical studies. *Virology* **124**:121–132.

Chapter 47

Application of Site-Directed Mutagenesis to the Study of Poliovirus Capsids: Myristylation of VP4 Is Required for Virion Stability, and a Sequence of 12 Amino Acids in VP1 Determines the Host Range of the Virus

Marc Girard, Daniel Marc, Annette Martin, Thérèse Couderc, Danièle Benichou, Adina Candrea, Radù Crainic, Florian Horaud, and Sylvie van der Werf

Polioviruses, of which there are three known serotypes (PV1, PV2, and PV3), are small, naked viruses of icosahedral symmetry. The three-dimensional structure of their capsid has been determined at a resolution of 2.9 Å (0.29 nm) (Hogle et al., 1985). The 60 copies of capsid polypeptide VP1 are assembled in pentamers around the fivefold axes of the icosahedron, whereas those of VP2 and VP3 alternate around its threefold axes. VP4 is located at the inner face of the virus shell. VP4, its precursor VP0, and the capsid protein P1 from which the four capsid polypeptides derive are myristylated on their N-terminal glycine by amide linkage (Chow et al., 1987). The N-terminal sequence of these proteins is Gly-Ala-Gln-Val-Ser-Ser, forming a consensus myristylation signal as reported by Paul et al. (1987) and Towler et al. (1988) (Table 1).

We wished to investigate the possible functions of the myristyl group in the different steps of poliovirus multiplication. Toward this aim, point mutations were generated in the nucleotide sequence coding for the myristylation signal of a PV1 (Mahoney) cDNA molecule that can be transcribed in vitro into an infectious genomic RNA (van der Werf et al., 1986). The properties of the mutated transcripts were then analyzed after their transfection into primate cells permissive for poliovirus replication. This approach allowed us to study the role of myristylation in the replication and assembly of poliovirus.

All three poliovirus serotypes are neurovirulent for monkeys, but only PV2, when adapted to mice, has been shown to be neurovirulent for these animals (Armstrong, 1939). When injected by the intracerebral (i.c.) route, as little as 10^4 PFU of PV2 (Lansing) will induce fatal paralysis in 50% of the animals. In contrast, the Mahoney strain of PV1 does not cause any disease, even after i.c. inoculation of as much as 10^7 PFU.

We previously reported the construction of a PV1-PV2 chimera, v510, in which VP1 amino acids 94 to 102 from neutralization antigenic site 1 of PV1 were replaced by those from PV2 (Martin et al., 1988a). The substitution was done by oligonucleotide cassette exchange in an infectious PV1 cDNA clone (Kean et al., 1986). The v510 chimera was shown to be neurovirulent for mice, whereas the PV1 parent was not (Martin et al., 1988b).

Only six amino acids, at VP1 residue positions 95, 97, and 99 to 102, differ between PV1 and v510. The question asked was whether all six amino acid changes were required and, if not, which substitutions of the six were critical for the mouse-adapted phenotype. To answer this question, two approaches were used: (i) PV1-PV2 chimeras with partial substitutions of neutralization antigenic site 1 were engineered, and (ii) v510 variants resistant to neutralization with two PV2-specific monoclonal antibodies (MAbs) that recognize antigenic site 1 of v510 were selected. We also investigated whether we

Marc Girard, Daniel Marc, Annette Martin, Danièle Benichou, and Sylvie van der Werf, Laboratory of Molecular Virology, Centre National de la Recherche Scientifique, URA 545, Pasteur Institute, 25, Rue du Dr. Roux, 75724 Paris Cedex 15, France. *Thérèse Couderc, Adina Candrea, Radù Crainic, and Florian Horaud,* Laboratory of Medical Virology, Pasteur Institute, 25, Rue du Dr. Roux, 75724 Paris Cedex 15, France.

TABLE 1
Sequence of the Amino Terminus of Known Myristylated Proteins[a]

Protein[b]	Amino acid position						
	1	2	3	4	5	6	7
Cyclic AMP-dependent protein kinase.............................Gly	Asn	Ala	Ala	Ala	Ala	Lys
Calcineurin B...Gly	Asn	Glu	Ala	Ser	Tyr	Pro
MuLV p15..Gly	Gln	Thr	Val	Thr	Thr	Pro
FeSV p15...Gly	Gln	Thr	Ile	Thr	Thr	Pro
HTLV-II p15 ....Gly	Gln	Ile	His	Gly	Leu	Ser
BLV p15...Gly	Asn	Ser	Pro	Ser	Tyr	Asn
RSV p60src ....Gly	Ser	Ser	Lys	Ser	Lys	Pro
HBV preS1 ....Gly	Gln	Asn	Leu	Ser	Thr	Ser
Poliovirus VP4 ....Gly	Ala	Gln	Val	Ser	Ser	Gln
SV40 VP2 ....Gly	Ala	Ala	Leu	Thr	Leu	Leu
Bovine G$_{o\alpha}$ GTP-binding protein ....Gly	Cys	Thr	Leu	Ser	Ala	Glu
Bovine G$_{s\alpha}$ GTP-binding protein ....Gly	Cys	Leu	Gly	Asn	Ser	Lys
SYN kinase ....Gly	Cys	Val	Gln	Cys	Lys	Asp
HIV-1 Nef protein ....Gly	Gly	Lys	Try	Ser	Lys	Ser
HIV-2 Nef protein ....Gly	Ala	Ser	Gly	Ser	Lys	Lys

[a] Myristylation is a cotranslational event that occurs after removal of the methionine residue at the N terminus of the nascent protein. (After Paul et al. [1987] and Towler et al. [1988].)

[b] Abbreviations: MuLV, murine leukemia virus; FeSV, feline sarcoma virus; HTLV-II, human T-cell leukemia virus type II; BLV, bovine leukemia virus; RSV, Rous sarcoma virus; HBV, hepatitis B virus; SV40, simian virus 40; SYN, src/yes-related novel kinase; HIV-1 and -2, human immunodeficiency virus types 1 and 2. For more details, see Semba et al. (1986).

could shorten the loop formed by neutralization antigenic site 1 and still keep virus infectivity. Finally, we used v510 as a model to study in mice the role of molecular determinants of attenuation in the PV1 genome.

SITE-DIRECTED MUTAGENESIS OF THE MYRISTYLATION SIGNAL OF VP4

The Lack of VP0 Myristylation Is Lethal

The mutations indicated in Table 2 were introduced in the region of the cDNA of PV1 (Mahoney) encoding the myristylation signal of VP4. This was achieved by oligonucleotide site-directed mutagenesis as described by Morinaga et al. (1984). Genomic RNA molecules transcribed in vitro with T7 RNA polymerase were transfected onto HeLa or Vero cell cultures in the presence of DEAE-dextran. The cells were then overlaid with Dulbecco modified Eagle medium and incubated at 37°C. Cultures were monitored daily for the appearance of cytopathic effect (CPE). Whereas transcripts from wild-type cDNA and from the Ser-5→Thr mutant were infectious ($\approx 10^5$ PFU/µg of RNA), leading to full CPE at 24 h after transfection of HeLa cells, those from the other mutants were not, as judged by absence of CPE after 2 days of incubation and by absence of visible plaques under agar (not shown).

At times, however, the occurrence of CPE in some of the plates of transfected cells could be observed after 3 to 5 days in culture. Virus stocks were prepared from each plate in which late CPE had occurred, and the resulting virus was analyzed by nucleotide sequencing in the region coding for the myristylation signal. Virus recovered after transfection with the Gly-1→Arg mutant showed reversion to the original Gly (Table 2). This corresponded to a CGC (Arg) to GGC (Gly) point mutation. The fact that CGC could also have reverted to AGC (Ser), UGC (Lys), CAC (His), CCC (Pro), or CUC (Leu), but that none of these reversions was observed, suggests that a Gly residue is required at this position for the virus to be viable.

Three types of reversions (Pro→Ala, Pro→Ser [the original sequence], and Pro→Thr) but none of the other theoretically possible reversions (Pro to Leu, Glu, or Arg) were observed at position 5, suggesting that only Ala, Ser, or Thr could restore virus viability. Similarly, residues found in the viable Pro-2 revertants were only Thr, Ser, Leu, and Ala (the original residue), not His or Arg, suggesting that the latter two were incompatible with virus viability. The resulting possible N-terminal sequence for PV1 is shown in Table 3. This sequence is in agreement with the consensus myristylation sequence of yeast and rat liver cells (Towler et al., 1987; Towler et al., 1988).

TABLE 2
PV1 Myristylation Mutants

Virus	VP0 amino acid sequence[a]						Infectivity of virus	Sequence of escape mutants[b]	
	1	2	3	4	5	6			
PV1	G	A	Q	V	S	S	+++		
1A1	R̲	-	-	-	-	-	−	CGC ⟶ (Arg)	GGC (Gly)
1A2	-	-	-	-	T̲	-	+++		
1A3	-	-	-	-	P̲	-	−	CCA ⟶ (Pro)	GCA (Ala)
									TCA (Ser)
									ACA (Thr)
1A4	-	P̲	-	-	-	-	−	CCT ⟶ (Pro)	ACT (Thr)
									TCT (Ser)
									GCT (Ala)
									CTT (Leu)

[a] The underlined mutations were generated in a cloned infectious PV1 cDNA molecule, and infectivity was determined by transfection (see text).
[b] Reverse mutants that spontaneously arose showed the indicated nucleotide sequence.

To confirm that the loss of viability induced by the mutation of VP0 amino acid 1, 2, or 5 was indeed due to the lack of myristylation of the protein, HeLa cells transfected with each of the mutated full-length transcripts (Table 2) were labeled from 3 to 7.5 h after transfection with either [^{35}S]methionine or [^{3}H]myristate, and cell extracts were prepared and immunoprecipitated with an anti-VP2 rabbit immune serum. Identical amounts of ^{35}S-labeled VP0 were observed in all extracts from cells transfected with either infectious or noninfectious RNA transcripts. In contrast, ^{3}H-labeled VP0 was detected readily only in extracts of cells transfected with wild-type PV1 transcripts or with the transcript from the

viable Ser-5→Thr mutant. Only faint bands of ^{3}H-labeled VP0 could be detected in the case of the Ala-2→Pro and Ser-5→Pro mutant-transfected cells, perhaps indicative of the early emergence of reversion mutants. No ^{3}H labeling was observed in the extracts from cells transfected with the Gly-1→Arg mutant (not shown). When the same experiment was repeated using HeLa cells infected with the revertant viruses, ^{3}H-labeled VP0 was readily detected in all cases (data not shown). Therefore, restoration of myristylation of VP0 was induced concomitantly with virus viability.

Role of Myristylation in the Replication of Viral RNA

To investigate at which step virus multiplication was blocked by the lack of VP0 myristylation, the ability of the mutated transcripts to replicate in HeLa cells was analyzed. Transfected HeLa cell monolayers were lysed with 0.5% Nonidet P-40 in 0.01 M Tris hydrochloride (pH 7.4)–1 mM EDTA–0.14 M NaCl buffer, cytoplasmic extracts were prepared, and the cytoplasmic RNAs were hybridized by dot blot hybridization with a ^{32}P-labeled viral RNA probe complementary to PV1 nucleotides 3417

TABLE 3
Poliovirus Consensus Myristylation Sequence

Amino acid position	Wild-type sequence	Other accepted amino acids
1	G	
2	A	T, S, L
3	Q	
4	V	
5	S	A, T
6	S	

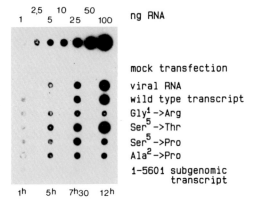

FIGURE 1. Effect of myristylation mutations on viral RNA replication. HeLa cells were transfected with the indicated RNAs or RNA transcripts. At the time indicated, the cell RNA was extracted, and 1 μg of total RNA was blotted onto a nylon membrane. Dot hybridization was with a ^{32}P-labeled cRNA (nucleotides 3417 to 4830).

to 4830 (Fig. 1). All of the genomic transcripts, whether infectious or not, were able to replicate in the transfected cells, as seen by the increase in the intensity of hybridization signals during the first 7.5 h after transfection. However, beyond 7.5 h posttransfection, the intensity of the hybridization signals increased only in the cell cultures transfected with infectious RNAs, not in those transfected with the noninfectious transcripts. These data show that the myristylation mutants are normally replicated after transfection but cannot initiate a second cycle of virus multiplication.

Role of Myristylation in Poliovirion Assembly

To investigate whether the block was at the stage of virus assembly, [^{35}S]methionine-labeled cell extracts were prepared at 7.5 h after transfection and analyzed by sucrose gradient centrifugation, using conditions under which the peak of 5S precursor protein P1 was in fractions 23 to 26 and that of the 14S pentamers was recovered in fractions 5 to 9 of the gradient. Faster-sedimenting material such as procapsids, provirions, or 150S virions were recovered in the pellet of the gradient.

The even fractions of the gradients were immunoprecipitated under native conditions with a serum directed against heat-treated ("C") particles before sodium dodecyl sulfate-polyacrylamide gel electrophoresis. The gradient pellet, once resuspended in lysis buffer, was

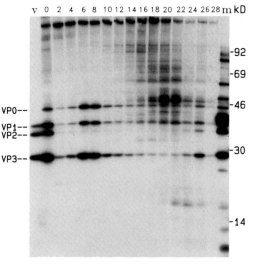

FIGURE 2. Recovery of 14S particles after transfection of HeLa cells with a wild-type RNA transcript. HeLa cells transfected with the RNA transcript from a PV1 (Mahoney) cDNA were labeled with [^{35}S]methionine from 3 to 7.5 h posttransfection, at which time an extract was prepared and fractionated by sucrose gradient centrifugation. Even fractions from the gradient were immunoprecipitated with an anti-C particle immune serum before analysis by sodium dodecyl sulfate-polyacrylamide gel electrophoresis and autoradiography. Lane v, Purified virus control; lane m, ^{35}S-labeled extract from PV1-infected cells. kD, Kilodaltons.

treated as fraction zero. Extracts from cells transfected with wild-type transcripts readily showed the presence of labeled VP0, VP1, and VP3 at the position of 14S pentamers (Fig. 2, lanes 6 to 8) as well as in the pelleted material (lane 0). In addition, the pellet showed the presence of labeled VP2, indicative of the cleavage of VP0 into VP4 plus VP2. This cleavage is known to occur at the last stage of virus assembly, when provirions mature into virions. A similar result was observed in cells transfected with RNA transcripts from the viable Ser-5→Thr mutant (not shown).

A remarkably different result was observed in the case of cells transfected with RNA transcripts from the nonviable Gly-1→Arg mutant (Fig. 3). These cells showed a total lack of labeled VP2 in the fast-sedimenting material in the pellet of the gradient. In contrast, labeled VP2 was found at the position of 14S pentamers together with VP1 and VP3, but only traces of labeled VP0 were detected. Thus, most of the 14S material in the Gly-1→Arg mutant-trans-

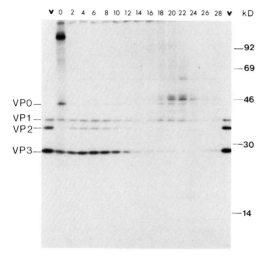

FIGURE 3. Recovery of 14S particles after transfection of HeLa cells with an RNA transcript from mutant Gly-1→Arg (mutant 1A1). For details, see legend to Fig. 2.

fected cells represented pentamers that had undergone cleavage of VP0 to VP2 plus VP4. Since this cleavage is known to take place only at the last stage of virion assembly, perhaps as a consequence of the entry of viral RNA (Arnold et al., 1987), it is most likely that pentamers with a cleaved VP0 can be generated only by dissociation of virions. This interpretation is strengthened by the absence of labeled virions in the pellet of the gradient in Fig. 3.

We conclude that the absence of VP0 myristylation resulting from the Gly-1→Arg mutation does not seem to prevent the assembly of the viral capsid polypeptides up to the stage of the virion, but the virions thus formed readily dissociate into 14S pentamers with the unusual polypeptide composition VP1-VP2-VP3. Whether they also contain VP4 remains to be determined. However, the possibility that dissociation of the virions results from the presence of the Arg residue at the N terminus of VP0 rather than from its lack of myristylation cannot be excluded at present. Analysis of the assembly of the two other nonviable mutants is in progress to test this hypothesis.

Together, the results presented here suggest that myristylation is required for the proper assembly and stability of virus particles and that in its absence, no stable poliovirions can be formed, although viral RNA replication, protein synthesis, and protein processing appear to be normal in transfected cells.

DETERMINANTS OF POLIOVIRUS NEUROVIRULENCE FOR MICE

The PV1-PV2 chimera v510, with VP1 amino acids 94 to 102 from PV2 (Lansing), was found to be neurovirulent for mice, which allowed us to conclude that VP1 amino acids 94 to 102 from PV2 (Lansing) were sufficient to confer mouse adaptation to the virus. Similar results have been obtained by others (Murray et al., 1988). To determine which residues of the six differing between PV1 and v510 in this region (Table 4) were critical for mouse adaptation, PV1-PV2 chimeras with limited substitutions of VP1 amino acids were constructed, using the oligonucleotide cassette exchange strategy that had been used to obtain chimera v510. No mouse neurovirulence (after injections at 10^6 PFU) could be observed when the Pro residue at position 95 in VP1 was replaced by the Asp residue found at the same position in PV2, generating vDL10, when both Pro-95 and Ser-97 were replaced by Asp and Pro, respectively, generating vDL20, or when residues at positions 100 (Asn→Arg) and 101 (Lys→Ala) were substituted, generating vDL40 (Table 4). Growth rate of the viruses on HeLa cells at 37°C, plaque size, heat stability of the virion at 45°C, and electrophoretic mobility of the hybrid VP1 polypeptides in sodium dodecyl sulfate-polyacrylamide gels remained unchanged in all three chimeras.

Next, vDL30, another chimera with a substitution of four amino acids (Pro-95→Asp, Ser-97→Pro, Asn-100→Arg, and Lys-101→Ala) was generated. This chimera was found to be partially neurovirulent for mice, as seen by the fact that a limited but reproducible number of animals were paralyzed after injection of 10^6 or 10^7 PFU of the virus. vDL30 showed a particular plaque phenotype. At 33°C it produced large plaques, like its PV1 parent and the other chimeras. At 37°C it produced significantly smaller plaques, and at 39°C it produced mainly minute plaques, among which a few large plaques could be found occasionally.

To investigate whether these large plaques resulted from the emergence of new variants, three large plaques from a virus titration at 39°C were picked up at random and plaque purified three times at 39°C, and stocks of the resulting viruses were prepared. The RNA of the three viruses was sequenced in the region of VP1. One of the large-plaque viruses exhibited a vDL30 sequence except for an Asp→Asn mutation at position 95, the second one showed a vDL30 sequence with an additional Ala→Val mutation at position 96, and the third had a vDL30 sequence with Ala instead of Asp at position 95

TABLE 4
Search for a Minimal Mouse Neurovirulence Sequence

Virus	VP1 amino acid sequence												Mouse neuro-virulence test[a] (no. paralyzed)
			95					100					
PV1	D	N	P	A	S	T	T	N	K	D	K	L	0
v510	-	-	D	-	P	-	K	R	A	S	-	-	6
vDL10	-	-	D	-	-	-	-	-	-	-	-	-	0
vDL20	-	-	D	-	P	-	-	-	-	-	-	-	0
vDL40	-	-	-	-	-	-	-	R	A	-	-	-	0
vDL30	-	-	D	-	P	-	-	R	A	-	-	-	2
vDL30 variants													
1	-	-	N	-	P	-	-	R	A	-	-	-	0
2	-	-	D	V	P	-	-	R	A	-	-	-	2
3	-	-	A	-	P	-	-	R	A	-	-	-	0

[a] Groups of six mice were injected i.c. with 10^6 PFU of chimeras vDL10, vDL20, and vDL40 and with 10^7 PFU of chimera vDL30 and vDL30 variants.

(Table 4). Inoculation of 10^7 PFU of each of these variants to groups of six mice by the i.c. route showed that variant 2 was the only one to exhibit the weak neurovirulence of its parent, vDL30. No mice were paralyzed after injection of variant 1 or 3, both of which are mutated at position 95. Thus, the Asp residue at position 95 seems to be critical (but not sufficient) for the host range phenotype of the chimera. The effect of additional substitutions, particularly at VP1 positions 99 and 102, will soon be investigated.

During the course of this study, we generated a PV1-PV2 chimera with the same six VP1 amino acid changes as in v510 but with an additional Asp→Val mutation at position 93 (Martin et al., 1988a; Martin et al., 1988b). This chimera, v410, was found to be devoid of virulence for mice. Comparison of v410 and v510 showed that the two chimeras differed by a variety of criteria (Table 5), such as plaquing efficiency in monkey cells, stability at 45°C, and neurovirulence for monkeys (courtesy of M. Arita, Japan National Institute of Health). Thus,

the mere substitution of a single amino acid at VP1 position 93 induced deep phenotypic changes in the host range and properties of the virus. The impact of the Asp-93→Val mutation on the three-dimensional structure of the viral capsid will be interesting to study by X-ray crystallography (Hogle et al., 1985).

We also asked whether we could shorten the loop formed by VP1 amino acids 94 to 102 and still retain virus infectivity. For that purpose, we first used the oligonucleotide cassette exchange strategy described above, with shorter oligonucleotides than the sequence to be substituted. Two deletions were thus engineered into a v510 cDNA molecule. One, Δ4, eliminated VP1 amino acids 95 to 99 and added a new Gly residue instead; a second, Δ5, eliminated amino acids 95 to 100 but also added a new Gly residue. A third deletion, Δ9, was next engineered by site-directed mutagenesis on an M13 molecule containing the appropriate PV1 cDNA fragment, using a 30-mer oligonucleotide with 15 bases complementary to the PV1 cDNA sequence on

TABLE 5
Comparison of PV1-PV2 Chimeras v410 and v510 and Parent PV1 (Mahoney)

Virus	VP1 amino acid 93	Relative titer (PFU)			Loss of titer after 5 min at 45°C (PFU)	Virulence for[a]:	
		HeLa	CV-1	Primary monkey kidney		Monkeys	Mice
v410	Val	100	1	0.5	$10^{3.0}$	Partial paralysis only	−
v510	Asp	200	20	100	$10^{1.5}$	Full neurovirulence	+
Mahoney	Asp	500	50	ND[b]	$10^{1.0}$	Full neurovirulence	−

[a] Monkeys were injected by the intraspinal route with 10^4 or 10^6 PFU. Mice received up to 10^7 PFU by the i.c. route.
[b] ND, Not determined.

TABLE 6
v510 Deletion Mutants[a]

Virus	VP1 amino acid sequence[b]											Mouse neurovirulence
	95					100						
v510	D	N	D	A	P	T	K	R	A	S	K	++
Δ4	–	–	G	⊢———————————⊣				R	A	S	K	–
Δ5	–	–	G	⊢———————————⊣					A	S	K	–
Δ9	–	⊢———————————————⊣									K	ND
Δ9-His	H	⊢———————————————⊣									K	ND

[a] Viable viruses with the indicated deletions in VP1 (lines) were recovered upon transfection of HeLa cells by the corresponding plasmids. Mouse neurovirulence was tested by injection of 10^7 PFU of virus i.c. ND, Not done.
[b] Underlined amino acids correspond to heterologous residues introduced at the mutagenesis step.

either side of the sequence to be deleted. The deletion thus created eliminated the entire amino acid loop from VP1 residues 94 to 102 (Table 6). By site-directed mutagenesis (Morinaga et al., 1984), we also constructed a Δ9 mutant with an additional Asp→His mutation at VP1 position 93 (Table 6). The Δ9 and Δ9-His mutants were viable and grew to high titers. These two mutants produced small plaques and their virions were heat labile at 45°C, but they did not show temperature sensitivity at 39°C. These viruses will presently be analyzed for monkey neurovirulence. Neither Δ4 nor Δ5 was able to induce paralysis upon i.c. inoculation into mice (Table 6).

Another approach to study the capsid amino acid sequence requirement for mouse neurovirulence was based on the observation that v510 could no longer be neutralized by PV1 antigenic site 1-specific MAb C3 but could be neutralized by two PV2-specific MAbs, HO2 and IIo (Martin et al., 1988b). We asked whether v510 escape mutants able to resist neutralization with either of these MAbs were still neurovirulent for mice. To do this, virus from plaques obtained in neutralization assays of v510 with either of the two MAbs was passaged once in the presence of the same MAb and then plaque purified. More than a dozen independent MAb-resistant mutants were thus generated. None of these escape mutants regained the ability to be neutralized by PV1-specific MAb C3. All HO2-resistant mutants were also IIo resistant, but the converse was not true. Most of these escape mutants were as stable at 45°C as their v510 parent, and most were rct40$^+$ (Table 7).

The escape mutants were sequenced by using appropriate oligonucleotides in the region of the RNA encoding VP1 amino acids 80 to 110. In parallel, their 50% lethal doses for mice were determined. Most of the escape mutants showed

loss of mouse neurovirulence (Table 7). This was true of the Asp→Gly and Asp→Asn mutations at positions 93 and 95 and of the Lys→Asn and Lys→Glu mutations at position 99. This was also true of an unexpected Leu→Arg mutation at position 104. However, neither the Arg→His mutation at position 100, selected for resistance to HO2, nor the Ala→Thr mutation at position 101, leading to resistance to IIo, had a pronounced effect on mouse neurovirulence. Perhaps the most surprising observation was that one of the HO2 escape mutants was a deletion mutant. The deletion eliminated all nine VP1 amino acids 94 to 102 (Table 7). This finding suggests that the foreign C3 loop may be unstable in the chimera and can be spontaneously deleted. It will be of interest to determine the frequency at which such an event occurs.

From this series of observations, it can be concluded that critical amino acid positions for mouse neurovirulence are VP1 residues 93, 95, 99, and 104. All mouse neurovirulent variants were rct40$^+$; but nonneurovirulent escape mutants could be either rct40$^+$ or rct$^-$. Also, it was possible to delete the complete C3 loop (VP1 amino acids 94 to 102) without altering virus viability.

Finally, we introduced into v510 a mutation from A to G at position 480 in the 5′ noncoding region of the viral genome. This mutation is linked to attenuation of monkey neurovirulence in PV1 (Omata et al., 1986; Kawamura et al., 1989). The resulting virus, v510-A, was injected into mice (10^6 PFU) by the i.c. route. No mice were paralyzed, showing that the attenuation mutation of the 5′ noncoding region worked in mice as well as in monkeys and suggesting that mouse neurovirulence of v510 resulted from the conjunction of two elements: (i) a host range mutation linked to VP1 amino acids 94 to 102, allowing the virus to grow in mice, and (ii) a

TABLE 7
v510 Neutralization Escape Mutants

Virus	VP1 amino acid sequence[a]												Heat stability[b] (15 min, 45°C)	RCT 40[c]	Mouse 50% lethal dose (PFU)
			95					100							
PV1	D	N	P	A	S	T	T	N	K	D	K	L	−	+	$>10^6$
v510	-	-	D	-	P	-	K	R	A	S	-	-	+	+	$10^{4.4}$
511D2	<u>G</u>	-	D	-	P	-	K	R	A	S	-	-	+	−	$>10^6$
512D1	<u>N</u>	-	D	-	P	-	K	R	A	S	-	-	+	+	$>10^6$
511D1	-	-	<u>G</u>	-	P	-	K	R	A	S	-	-	++	+	$>10^6$
515D2	-	-	<u>N</u>	-	P	-	K	R	A	S	-	-	+	+	$>10^6$
511H3	-	-	D	-	P	-	<u>N</u>	R	A	S	-	-	+	+	$>10^6$
513H1	-	-	D	-	P	-	<u>E</u>	R	A	S	-	-	+	−	$>10^6$
515H1	-	-	D	-	P	-	K	<u>H</u>	A	S	-	-	++	+	$10^{5.5}$
513D1	-	-	D	-	P	-	K	R	<u>T</u>	S	-	-	−	+	$10^{5.4}$
516H1	-	├──── 9-amino-acid deletion ────┤									-	-	++	−	$>10^6$
516D1	-	-	D	-	P	-	K	R	A	S	-	<u>R</u>	+	+	$>10^6$

[a] Underlined amino acids correspond to the mutations found in the genomes of the indicated escape mutants.
[b] Loss of infectivity after 15 min of incubation at 45°C was scored as follows: −, less than 10^1; +, 10^1 to 10^2; ++, more than 10^2.
[c] Reproductive capacity at supraoptimal temperature (RCT) was determined as the difference of the virus titer (PFU) at optimal (37°C) and supraoptimal (40°C) temperatures and was scored as +, less than 10^2; −, more than 10^2.

virulent phenotype inherited from the Mahoney strain of poliovirus and linked to the nucleotide sequence in the 5′ noncoding region of the genome.

ACKNOWLEDGMENTS. We thank M. Arita for performing the monkey neurovirulence tests on v510 and v410 and Jim Hogle for continuous interest and helpful suggestions.

This work was supported in part by the Centre National de la Recherche Scientifique (ERA 1030, AIP 955198, and URA 545), by World Health Organization project V26/181/3 awarding T.C. a doctoral fellowship, and by grants 86.3003 and 88.1006 from the Institut National de la Santé et de la Recherche Médicale.

REFERENCES

Armstrong, C. 1939. Successful transfer of the Lansing strain of poliomyelitis virus from the cotton rat to the white mouse. *Public Health Rep.* **54**:2302–2305.

Arnold, E., M. Luo, G. Vriend, M. G. Rossmann, A. C. Palmenberg, G. D. Parks, M. J. H. Nicklin, and E. Wimmer. 1987. Implications of the picornavirus capsid structure for polyprotein processing. *Proc. Natl. Acad. Sci. USA* **84**:21–25.

Chow, M., J. F. E. Newman, D. J. Filman, J. M. Hogle, D. J. Rowlands, and F. Brown. 1987. Myristylation of picornavirus capsid protein VP4 and its structural significance. *Nature* (London) **327**:482–486.

Hogle, J. M., M. Chow, and D. J. Filman. 1985. Three-dimensional structure of poliovirus at 2.9 Å resolution. *Science* **229**:1358–1365.

Kawamura, N., M. Kohara, S. Abe, T. Komatsu, K. Tago, M. Arita, and A. Nomoto. 1989. Determinants in the 5′ noncoding region of poliovirus Sabin 1 RNA that influence the attenuation phenotype. *J. Virol.* **63**:1302–1309.

Kean, K. M., C. Wychowski, H. Kopecka, and M. Girard. 1986. Highly infectious plasmids carrying poliovirus cDNA are capable of replication in transfected simian cells. *J. Virol.* **59**:490–493.

Martin, A., C. Wychowski, D. Bénichou, R. Crainic, and M. Girard. 1988a. Construction of a chimaeric type1/type2 poliovirus by genetic recombination. *Ann. Virol.* **139**:79–88.

Martin, A., C. Wychowski, T. Couderc, R. Crainic, J. Hogle, and M. Girard. 1988b. Engineering a poliovirus type 2 antigenic site on a type 1 capsid results in a chimaeric virus which is neurovirulent for mice. *EMBO J.* **7**:2839–2847.

Morinaga, Y., T. Franceschini, B. Inouye, and M. Inouye. 1984. Improvement of oligonucleotide directed site specific mutagenesis using double stranded plasmid DNA. *Bio/Technology* **2**:636–639.

Murray, M. G., J. Bradley, X. F. Yang, E. Wimmer, E. G. Moss, and V. R. Racaniello. 1988. Poliovirus host range is determined by a short amino acid sequence in neutralization antigenic site 1. *Science* **241**:213–215.

Omata, T., M. Kohara, S. Kuge, T. Komatsu, S. Abe, B. L. Semler, A. Kameda, H. Itoh, M. Arita, E. Wimmer, and A. Nomoto. 1986. Genetic analysis of the attenuation phenotype of poliovirus type 1. *J. Virol.* **58**:348–358.

Paul, A. V., A. Shultz, S. E. Pincus, S. Oroszlan, and E. Wimmer. 1987. Capsid protein VP4 of poliovirus is N-myristoylated. *Proc. Natl. Acad. Sci. USA* **84**:7827–7831.

Semba, K., M. Nishizawa, N. Miyajima, M. C. Yoshida, J. Sukegawa, Y. Yamanashi, M. Sasaki, T. Yamamoto, and K. Toyoshima. 1986. Yes-related protooncogene, syn, belongs to the protein-tyrosine kinase family. *Proc. Natl. Acad. Sci. USA* **83**: 5459–5463.

Towler, D. A., S. P. Adams, S. R. Eubanks, D. S. Towery, E. Jackson-Machelski, L. Glaser, and J. I. Gordon. 1987. Purification and characteriza-tion of yeast myristoyl CoA:protein N-myristoyl-transferase. *Proc. Natl. Acad. Sci. USA* **84**:2708–2712.

Towler, D. A., S. P. Adams, S. R. Eubanks, D. S. Towery, E. Jackson-Machelski, L. Glaser, and J. I. Gordon. 1988. Myristoyl CoA:protein N-myristoyl-transferase activities from rat liver and yeast pos-sess overlapping yet distinct peptide substrate spec-ificities. *J. Biol. Chem.* **263**:1784–1790.

van der Werf, S., J. Bradley, E. Wimmer, W. F. Studier, and J. Dunn. 1986. Synthesis of infectious poliovirus RNA by purified T7 RNA polymerase. *Proc. Natl. Acad. Sci. USA* **83**:2330–2334.

Chapter 48

Mapping Neurovirulence in the Theiler's Murine Encephalomyelitis Virus Genome

M. A. Calenoff, K. S. Faaberg, and H. L. Lipton

Theiler's murine encephalomyelitis viruses (TMEV) are naturally occurring enteric pathogens of mice that can be divided into two groups on the basis of their neurovirulence after intracerebral (i.c.) inoculation (Lipton, 1975; Ohara and Roos, 1987). The first group consists of two highly virulent viruses, GDVII and FA, which produce a rapidly fatal encephalitis in adult mice. All of the remaining isolates, referred to as TO strains because they resemble Theiler's original isolates (Theiler, 1937), belong to the second group. These viruses are much less virulent than GDVII and FA but still cause central nervous system disease in the form of poliomyelitis (early onset), followed by a chronic, inflammatory demyelinating disease (late onset) that is due to viral persistence (Lipton, 1975; Clatch et al., 1986). The difference in virulence between the two groups is substantial, on the order of magnitude $\geq 10^5$ PFU per 50% lethal dose (LD_{50}) (Lipton, 1980). Therefore, the existence of two distinct neurovirulence groups makes the TMEV particularly useful for studying the molecular pathogenesis of picornaviruses.

The TMEV constitute a separate serological group within the picornavirus family (Lipton, 1978). On the basis of the complete nucleotide sequence and genome organization, the TMEV have been unofficially classified as cardioviruses, i.e., with encephalomyocarditis virus and mengovirus (Pevear et al., 1987; Ohara et al., 1988). Recently, we have sequenced strains representing both neurovirulence groups: GDVII virus, which is highly virulent, and BeAn virus, which is less virulent (Pevear et al., 1987; Pevear et al., 1988a). Computer-generated comparisons of GDVII and BeAn showed that they are 90.4%

identical at the nucleotide level and 95.7% identical at the amino acid level (Pevear et al., 1988b). Because the 775 nucleotide and 99 amino acid differences are dispersed throughout the genome, sequence analysis alone is not sufficient to identify the genomic regions or gene products responsible for pathogenetic properties, such as virulence and attenuation.

We have now developed a genetic system by constructing full-length cDNA copies of the highly virulent GDVII and less virulent BeAn virus RNAs and produced infectious transcripts in vitro. To identify the sites of neurovirulence determinants, recombinant chimeras were made between the two parental cDNAs, and the phenotype of each chimera was characterized in vitro and after i.c. inoculation of mice.

INFECTIVITY OF FULL-LENGTH PARENTAL RNA TRANSCRIPTS

BeAn virus cDNA was originally cloned into the *Pst*I site of pBR322 (Pevear et al., 1987), and a full-length clone was constructed from five subgenomic clones (unpublished data). After mutation of an internal *Pst*I site at nucleotide 6694, the full-length clone was inserted into the *Pst*I site of the transcription vector pGEM3 downstream of the T7 RNA polymerase promoter. Similarly, a full-length GDVII virus cDNA clone was assembled in pGEM3 from six subgenomic fragments after the initial cloning in pBR322 (Pevear et al., 1988a; manuscript in preparation).

cDNA clones, oriented with the 5' genomic end downstream of the T7 promoter, were lin-

M. A. Calenoff and K. S. Faaberg, Department of Neurology, Northwestern University Medical School, Chicago, Illinois 60611. *H. L. Lipton,* Department of Neurology, University of Colorado School of Medicine, 4200 East 9th Avenue, Denver, Colorado 80262.

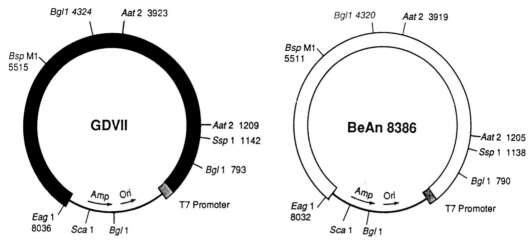

FIGURE 1. Final constructs of full-length cDNA clones of BeAn and GDVII in pGEM3. These two constructs are oriented with the 5' viral sequences oriented toward the T7 RNA polymerase promoter. The conserved restriction endonuclease sites used to make recombinant clones are indicated.

earized at the *Sph*I site within the vector, and positive-strand RNA transcripts were synthesized in vitro. In most instances, the entire reaction mixture was used to transfect BHK-21 cells without further purification, using DEAE-dextran as the facilitator (Vaheri and Pagano, 1965). The monolayers were incubated at 33°C, and by 3 to 5 days after transfection, focal areas of cytopathology (plaques) appeared. Progeny viral stocks, which were tested for in vitro growth phenotype and mouse neurovirulence, were prepared after one additional passage in BHK-21 cells.

The infectivity of positive-strand virus RNA transcripts was tested by transfection of BHK-21 cells and compared with that of the virion RNAs. Both GDVII and BeAn 8386 RNA transcripts produced 10^2 to 10^3 PFU/μg of RNA transfected (data not shown). In contrast, the virion RNAs had an infectivity of approximately 10^5 PFU/μg. Both TMEV cDNA templates (Fig. 1) contain extra sequences between the T7 promoter in the vector and the first viral 5' nucleotide; e.g., BeAn cDNA has 53 nucleotides of the vector in addition to 14 oligo(dG)-oligo(dC) tails. Twenty-two extra nucleotides are present at the 3' end downstream of the poly(A) tail. These additional sequences are known to reduce the efficiency of transfection of picornavirus RNA transcripts (Sarnow, 1989; van der Werf et al., 1986), and they are likely to be responsible for the lower efficiency of transfection that we observed.

PHENOTYPIC CHARACTERISTICS OF VIRUS PROGENY DERIVED FROM PARENTAL cDNA CLONES

Plaque size and temperature sensitivity of the viruses derived from RNA transfection experiments were determined. The BeAn parent had a titer of 5.0×10^7 PFU/ml, produced small plaques (1 mm in diameter), and was temperature sensitive, and $>10^6$ PFU inoculated i.c. was not lethal for adult SJL/J mice (LD_{50}, $\geq 3 \times 10^6$ PFU) (Table 1 and Fig. 2). In contrast, the GDVII parent had a titer of 2.8×10^8 PFU/ml, produced large plaques (3 to 5 mm in diameter), and was not temperature sensitive, and <20 PFU was lethal for SJL/J mice (LD_{50}, 17 PFU).

CONSTRUCTION OF RECOMBINANT TMEV

Recombinant viruses constructed with the two cDNA clones were made by using the restriction endonuclease sites for *Ssp*I, *Bgl*I, *Aat*II, *Bsp*MI and *Eag*I, as well as the *Sca*I site in pGEM3. Most of the constructs required partial digestions of these conserved restriction sites. As a result, seven different cDNA clones from corresponding segments of the BeAn and GDVII genomes were constructed and given a numerical designation (Fig. 3). Since there is a 10% difference in the nucleotide sequences of the two parental viruses (Pevear et al., 1988a), the origin of all exchanged genomic parts could be confirmed by cleavage with restriction endonucleases.

TABLE 1
In Vitro Growth Characteristics of Parental and Chimeric TMEV
Derived from RNA Transcripts

Virus[a]	Titer (PFU/ml)	Plaque size (mm)	Plaquing efficiency[b]
BeAn 8386 parent	5.0×10^7	1	$<1.0 \times 10^{-6}$
GDVII parent	2.8×10^8	3–5	0.46
Chimeras			
1	1.7×10^8	1	$<2.0 \times 10^{-4}$
6	7.6×10^7	1	$<1.3 \times 10^{-6}$
3	2.2×10^7	1	$<4.5 \times 10^{-6}$
7	2.3×10^8	3–5	$<9.1 \times 10^{-5}$
9	7.0×10^7	1	1.0×10^{-3}
10	7.5×10^7	5	0.20
13	1.0×10^8	1	$<1.0 \times 10^{-5}$

[a] BHK-21 cells were transfected with 10 to 20 μg of transcript RNA, and progeny viruses were prepared after one additional cell passage.
[b] Ratio of the virus titer at 39.8°C to that at 33°C.

PHENOTYPIC CHARACTERISTICS OF RECOMBINANT TMEV

Seven recombinant viruses from the highly virulent GDVII and less virulent BeAn parental

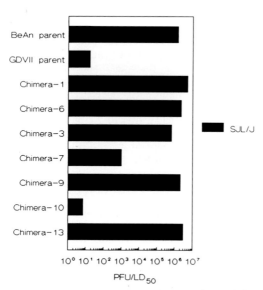

FIGURE 2. Comparative neurovirulence for TMEV parental and chimeric viruses after i.c. inoculation of mice, determined by the mean LD_{50}. Parental and chimeric viral stocks were derived from infectious transcript RNAs. Adult SJL/J mice were inoculated i.c. with 10-fold dilutions of virus stocks in 30 μl and observed daily for signs of illness and mortality. All control mice that were inoculated with DMEM survived (data not shown).

viruses were constructed in vitro from cDNA clones. Progeny viruses derived from these recombinants after transfection were used to analyze the effect of nucleotide and amino acid changes on the phenotype of the virus. Five restriction sites marked the ends of the exchanged DNA fragments that approximated the five cardiovirus genomic regions (Fig. 1; Rueckert and Wimmer, 1984). The following genomic exchanges were made: the 5′ noncoding regions (chimeras 1 and 6), the leader (L) and P1 region (chimeras 3, 7, and 9), both the 5′ noncoding and L-P1 regions (chimera 10), and the P3 region (chimera 13) (Fig. 3). The L-P1 genome sequences derived from the parental strain were the determinant of plaque size (Table 1). Chimeras with the GDVII L-P1 region produced large plaques in vitro, whereas chimeras with the BeAn L-P1 region produced small plaques. The mutations that influence TMEV temperature sensitivity have not been defined. Assaying the recombinants for their temperature-sensitive phenotype, we found a correlation between temperature sensitivity and the origin of the L-P1 region. Recombinants with the BeAn L-P1 region were temperature sensitive. Chimera 7, which contains 279 nucleotides of the 5′ noncoding region upstream of the polyprotein initiation codon, protein 2A, and 20 amino acids of 2B, in addition to the L-P1 region derived from GDVII, is also temperature sensitive. The fact that chimera 7 is temperature sensitive and possesses the GDVII L-P1 region indicates that additional temperature-sensitive mutation sites are present elsewhere in the genome. One of the sites of joining in chimera 7 was in the 5′ noncoding

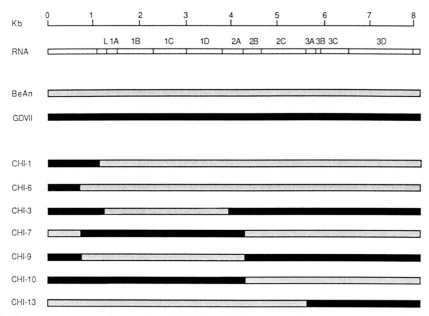

FIGURE 3. Genome structures of recombinant TMEVs, with the expected structures shown by the combination of BeAn and GDVII virus sequences. The genomic RNA and its organization are shown below in the standard nomenclature (Rueckert and Wimmer, 1984).

region, and alterations in the RNA structure in this location have been shown to produce temperature sensitivity with poliovirus mutants (Dildine and Semler, 1989; Racaniello and Meriam, 1986).

MOUSE NEUROVIRULENCE OF RECOMBINANT VIRUSES

Six- to 8-week-old SJL/J (Jackson Laboratories) female mice were inoculated in the right cerebral hemisphere with 10-fold dilutions of parental and chimeric viruses in 30 μl (5 to 10 mice per dilution), and all mice were observed daily for signs of illness and death. The LD_{50} was calculated by the method of Reed and Muench (1938), and critical LD_{50} determinations were repeated.

All of the recombinant viruses having the BeAn region encoding the L and P1 proteins were not lethal for mice, similar to parental BeAn virus (Fig. 2). In contrast, chimeras 7 and 10, which had the L-P1 region of GDVII, were highly lethal for mice. Because of the large LD_{50} differences obtained between the two groups, none of the virus stocks of the less virulent phenotype were concentrated to obtain an exact LD_{50} endpoint. These results indicate that the genomic region encoding the viral L and capsid proteins contain determinants responsible for

neurovirulence. In addition, another determinant(s) enhancing neurovirulence was detected in the 5' noncoding region. Chimera 10, which has the GDVII nucleotides 1 to 793, was 140 times more virulent than chimera 7, in which these nucleotides are derived from BeAn (Fig. 2), and was equivalent in neurovirulence to parental GDVII virus.

SUMMARY AND CONCLUSIONS

Full-length cDNA clones of the highly virulent GDVII and less virulent BeAn viruses, representing the two TMEV neurovirulence groups, were constructed in the bacterial plasmid pGEM3. RNA transcripts made from cDNAs were transfected into BHK-21 cells, yielding progeny viruses that had the exact in vitro growth phenotype and mouse neurovirulence pattern of the respective parent. Seven recombinant viruses were constructed by exchanging corresponding genomic regions (5' noncoding, L-P1, P2, P3, and 3' noncoding) between the parental cDNAs. RNA transcripts from the recombinant cDNAs were infectious and enabled analysis of their in vitro phenotypic characteristics, such as plaque size and temperature sensitivity, and mouse neurovirulence.

A major determinant of neurovirulence was found to lie in the L-P1 region of the genome, since the presence of the L-P1 region from the less virulent BeAn strain always resulted in viruses that were not lethal. The presence of the corresponding region from the highly virulent GDVII strain was associated with recombinants that produced a rapidly fatal encephalitis. For chimeras 3, 7, and 9, part of the genome encoding the L protein was exchanged along with the P1 region; therefore, one cannot exclude a role for the leader in neurovirulence on the basis of these constructs. Chimera 3, in which the P1 region came from the BeAn strain but most of the L sequence (N-terminal 46 amino acids) was derived from the GDVII strain, was in fact attenuated. These results suggest that neurovirulence maps to the coat proteins rather than to the L protein. The coat proteins encapsidate the single-stranded RNA genome and are responsible for the surface antigenic and receptor attachment sites. The function of the cardiovirus L protein remains to be elucidated.

Virion coat protein changes may alter virus receptor activity and antigenicity, as well as virion stability and possibly other virus processes, such as RNA encapsidation and uncoating. It does not seem plausible that merely altering antigenicity could account for the large difference in neurovirulence that exists between the two TMEV groups. Moreover, both groups elicit strong humoral and cellular immune responses in mice (Clatch et al., 1986; Ohara and Roos, 1987). It seems more likely that the GDVII L-P1 region, which correlates with enhanced neurovirulence, is due to virion receptor binding to host cells. In this regard, the neuronal tropisms of the two groups of TMEV differ (Lipton, 1975; Sethi and Lipton, 1981; Stroop et al., 1981). GDVII virus infects neocortical, hippocampal, and motor neurons; in contrast, BeAn virus principally grows in motor neurons in the brain stem and spinal cord and, to some extent, in neurons in the hippocampus. Thus, the increased virulence of GDVII virus correlates with its ability to infect neurons more extensively in the mouse neocortex, which could be the function of its receptor. Interestingly, 2 of the 30 amino acid residues in the putative TMEV receptor-binding site differ between the BeAn and GDVII viruses (Pevear et al., 1988b). It is intriguing to speculate that by changing either or both of these residues, one may alter the TMEV host range (neurotropism).

In support of this notion are the recent findings of Murray et al. (1988) and Martin et al. (1988). Both groups have shown that replacement of the amino acids of neutralizing immunogenic site I of the Mahoney strain of poliovi-

rus type 1 with the corresponding sequence of the Lansing strain of poliovirus type 2 enabled the hybrid Mahoney virus to infect the mouse central nervous system and cause poliomyelitis. Because antigenic site I is located near the virion receptor attachment site (Rossmann and Palmenberg, 1988), it has been postulated that site I is involved in attachment of poliovirus to mouse neurons. Zurbriggen and Fujinami (1989) also reported that a neutralizing monoclonal antibody to VP1 of the DA strain (less virulent TMEV group) selected a viral escape mutant of reduced virulence that lost its ability to persist in the central nervous system. This outcome is somewhat similar to the situation of reduced mouse neurovirulence of poliovirus type 2 variants selected with monoclonal antibodies to antigenic site I (La Monica et al., 1987). In both instances, a change in the virion receptor attachment site may be responsible for attenuation.

ACKNOWLEDGMENTS. This research was supported by Public Health Service grants NS 21913 and NS 23349 from the National Institutes of Health and by National Multiple Sclerosis Society grant RG 1815 B-7. K.S.F. is a postdoctoral fellow of the National Multiple Sclerosis Society.

We thank Jay Desai and Robert Morrison for excellent technical assistance.

REFERENCES

Clatch, R. J., H. L. Lipton, and S. D. Miller. 1986. Characterization of Theiler's murine encephalomyelitis virus (TMEV)-specific delayed-type hypersensitivity responses in TMEV-induced demyelinating disease: correlation with clinical signs. *J. Immunol.* **136:**920–927.

Dildine, S. L., and B. L. Semler. 1989. The deletion of 41 proximal nucleotides reverts a poliovirus mutant containing a temperature-sensitive lesion in the 5' noncoding region of genomic RNA. *J. Virol.* **63:**847–862.

La Monica, N., W. J. Kupsky, and V. R. Racaniello. 1987. Reduced mouse neurovirulence of poliovirus type 2 Lansing antigenic variants selected with monoclonal antibodies. *Virology* **161:**429–437.

Lipton, H. L. 1975. Theiler's virus infection in mice: an unusual biphasic disease process leading to demyelination. *Infect. Immun.* **11:**1147–1155.

Lipton, H. L. 1978. Characterization of the TO strains of Theiler's mouse encephalomyelitis viruses. *Infect. Immun.* **20:**869–872.

Lipton, H.L. 1980. Persistent Theiler's murine encephalomyelitis virus infection in mice depends on plaque size. *J. Gen. Virol.* **46:**169–177.

Martin, A., C. Wychowski, T. Couderc, R. Crainic, J. Hogle, and M. Girard. 1988. Engineering a poliovirus type 2 antigenic site on a type 1 capsid results in a chimaeric virus which is neurovirulent for mice. *EMBO J.* **7**:2839–2847.

Murray, M. G., J. Bradley, X.-F. Yang, E. Wimmer, E. G. Moss, and V. R. Racaniello. 1988. Poliovirus host range is determined by a short amino acid sequence in neutralization antigenic site I. *Science* **241**:213–216.

Ohara, Y., and R. Roos. 1987. The antibody response in Theiler's virus infection: new perspectives on multiple sclerosis. *Prog. Med. Virol.* **34**:156–179.

Ohara, Y., S. Stein, J. Fu, L. Stillman, L. Klaman, and R. P. Roos. 1988. Molecular cloning and sequence determination of DA strain of Theiler's murine encephalomyelitis viruses. *Virology* **164**:245–255.

Pevear, D. C., J. Borkowski, M. Calenoff, C. K. Oh, O. Ostrowski, and H. L. Lipton. 1988a. Insights into Theiler's virus neurovirulence based on a genomic comparison of the neurovirulent GDVII and less virulent BeAn strains. *Virology* **165**:1–12.

Pevear, D. C., M. Calenoff, E. Rozhon, and H. L. Lipton. 1987. Analysis of the complete nucleotide sequence of the picornavirus Theiler's murine encephalomyelitis virus indicates that it is closely related to cardioviruses. *J. Virol.* **61**:1507–1516.

Pevear, D. C., M. Luo, and H. L. Lipton. 1988b. Three-dimensional model of the capsid proteins of two biologically different Theiler's virus strains: clustering of amino acid differences identifies possible locations of immunogenetic sites on the virion. *Proc. Natl. Acad. Sci. USA* **85**:4496–4500.

Racaniello, V. R., and C. Meriam. 1986. Poliovirus temperature-sensitive mutant containing a single nucleotide deletion in the 5'-noncoding region of the viral RNA. *Virology* **155**:498–507.

Reed, L. J., and H. A. Muench. 1938. A simple method of estimating fifty percent endpoints. *Am. J. Hyg.* **27**:493–497.

Rossmann, M. G., and A. C. Palmenberg. 1988. Conservation of the putative receptor attachment site in picornaviruses. *Virology* **164**:373–382.

Rueckert, R. R., and E. Wimmer. 1984. Systemic nomenclature of picornaviral proteins. *J. Virol.* **50**: 957–959.

Sarnow, P. 1989. Role of 3'-end sequences in infectivity of poliovirus transcripts made in vitro. *J. Virol.* **63**:467–470.

Sethi, P., and H. L. Lipton. 1981. The growth of four human and animal enteroviruses in the central nervous system of mice. *J. Neuropathol. Exp. Neurol.* **40**:258–270.

Stroop, W. G., J. R. Baringer, and M. Brahic. 1981. Detection of Theiler's virus RNA in mouse central nervous system by in-situ hybridization. *Lab. Invest.* **45**:504–509.

Theiler, M. 1937. Spontaneous encephalomyelitis of mice, a new virus disease. *J. Exp. Med.* **65**:705–719.

Vaheri, A., and J. S. Pagano. 1965. Infectious poliovirus RNA: a sensitive method of assay. *Virology* **27**:434–436.

van der Werf, S., J. Bradley, E. Wimmer, F. W. Studier, and J. J. Dunn. 1986. Synthesis of infectious poliovirus RNA by purified T7 RNA polymerase. *Proc. Natl. Acad. Sci. USA* **83**:2330–2334.

Zurbriggen, A., and R. S. Fujinami. 1989. A neutralization-resistant Theiler's virus variant produces an altered disease pattern in the mouse central nervous system. *J. Virol.* **63**:1505–1513.

Chapter 49

Studies of Alphavirus Virulence Using Full-Length Clones of Sindbis and Venezuelan Equine Encephalitis Viruses

R. E. Johnston, N. L. Davis, J. M. Polo, D. L. Russell, D. F. Pence, W. J. Meyer, D. C. Flynn, L. Willis, S.-C. Lin, and J. F. Smith

The study of viral pathogenesis is basically the study of the interaction of the virus with cells of the host. One aspect of this interaction, and a major determinant of pathogenicity, is the specificity with which a virus selectively targets cells and tissues in vivo. We have examined (i) early interactions between the alphaviruses Sindbis virus and Venezuelan equine encephalitis virus (VEE) and their host cells in culture and (ii) how mutations affecting these interactions influence the course of disease in experimental animals. In previous work, we selected virus mutants capable of rapid penetration into baby hamster kidney cells, penetration being defined operationally as the point at which infectious virions are no longer susceptible to antibody neutralization at the cell surface or to removal from the cell surface by trypsin (Baric et al., 1981; Olmsted et al., 1984). Many of these mutants display an attenuated phenotype in vivo. For instance, of eight rapidly penetrating mutants of VEE, seven were avirulent in mice (Johnston and Smith, 1988). These mutations are not detrimental to virus growth in susceptible cells, and their inability to cause disease cannot be ascribed to being "weakened" in a general sense. Rather, such mutations appear to affect specific functional domains in the E2 glycoprotein that are involved in the disease process.

The relationship of rapid penetration in cell culture to reduced virulence in vivo is not im-mediately obvious. We hypothesize that both penetration in cultured cells and targeting of critical tissues in vivo (and hence pathogenesis) are related functions of the viral glycoproteins. If a glycoprotein domain important in the penetration event overlaps a domain having a major influence on pathogenicity, then selection for an altered penetration rate (whether faster or slower) is likely to coselect for an alteration in pathogenicity. In this context, therefore, rapid penetration is simply a trait for which a convenient positive selective pressure can be applied.

The overlapping-domain hypothesis has been tested in the Sindbis virus system. Sequencing the glycoprotein genes of Sindbis virus penetration-attenuation mutants revealed a single-site mutation at E2 residue 114, where an Arg in an attenuated mutant (SB-RL) had been substituted for a wild-type Ser (Davis et al., 1986). The putative causative mutation at E2 residue 114 was introduced into the full-length Sindbis virus clone, Toto1101 (Polo et al., 1988). Using two constructs that differed only at E2 codon 114, it was demonstrated that the Ser-to-Arg mutation is sufficient for both the attenuation and penetration phenotypes. In reference to the overlapping-domain model, E2 residue 114 must participate in both penetration and pathogenesis domains.

We have continued to explore the genetics of alphavirus pathogenesis and penetration by

R. E. Johnston, N. L. Davis, J. M. Polo, D. F. Pence, W. J. Meyer, and S.-C. Lin, Department of Microbiology and Immunology, University of North Carolina, Chapel Hill, North Carolina 27599-7290.   *D. L. Russell,* Department of Biology, Washington and Lee University, Lexington, Virginia 24450.   *D. C. Flynn,* Department of Microbiology, University of Virginia, School of Medicine, Charlottesville, Virginia 22908.   *L. Willis,* Department of Biochemistry, North Carolina State University, Raleigh, North Carolina 27695.   *J. F. Smith,* Division of Virology, U.S. Army Medical Research Institute of Infectious Diseases, Frederick, Maryland 21701.

using monoclonal antibody (MAb) escape mutants and site-directed mutagenesis of full-length cDNA clones. This chapter will address (i) the nature of the pathogenesis domain encompassing E2 residue 114, (ii) the early cell surface events that have been altered in the penetration mutants, (iii) the assembly of a model alphavirus vaccine by using multiple attenuating mutations placed into a full-length clone of Sindbis virus, and (iv) the application of this information to the generation of a new live attenuated vaccine for VEE.

MAPPING AN E2 PATHOGENESIS SITE

The pathogenesis domain that includes E2 residue 114 was defined by using E2-specific MAbs. In addition to changes in penetration rate and pathogenesis, the attenuating Ser-to-Arg mutation at E2 residue 114 also alters the antibody neutralization phenotype of the virus (Olmsted et al., 1986; Polo et al., 1988). In biological mutants as well as virus derived from substituted cDNA clones, strains having E2 Arg-114 are more efficiently neutralized by the E2c class of MAbs than are strains having E2 Ser-114. This finding suggested that the E2c neutralizing antigenic site might be related to the pathogenesis domain of which E2 residue 114 is a part. Accordingly, MAb escape mutants were selected by using the two MAbs of the E2c class, R6 and R13. One panel of escape mutants was selected from Sindbis virus strain AR339 (SB), our prototype virulent strain (E2 Ser-114). Another panel was selected from SB-RL, the prototype attenuated, rapidly penetrating strain (E2 Arg-114) which itself was derived from SB. Whether selected with MAb R6 or R13, none of the mutants were capable of binding either of the E2c MAbs. The glycoprotein genes of each mutant were sequenced, and each mutant harbored a single coding change in one of three E2 codons, 62, 96, or 159. These data suggested that E2c is a conformational antigenic site composed of amino acid residues that are dispersed along the primary sequence of the protein but which are nevertheless closely associated within the folded E2 molecule.

The phenotypic effect of amino acid substitutions at E2 positions 62, 96, and 159 depended not only on the specific substitution but also on the residue at 114. One effect was suppression of the attenuated phenotype of SB-RL. In the SB-RL background (E2 Arg-114), an Asp-for-Asn substitution at E2 62 or a Glu-for-Lys substitution at E2 residue 159 suppressed the attenuating effect of E2 Arg-114, giving rise to a

virulent strain. At present, the mechanism of the intragenic suppression is a matter of speculation. For example, it is of interest that both of the intragenic suppressor mutations encoded an acidic amino acid. In the virulent SB strain, E2 Ser-114 is in the center of a small region of hydrophobic and uncharged amino acids. Introduction of an Arg at this position considerably decreases the predicted hydrophobicity of the area. If E2 residues 62 and 159 are in close proximity to E2 residue 114, then it is possible that an acidic residue at E2 position 62 or 159 could neutralize the charged E2 Arg-114 and perhaps abrogate its effect on local hydrophobicity and its attenuating phenotype. Consistent with this hypothesis is the finding that an Asn substitution at E2 position 159 did not suppress the attenuating phenotype specified by E2 Arg-114, nor did an analogous mutation selected in the SB background alter the virulent phenotype specified by E2 Ser-114. We are exploring the intragenic suppression further by characterizing appropriate site-directed mutants placed in the full-length Sindbis virus cDNA clone.

A second effect was attenuation of virulence in the SB background (E2 Ser-114). In SB, a His-for-Tyr substitution at E2 position 96 was attenuating. This finding, in combination with data from mutants derived from SB-RL, shows that a mutation at E2 position 62, 96, 114, or 159 affects E2c MAb binding or activity as well as pathogenesis in neonatal mice. This suggests that these mutations not only define the E2c antigenic site but also define a pathogenesis domain on the glycoprotein spike which governs the course of disease in vivo.

A pathogenesis domain analogous to that identified by the E2c MAbs may be present in other alphaviruses. Vrati et al. (1986) isolated an attenuated mutant of Ross River virus containing an in-frame deletion of E2 residues 55 to 61. Griffin and Johnson (1977) isolated a variant of Sindbis virus AR339 that is neurovirulent in adult mice inoculated intracerebrally (i.c.), whereas its AR339 parent is avirulent in this model. By using recombinants constructed in vitro, it was determined that acquisition of the adult virulence phenotype requires two mutations, one at E1 position 313 and another at E2 position 55 (Lustig et al., 1988). In Sindbis virus, this pathogenesis domain may include additional residues that were not represented among the E2c MAb escape mutants. Russell et al. (1989) isolated several rapidly penetrating mutants of Sindbis virus strain S.A.AR86, a natural isolate virulent for adult mice by the i.c. route. Among these mutants was one that was rapidly penetrating but virulent in both neonatal and adult mice (S.-C. Lin, D. L. Russell, and R. E. Johnston, manuscript in preparation). Sequencing suggested that the rapid penetration phenotype resulted from a mutation at E2 position 70, and introduction of this mutation into a full-length clone of Sindbis virus strain AR339 produced a virus that was rapidly penetrating. Although this mutation specified a rapidly penetrating, virulent phenotype in the S.A.AR86 background, in the AR339 background the same mutation specified attenuation. These observations provide additional evidence of an alphavirus pathogenesis domain in the vicinity of E2 residues 55 to 70. However, determination of the spatial relationships between these E2 residues and those identified among the E2c MAb escape mutants must await the solution of the three-dimensional structure of the Sindbis virus glycoprotein spike.

The relationship of the pathogenesis domain to a putative overlapping domain important for penetration was explored by examining the penetration phenotype of each of the E2c MAb-selected mutants. Although specific mutations at E2 residues 62 and 159 were intragenic suppressors of the SB-RL attenuation phenotype, only the E2 position 62 mutation also suppressed the SB-RL rapid-penetration phenotype. In conjunction with E2 Arg-114, which specifies rapid penetration and attenuation in SB-RL, the Glu substitution at E2 position 159 produced a rapidly penetrating, virulent phenotype. The His-for-Tyr substitution at E2 position 96, in conjunction with E2 Ser-114 in the SB background, produced a slowly penetrating, attenuated phenotype. From these data with a limited variety of amino acid substitutions at each critical residue, it appears that the Sindbis virus glycoprotein domains influencing pathogenesis and penetration do overlap but may not necessarily be coincident.

REARRANGEMENT OF THE GLYCOPROTEIN SPIKE AT THE CELL SURFACE

We have examined the early events in Sindbis virus infection to identify the specific stage at which the penetration of the prototype attenuated mutant, SB-RL, is accelerated. The results of these studies suggest that after attachment of Sindbis virus to cells, and either immediately before or coincident with internalization of infecting virions, the E1-E2 glycoprotein spikes undergo a programmed rearrangement. As a result of this rearrangement, previously unexposed epitopes become accessible to their cognate MAbs. It is this early step which appears to be accelerated in the rapidly penetrating mutants. The data supporting these conclusions are summarized below.

The rearrangement was demonstrated by using MAbs as specific probes of virion structure. Each of the MAbs used bound to virions in enzyme-linked immunosorbent assays (ELISA) in which purified virus had been disrupted by nonspecific adsorption to the wells of an ELISA plate. The MAbs were categorized further on the basis of binding or lack of binding to intact virions in solution. Some recognized external epitopes accessible at the virion surface, whereas others recognized inaccessible or internal epitopes. We reasoned that an alteration in virion structure might be detectable as a change in epitope accessibility. Virus-cell complexes were established by allowing freshly grown and gradient-purified Sindbis virions to attach to baby hamster kidney cells at 4°C, a temperature at which attachment occurs but penetration is inhibited. Binding of MAb to such complexes was detected with ^{125}I-goat anti-mouse immunoglobulin. Attached virions at 4°C displayed the same epitope accessibility as did virions in solution. However, when the virus-cell complexes, in the presence of the test MAb, were shifted to 37°C to allow the process of infection to proceed, binding of a subset of E1 and E2 MAbs to previously inaccessible epitopes was detected. We refer to the newly accessible epitopes as transitional epitopes. This result suggested that a significant rearrangement of the glycoprotein spike had occurred as a consequence of interactions at the cell surface and that the exposure of these transitional epitopes might mediate subsequent events in the normal entry pathway.

The rearrangement appeared to occur at the cell surface on particles that closely resemble virions. In electron microscope studies, gold-tagged MAbs to one of the transitional epitopes bound to structures at the cell surface which were indistinguishable from virions. Morphometric analysis showed that virtually all of the specific labeling was associated with such particles rather than the plasma membrane itself.

The rearrangement required no new viral or cellular protein synthesis, as it occurred normally in the presence of cycloheximide. The rearrangement did not appear to be induced by transit of virions through a low-pH compartment in that (i) it was not inhibited by NH_4Cl and (ii) it occurred normally at both permissive and nonpermissive temperatures in Chinese hamster ovary cells temperature sensitive for endosomal acidification.

Several lines of evidence suggested that the rearranged virions are intermediates in the normal infection pathway. First, the rearrangement was detected earlier with a rapidly penetrating mutant. Second, the time course and temperature dependence of the rearrangement were indistinguishable from the time course and temperature dependence of penetration. Third, MAb R12, which is specific for one of the transitional E2 epitopes exposed on rearranged particles, retarded penetration of infectious virions. In this experiment, virus was allowed to attach to cells at 4°C, and then the virus-cell complexes were shifted to 30°C to allow penetration. In a control experiment, infected cultures were treated with MAb R15 after attachment but before allowing penetration. MAb R15 is a neutralizing MAb directed to an external epitope on E2 and therefore neutralized 88% of the attached PFU. When MAb R15 was added after allowing 45 min at 30°C for penetration, only 5% of the PFU were still susceptible to R15 neutralization. However, if MAb R12 was added after attachment and was present during incubation at 30°C, 32% of the PFU remained susceptible to R15 neutralization after the penetration period. Our interpretation of this result is that when the attached virions rearranged, MAb R12 bound to the rearranged intermediates and slowed the rate at which they penetrated the cell, leaving a larger proportion of virions at the cell surface, where they were susceptible to neutralization by MAb R15 added at the end of the penetration period. MAb R12 was a specific probe for rearranged virions, and penetration in this experiment was measured by assay of residual infectivity. Therefore, the observed effect of MAb R12 on penetration strongly supports the notion that rearranged virions are intermediates in the normal infection pathway.

ASSEMBLY OF A MODEL ALPHAVIRUS VACCINE

The principal disadvantage of live virus vaccines is the potential for reversion to virulence. Theoretically, this disadvantage could be overcome by the presence of multiple, independently attenuating mutations in the vaccine strain. This appears to be the case with the Sabin type 1 poliovirus vaccine, which has a low reversion frequency (Omata et al., 1986). Two developments facilitated the assembly of a model alphavirus vaccine to test this hypothesis. First, the finding that selection for rapid penetration tends to coselect for attenuated mutants allowed the identification of specific mutations that reduce virulence yet do not affect the ability of the virus to grow in susceptible cells. Second, the construction of a full-length Sindbis virus cDNA clone (Toto1101) (Rice et al., 1987), which could be transcribed into an infectious RNA, provided a workable genetic system in which to assess the effects of attenuating mutations individually and in combination. The goal of the experiments outlined below was to devise a model alphavirus vaccine containing multiple, independently attenuating mutations. By combining an attenuating E1 gene from Toto1101 with the Arg-for-Ser substitution found in the SB-RL E2 gene, we have produced a virus that is considerably more attenuated than either of the single mutants and which uniformly protects against high-dose challenge.

We have shown previously that virus derived from the original Rice clone of Sindbis virus (Toto1101) is attenuated in neonatal mice (20 to 30% mortality after subcutaneous [s.c.] inoculation of 100 PFU) (Polo et al., 1988). When the 6K and E1 genes of Toto1101 are replaced with the 6K and E1 genes cloned from Sindbis virus AR339, virulence is restored (100% mortality). The two 6K-E1 regions differ at three loci, with the Toto1101 residues E1 Ala-72, Gly-75, and Ser-237 corresponding to AR339 E1 Val-72, Asp-75, and Ala-237. The contribution of each locus to the pathogenesis phenotype was determined by using site-directed mutagenesis to replace E1 residue 72, 75, or 237 in Toto1101 with the corresponding amino acid from AR339. Replacement of either E1 residue 75 or 237 increased the mortality induced by Toto1101-derived virus from 20 to 30% to 60 to 65%, whereas replacement at Toto1101 E1 residue 72 had no significant effect. Combining the E1 position 75 and 237 replacements in Toto1101 increased the mortality to 98%. However, all three replacements appeared to be required to induce 100% mortality.

Attenuating E2 and E1 mutations were combined in the following experiments. TR2100 is a virulent construct containing E2 Ser-114 and the virulent E1 gene from AR339 (Polo et al., 1988). At an s.c. dose of 100 PFU, virus from this construct induced 100% mortality, with an average survival time of 4 days. Peak viremia of 10^9 PFU/ml was attained at 24 h postinoculation. Virus titer in the brain rose steadily from 10^8 PFU/g at 24 h to exceed 10^{10} PFU/g by 4 days, after which all animals had died. TR2200 is an attenuated construct that differs from TR2100 by a single coding change to E2 Arg-114. The s.c.

and i.c. 50% lethal doses (LD_{50}s) of TR2200 were $>10^7$ and 1.4×10^6 PFU, respectively. In animals inoculated s.c. with 100 PFU, a peak viremia of 10^7 PFU/ml was obtained, and the maximum occurred at 3 days postinoculation. Likewise, the invasion of the brain appeared to be delayed. No virus was detected in the brain before day 2, and the maximum titer of $10^{8.5}$ PFU/g was not reached until day 6. Combining the E2 Arg-114 mutation with the attenuating E1 gene from Toto1101 in TR1000 raised the s.c. and i.c. LD_{50}s to $>10^7$ PFU. First detection of virus in the brain was delayed to day 3 after inoculation with 100 PFU s.c. The maximum titer of virus in the brain was reduced to approximately $10^{6.5}$ PFU/g, which occurred on day 9 postinoculation. The s.c. and i.c. LD_{50}s of virus derived from Toto1101 were 3×10^5 and $<10^4$ PFU, respectively. Therefore, virus containing attenuating mutations in both of the glycoprotein genes was more attenuated than either of the mutations alone but nevertheless was sufficiently immunogenic to protect against an i.c. challenge of 500 LD_{50} of S.A.AR86, a Sindbis virus strain virulent in adult mice.

GENERATION OF AN ENGINEERED VEE VACCINE

On the basis of the information obtained with the Sindbis virus model system, we have begun the initial experiments required to produce an engineered live attenuated vaccine for VEE. As a first step in this process, a cDNA clone of the VEE genome was constructed and placed downstream from a T7 promoter such that infectious RNA transcripts are derived from the clone by in vitro transcription (Davis et al., 1989). Second, rapidly penetrating, mouse avirulent mutants of VEE were isolated (Johnston and Smith, 1988). By sequencing the glycoprotein genes of the mutants, putative attenuating mutations have been mapped to E2 positions 76, 120, and 209. It is of interest that the mutation in one of the rapidly penetrating mutants was at E2 position 120, a substitution of Lys for the parental Thr. A similar mutation was found at E2 position 120 in the experimental live VEE vaccine TC-83, a substitution of Arg for Thr (Kinney et al., 1986). Each of these mutations, individually and in combination, is being placed into the full-length cDNA clone by site-directed mutagenesis. The individual constructs will be used to confirm the attenuating effect of each of the mutations by comparing the pathogenesis of virus derived from clones that differ by a single nucleotide change. Virus from constructs containing multiple mutations will be used to (i) determine potential additive attenuating effects, (ii) assess the influence of multiple mutations on reversion in vivo, and (iii) document the effectiveness of such engineered strains as vaccines by challenge with virulent wild-type VEE.

SUMMARY

We have used a genetic approach to examine alphavirus virulence in animals, penetration in cell culture, and the relationship between these two phenotypes. Each phenotype is strongly influenced by a specific domain on an exposed face of glycoprotein E2. The domain associated with pathogenesis overlaps the domain which affects penetration such that genetic selection for an altered penetration phenotype often coselects for altered virulence in vivo. With respect to penetration of Sindbis virus, at least one class of mutations that accelerate penetration also facilitates the generation of rearranged virions at the cell surface. These rearranged particles have been identified as intermediates in the normal infection process and are formed after attachment but before internalization. With respect to Sindbis virus pathogenesis, we have mapped portions of an E2 pathogenesis domain by using neutralization escape mutants. Certain individual attenuating mutations retard invasion of and reduce replication in the brain. Combining such mutations appears to produce an additive attenuating effect on neuroinvasiveness and neurovirulence. Guided by the Sindbis virus model system, we have generated a full-length cDNA clone of VEE that can be transcribed in vitro into an infectious RNA and have begun to insert into this clone putative attenuating mutations identified by sequencing rapidly penetrating, attenuated mutants of VEE. It is our hope that combinations of attenuating mutations will produce an effective live attenuated vaccine for VEE having a low rate of reversion to virulence.

ACKNOWLEDGMENTS. This work was supported by Public Health Service grants AI22186 and NS26681 and by the U.S. Army Medical Research and Development Command (DAMD 17-87-C-7259).

REFERENCES

Baric, R. S., D. W. Trent, and R. E. Johnston. 1981. A Sindbis virus variant with a cell determined latent period. *Virology* **110**:237–242.

Davis, N. L., F. J. Fuller, W. G. Dougherty, R. A. Olmsted, and R. E. Johnston. 1986. A single nucleotide change in the E2 glycoprotein of Sindbis virus affects penetration rate in cell culture and virulence in neonatal mice. *Proc. Natl. Acad. Sci. USA* **83:** 6771–6775.

Davis, N. L., L. V. Willis, J. F. Smith, and R. E. Johnston. 1989. In vitro synthesis of infectious Venezuelan equine encephalitis virus RNA from a cDNA clone: analysis of a viable deletion mutant. *Virology* **171:**189–204.

Griffin, D. E., and R. T. Johnson. 1977. Role of the immune response in recovery from Sindbis virus encephalitis in mice. *J. Immunol.* **118:**1070–1075.

Johnston, R. E., and J. F. Smith. 1988. Selection for accelerated penetration rate in cell culture coselects for attenuated mutants of Venezuelan equine encephalitis virus. *Virology* **162:**437–443.

Kinney, R. M., B. J. B. Johnson, V. L. Brown, and D. W. Trent. 1986. Nucleotide sequence of the 26S mRNA of the virulent Trinidad donkey strain of VEE virus and deduced sequence of the encoded structural protein. *Virology* **152:**400–413.

Lustig, S., A. C. Jackson, C. S. Hahn, D. E. Griffin, E. G. Strauss, and J. H. Strauss. 1988. Molecular basis of Sindbis virus neurovirulence in mice. *J. Virol.* **62:**2329–2336.

Olmsted, R. A., R. S. Baric, B. A. Sawyer, and R. E. Johnston. 1984. Sindbis virus mutants selected for rapid growth in cell culture display attenuated virulence in animals. *Science* **225:**424–427.

Olmsted, R. A., W. J. Meyer, and R. E. Johnston. 1986. Characterization of Sindbis virus epitopes important for penetration in cell culture and pathogenesis in animals. *Virology* **148:**245–254.

Omata, T., M. Kohara, S. Kuge, T. Komarsu, S. Abe, B. L. Semler, A. Kameda, H. Itoh, M. Arita, E. Wimmer, and A. Nomoto. 1986. Genetic analysis of the attenuation phenotype of poliovirus type 1. *J. Virol.* **58:**348–358.

Polo, J. M., N. L. Davis, C. M. Rice, H. V. Huang, and R. E. Johnston. 1988. Molecular analysis of Sindbis virus pathogenesis in neonatal mice by using virus recombinants constructed in vitro. *J. Virol.* **62:** 2124–2133.

Rice, C. M., R. Levis, J. H. Strauss, and H. V. Huang. 1987. Production of infectious transcripts from Sindbis virus cDNA clones: mapping of lethal mutations, rescue of a temperature-sensitive marker, and in vitro mutagenesis to generate defined mutants. *J. Virol.* **61:**3809–3819.

Russell, D. L., J. M. Dalrymple, and R. E. Johnston. 1989. Sindbis virus mutations which coordinately affect glycoprotein processing, penetration, and virulence in mice. *J. Virol.* **53:**1619–1629.

Vrati, S. V., S. G. Faragher, R. C. Weir, and L. Dalgarno. 1986. Ross River virus mutant with a deletion in the E2 gene: properties of the virion, virus-specific macromolecule synthesis, and attenuation of virulence for mice. *Virology* **151:**222–232.

Chapter 50

Molecular Studies on Enteroviral Heart Disease

Reinhard Kandolf, Annie Canu, Karin Klingel, Philip Kirschner, Heike Schönke, Jürgen Mertsching, Roland Zell, and Peter Hans Hofschneider

Enteroviruses of the family *Picornaviridae*, such as coxsackieviruses of group B types 1 to 5, are generally considered to be the most common agents of viral myocarditis (Abelmann, 1973; Woodruff, 1980; Johnson and Palacios, 1982; Reyes and Lerner, 1985; Kandolf, 1988). Other members of the human enterovirus group, comprising at present over 70 serotypes (e.g., various group A coxsackieviruses and echoviruses), have also been associated with human viral heart disease. Enterovirus-induced myocarditis may be acute, subacute, or chronic. The patient may experience full recovery, a chronic progressive disease, or a fulminant disease resulting in early death.

The structure and molecular genetics of the enteroviruses are well understood, chiefly by analogy with the extensively studied poliovirus (for a review, see Wimmer et al. [1987]). The genetic material of enteroviruses is contained in a single-stranded RNA molecule of about 7,500 nucleotides of positive polarity, covalently linked at the 5' end to a small virus-encoded protein, VPg (3B; for a systematic nomenclature of picornavirus proteins, see Rueckert and Wimmer [1984]). The viral RNA, which is infectious, is polyadenylated at the 3' end and serves as messenger for the synthesis of virus-directed proteins and as template for replication. Four structural proteins, VP1 to VP4 (1A to 1D), are generated from the amino-terminal part of the precursor polyprotein, whereas the nonstructural proteins like the polymerase ($3D^{pol}$), VPg (3B), two proteinases ($2A^{pro}$ and $3C^{pro}$), and several polypeptides of as yet unknown function are released from the carboxy-terminal part.

Diagnosing viral heart disease by clinical features is an exercise of uncertain validity. In practice, the measurement of serum antibody titers in suspected enteroviral heart disease seldom proves useful diagnostically, except for cases that occur as part of an epidemic (Dec et al., 1985). In addition, the histologic features of biopsy specimens of patients with a clinical suspicion of viral heart disease are usually not virus specific and are indistinguishable from those seen in inflammatory heart disease of nonviral origin, e.g., bacterial infections or connective tissue diseases (Wynne and Braunwald, 1980). Thus, we lack information on how often viruses are actually the cause of acute dilated cardiomyopathy, what the prognosis is in such cases, and whether specific therapy can affect prognosis. The reason for this fundamental lack of knowledge is the difficulty of establishing an unequivocal diagnosis of viral heart disease. Confirmation of the clinical suspicion of viral heart disease requires the demonstration of replicating virus inside myocardial cells, which is exceedingly difficult by conventional methods, such as isolation of infectious virus or immunofluorescent staining of virus-specific antigens.

Molecular genetic techniques have now provided investigators with powerful tools to define the role of enteroviruses in the induction of the disease (Hyypiä et al., 1984; Tracy, 1984; Kandolf and Hofschneider, 1985; Rotbart et al., 1985; Bowles et al., 1986) as well as to study the molecular basis of pathogenicity. This review will focus on the development of an enterovirus group-specific in situ hybridization technique (Kandolf et al., 1987a; Kandolf, 1988; Kandolf and Hofschneider, 1989), which is the method of choice for diagnosing enteroviral heart disease by using endomyocardial biopsy samples. In addition, antisera raised against bacterially syn-

Reinhard Kandolf, Annie Canu, Karin Klingel, Philip Kirschner, Heike Schönke, Jürgen Mertsching, Roland Zell, and Peter Hans Hofschneider, Department of Virus Research, Max Planck Institute for Biochemistry, D-8033 Martinsried, Federal Republic of Germany.

thesized coxsackievirus B3 (CVB3) proteins are described, and they reveal a broad spectrum of cross-reactivity within the enteroviruses (Werner et al., 1988). Visualizing both nucleic acid and the expression of distinct viral proteins on the same section will be enormously useful in future studies of the molecular basis of persistent enterovirus infection. Finally, we review in vitro experiments carried out in the absence of effective antiviral therapy, demonstrating the high antiviral activity of human natural fibroblast interferon (beta interferon [IFN-β]) in cultured human heart cells.

CLONED CVB3 cDNA AS A DIAGNOSTIC REAGENT FOR THE DETECTION OF ENTEROVIRUSES

One major prerequisite for the introduction of in situ hybridization as a diagnostic tool in suspected enteroviral heart disease was the molecular cloning and characterization of the single-stranded genomic RNA of a cardiotropic CVB3 (Kandolf and Hofschneider, 1985) that had been propagated in cultured human heart cells (Kandolf et al., 1985). Full-length, reverse-transcribed CVB3 cDNA generated replication-competent CVB3 upon transfection of recombinant viral cDNA into mammalian cells, offering unique opportunities for the genetic analysis of this virus. As expected, in vitro-synthesized CVB3 RNA transcripts were also found to be infectious. From the diagnostic point of view, cloned CVB3 cDNA offers the unique possibility for a group-specific diagnosis of enterovirus infections. Because of the high degree of nucleic acid sequence identity shared among the numerous serotypes of the human enterovirus group, detection of the most commonly associated agents of human viral heart disease, including the group A and B coxsackieviruses and the echoviruses, is possible in a single hybridization assay. The broad spectrum of enteroviruses detected by cloned CVB3 cDNA greatly facilitates diagnosis of enteroviral heart disease, since from the clinical point of view, the antigenic typing of an etiologically implicated enterovirus serotype appears to be of secondary importance and can be carried out later—for example, by standard virological techniques or by hybridization with serotype-specific cDNA fragments available to date only for a limited number of enterovirus serotypes.

Highly specific hybridization conditions have been established for the detection of enteroviruses. When radioactively labeled cloned CVB3 cDNA corresponding to 95.4% of the viral genome (nucleotides 66 to 7128) was hybridized to electrophoretically resolved CVB3 RNA and to total RNA from cultured human heart cells, specific hybridization was found for the viral RNA and not for human RNA (Kandolf and Hofschneider, 1985). In addition, no hybridization was detected with restriction enzyme-digested human cellular DNA under conditions appropriate for the detection of single-copy genes.

IN SITU DETECTION OF ENTEROVIRUS RNA IN INFECTED CELLS BY NUCLEIC ACID HYBRIDIZATION

The feasibility of using the in situ hybridization technique (Wolf et al., 1973; Haase et al., 1984) to detect enterovirus RNA was first established in cell culture systems and then applied to myocardial tissue of athymic mice persistently infected with CVB3 (Kandolf et al., 1987a). Uninfected Vero cells exhibited essentially no grains when hybridized to the ^{3}H-labeled or ^{35}S-labeled CVB3 cDNA probe. By contrast, highly significant labeling was achieved in Vero cells infected with various enteroviruses, for example, CVB1, CVB3, CVB5, or echovirus 11.

Using myocardial tissue sections from CVB3-infected mice, in situ hybridization proved to be a powerful tool not only with respect to establishing an unequivocal diagnosis of myocardial infection but also with respect to understanding its pathogenesis (Fig. 1). The autoradiographic silver grains, which indicate hybridization between viral RNA and the radiolabeled CVB3 cDNA probe, are clearly localized to individual infected myocytes (Fig. 1A). These cells are easily identified by interference contrast microscopy in unstained sections because of their characteristic size and morphology. In this model system of enteroviral heart disease, myocardial infection was found to be multifocal and randomly distributed in the heart muscle (Fig. 1B). Myocardial cross sections revealed a transmural disseminated infection of the myocardium, as demonstrated in Fig. 1C for the left ventricle. Infected myocytes were often found in clusters within areas of severe myocardial lesions and fibrosis (Fig. 1D). Furthermore, progression of the infection could be observed from areas with myocardial fibrosis to as yet uninfected myocytes (Fig. 1E and F), demonstrating the possible cell-to-cell spread of the virus. Viral RNA, however, was also found in isolated myocytes in apparently normal myocardial tissue, which was primarily observed in the early stage of the infection (Fig. 1G). In addition, viral RNA

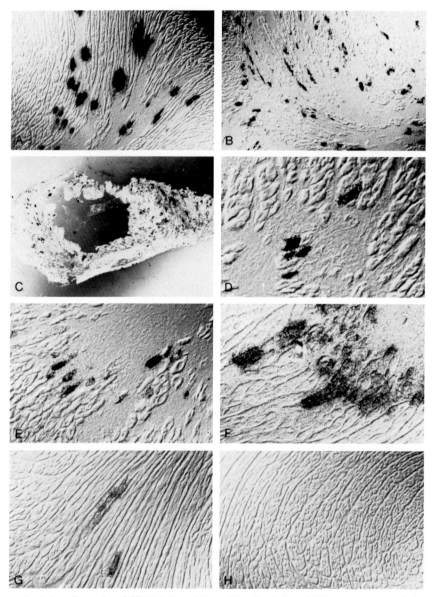

Figure 1. Autoradiographs of CVB3-infected (A to G) and uninfected (H) mouse myocardial tissue hybridized in situ with ^{3}H-labeled (B to E) or ^{35}S-labeled (A, F, G, and H) cloned CVB3 cDNA. Days after infection of athymic mice were 56 (A), 23 (B to E), 42 (F), and 4 (G). Exposure times were 9 days (A and H), 6 weeks (B to E), 4 days (F), and 2 days (G). Note that since silver grains are positioned at various levels within the photoemulsion, some grains are not observed and appear out of focus in photography. Magnifications for interference contrast microscopy of unstained sections: ×200 (A, G, and H), ×100 (B), ×25 (C), ×400 (D and E), and ×650 (F). (From Kandolf et al. [1987a].)

also appeared to be located within the small interstitial myocardial cells (Fig. 1B and E). Labeled myocardial cells were not found when myocardial tissue sections were probed with the radiolabeled plasmid p2732B control DNA

(Kandolf and Hofschneider, 1985) or with the cloned *Eco*RI J fragment of the genetically unrelated cytomegalovirus (Nelson et al., 1982). In addition, no labeled myocardial cells were found when myocardial tissue sections of uninfected mice were probed with the radiolabeled CVB3 cDNA probe (Fig. 1H).

A high sensitivity of detection was made possible by several improvements in methodology, including optimized hybridization conditions to prevent nonspecific binding and the use of radiolabeled cloned cDNA fragments about 100 nucleotides in length (Kandolf et al., 1987a). Quantification of the CVB3 copy number in infected Vero cells by RNA blot analysis and comparison with the results of in situ hybridization to cells from the same culture indicated that as few as 20 viral copies are easily detectable within 2 weeks of autoradiographic exposure. Clearly positive hybridization signals with infected mouse myocardial tissue were observed after only 2 days of exposure to the ^{35}S-labeled cDNA probe, indicating a high copy number of replicating viral genomes in myocardial cells (Fig. 1G). Furthermore, overexposed myocardial slide preparations showed extremely low background signals, which confirmed the high specificity of in situ hybridization for the detection of enterovirus RNA.

IN SITU DETECTION OF ENTEROVIRUS RNA IN THE MYOCARDIUM OF PATIENTS WITH MYOCARDITIS AND DILATED CARDIOMYOPATHY

In situ hybridization has already proved to be a valuable tool to assess the presence of enterovirus RNA in endomyocardial biopsy samples obtained from patients with suspected myocarditis or dilated cardiomyopathy (Kandolf, 1988; Kandolf and Hofschneider, 1989). Replicating enterovirus RNA was found to be present in 23 of 95 patients with a clinical suspicion of acute myocarditis, including 10 of 33 patients with dilated cardiomyopathy of recent onset. All 53 patients of a pathological control group with other specific heart muscle diseases not consistent with a primary viral etiology (e.g., ischemic, hypertrophic, or metabolic cardiomyopathies) were negative when myocardial tissue was examined by using in situ hybridization.

Moreover, replicating enterovirus RNA was detected not only at an early stage of clinically obvious myocarditis but also in chronic dilated cardiomyopathy, indicating persistence of the virus in the human heart. Of 48 patients with chronic dilated cardiomyopathy, 8 patients, including 4 of 19 who had undergone heart transplantation because of end-stage dilated cardiomyopathy, were found to have myocardial enterovirus infection. The concept of

possible enterovirus persistence in chronic dilated cardiomyopathy, evolving from acute and possibly subacute infections, is currently being substantiated by our finding of enterovirus persistence in subsequent biopsies obtained from patients with ongoing disease. The negative results obtained in the control group of patients with other specific heart muscle diseases clearly assign an etiologic role to the enteroviruses in patients with myocarditis and dilated cardiomyopathy and suggest that a preexisting cardiomyopathy is not a predisposing factor for myocardial infection.

The number of infected myocardial cells appears to be related to the severity of clinical symptoms. In patients with mild perimyocarditis or healing myocarditis, only a few myocardial cells were found to express enterovirus RNA, but numerous infected cells were found in patients with severe dilated-type cardiomyopathy. As expected, the most extensive patterns of myocardial infections were observed in fulminant cases of the disease resulting in early death. Figure 2 shows, as an example, the pattern of acute enterovirus infection in the autopsy heart of an infant who died of fulminant CVB4 myocarditis. The autoradiographic silver grains, which indicate hybridization between viral RNA and the radiolabeled cloned cDNA probe, are clearly localized to numerous infected myocardial cells, thereby providing an unequivocal diagnosis of myocardial infection (Fig. 2A). Infected myocytes are easily identified because of their characteristic size and morphology. Infected myocytes are also seen adjacent to inflammatory cells, which provide the basis for the histopathological diagnosis of myocarditis. No labeled cells were observed when myocardial tissue sections were probed with the radiolabeled plasmid vector control DNA probe (Fig. 2B).

Another important observation was that replicating enterovirus RNA was found to be present not only in myocytes but also in small interstitial cells, possibly myocardial fibroblasts, which agrees with the previous in vitro findings in cultured human heart cells and persistently infected human myocardial fibroblasts (Kandolf et al., 1985).

GENERATION OF ENTEROVIRUS GROUP-SPECIFIC ANTISERA

Although there are a number of interrelationships between different enteroviruses (Melnick et al., 1979; Yolken and Torsch, 1981), the applicability of group-specific antigen detection

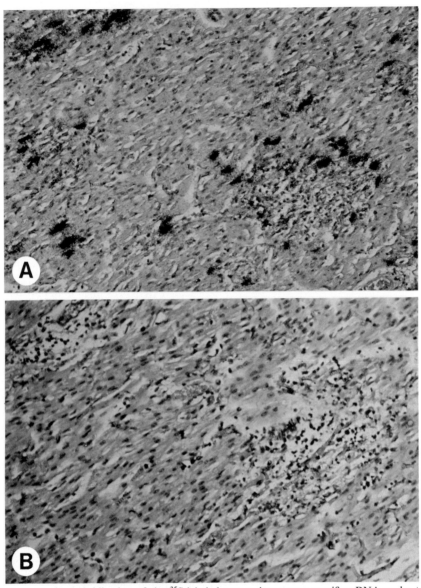

FIGURE 2. In situ hybridization of the ^{35}S-labeled enterovirus group-specific cDNA probe to the paraffin-embedded autopsy heart of an infant who died of acute CVB4 infection. Autoradiographic silver grains (A) can be clearly localized to distinct infected myocytes, thereby providing the possibility for an unequivocal diagnosis of the enterovirus infection. Note that infected myocytes are seen adjacent to areas of inflammation but also in areas without inflammation. Labeled myocardial cells were not observed when myocardial tissue sections were hybridized with the ^{35}S-labeled plasmid vector p2732B (B; control), demonstrating the specificity of in situ hybridization. Magnification for hematoxylin-eosin stains: ×75 (A) and ×85 (B).

has met only modest success in the identification of cardiotropic enteroviruses because of the antigenic heterogeneity among and within the serotypes. Compared with hyperimmune antisera obtained by immunization of rabbits with purified enteroviral virions (Mertens et al., 1983), antisera raised against bacterially synthesized CVB3 proteins revealed a much broader spectrum of cross-reactivity within the enteroviruses (Werner et al., 1988). Several subgenomic fragments from infectious recombinant CVB3 cDNA were inserted into the expression plasmid pPLc24 (Remaut et al., 1983) within the coding sequence of the replicase gene of bacteriophage MS2, thereby achieving a fusion of both reading frames. With this approach, plasmids were constructed that expressed either the structural proteins VP4, VP2, and VP3 (p1A to p1C) or VP1 (p1D) or the RNA-dependent RNA polymerase (p3D^{pol}) of CVB3. Polyclonal antisera raised in rabbits against the purified expression products of structural proteins offer the unique possibility for enterovirus-specific identification of various commonly implicated agents of viral heart disease, including group A and group B coxsackieviruses and echoviruses. This broad detection spectrum is expected to facilitate rapid identification of enterovirus infection by antigen detection. The use of these antibodies in combination with in situ hybridization is currently being pursued. With a double-labeling assay (Brahic et al., 1984), the simultaneous in situ detection of enterovirus RNA and the demonstration of distinct viral proteins will now become possible at the single-cell level.

ANTIVIRAL ACTIVITY OF HUMAN FIBROBLAST IFN IN CULTURED HUMAN HEART CELLS

Effective antiviral therapy has not yet been established in the treatment of viral heart disease. The optimal goal of antiviral treatment, to restore function to the infected cell, usually appears unattainable. A realistic goal would be to inhibit viral replication and thus prevent virus spread to as yet uninfected cells. In many natural infections, this is achieved in part by the endogenous IFN system (Joklik, 1986). With respect to the potential clinical application of exogenous IFN in viral heart disease, one major prerequisite is the demonstration that the virus to be treated is susceptible in vitro to the action of a given type of IFN in the specific host cell.

The protective role of natural human fibroblast interferon (IFN-β) in CVB3-infected, enriched human fetal myocytes has been estab-

lished (Kandolf et al., 1985). Myocytes protected by IFN-β continued to beat rhythmically as long as IFN-β was given with exchanges of medium every 2 days. By contrast, virus replication in unprotected cultures was accompanied by loss of spontaneous contractility within 9 h, followed by complete lysis of myocytes within 20 h of infection.

The discovery of persistently infected, CVB3 carrier cultures of human myocardial fibroblasts (Kandolf et al., 1985) provided another useful test system for studying the activity of antiviral agents. In this type of infection, only a small proportion of the cell population is productively infected. Virus titers of 10^7 PFU of infectious CVB3 per ml of culture medium are obtained. Treatment of these cultures with IFN-β at a dose of 300 IU/ml of culture medium every 24 h eliminated infectious progeny virus within 11 to 13 days. An important finding is the observation that this potent in vitro activity of IFN-β is completely blocked in the presence of 10 μM prednisolone (Kandolf et al., 1987b). By contrast, azathioprine as well as cyclosporine do not interfere with the antiviral activity of IFN-β.

By using in situ hybridization, the quantitative impact of therapeutic interventions on viral replication can now be measured at the nucleic acid level. Preliminary results indicate that after IFN-β treatment, small amounts of CVB3 RNA are still detectable within some myocardial fibroblasts although infectious virus is no longer detectable by biologic tests. The molecular basis of this type of enterovirus persistence with reduced or even undetectable infectious virus remains of focal interest.

COMMENT AND OUTLOOK

Recombinant DNA techniques not only have allowed an important breakthrough in the study of enterovirus genetics but have already contributed significantly to an improved diagnosis of enterovirus infections. To date, there is firm evidence from in situ nucleic acid hybridization that enterovirus infections of the human heart lead to a significant number of patients presenting with clinical signs and symptoms of myocarditis or dilated cardiomyopathy. In patients with acute onset of dilated cardiomyopathy, which is (apart from sudden death) the most dramatic manifestation of myocarditis, the incidence of myocardial enterovirus infections was found to be approximately 30%. Particularly intriguing is the concept of possible enterovirus persistence in chronic dilated cardiomyopathy, a condition which in its end stage can be treated only by heart transplantation.

Presumably, enterovirus persistence in myocardial cells is the result of continued synthesis of the viral precursor polyprotein and correct processing, at least of the virus-encoded polymerase. However, a lack of correct processing of other viral gene products (for example, coat proteins) could restrict replication. Restricted replication could explain the common failure to isolate infectious virus from patients after the acute phase of virus replication. In addition, reactivation of infectious virus replication could explain the clinical phenomenon of recurrence and progression to long-term cardiac disease.

To investigate further the molecular mechanisms of enterovirus persistence, the infectious CVB3 cDNA is used as a source for the generation of distinct viral proteins in different expression systems. Polyclonal antisera raised against bacterially synthesized CVB3 structural proteins offer the possibility of enterovirus group-specific identification of the various commonly implicated agents of viral heart disease by antigen detection. If cloned CVB3 cDNA is used in combination with enterovirus antisera, simultaneous in situ detection of viral RNA and proteins will provide a powerful means to address questions concerning the molecular basis of enterovirus persistence. The combination of in situ hybridization plus immunohistochemistry is also expected to provide a valuable tool in the study of different immunological concepts of virus-induced autoimmunity (Weinstein and Fenoglio, 1987). Moreover, new constructions involving eucaryotic expression vectors are being designed in order to study the interaction between distinct viral polypeptides and cellular mechanisms.

In addition to in situ hybridization, the recently developed polymerase chain reaction (Saiki et al., 1988) makes possible the in vitro and in situ amplification of distinct enteroviral gene regions after cDNA synthesis. So far, we have selected primers corresponding to highly conserved sequences of the 5' untranslated region of various enteroviruses. The enterovirus polymerase chain reaction was found to be sensitive, and specificity of amplified sequences was shown by hybridization to an internal, virus-specific oligonucleotide (unpublished results). Direct sequencing of enzymatically amplified gene regions (Innis et al., 1988) will allow typing of an etiologically implicated enterovirus strain and the identification of specific mutations that might be implicated in enterovirus persistence.

Besides providing an unequivocal diagnosis of myocardial infection, there are therapeutic implications from the demonstration of replicating enterovirus RNA in myocardial cells. Clearly, corticosteroids should be proscribed in the presence of myocardial infection, since increased viral replication (Kilbourne and Horsfall, 1951) and inhibition of the endogenous IFN system may follow their use (Kandolf et al., 1987b).

The most appealing treatment for enteroviral heart disease, which in some cases may prevent the development of chronic dilated cardiomyopathy, appears to be antiviral therapy. The potent antiviral activity of IFN-β in CVB3-infected cultured human heart cells has been demonstrated. Although the relationship between the in vitro effects of IFN-β and its potential in vivo activities remains to be determined, our in vitro observations are an important prerequisite for correct use of exogenous IFN-β in enterovirus infections. When applied at an early stage of the disease, IFN-β would probably reduce the spread of the virus in the myocardium by protecting uninfected cells and thereby limit the disease. However, care should be taken with patients with severely impaired hemodynamic functions, who might not tolerate the increased oxygen demand induced by fever and tachycardia, which are common side effects of IFN treatment. Our in vitro results with cultured human beating-heart cells show an IFN-β-mediated dose-dependent increase of the beat frequency at the high dose of 10^3 IU/ml of culture medium, which was not found at a dose of 300 IU/ml (Kandolf et al., 1987b). Nonetheless, this finding indicates the need for careful evaluation of potential in vivo activities of IFN-β in cardiac patients.

ACKNOWLEDGMENTS. We thank H. Riesemann and B. Schwaiger for excellent technical assistance. Endomyocardial biopsies were obtained from E. Erdmann and H. P. Schultheiss, Department of Internal Medicine I, University of Munich. Myocardial tissue from explanted hearts was obtained from B. Kemkes, Department of Cardiac Surgery, University of Munich. Myocardial tissue from autopsy hearts was kindly provided by A. Foulis, Department of Pathology, University of Glasgow.

This work was supported in part by grant Ka 593/2-2 from the Deutsche Forschungsgemeinschaft and by grant 321-7291-BCT-0370 "Grundlagen und Anwendungen der Gentechnologie" from the German Ministry for Research and Technology. R.K. is a Hermann and Lilly Schilling Professor of Medical Research.

REFERENCES

Abelmann, W. H. 1973. Viral myocarditis and its sequelae. *Annu. Rev. Med.* **24:**145–152.

Bowles, N. E., P. J. Richardson, E. G. J. Olsen, and L. C. Archard. 1986. Detection of coxsackie B virus-specific RNA sequences in myocardial biopsy samples from patients with myocarditis and dilated cardiomyopathy. *Lancet* **i:**1120–1123.

Brahic, M., A. T. Haase, and E. Cash. 1984. Simultaneous *in situ* detection of viral RNA and antigens. *Proc. Natl. Acad. Sci. USA* **81:**5445–5448.

Dec, G. W., I. F. Palacios, J. T. Fallon, H. T. Aretz, J. Mills, D. C.-S. Lee, and R. A. Johnson. 1985. Active myocarditis in the spectrum of acute dilated cardiomyopathy. *N. Engl. J. Med.* **312:**885–890.

Haase, A., M. Brahic, L. Stowring, and H. Blum. 1984. Detection of viral nucleic acids by *in situ* hybridization. *Methods Virol.* **7:**189–226.

Hyypiä, T., P. Stålhandske, R. Vainionpää, and U. Petterson. 1984. Detection of enteroviruses by spot hybridization. *J. Clin. Microbiol.* **19:**436–438.

Innis, M. A., K. B. Myambo, D. H. Gelfand, and M. A. D. Brow. 1988. DNA sequencing with *Thermus aquaticus* DNA polymerase and direct sequencing of polymerase chain reaction-amplified DNA. *Proc. Natl. Acad. Sci. USA* **85:**9436–9440.

Johnson, R. A., and I. Palacios. 1982. Dilated cardiomyopathies of the adult. *N. Engl. J. Med.* **307:**1119–1126.

Joklik, W. K. 1986. Interferons, p. 281–307. *In* B. N. Fields and D. M. Knipe (ed.), *Fundamental Virology.* Raven Press, New York.

Kandolf, R. 1988. The impact of recombinant DNA technology on the study of enterovirus heart disease, p. 293–318. *In* M. Bendinelli and H. Friedman (ed.), *Coxsackieviruses—A General Update.* Plenum Publishing Corp., New York.

Kandolf, R., D. Ameis, P. Kirschner, A. Canu, and P. H. Hofschneider. 1987a. *In situ* detection of enteroviral genomes in myocardial cells by nucleic acid hybridization: an approach to the diagnosis of viral heart disease. *Proc. Natl. Acad. Sci. USA* **84:**6272–6276.

Kandolf, R., A. Canu, and P. H. Hofschneider. 1985. Coxsackie B3 virus can replicate in cultured human foetal heart cells and is inhibited by interferon. *J. Mol. Cell. Cardiol.* **17:**167–181.

Kandolf, R., and P. H. Hofschneider. 1985. Molecular cloning of the genome of a cardiotropic coxsackie B3 virus: full-length reverse-transcribed recombinant cDNA generates infectious virus in mammalian cells. *Proc. Natl. Acad. Sci. USA* **82:**4818–4822.

Kandolf, R., and P. H. Hofschneider. 1989. Enteroviral heart disease. *Springer Semin. Immunopathol.* **11:**1–13.

Kandolf, R., P. Kirschner, D. Ameis, A. Canu, and P. H. Hofschneider. 1987b. Cultured human heart cells: a model system for the study of the antiviral activity of interferons. *Eur. Heart J.* **8**(Suppl. J)**:**453–456.

Kilbourne, E., and F. Horsfall. 1951. Lethal infections with coxsackievirus of adult mice given cortisone. *Proc. Soc. Exp. Biol. Med.* **77:**135–138.

Melnick, J. L., H. A. Wenner, and C. A. Phillips. 1979. Enteroviruses, p. 471–534. *In* E. H. Lennette and N. J. Schmidt (ed.), *Diagnostic Procedures for Viral, Rickettsial and Chlamydial Infections,* 5th ed. American Public Health Association, Washington, D.C.

Mertens, T., U. Pika, and H. J. Eggers. 1983. Cross antigenicity among enteroviruses as revealed by immunoblot technique. *Virology* **129:**431–442.

Nelson, J. A., B. Fleckenstein, D. A. Galloway, and J. K. McDougall. 1982. Transformation of NIH 3T3 cells with cloned fragments of human cytomegalovirus strain AD169. *J. Virol.* **43:**83–91.

Remaut, E., H. Tsao, and W. Fiers. 1983. Improved plasmid vectors with a thermoinducible expression and temperature-regulated runaway replication. *Gene* **22:**103–133.

Reyes, M. P., and A. M. Lerner. 1985. Coxsackievirus myocarditis—with special reference to acute and chronic effects. *Prog. Cardiovasc. Dis.* **27:**373–394.

Rotbart, H. A., M. J. Levin, L. P. Villarreal, S. M. Tracy, B. L. Semler, and E. Wimmer. 1985. Factors affecting the detection of enteroviruses in cerebrospinal fluid with coxsackievirus B3 and poliovirus 1 cDNA probes. *J. Clin. Microbiol.* **22:**220–224.

Rueckert, R. R., and E. Wimmer. 1984. Systematic nomenclature of picornavirus proteins. *J. Virol.* **50:**957–959.

Saiki, R. K., D. H. Gelfand, S. Stoffel, S. J. Scharf, R. Higuchi, G. T. Horn, K. B. Mullis, and H. A. Erlich. 1988. Primer-directed enzymatic amplification of DNA with a thermostable DNA polymerase. *Science* **239:**487–491.

Tracy, S. 1984. A comparison of genomic homologies among the coxsackievirus B group: use of fragments of the cloned coxsackievirus B3 genome as probes. *J. Gen. Virol.* **65:**2167–2172.

Weinstein, C., and J. J. Fenoglio. 1987. Myocarditis. *Hum. Pathol.* **18:**613–618.

Werner, S., W. M. Klump, H. Schönke, P. H. Hofschneider, and R. Kandolf. 1988. Expression of coxsackievirus B3 proteins in *Escherichia coli* and generation of virus-specific antisera. *DNA* **7:**307–316.

Wimmer, E., R. J. Kuhn, S. Pincus, C. F. Yang, H. Toyoda, M. J. H. Nicklin, and N. Takeda. 1987. Molecular events leading to picornavirus genome replication. *J. Cell Sci.* Suppl. **7:**251–276.

Wolf, H., H. zur Hausen, and V. Becker. 1973. EB viral genomes in epithelial nasopharyngeal carcinoma cells. *Nature* (London) *New Biol.* **244:**245–247.

Woodruff, J. F. 1980. Viral myocarditis—a review. *Am. J. Pathol.* **101:**427–479.

Wynne, J., and E. Braunwald. 1988. The cardiomyopathies and myocarditides, p. 1410–1469. *In* E. Braunwald (ed.), *Heart Disease. A Textbook of Cardiovascular Medicine*, 3rd ed. The W. B. Saunders Co., Philadelphia.

Yolken, R. H., and V. M. Torsch. 1981. Enzyme-linked immunosorbent assay for detection and identification of coxsackieviruses A. *Infect. Immun.* **31:**742–750.

IX. STRATEGIES FOR CONTROL OF VIRUS DISEASE

Chapter 51

Satellite RNA of Cucumber Mosaic Virus: a Pathogenic RNA

David Baulcombe, Martine Devic, and Martine Jaegle

SATELLITE RNA AS A MOLECULAR PARASITE

Many viruses are associated with small, extragenomic RNA molecules that may be considered as molecular parasites. These RNA species, known as satellite RNA (satRNA), are parasites in the sense that they are not required by the virus but are themselves dependent on the virus for all functions necessary for propagation through the infected plant and transmission from plant to plant (Francki, 1985). A previous review suggested that an important feature of satRNA is a lack of extensive homology with the helper virus (Murant and Mayo, 1982). However, since satRNAs are recognized by replicase and other functions of the helper virus, it is likely, even if there is no homology in the primary structure, that features of the secondary or tertiary structure are shared with the helper virus.

Like other parasites, satRNA may modify its host. In the instance of satRNA associated with cucumber mosaic virus (CMV), this modification is observed as an effect on the symptoms produced by the infected plant. Depending on the strain of satRNA, the genotype of the host plant and the strain of helper virus, the modification is either an amelioration of the viral symptoms or the induction of severe symptoms. The severe symptoms have sometimes been described as exacerbated viral symptoms. However, as they are quite distinct in nature from the viral symptoms, it is more useful to consider the satRNA-induced symptoms as quite different from the symptoms produced in a plant that is infected with virus alone. This point is illus-trated by the effects of CMV infection on tobacco or tomato when the satRNA of CMV strain Y (Takanami, 1981) is added to the inoculum. In the absence of satRNA, the virus produces mild mosaic symptoms on both plants and a fern leaf effect on tomato. The addition of satRNA Y to the inoculum results in a bright yellow mosaic symptom on tobacco and a systemic necrosis on tomato. SatRNA that induces severe symptoms is referred to here as virulent; satRNA that ameliorates symptoms is referred to as benign.

It was demonstrated previously in transgenic plants transformed with a DNA copy of satRNA from CMV, coupled to a suitable promoter, that a polyadenylated form of the satRNA was produced (Baulcombe et al., 1986). This transcript had very little biological activity in its transcribed form, since the satRNA sequence was surrounded by nonviral sequence at the 5' and 3' ends (Fig. 1) (unpublished data). However, when the transgenic plants were infected with CMV, the transcribed satRNA was replicated and accumulated to a high levels. This replicated satRNA had the full biological activity of the natural satRNA, including the ability to attenuate symptoms of CMV and also a related virus, tomato aspermy virus (Harrison et al., 1987). Gerlach et al. (1987) have shown similar resistance against tobacco ringspot nepovirus resulting from expression of its satRNA in transgenic plants. Therefore, this use of satRNA is a second example, together with the use of viral coat protein genes (Powell Abel et al., 1986), in which expression of viral RNA in transgenic plants produces resistance to the effects of viral infection. The properties of the

David Baulcombe and Martine Devic, The Sainsbury Laboratory, Colney Lane, Norwich NR4 7UH, United Kingdom. *Martine Jaegle,* Department of Molecular Biology, Institute for Plant Science Research, Maris Lane, Trumpington, Cambridge CB2 2LQ, United Kingdom.

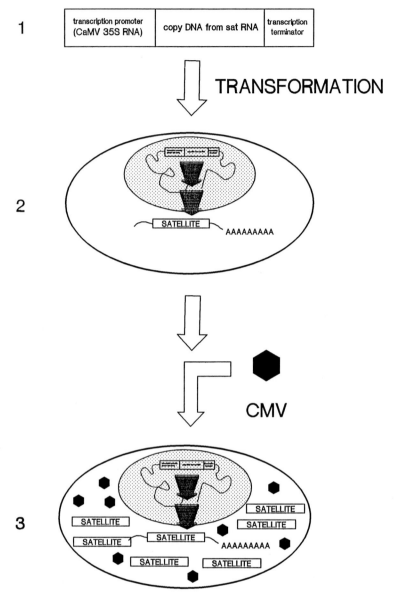

FIGURE 1. Plant transformation to express biologically active satellite RNA. The figure illustrates steps described in detail by Baulcombe et al. (1986). In step 1, a DNA copy of CMV satRNA was coupled to a cauliflower mosaic virus (CaMV) promoter and a transcription terminator. Step 2 was the transfer, by *Agrobacterium*-mediated transformation, of the DNA into the nuclear genome of tobacco plants. The biological activity of the transcripts produced from the integrated DNA was demonstrated when the transformed plants were infected with CMV (step 3). The satRNA replicated to a high level and was also encapsidated in virus particles.

satRNA effect are complementary to those of coat protein-mediated resistance, so that these two types of viral sequence may be a particularly effective combination when expressed in crop plants. For example, the satRNA effect differs from coat protein-mediated resistance in that no new protein is produced and in its lack of sensitivity to both the level of satRNA expression in the plant and the concentration of inoculum (Harrison et al., 1987). However, since the

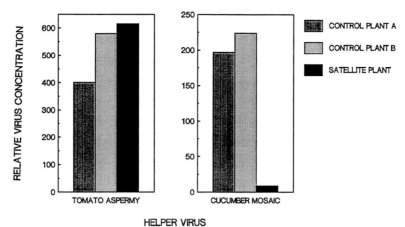

FIGURE 2. Effect of satRNA in transgenic plants on virus accumulation. Transgenic plants either expressing (satellite plant) or not expressing (control plants A and B) satRNA were inoculated with satellite-free helper virus. Virus concentration in systemically infected leaves was measured by inoculation of extracts onto a local lesion host (*Chenopodium quinoa*).

virulent species of satRNA have a nucleotide sequence very similar to that of the benign satRNAs (Palukaitis, 1988), it is not yet possible to use this genetically engineered resistance in the field. It is necessary to identify functional domains within the satRNA so that a modified form can be expressed in plants without the risk of mutation or transmission from the protected plants to nearby nontransformed plants.

A modified satRNA will be useful also as a component of a viral "vaccine" that can be applied to nontransformed plants. Tien et al. (1987) have shown already that a natural isolate of benign satRNA of CMV is effective as the active component of a vaccine used to protect field-grown pepper plants against the effects of CMV.

BENIGN satRNA

A model for the mechanism of satellite-mediated suppression of viral symptom production proposes a competition between the satRNA and the helper virus for limiting amounts of replicase enzyme. Evidence for this model is based on analysis of the kinetics of helper virus RNA synthesis in cultured cells inoculated in the presence or the absence of satRNA (Piazzolla et al., 1982). It was apparent that viral RNA synthesis was inhibited in the presence of satRNA. There is also evidence from the analysis of steady-state levels of both virus and viral RNA showing that satRNA inhibits viral accumulation (Harrison et al., 1987). However, it is unlikely

that this is the simple explanation of the amelioration effect. There are several examples, including tomato aspermy virus, which can be attenuated by satRNA of CMV, in which symptom severity is not directly related to the amount of virus. Transgenic plants expressing satRNA were capable of effective inhibition of tomato aspermy virus symptoms despite there being no reduction of the amount of viral RNA compared with control plants that were not producing satRNA (Harrison et al., 1987). Similarly, there was no effect on production of capsid protein (Harrison et al., 1987) (Fig. 2). It is likely, therefore, that the satRNA can interfere with the process of symptom induction either by the virus or by the plant.

The satRNA of tobacco ringspot virus also has the ability to attenuate the effects of viral infection without necessarily inhibiting the helper virus replication (Ponz et al., 1987). In this instance, the effect is even more striking than with the satRNAs of CMV and tomato aspermy virus, as the attenuated virus (cherry leafroll nepovirus) cannot replicate the satRNA of tobacco ringspot virus.

Benign satRNA of CMV can also protect against the effects of a virulent satRNA under certain conditions. This was demonstrated by inoculating plants first with a viral inoculum containing a benign satRNA and, after this infection became established, with a viral inoculum supplemented with a virulent satRNA. An experiment designed to test whether the transformed plants expressing satRNA could resist

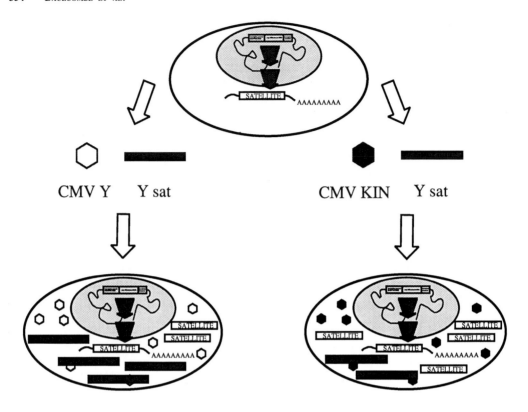

YELLOW MOSAIC SYMPTOMS   ATTENUATED SYMPTOMS

FIGURE 3. Effect of satRNA in transgenic plants on infections with a virulent satRNA. The diagram illustrates schematically the accumulation of satRNA and symptom production when the virulent Y satRNA was introduced with different helper virus strains.

the virulent satellite in the same way is illustrated in Fig. 3. The virulent satRNA was from the Y strain of CMV (Takanami, 1981), and its effects could be diagnosed by the production of a yellow mosaic symptom on the systemically infected leaves. The strain of helper virus used in the experiment was either the Y strain of CMV, which is the natural helper of the Y satRNA, or CMV KIN, which was not associated with a satRNA when it was isolated (Harrison et al., 1987). The results summarized in Fig. 3 showed different results for the two types of helper strain. When CMV KIN was used as helper, there was inhibition of Y satRNA at the levels of both symptoms and satRNA accumulation. This result suggests that the transcribed satRNA is at an advantage over the satRNA in the challenge inoculum, possibly because it is present in cells before the arrival of the virus. However, this advantage was overridden when the helper virus was the Y strain of CMV (Fig. 3). A likely explanation for this second result

proposes that an affinity between the CMV Y strain and the Y satRNA gives the Y satRNA an advantage in the competition with the satRNA produced in transgenic plants. At the moment, there is no obvious suggestion for the nature of this affinity. Perhaps an antisense interaction, which has been described Rezaian and Symons (1986) (Fig. 4), is stronger in the homologous combination of helper and satRNA so that the Y satRNA is encapsidated and transported from cell to cell more efficiently than with CMV KIN. Alternatively, the viral replicase may operate more efficiently in the homologous combination.

VIRULENT satRNA

It is clear that the formation of symptoms in response to the presence of satRNA in an inoculum of CMV involves an interaction, either direct or indirect, between the satRNA and components of the host plant. As different spe-

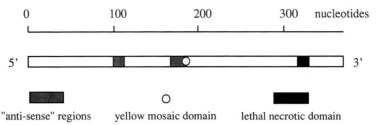

FIGURE 4. Domain structure of CMV Y satRNA. Shown are the locations of antisense, yellow mosaic-inducing, and lethal necrotic domains in the nucleotide sequence of CMV Y satRNA. The antisense domain was defined by Rezaian and Symons (1986); the symptom domains were defined by Devic et al. (in press) and M. Devic, M. Jaegle, and D. Baulcombe (unpublished data).

cies of satRNA induce different responses in the host plant, it is likely that this interaction involves variable parts of the satRNA molecule. These domains have been located in the satRNA by creating hybrid molecules and correlating the phenotype with the presence of domains from a virulent satRNA. The virulent satRNA used in this study was the Y satRNA described by Takanami (1981) as producing a lethal necrotic disease on tomato and a bright yellow mosaic on tobacco.

A series of hybrid molecules was produced, and from their phenotype it was concluded that the domain responsible for induction of the yellow mosaic disease is located centrally in the molecule. The lethal necrotic domain was separate and toward the 3' end (Devic et al., in press).

It was not too surprising that the yellow mosaic domain was located within the central part of the molecule. The Y satRNA has the unique property among known satRNA species of including the yellow mosaic disease and has a similarly unique nucleotide sequence in the central region of the molecule. A better resolution of the yellow mosaic domain was obtained by a mutational analysis that identified a site where changing two nucleotides destroyed the yellow mosaic-inducing property of the molecule.

The localization of the domain responsible for the induction of the lethal necrotic disease was less obvious, as there was no particular area in which a sequence motif was unique to the virulent form of the satRNA. However, there was a region in which all of the necrosis-inducing forms of satRNA were identical and where there was sequence variation in the benign forms. However, there was not a simple correlation between the presence of this sequence and the symptom-inducing phenotype, as each benign form varied from the virulent form at different nucleotide positions. The complexity of the situation was confirmed when

various defined mutations were created in that region of the molecule. The simplest interpretation of the data is that the necrosis domain extends over several nucleotides and variation at several positions in that region has the potential to destroy the necrosis-inducing capability of the satRNA.

FUTURE DEVELOPMENTS

The domain structure of satRNA is summarized in Fig. 4. This information will be used in the development of satRNA for use in crop protection and for more fundamental investigation of the mechanism of symptom induction by satRNA.

In the applied context, the new information suggests regions of the molecule that could be modified to produce a satRNA that is not capable of inducing symptoms. Ideally, the modified satRNA would differ from the virulent forms at several positions, so that back-mutation would not occur readily. In the course of this work, several mutant forms of the satellite have been created, illustrating that modified forms of the satRNA are replicable, can move systemically in a plant, and can be transmitted by aphids from plant to plant. One mutation (M. Jaegle, unpublished data) involved a large deletion within the central part of the molecule. Surprisingly, a smaller deletion from within the region of the larger deletion did not accumulate when inoculated onto plants together with helper virus. It appears, therefore, that any modifications must leave features of secondary or higher-order structure of the satRNA molecule intact. Unfortunately, understanding of higher-order structures involving the satRNA molecule is not very sophisticated. Current models of satRNA structure do not explain why the smaller deletion described above does not accumulate on plants, and the possibility remains that interactions of satRNA with other viral or host plant factors determine the structure of satRNA in vivo.

In principle, the satRNA could induce symptoms via the action of small satRNA-encoded peptides or directly as an RNA molecule. Currently, it appears that the latter is the more likely alternative, as various mutations have been created which interrupt the small open reading frames and which have no effect on the ability of the satRNA to induce symptoms. How RNA molecules might induce symptoms is not immediately obvious. A feasible model proposes that satRNA interferes in RNA-mediated processes of the healthy cell. It is now clear that there are several potential targets in this way, since RNA is known to be a component in RNases and protein transport complexes as well as in the protein synthetic apparatus.

In future work, we shall investigate these RNA-mediated processes and look for proteins and other components that bind to satRNA of CMV. The outcome will be a more profound understanding of the molecular biology of satRNA so that it can be used safely in crop protection. It is also anticipated that detailed knowledge of the pathogenic process will provide information about the healthy cell and suggest new methods of engineering disease resistance in plants.

REFERENCES

Baulcombe, D. C., G. R. Saunders, M. W. Bevan, M. A. Mayo, and B. D. Harrison. 1986. Expression of biologically active viral satellite RNA from the nuclear genome of transformed plants. *Nature* (London) 321:446–449.

Devic, M., M. Jaegle, and D. C. Baulcombe. Symptom production on tobacco and tomato is determined by two distinct domains of the satellite RNA of cucumber mosaic virus (strain Y). *J. Gen. Virol.*, in press.

Francki, R. I. B. 1985. Plant virus satellites. *Annu. Rev. Microbiol.* 39:151–174.

Gerlach, W. L., D. Llewellyn, and J. Haseloff. 1987. Construction of a plant disease resistance gene from the satellite RNA of tobacco ringspot virus. *Nature* (London) 328:802–805.

Harrison, B. D., M. A. Mayo, and D. C. Baulcombe. 1987. Virus resistance in transgenic plants that express cucumber mosaic virus satellite RNA. *Nature* (London) 328:799–802.

Murant, A. F., and M. A. Mayo. 1982. Satellites of plant viruses. *Annu. Rev. Phytopathol.* 20:49–70.

Palukaitis, P. 1988. Pathogenicity regulation by satellite RNAs of cucumber mosaic virus: minor nucleotide sequence changes alter host responses. *Mol. Plant-Microbe Interact.* 1:175–181.

Piazzolla, P., M. E. Tousignant, and J. M. Kaper. 1982. Cucumber mosaic virus-associated RNA 5—the overtaking of viral RNA synthesis by CARNA 5 and dsCARNA 5 in tobacco. *Virology* 122:147–157.

Ponz, F., A. Rowhani, S. M. Mircetich, and G. Bruening. 1987. Cherry leafroll virus infections are affected by a satellite RNA that the virus does not support. *Virology* 160:183–190.

Powell Abel, P., R. S. Nelson, N. De, N. Hoffmann, S. G. Rogers, R. T. Fraley, and R. N. Beachy. 1986. Delay of disease development in transgenic plants that express the tobacco mosaic virus coat protein gene. *Science* 232:738–743.

Rezaian, M. A., and R. H. Symons. 1986. Anti-sense regions in satellite RNA of cucumber mosaic virus form stable complexes with the viral coat protein gene. *Nucleic Acids Res.* 14:3229–3239.

Takanami, Y. 1981. A striking change in symptoms on cucumber mosaic virus-infected tobacco plants induced by a satellite RNA. *Virology* 109:120–126.

Tien, P., X. Zhang, B. Qiu, B. Qin, and G. Wu. 1987. Satellite RNA for the control of plant diseases caused by cucumber mosaic virus. *Ann. Appl. Biol.* 111:143–152.

Chapter 52

Genotypic Relationships among Wild Polioviruses from Different Regions of the World

Olen M. Kew, Mark A. Pallansch, Baldev K. Nottay, Rebeca Rico-Hesse, Lina De,
and Chen-Fu Yang

For a generation of virologists, polioviruses (and other picornaviruses) have been best known as prototype positive-stranded RNA viruses. Many of the concepts and methodologies central to modern virology were first developed in this system. Indeed, the remarkable homologies in virion structure, genomic organization, and macromolecular processing that reflect common evolutionary roots of positive-stranded RNA viruses infecting animals and plants have been the dominant theme of these symposia (Rueckert and Brinton, 1986; this volume).

Polioviruses were first known as the agents of a severe acute paralytic disease (Paul, 1971). Polioviruses were discovered 80 years ago in Vienna by Landsteiner and Popper (1909), at a time when poliomyelitis was endemic throughout the world. Widespread immunization over the last 3 decades has virtually eliminated poliomyelitis from developed countries (Assaad and Ljungars-Esteves, 1984). Nearly all of the very few remaining cases in these countries are associated with reversion of the oral poliovaccines to virulence (Nkowane et al., 1987), a problem currently under intensive molecular investigation (Kawamura et al., 1989; Pilipenko et al., 1989; Skinner et al., 1989). The present situation, more than 30 years after development of effective poliovaccines (Salk and Salk, 1977; Sabin, 1985), is much less favorable in most developing countries, where nearly 300,000 cases, primarily in children under 2 years of age, occurred in 1988 (World Health Organization, 1989). The incidence of poliomyelitis has declined sharply in those regions (particularly in the Americas, China, and the eastern Mediterranean) where comprehensive national immunization programs have been implemented (Henderson et al., 1988). Even so, pockets of wild-virus transmission persist in five continents.

In response to global initiatives to control, and possibly eradicate, poliomyelitis (Pan American Health Organization, 1985; World Health Assembly, 1988), we have applied several molecular methods to follow the patterns of wild-poliovirus transmission in nature (Rico-Hesse et al., 1987; Kew et al., in press). The main objectives of our studies are to determine the epidemiologic relationships among poliomyelitis cases and outbreaks occurring in different regions of the world and to identify the principal reservoirs sustaining wild-poliovirus endemicity. It is hoped that findings from these studies will help promote formulation of more effective strategies for control of wild-poliovirus infections.

Precise monitoring of the transmission of wild polioviruses is possible because the virus genome evolves rapidly during replication in humans (Nottay et al., 1981; Minor et al., 1986b). Mutations, primarily (>97%) generating synonymous codons (Rico-Hesse et al., 1987), accumulate at rates of up to one to two nucleotide substitutions per week (Nottay et al., 1981). Since the potential combinations of synonymous codons within even small genomic intervals is very large, it is possible to obtain a reasonable survey of the natural distribution of poliovirus genotypes from quite limited sequence data

Olen M. Kew, Mark A. Pallansch, Baldev K. Nottay, Lina De, and Chen-Fu Yang, Division of Viral Diseases, Center for Infectious Diseases, Centers for Disease Control, Atlanta, Georgia 30333.   *Rebeca Rico-Hesse,* Department of Epidemiology and Public Health, Yale University School of Medicine, P.O. Box 3333, New Haven, Connecticut 06510.

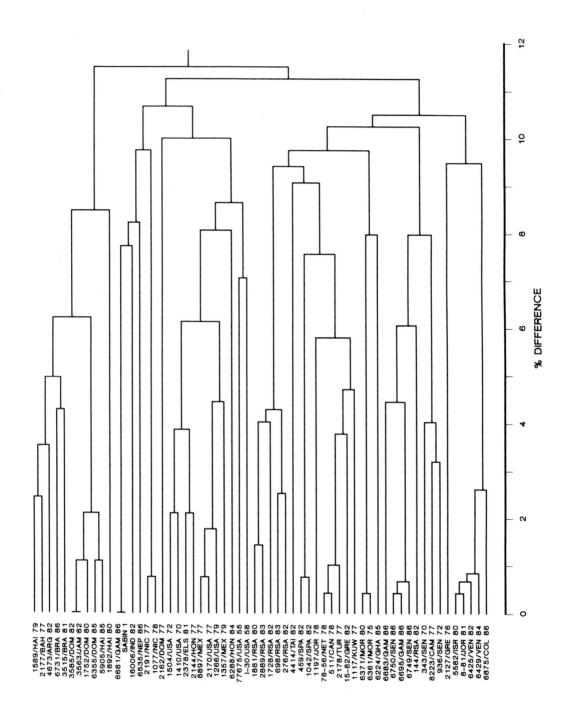

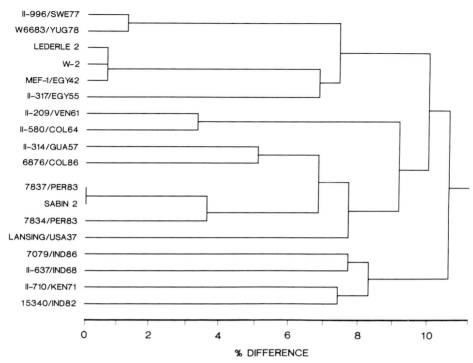

FIGURE 2. Sequence relationships (VP1-2A region; nucleotides 3295 to 3444; numbered by alignment with the reference Sabin 2 sequence of Toyoda et al. [1984]) among 18 type 2 polioviruses. Isolate PV2/7837/PER84 is vaccine derived.

(e.g., 2% of the total genome). For our initial comparisons, we selected an interval at the junction of the capsid and noncapsid domains (90 nucleotides encoding VP1 and 60 encoding the protease 2A). Since few residues encoded within this interval reside on the virion surface (Hogle et al., 1985; Minor et al., 1986a; Page et al., 1988), we reasoned that fixation of the observed mutations probably occurred primarily by genetic drift rather than by immune selection.

Sequence comparisons (VP1-2A region) among many independent isolates are summarized in dendrograms (Fitch and Margoliash, 1967; Fig. 1 to 3). The limit divergence observed within each serotype (type 1, 23%; type 2, 23%; type 3, 26%) approached the 29% divergence across the same interval for the three serotypes of Sabin poliovaccine strains (Toyoda et al., 1984). The genomes of all wild isolates clustered

by serotype in dendrograms (data not shown). Except for sequences encoding virion surface amino acids, the extent of nucleotide divergence within the VP1-2A interval is nearly equivalent within and across serotypes.

We have previously defined a genotype as a group of polioviruses having no more than 15% genomic divergence within the 150-nucleotide VP1-2A interval (Rico-Hesse et al., 1987). This demarcation, although somewhat subjective, is consistent with associations confirmable by independent methods and includes most links expected on epidemiologic grounds while excluding nearly all spurious associations. Confidence in the relationships given near the right sides of the dendrograms diminishes as sequence divergence exceeds 15%. Inclusion of additional sequence information per genome generally results in improving dendrogram resolution among

FIGURE 1. Dendrogram summarizing genomic sequence relatedness among 60 type 1 polioviruses across the interval 3296 to 3445 (VP1-2A region; nucleotide positions are numbered according to Nomoto et al. [1982]). Isolate PV1/6681/GAM86 is vaccine derived; all other isolates are wild. The percent nucleotide sequence divergence between any pair of isolates is twice the distance along the abscissa to the connecting node.

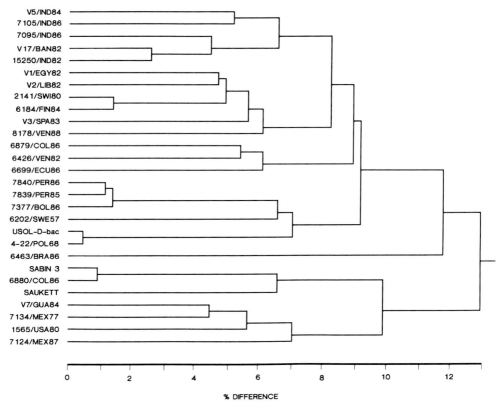

FIGURE 3. Sequence relationships among 28 type 3 polioviruses across the interval of nucleotides 3287 to 3436 (VP1-2A region; nucleotides are numbered according to Stanway et al. [1984]). Isolate PV3/6880/COL86 is vaccine derived.

closely related viruses but has little effect on the overall structure of the main branches (the exceptions are recombinant genomes; see below). The evolutionary separation of wild genotypes occurs so rapidly that they all appear to be nearly equally divergent, and no distinct "ancestral" poliovirus genomes can be discerned. Analyses more refined than the methods used here (such as assigning specific statistical weights to the less frequent transversion mutations) will be required if broader evolutionary questions are to be addressed.

GENOTYPIC RELATIONSHIPS AMONG CONTEMPORARY WILD POLIOVIRUSES

Genetically related polioviruses cluster geographically. This is easily visualized by assigning patterns to each genotype and mapping their distribution. We have sequenced many isolates obtained since 1980 from the American region,

all of which distributed into a small number (three type 1, two type 2, four type 3) of endemic genotypes (Fig. 4). Intensive investigation of recent cases in the Americas has failed to reveal the existence of any additional indigenous genotypes. Although information for other regions is still fragmentary, it is clear that many separate genotypes currently coexist worldwide (Fig. 1 to 3).

The existence of multiple genotypes throughout the world has important implications for poliomyelitis control. The geographic clustering of genotypes clearly suggests that local conditions (such as immunization coverages and seroconversion rates), rather than the particular properties of each virus (such as antigenicity), largely determine the extent of endemicity. The global picture of wild-poliovirus epidemiology is thus quite different from the well-documented example of influenza A viruses, in which selection of antigenic variants in one region may be associated with their rapid spread in global pan-

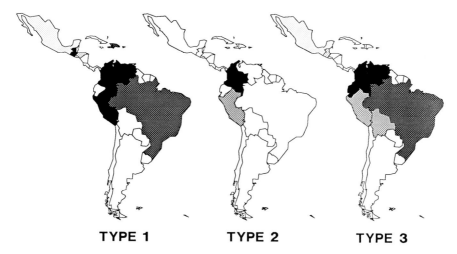

TYPE 1   TYPE 2   TYPE 3

FIGURE 4. Geographic distribution of wild-poliovirus genotypes obtained as clinical isolates in the American region since 1980. Each genotype (a group of polioviruses sharing >85% nucleotide homologies within the VP1-2A interval) is assigned a pattern according to the most populous country to which it was endemic during this period. Most assignments are based on analyses of several independent case isolates from each country. Large declines in poliomyelitis incidence have occurred in the Americas since 1980. In 1989, cases associated with wild viruses were largely restricted to the west coast of Mexico (type 3), some coastal communities in western South America (types 1 and 3), and northeastern Brazil (types 1 and 3). No wild type 2 polioviruses have been detected in the Americas since 1986. Since 1980, cases and outbreaks associated with imported genotypes occurred in the United States (1980; type 3, Mexican genotype), Argentina (1982; type 1), Jamaica (1982; type 1, Dominican Republic genotype), Guatemala (1986; type 1, Andean genotype), and Canada (1988; type 1, south Indian genotype).

demics (Buonagurio et al., 1986). Intratypic cross-neutralization among polioviruses is sufficient to permit the poliovaccines to confer protective immunity to all known genotypes (World Health Organization, 1989). Therefore, complete control of wild-poliovirus transmission appears attainable for every region of the world.

SOME MOLECULAR EPIDEMIOLOGIC VIGNETTES

Considerable detailed epidemiologic information may be inferred from the structures of the dendrograms. Links between cases can be independently and unequivocally established, and possible alternative transmission pathways can be distinguished. A few examples are presented to illustrate how sequence information has provided new insights into poliovirus epidemiology.

Distinguishing Epidemicity from Endemicity

Local poliomyelitis epidemics typically involve the clonal spread over short time periods of a specific genotypic variant. Isolates obtained during epidemics generally show very limited sequence heterogeneity. Viruses isolated during interepidemic periods are typically more divergent, reflecting the existence of independent pockets of wild-virus circulation. Countries with low immunization coverages generally have multiple reservoirs of endemic transmission, such that cases in one part of the country may have no direct relationship to cases elsewhere in the country. For example, type 1 viruses obtained over a 3-year period in South Africa were very heterogeneous in sequence and belonged to two genotypic groups (Fig. 1). Similarly, the genomes of type 3 polioviruses from India and Bangladesh isolated in 1982 to 1986 diverged (in the VP1-2A interval) by up to 14% (Fig. 3). Before the recent nationwide immunization campaigns in Mexico, endemicity of type 1 polioviruses was maintained for at least 8 years within several independent reservoirs (unpublished results).

Identification of the Sources of Imported Polioviruses

Gaps in population immunity may permit occurrence of cases associated with imported genotypes in countries where wild polioviruses are no longer endemic. Before the availability of molecular methods, the endemic origins of the imported viruses often could not be reliably located. Since the degree of nucleotide sequence divergence between two polioviruses is an approximate measure of the extent of their epidemiologic separation, otherwise unsuspected links are evident from the dendrograms (Rico-Hesse et al., 1987). For example, the most common sources of wild polioviruses imported into the United States have been Mexico and Central America (Fig. 1 and 3). Comparisons of sequences of type 1 isolates from sporadic cases in the United States with those of wild polioviruses from several regions of Mexico suggested the independent importation of virus from different Mexican states (unpublished results).

Cases and outbreaks in Europe (Schaap et al., 1984; Poyry et al., 1988) are usually associated with viruses related to those endemic to the eastern Mediterranean countries (Fig. 1 and 3). An exception was the 1967-to-1968 epidemic in Poland, which immediately followed the field testing of an experimental type 3 oral poliovaccine strain, USOL-D-bac (Kostrzewski et al., 1970). Case isolates, among which 4-22/POL68 is representative, were close derivatives of USOL-D-bac (Fig. 3). Interestingly, the only documented example of importation of an eastern Mediterranean genotype into Canada and the United States in 1978 to 1979 occurred by way of an earlier epidemic in the Netherlands (Nottay et al., 1981; Fig. 1).

Displacement and Elimination of Genotypes

Currently, type 1 polioviruses are the most frequent cause of poliomyelitis, and type 2 strains are the least frequent. In the prevaccine era, all three serotypes had worldwide endemicity. The indigenous type 2 genotypes appear to have been eliminated from Europe, Japan, North America, and possibly South America. Type 2 is the serotype most effectively controlled by immunization, and the presence of wild type 2 polioviruses in a community is an indication of possibly serious deficiencies in vaccine coverages. Experience in the Americas

suggests that wild type 3 polioviruses are usually the last to be eliminated. The apparent order of elimination of the three serotypes (2, 1, 3) may relate to the relative efficiencies by which the poliovaccines induce neutralizing antibodies to each serotype (Patriarca et al., 1988).

Endemicity can be reestablished by imported genotypes if effective vaccine coverages are not maintained. For example, a type 1 genotype thought to have originated in the northern Andean countries is currently endemic to both South America and the eastern Mediterranean. The close sequence correspondence among isolates from Venezuela (1980; not shown), Israel (1980), Jordan (1981), and Venezuela (1982) suggests that introduction occurred about 1980 (Fig. 1). Later isolates from both regions form two geographically defined subgenotypes (unpublished results). Endemicity has been restored to other parts of the Middle East by viruses imported from the Indian subcontinent.

Identification of Recombinant Genomes

The genomes of some case isolates cluster anomalously in dendrograms comparing related viruses from the same locales. Several of these viruses appear not to represent imported genotypes but instead have nucleotide sequences expected of recombinants (Rico-Hesse et al., 1987). Several presumptive recombinant genomes have been partially characterized, and one has been sequenced completely (Fig. 5). The probable recombinant genomes differ from those of most wild isolates by having their capsid regions, which are highly homologous to those of other local isolates, sharply bounded by long intervals of highly divergent sequences. In the best-characterized example, the capsid region genes appear to have been exchanged as a cassette into an alternate genetic background. This process may be analogous to the antigenic shifts of influenza viruses (Young and Palese, 1979). The existence of natural wild-virus recombinants requires that the dendrograms be interpreted with care, as different genomic domains may descend from separate lineages.

High-Resolution Molecular Epidemiology

Distinctions between isolates increase with inclusion of longer genomic intervals for comparison. Details of the pathways of virus transmission during the 1988 epidemic in Israel were inferred from the patterns of fixation of mutations into the 906-nucleotide VP1 genes of representative case isolates. Contrary to expectations, the sequence data strongly suggested that

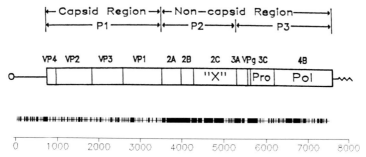

FIGURE 5. Diagrammatic representation of the genome of a probable natural wild-poliovirus recombinant. Locations of nucleotide differences between the type 1 isolate and the reference type 1 Mahoney strain are indicated by vertical lines. The capsid regions are about 3% divergent and extensive intervals within the 5' untranslated and noncapsid regions are highly (up to 20%) divergent, consistent with the view that the different genomic domains were derived from separate virus lineages.

the index epidemic case occurred 6 months before the epidemic was first recognized and that the virus had spread within the country along two major pathways (O. Kew et al., manuscript in preparation).

Reconstruction of past epidemics is of less practical importance than determination of ongoing transmission patterns. As immunization programs develop, the genetic heterogeneity among contemporary case isolates diminishes as many lineages are eliminated. Sequence comparisons of isolates óbtained throughout a region may facilitate identification of the surviving reservoirs of wild-virus transmission.

Rapid Methods for Detecting Wild Polioviruses

Detection of wild polioviruses is most critical for those developing countries where poliomyelitis is still endemic. To simplify identification of wild isolates, we have prepared Sabin vaccine strain- and wild-genotype-specific oligodeoxynucleotide probes. The probes are in routine use in Brazil and other countries (da Silva et al., 1990). Similarly, specific polymerase chain reaction primer pairs (Saiki et al., 1988) have been used for the rapid (<8 h) identification of polioviruses isolated in cell culture or present in stool suspensions. Since regions of wild-poliovirus genomes can be selectively amplified in the presence of large excesses (10^6-fold) of Sabin vaccine strain RNAs (C. F. Yang, manuscript in preparation), it may become possible through environmental sampling (Poyry et al., 1988) to detect circulation of wild polioviruses in communities having no cases of clinical disease. Sampling in critical areas may provide early warning of conditions predictive of outbreaks.

CONCLUSIONS

Poliomyelitis remains a major health problem in the developing world. We have used comparative genomic sequencing to monitor the distribution of wild polioviruses in nature. The relationships between cases can be precisely determined because the viral genome evolves rapidly during natural transmission. Sequence data have been used in the design of wild-genotype-specific hybridization probes and polymerase chain reaction primer pairs for rapid detection of wild viruses in endemic areas. It is hoped that surveillance based on molecular methods will facilitate efforts to control, and even eradicate, wild polioviruses.

REFERENCES

Assaad, F., and K. Ljungars-Esteves. 1984. World overview of poliomyelitis: regional patterns and trends. Rev. Infect. Dis. 6(Suppl. 2):S302–S307.

Buonagurio, D. A., S. Nakada, J. D. Parvin, M. Krystal, P. Palese, and W. M. Fitch. 1986. Evolution of human influenza A viruses over 50 years: rapid, uniform rate of change in NS gene. Science 232:980–982.

da Silva, E. E., H. G. Schatzmayr, and O. M. Kew. 1990. Nucleotide sequences of the VP1 capsid proteins of wild polioviruses types 1 and 3 from epidemic areas of Brazil. Braz. J. Med. Biol. Res. 23:1–5.

Fitch, W. M., and E. Margoliash. 1967. Construction of phylogenetic trees. Science 155:279–284.

Henderson, R. H., J. Keja, G. A. Hayden, A. Galazka, J. Clements, and C. Chan. 1988. Immunizing the children of the world: progress and prospects. *Bull. WHO* 66:535–543.

Hogle, J. M., M. Chow, and D. J. Filman. 1985. Three-dimensional structure of poliovirus at 2.9 Å resolution. *Science* 229:1353–1365.

Kawamura, N., M. Kohara, S. Abe, T. Komatsu, K. Tago, M. Arita, and A. Nomoto. 1989. Determinants in the 5′ noncoding region of poliovirus Sabin 1 RNA that influence the attenuation phenotype. *J. Virol.* 63:1302–1309.

Kew, O. M., B. K. Nottay, R. R. Rico-Hesse, and M. A. Pallansch. Molecular epidemiology of wild poliovirus transmission. *Appl. Virol. Res.*, in press.

Kostrzewski, J., A. Kulesza, and A. Abgarowicz. 1970. Type 3 poliomyelitis epidemic in Poland in 1968. *Epidemiol. Rev.* 34:89–103.

Landsteiner, K., and E. Popper. 1909. Uebertragung der Poliomyelitis acuta auf Affen. *Z. Immunitaetsforsch.* 2:375–391.

Minor, P. D., M. Ferguson, D. M. A. Evans, J. W. Almond, and J. P. Icenogle. 1986a. Antigenic structure of polioviruses of serotypes 1, 2, and 3. *J. Gen. Virol.* 67:1283–1291.

Minor, P. D., A. John, M. Ferguson, and J. P. Icenogle. 1986b. Antigenic and molecular evolution of the vaccine strain of type 3 poliovirus during the period of excretion by a primary vaccinee. *J. Gen. Virol.* 67:693–706.

Nkowane, B. M., S. G. F. Wassilak, W. A. Orenstein, K. J. Bart, L. B. Schonberger, A. R. Hinman, and O. M. Kew. 1987. Vaccine-associated paralytic poliomyelitis, U.S.A., 1973–1984. *J. Am. Med. Assoc.* 257:1335–1340.

Nomoto, A., T. Omata, H. Toyoda, S. Kuge, H. Horie, Y. Kataoka, Y. Genba, Y. Nakano, and N. Imura. 1982. Complete nucleotide sequence of the attenuated poliovirus Sabin 1 strain genome. *Proc. Natl. Acad. Sci. USA* 79:5793–5797.

Nottay, B. K., O. M. Kew, M. H. Hatch, J. T. Heyward, and J. F. Obijeski. 1981. Molecular variation of type 1 vaccine-related and wild polioviruses during replication in humans. *Virology* 108:405–423.

Page, G. S., A. G. Mosser, J. M. Hogle, D. J. Filman, R. R. Rueckert, and M. Chow. 1988. Three-dimensional structure of poliovirus serotype 1 neutralizing determinants. *J. Virol.* 62:1781–1794.

Pan American Health Organization. 1985. *Eradication of Indigenous Transmission of the Wild Polio Virus from the Americas. Plan of Action.* Pan American Health Organization, Washington, D.C.

Patriarca, P., F. Laender, G. Palmeira, M. J. Couto Oliveira, I. Lima Filho, M. C. de Souza Dantes, M. Tenorio Cordeiro, J. B. Risi, and W. A. Orenstein.

1988. Randomised trial of alternative formulations of oral poliovaccine in Brazil. *Lancet* i:429–432.

Paul, J. R. 1971. *A History of Poliomyelitis.* Yale University Press, New Haven, Conn.

Pilipenko, E. V., V. M. Blinov, L. I. Romanova, A. N. Sinyakov, S. V. Maslova, and V. I. Agol. 1989. Conserved structural domains in the 5′-untranslated region of picornaviral genomes: an analysis of the segment controlling translation and neurovirulence. *Virology* 168:201–209.

Poyry, T., M. Stenvik, and T. Hovi. 1988. Viruses in sewage waters during and after a poliomyelitis outbreak and subsequent nationwide oral poliovaccination campaign in Finland. *Appl. Environ. Microbiol.* 54:371–374.

Rico-Hesse, R., M. A. Pallansch, B. K. Nottay, and O. M. Kew. 1987. Geographic distribution of wild poliovirus type 1 genotypes. *Virology* 160:311–322.

Rueckert, R. R., and M. A. Brinton. 1986. *Positive Strand RNA Viruses.* Alan R. Liss, Inc., New York.

Sabin, A. B. 1985. Oral poliovirus vaccine: history of its development and use, and current challenge to eliminate poliomyelitis from the world. *J. Infect. Dis.* 151:420–436.

Saiki, R. K., D. H. Gelfand, S. Stoffel, S. J. Scharf, R. Higuchi, G. T. Horn, K. B. Mullis, and H. A. Ehrlich. 1988. Primer-directed enzymatic amplification of DNA with a thermostable DNA polymerase. *Science* 239:487–491.

Salk, J., and D. Salk. 1977. Control of poliomyelitis and influenza with killed virus vaccines. *Science* 195:834–847.

Schaap, G. J. P., H. Bijkerk, R. A. Coutinho, J. G. Kapsenberg, and A. L. van Wezel. 1984. The spread of wild poliovirus in the well-vaccinated Netherlands in connection with the 1978 epidemic. *Prog. Med. Virol.* 29:124–140.

Skinner, M. A., V. R. Racaniello, G. Dunn, J. Cooper, P. Minor, and J. W. Almond. 1989. New model for the secondary structure of the 5′ non-coding RNA of poliovirus is supported by biochemical and genetic data that also show that RNA secondary structure is important in neurovirulence. *J. Mol. Biol.* 207:379–392.

Stanway, G., P. J. Hughes, R. C. Mountfort, P. Reeve, P. D. Minor, G. C. Schild, and J. W. Almond. 1984. Comparisons of the complete nucleotide sequences of the genomes of the neurovirulent poliovirus P3/Leon/37 and its attenuated Sabin vaccine derivative P3/Leon 12 a_1b. *Proc. Natl. Acad. Sci. USA* 81:1539–1543.

Toyoda, H., M. Kohara, Y. Kataoka, T. Suganuma, T. Omata, N. Imura, and A. Nomoto. 1984. Complete nucleotide sequences of all three poliovirus serotype genomes: implication for genetic relationship, gene function and antigenic determinants. *J. Mol. Biol.* **174:**561–585.

World Health Assembly. 1988. Global eradication of poliomyelitis by the year 2000. Resolution WHA 41.28. World Health Assembly, Geneva.

World Health Organization. 1989. *Expanded Pro-gramme on Immunization. Global Situation—Polio-myelitis.* World Health Organization, Geneva.

Young, J. F., and P. Palese. 1979. Evolution of influenza viruses in nature: recombination contributes to genetic variation of H1N1 strains. *Proc. Natl. Acad. Sci. USA* **76:**6547–6551.

Chapter 53

Rational Design of Antipicornavirus Agents

Mark A. McKinlay, Frank J. Dutko, Daniel C. Pevear, Maureen G. Woods, Guy D. Diana,
and Michael G. Rossmann

The family *Picornaviridae* is a major cause of virus-associated morbidity in humans. The picornavirus family consists of two major groups of viruses associated with human disease: the rhinoviruses and enteroviruses. The approximately 100 serotypes of rhinoviruses are the etiologic agents in the majority of the mild upper-respiratory illnesses known as the common cold (Couch, 1984). The 68 serotypes of enteroviruses, consisting of the polio-, coxsackie-, and echoviruses, cause a spectrum of clinical syndromes ranging in severity from mild upper-respiratory disease with or without myalgia and fever to myocarditis, aseptic meningitis, and neonatal sepsis.

It is estimated that, on average, preschool children experience 6 to 10 colds per year and adults have 2 to 5 colds annually (Gwaltney, 1985). The limited epidemiologic data available indicate that 5 million to 15 million enteroviral infections occur each year in the United States (Kogon et al., 1969). These data provided the incentive for pharmaceutical firms to pursue the development of antiviral drugs effective in inhibiting picornavirus replication. Most of the synthetic agents (Sperber and Hayden, 1988) emerging from broad-based screens have been found to stabilize the virion capsid, thereby preventing virion uncoating and, in some cases, adsorption to the host cell (Pevear et al., 1989). Representatives of the structurally diverse class of capsid-stabilizing agents studied most extensively to date include dichloroflavan (Bauer et al., 1981; Tisdale and Selway, 1983, 1984), Ro 09-0410 (Ishitsuka et al., 1982; Ninomiya et al., 1984), RMI 15,731 (Ash et al., 1979), R 61,837 (Al-NaKib and Tyrell, 1987), and disoxaril (Otto et

al., 1985; McKinlay, 1985; Fox et al., 1986; Zeichhardt et al., 1987) (Fig. 1).

All of these compounds were developed by using conventional medicinal chemical approaches. Rational drug design, defined as the directed synthesis of new compounds on the basis of an understanding of a prototype drug-virus structural or functional protein interaction at the atomic level, has been made possible only in the past few years as a result of breakthroughs in the area of X-ray crystallography of animal viruses. This chapter describes how rational drug design is influencing the development of new agents with enhanced potency and spectrum and the general lessons that have been learned about the value and potential shortcomings of this approach to drug design.

RESULTS OF CONVENTIONAL DESIGN APPROACH

Before the availability of X-ray crystallographic data on the site of binding of capsid-stabilizing agents at atomic resolution (Smith et al., 1986), considerable progress had been made toward the synthesis of potent compounds with broad spectra of activity by using conventional medicinal chemical approaches. Through the determination of structure-activity relationships for a large series of molecules, researchers at the Sterling Research Group in Rensselaer were able to synthesize compounds with improved activity profiles. By shortening the aliphatic chain of disoxaril (Fig. 1) from seven methylene units to five and adding a halogen substituent on the phenyl ring, the spectrum and potency

Mark A. McKinlay, Frank J. Dutko, Daniel C. Pevear, and Maureen G. Woods, Department of Virology and Oncopharmacology, Sterling Research Group, Rensselaer, New York 12144. *Guy D. Diana,* Department of Medicinal Chemistry, Sterling Research Group, Rensselaer, New York 12144. *Michael G. Rossmann,* Department of Biological Sciences, Lilly Hall of Life Sciences, Purdue University, West Lafayette, Indiana 47907.

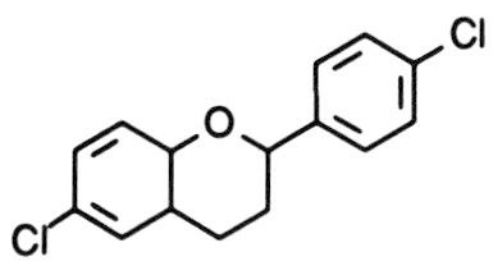

dichloroflavan
(BW683c)

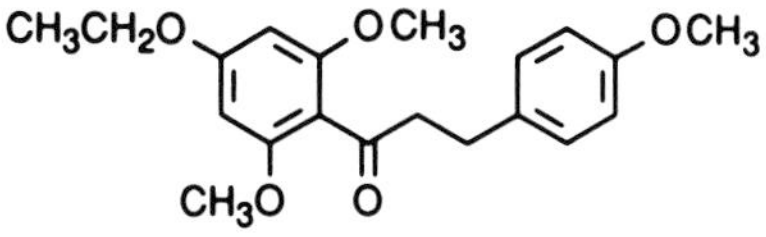

Ro 09-0410

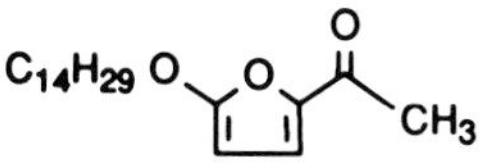

RMI 15,731

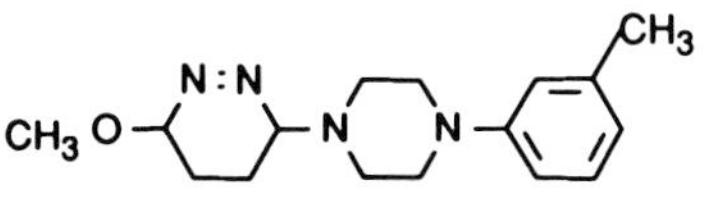

R 61-837

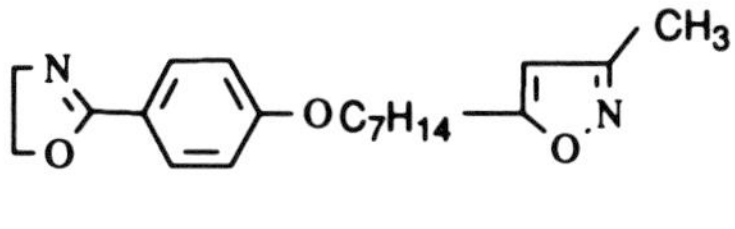

disoxaril
(Win 51711)

FIGURE 1. Picornavirus capsid-stabilizing agents.

against 48 rhinovirus serotypes were markedly enhanced (Fig. 2). The addition of a second chloro group on the phenyl ring resulted in a compound, Win 54954, with excellent activity against both the rhinoviruses and enteroviruses (Fig. 3). Win 54954 was also orally effective in preventing paralysis in animals infected with coxsackievirus A9 (Fig. 4). A study of the pharmacokinetics in mice suggested that serum levels of Win 54954 had to exceed the MIC for the

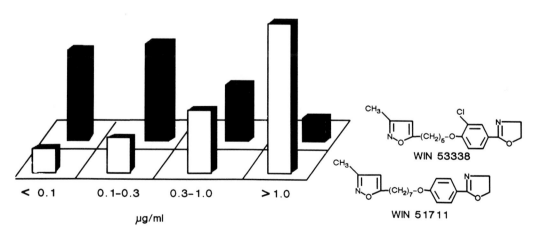

WIN 53338

WIN 51711

FIGURE 2. Structure-activity relationships of two antipicornaviral drugs. Human rhinovirus serotypes ($n = 47$) were tested against Win 51711 or 53338; shown is the percentage of serotypes inhibited at the indicated drug concentrations.

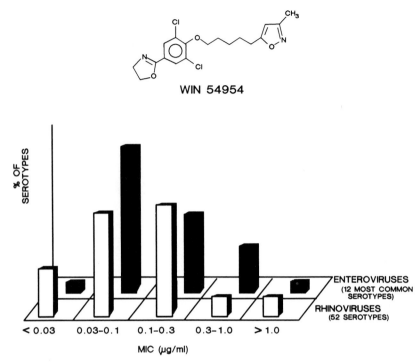

FIGURE 3. Spectrum of activity of Win 54954. Shown is the percentage of rhinovirus or enterovirus serotype inhibited at the indicated drug concentrations.

infecting virus for only a portion of the dosing interval in order for efficacy to be observed (data not shown). These results are encouraging and suggest that Win 54954 may be a clinically useful agent in the treatment of human picornaviral infections.

RATIONAL DRUG DESIGN

The most significant technical advance, which is beginning to make rational antiviral drug design a reality, has been in the area of X-ray crystallography of macromolecules ranging in size up to spherical viruses with a diameter of 300 Å (30 nm). After years of computational and technical development, the structure of the first spherical plant virus, tomato bushy stunt virus, was solved at 2.9-Å resolution in 1978 (Harrison et al., 1978). It was not until 1985 that the structure of the first animal virus, human rhinovirus 14 (HRV-14), was solved (Rossmann et al., 1985). This significant advance was followed soon thereafter by the determination of the structures of human poliovirus (Hogle et al., 1985) and mengovirus (Luo et al., 1987). It is likely that the structure of canine parvovirus

(Luo et al., 1988) will be solved in the near future.

All of the picornavirus capsid-stabilizing agents are extremely hydrophobic in nature such that in the cases of dichloroflavan and Ro 09-0410, they can be dissociated from the virion only by nonpolar solvents such as chloroform (Tisdale and Selway, 1983; Ninomiya et al., 1984). The nature and specificity of the presumed hydrophobic binding of these compounds to the virion and how the compound-virion interaction results in uncoating inhibition were unknown until the structure of an inhibitor bound to HRV-14 was solved to atomic resolution by X-ray crystallography (Smith et al., 1986). This pivotal study pinpointed the binding site of a disoxaril homolog containing a methyl group on the oxazoline ring, within a hydrophobic pocket inside VP1, one of the four viral capsid proteins (Fig. 5). The availability of this pocket suggests that the pocket plays some functional role such as providing the flexibility in the capsid necessary to allow disassembly or uncoating to occur. The binding of a hydrophobic molecule in the pocket may serve to make the capsid structure more rigid and resistant to uncoating forces. Alternatively, the compounds

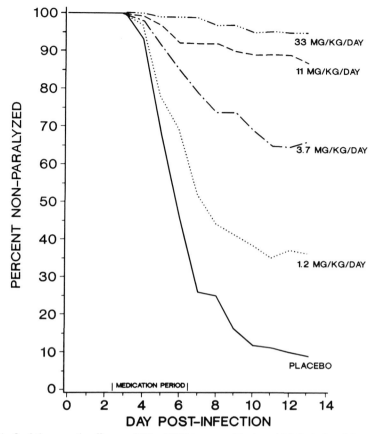

FIGURE 4. Oral therapeutic efficacy of Win 54954 in coxsackievirus A9-infected suckling mice. Mice were medicated once per day for 5 days beginning 2.5 days after infection with 1.2 (n = 90), 3.7 (n = 89), 11 (n = 90), or 0 and 33 (n = 60) mg per kg of body weight per day or received placebo (n = 88).

may act to hold the capsid together via the strength of the hydrophobic interactions.

Recent results indicate that for HRV-14 (major receptor-binding group of rhinoviruses), binding of compounds in the pocket causes conformational changes of up to 5.5 Å in the floor of the canyon (putative receptor-binding site) and results in inhibition of virion attachment to intracellular adhesion molecule 1 (ICAM-1), the cellular receptor (Pevear et al., 1989). The conformational changes induced upon drug binding may sterically prevent interactions between the virus and ICAM-1 necessary for attachment to occur.

An examination of the hydrophobic binding site revealed two features important for drug design. First, the compounds appear to form only one weakly directional hydrogen bond between Asn-219 of VP1 and a heteronitrogen of the compound, with the remainder of the binding energy apparently contributed through hydro-

phobic interactions. The design of agents to increase binding affinity presents a novel set of problems in that no other reported drug-"receptor" interactions are so highly dependent on hydrophobic binding. The initial attempts at drug design have therefore focused on ways to more completely fill the binding pocket. In this regard, enantiomeric effects of substituents on the oxazoline ring of a disoxaril homolog on in vitro antiviral potency have been described. The enhanced potency of the S enantiomers has been explained on the basis of improvements in the hydrophobic interactions of alkyl substituents and Leu-106 and Ser-107 of VP1 (Diana et al., 1988).

In an effort to more fully understand the nature of the drug-virus interactions, viruses resistant to the Win compounds have been isolated and characterized (Badger et al., 1988; Heinz et al., 1989). Most of the highly resistant mutants have been mapped to two sites, Val-188

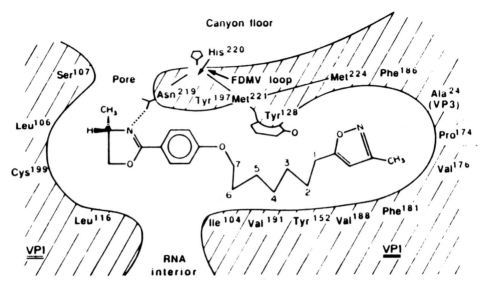

FIGURE 5. Diagrammatic view of the binding site of Win 52084 within VP1 of HRV-14.

and Cys-199, within the drug-binding pocket in VP1. In all cases, the substitutions found in the highly drug-resistant mutants have been to amino acids with larger side chains. In the case of the Val-188 → Leu mutation, drug binding is significantly reduced, which correlates with the decrease in antiviral potency. These results are consistent with the hypothesis that resistance to high concentrations of drug is a result of steric hindrance to drug binding.

HRV-14 mutants with resistance to low concentrations of compound have been isolated and sequenced and have been found to map to numerous sites more than 3 Å from the drug-binding pocket (Heinz et al., 1989). How these amino acid substitutions alter drug sensitivity is the subject of current studies. Since the binding of HRV-14 to the cellular receptor is blocked by these compounds, it is possible that the low-resistance mutations occur in the receptor-binding site, resulting in higher binding affinity of the mutant to the cellular receptor. The higher binding affinity could overcome the adsorption block seen with wild-type HRV-14. A second possibility is that these mutations result in changes in capsid conformation required for the compounds to enter the pocket, thereby decreasing binding affinity. Alternatively, the mutations may affect protomer-protomer interactions, reducing the inherent stability of the virion such that the binding of antiviral agents within the capsid does not stabilize the virion sufficiently to prevent uncoating. Whatever the results show, these mutants will be extremely informative with respect to understanding the

picornavirus-uncoating process and how these agents work to prevent it.

To date, only the Win compounds have been studied extensively by X-ray crystallography, although work has begun with the compounds from Janssen Pharmaceuticals. The preliminary data confirm, as expected, that R 61,837 binds in the same hydrophobic pocket within HRV-14.

DRUG DESIGN LESSONS

A number of lessons have been learned from the work with the picornavirus capsid-stabilizing compounds. First, the structure generated by X-ray crystallography is only a static representation of a macromolecule that obviously exists in nature as a dynamic entity. This fact is illustrated in the picornavirus studies: examination of the native structure would not suggest that molecules such as disoxaril would bind, since Met-221 of VP1 blocks access to the pocket. Design of an antiviral agent to block picornavirus uncoating simply by using the structure of the native virus would, therefore, have been virtually impossible. This result illustrates the need to solve the structure of the compound bound to the receptor, enzyme, or virus target of interest before initiating drug design.

A second observation from the picornavirus work of relevance to any antiviral molecular target is the difficulty in relating the affinity of binding (K_d) to the target with the antiviral efficacy observed in vitro. With the Win compounds, preliminary results suggest that the K_d for a number of radiolabeled compounds does not correlate perfectly with the in vitro MIC when one moves between chemical classes. This result is to be expected, since factors such as penetration into the cell and metabolism of the compound by the host cells used in the 48- to 72-h assay can affect MIC values. For this reason, the K_d values, which are of great importance in directing the design and synthesis of new compounds, must be considered as only one parameter in directing the synthesis of new molecules. Going one step further, the obvious goal of any drug discovery program is to produce an agent that is bioavailable. Certainly K_d alone will rarely correlate with the in vivo efficacy of an agent. In the disoxaril series, in which oral activity can be readily assessed in animal models of human infection (McKinlay et al., 1986; McKinlay and Steinberg, 1986), this problem is especially acute because the hydrophobic requirements for drug binding are at odds with the need for aqueous solubility in developing an oral dosage form.

Two drug design problems are particular to the picornavirus capsid-stabilizing agents. The first is the multitude of serotypes that must be inhibited by an agent before that agent is a clinically useful product. For all of the viruses whose RNA has been sequenced to date, amino acid substitution can be seen within the drug-binding pocket even though this region is the most conserved part of the protein coat. The design of agents to maximize hydrophobic interactions within the binding pocket of one serotype could in fact reduce or eliminate activity against a serotype with a different complement of amino acids in the pocket. Therefore, structural data are necessary on serotypes whose sensitivity to the class of compounds is most predictive of the sensitivity of the entire spectrum of serotypes. In this regard, HRV-14 appears to be of limited use in drug design, since the structural modifications that improve activity against HRV-14 decrease activity against most other rhinovirus serotypes.

The second problem affecting drug design in the picornavirus area is the hydrophobic nature of the binding. An illustration of the difficulties that are realized in trying to design a "piece of grease" to fit into a "greasy pocket" is the flip-flop in orientation that was seen when the alkyl chain length of the compounds was reduced from seven methylene units to five. In this case, the surprising 180° change in orientation still results in the same conformational change in VP1 and the conservation of a probable hydrogen bond between Asn-219 and the nitrogen on the isoxazole ring. In the absence of strong hydrogen bonds to anchor the compounds to the pocket, the orientation of any designed compound will have to be examined crystallographically.

FUTURE PROSPECTS

The most important component of a drug design program is the obvious need to predict the activity of a molecule before investing the effort in its synthesis. Although a certain amount of success can be had by viewing a ligand-protein structure on a graphics terminal, the multiplicity of possible conformations that the ligand and protein can assume renders it exceedingly difficult to make accurate predictions with any regularity. Recent developments in the use of thermodynamic calculations and molecular dynamics simulations permit consideration of previously impossible computational problems through the use of supercomputers and new computational approaches (McCammon, 1987; McCammon and Harvey, 1987).

The examples discussed above point out some of the shortcomings of the evolving field of rational drug design. Despite the current limitations of the technology, the availability of this new tool provides medicinal chemists with valuable information that should result in the more rapid development of compounds with enhanced potency.

REFERENCES

Al-NaKib, W., and D. A. J. Tyrell. 1987. A "new" generation of more potent synthetic antirhinovirus compounds: comparison of their MICs and their synergistic interactions. *Antiviral Res.* **8:**179–188.

Ash, R. J., R. A. Parker, A. C. Hagan, and G. D. Mayer. 1979. RMI 15,731(1-[5-tetradecyloxy-2-furanyl]-ethanone), a new antirhinovirus compound. *Antimicrob. Agents Chemother.* **16:**301–305.

Badger, J., I. Minor, M. J. Kremer, M. A. Oliviera, T. J. Smith, M. P. Griffith, D. M. Guerin, S. Krishnaswamy, M. Luo, M. G. Rossmann, M. A. McKinlay, G. D. Diana, F. J. Dutko, M. Fancher, R. Rueckert, and B. A. Heinz. 1988. Structural analysis of a series of antiviral agents complexed with human rhinovirus 14. *Proc. Natl. Acad. Sci. USA* **85:**3304–3308.

Bauer, D. J., J. W. T. Selway, J. F. Batchelor, M. Tisdale, I. C. Caldwell, and D. A. B. Young. 1981. 4',6-Dichloroflavan (BW683c), a new antirhinovirus compound. *Nature* (London) **292:**369–370.

Couch, R. B. 1984. The common cold: control? *J. Infect. Dis.* **150**:167–173.

Diana, G. D., M. J. Otto, A. Treasurywala, M. A. McKinlay, R. C. Oglesby, E. G. Maliski, M. G. Rossmann, and T. J. Smith. 1988. Enantiomeric effects of homologues of disoxaril on the inhibitory activity against human rhinovirus-14. *J. Med. Chem.* **31**:540–544.

Fox, M. P., M. J. Otto, and M. A. McKinlay. 1986. Prevention of rhinovirus and poliovirus uncoating by WIN 51711, a new antiviral drug. *Antimicrob. Agents Chemother.* **30**:110–116.

Gwaltney, J. M., Jr. 1985. The common cold, p. 351–355. *In* G. L. Mandel, R. G. Douglas, and J. E. Bennett (ed.), *Principles and Practices of Infectious Diseases*, 2nd ed. John Wiley & Sons, Inc., New York.

Harrison, S. C., A. J. Olson, C. E. Schutt, F. K. Winkler, and G. Bricogne. 1978. Tomato bushy stunt virus at 2.9 Å resolution. *Nature* (London) **276**:368–373.

Heinz, B. A., R. R. Rueckert, D. A. Shepard, F. J. Dutko, M. A. McKinlay, M. Fancher, M. G. Rossmann, J. Badger, and T. J. Smith. 1989. Genetic and molecular analysis of spontaneous mutants of human rhinovirus 14 resistant to an antiviral compound. *J. Virol.* **63**:2476–2485.

Hogle, J. M., M. Chow, and D. J. Filman. 1985. Three-dimensional structure of poliovirus at 2.9 Å resolution. *Science* **229**:1358–1365.

Ishitsuka, H., Y. T. Ninomiya, C. Ohsawa, and M. Fujiu. 1982. Direct and specific inactivation of rhinovirus by chalcone Ro 09-410. *Antimicrob. Agents Chemother.* **22**:617–621.

Kogon, A., I. Spigland, T. E. Frothingham, L. Eleveback, C. Williams, C. E. Hall, and J. P. Fox. 1969. The Virus Watch Program: a continuing surveillance of viral infections in metropolitan New York families. VII. Observations on viral excretion, seroimmunity, intrafamilial spread and illness association in coxsackie and echovirus infections. *Am. J. Epidemiol.* **89**:51–61.

Luo, M., J. Tsao, M. G. Rossmann, S. Basak, and R. W. Compans. 1988. Preliminary x-ray crystallographic analysis of canine parvovirus crystals. *J. Mol. Biol.* **200**:209–211.

Luo, M., G. Vriend, G. Kamer, I. Minor, E. Arnold, M. G. Rossmann, V. Boege, D. G. Scraba, G. M. Duke, and A. C. Palmenberg. 1987. The structure of mengo virus at atomic resolution. *Science* **235**:182–191.

McCammon, J. A. 1987. Computer-aided molecular design. *Science* **238**:486–491.

McCammon, J. A., and S. C. Harvey. 1987. *Dynamics of Proteins and Nucleic Acids*. Cambridge University Press, Cambridge.

McKinlay, M. A. 1985. WIN 51711: a new systemically active broad-spectrum antipicornavirus agent. *J. Antimicrob. Chemother.* **16**:284–286.

McKinlay, M. A., J. A. Frank, and B. A. Steinberg. 1986. Use of WIN 51711 to prevent echovirus type-9 induced paralysis in suckling mice. *J. Infect. Dis.* **154**:676–681.

McKinlay, M. A., and B. A. Steinberg. 1986. Oral efficacy of WIN 51711 in mice infected with human poliovirus. *Antimicrob. Agents Chemother.* **29**:30–31.

Ninomiya, Y., M. Ohsawa, I. Aoyama, Y. Umeda, Y. Suhara, and H. Ishitsuka. 1984. Antivirus agent, Ro-09-0410, binds to rhinovirus specifically and stabilizes the virus conformation. *Virology* **134**:269–276.

Otto, M. J., M. P. Fox, M. J. Fancher, M. F. Kuhrt, G. D. Diana, and M. A. McKinlay. 1985. In vitro activity of WIN 51711: a new broad-spectrum antipicornavirus drug. *Antimicrob. Agents Chemother.* **27**:883–886.

Pevear, D. C., M. J. Fancher, P. J. Felock, M. G. Rossmann, M. S. Miller, A. Treasurywala, M. A. McKinlay, and F. J. Dutko. 1989. Conformational change in the floor of the human rhinovirus canyon blocks adsorption to HeLa cell receptors. *J. Virol.* **63**:2002–2007.

Rossmann, M. G., E. Arnold, J. W. Erickson, E. A. Frankenberger, M. P. Griffith, H. J. Hecht, J. E. Johnson, G. Kamer, M. Luo, A. G. Mosser, R. R. Rueckert, B. Sherry, and G. Vriend. 1985. Structure of a human common cold virus and functional relationship to other picornaviruses. *Nature* (London) **317**:145–153.

Smith, T. J., M. J. Kremer, M. Luo, G. Vriend, E. Arnold, G. Kamer, M. G. Rossmann, M. A. McKinlay, G. D. Diana, and M. J. Otto. 1986. The site of attachment in human rhinovirus-14 for antiviral agents that inhibit uncoating. *Science* **223**:1286–1293.

Sperber, S. J., and F. G. Hayden. 1988. Chemotherapy of rhinovirus colds. *Antimicrob. Agents Chemother.* **32**:409–419.

Tisdale, M., and J. W. T. Selway. 1983. Inhibition of an early stage of rhinovirus replication by dichloroflavan (BW683c). *J. Gen. Virol.* **64**:795–803.

Tisdale, M., and J. W. T. Selway. 1984. Effect of dichloroflavan (BW683c) on the stability and uncoating of rhinovirus type 1B. *J. Antimicrob. Chemother.* **14**(Suppl. A):97–105.

Zeichhardt, H., M. J. Otto, M. A. McKinlay, P. Willingman, and K. O. Habermehl. 1987. Inhibition of poliovirus uncoating by disoxaril (WIN 51711). *Virology* **160**:281–285.

Chapter 54

Antiviral Compounds Distinguish between Two Groups of Rhinoviruses with Different Biological Properties

Koen Andries, Bart Dewindt, Jerry Snoeks, Henri Moereels, and Paul J. Lewi

A variety of chemically different compounds inhibit the replication of several serotypes of rhinoviruses (Andries et al., 1989). We noticed that one of these compounds, Win 51711, had an antiviral spectrum clearly different from a consensus spectrum identified for other capsid-binding compounds, such as sodium dodecyl sulfate (SDS), dichloroflavan, chalcone, MDL 20,610, and R 61837 (Table 1).

On the other hand, mutants resistant to R 61837 were shown to be cross-resistant to all other capsid-binding compounds, including Win 51711 (Andries et al., in press). This finding indicated that compounds sharing the same binding site could nevertheless have different spectra of antiviral activity.

In an attempt to understand why compounds binding to a similar site can display essentially different spectra of antiviral activity, we systematically tested the antiviral sensitivities of all 100 typed rhinoviruses, 3 poliovirus serotypes, and coxsackievirus A21 against a panel of 15 antiviral compounds belonging to structurally different chemical classes but all sharing the binding site of Win 51711. Compounds included reference antiviral agents such as chalcone, dichloroflavan, MDL 20,610, RMI 15,731, Win 51711, a Sandoz antirhinovirus compound (EP 187618), and SDS, as well as the Janssen pyridazinamines R 61837, R 60164, R 62025, R 62827, R 66703, R 66933, R 67041, and R 72440 (Andries et al., 1989).

MULTIVARIATE ANALYSIS

Results of MIC tests (Andries et al., 1988) were analyzed by using spectral map analysis (Lewi, 1976). First, logarithms were taken from the reciprocal MIC values. Next, we computed from the transformed data the mean sensitivity (to the 15 compounds) of each serotype as well as the mean potency of each antiviral compound (in the 100 serotypes). Then, we subtracted from each value in the transformed data table the corresponding mean sensitivity and mean potency. This operation is called double centering. If double centering were not applied, the intrinsic interactions between serotypes and compounds would be confounded by differences in sensitivity of the serotypes and by differences in potency of the compounds. Principal-components analysis of the logarithmically transformed and double-centered data was then applied. In our application, we found that three dimensions accounted for 91% of the information in the logarithmically transformed data table. These computed dimensions are called principal components. In this way, we reduced the original 15 dimensions to three factors. The result of principal-components analysis is in the form of a table of coordinates of the 15 compounds and 100 serotypes in three-dimensional space. Figure 1 represents all serotypes as points in a two-dimensional plane defined by the two most important factors.

RHINOVIRUS GROUPING ON THE BASIS OF ANTIVIRAL SENSITIVITY

Multivariate analysis of MIC data clustered viruses with similar susceptibilities to antiviral compounds. Hierarchical and nearest-neighbor cluster analysis confirmed the existence of two groups of rhinoviruses. The major group, which

Koen Andries, Bart Dewindt, Jerry Snoeks, Henri Moereels, and Paul J. Lewi, Janssen Research Foundation, Turnhoutseweg 30, B-2340 Beerse, Belgium.

TABLE 1

MICs of Antiviral Compounds Tested against Human Rhinoviruses

Human rhino-virus	MIC (μg/ml)					
	Sodium dodecyl sulfate	Dichloro-flavan	Chalcone	MDL 20,610	R 61837	Win 51711
1B	5.200	0.018	0.400	0.100	0.726	4.600
9	0.638	0.125	0.097	0.031	0.006	3.500
74	1.400	0.107	0.071	0.052	0.203	2.275
82	0.500	0.163	0.090	0.150	0.036	2.000
29	11.200	0.002	0.130	0.007	0.047	1.650
11	1.200	0.023	0.016	0.002	0.016	1.125
69	32.000	16.000	16.000	32.000	28.000	0.213
14	32.000	16.000	0.600	32.000	11.700	0.175
4	32.000	16.000	0.600	32.000	11.700	0.175
97	21.600	6.600	32.000	4.000	1.250	0.063
6	32.000	16.000	16.000	32.000	32.000	0.039
70	27.000	16.000	32.000	32.000	12.000	0.016

we will call antiviral group B, consists of viruses susceptible to antiviral agents such as chalcone, dichloroflavan, MDL 20,610, and R 61837. Antiviral group A consists of viruses with a more than average susceptibility to compounds such as Win 51711. The division of the rhinoviruses into two groups is not arbitrary but based on cluster analysis of the three-dimensional result of the multivariate analysis as described above. Antiviral group B of rhinoviruses contains twice as many ($n = 67$) serotypes as antiviral group A ($n = 33$).

RELATIONSHIP OF ANTIVIRAL GROUPS TO RECEPTOR GROUPS

The antiviral groups hypothesis is not in contradiction with the receptor groups hypothesis (Abraham and Colonno, 1984). In our model, viruses belonging to the minor receptor group are not randomly distributed but constitute a subgroup of antiviral group B (squares in Fig. 1). If anything can be learned from our study about viral evolution, we can hypothesize that the mutual divergence of receptor groups took place after that of the antiviral groups.

RELATIONSHIP OF ANTIVIRAL GROUPS TO AMINO ACID SEQUENCES

In a further approach, we studied the correlation between our multivariate analysis and protein sequence information by considering the VP1 amino acid alignment of Palmenberg (1989). At first, we selected the 17 amino acids that correspond to the lining of the hydrophobic pocket in human rhinovirus 14. Degrees of similarity between sequenced rhinovirus serotypes obtained from the multivariate analysis were compared with the percentages of pocket amino acid identity of the same serotypes. The two data sets revealed a high and significant correlation ($r = 0.89$, $P < 0.001$, Spearman). This correlation indicates that the antiviral compounds were really used as a panel of small molecular probes to characterize the composition of the antiviral binding site, which is a process very similar to the characterization of epitopes by using a panel of monoclonal antibodies.

When we compared similarities obtained from the multivariate analysis with percentage amino acid identity of the whole VP1 proteins of the same serotypes, we still found a very high correlation ($r = 0.87$ for degrees of similarity compared with VP1 amino acid identities; $P < 0.001$). This relationship exists because there is a strong correlation between the amino acid sequence of the pocket and the amino acid sequence of VP1. Indeed, we found a correlation of 0.84 for pocket amino acid identities compared with VP1 amino acid identities ($P < 0.001$). Our data suggest that the homologies found in the amino acid composition of the pocket are predictive of the homology in the amino acid composition of the whole capsid.

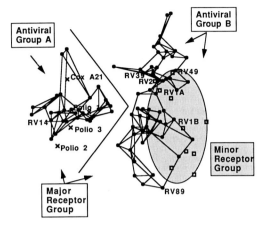

FIGURE 1. Division of rhinoviruses into two antiviral groups by multivariate analysis of antiviral tests. Positions of viruses on the map are the result of interactions between serotypes and compounds. Two serotypes occupy the same position if their sensitivities to the 15 compounds are proportional (although the two serotypes may possess different average sensitivities). Likewise, two compounds have the same positions (positions of compounds are not shown) if their activities against the 100 serotypes are proportional (although the compounds may differ in average potency). If a serotype presents a sensitivity to a compound above the average, then the serotype and compound attract each other. On the other hand, if a serotype shows less than average sensitivity for a compound, then they repel each other. These mutual attractions and repulsions determine a visible pattern of interactions between serotypes and compounds. The two-group pattern of serotypes that appears from the figure has been established by cluster analysis. Symbols: ●, rhinoviruses belonging to the major receptor group; □, rhinoviruses belonging to the minor receptor group; ×, enteroviruses. Only sequenced serotypes are labeled with their serotype numbers. Abbreviations: RV, rhinovirus; Polio, poliovirus; Cox, coxsackievirus.

RELATIONSHIP OF ANTIVIRAL GROUPS TO OCCURRENCE OF CLINICAL COLDS

We used our data to analyze available epidemiological data on rhinoviruses. In total, 1,205 rhinovirus isolates were collected from people suffering from a cold and serotyped by two independent groups over three time periods of about 5 years each (Fox et al., 1985; Monto et al., 1987). Given the fact that antiviral group B

contains twice as many serotypes as antiviral group A, one would expect serotypes of that group to account for exactly twice as many colds. However, this was not the case. Serotypes of antiviral group B accounted for five times as many colds as serotypes of antiviral group A. The chance that this was just a coincidence was calculated to be less than 1 in 10,000 (randomization test). Within antiviral group B viruses, we could not identify differences in pathogenicity among serotypes belonging to the different receptor groups ($P = 0.15$). We therefore assume that something in the structure of group B rhinoviruses endows them with a higher pathogenicity.

In summary, we identified two groups of rhinoviruses by analyzing the binding characteristics of small molecular probes for the various serotypes. Additional evidence of the existence of two groups of rhinoviruses was obtained by comparing our results with the distribution of receptor groups and with sequence data of amino acids of VP1 and finally by the observation of a difference in epidemiological profile.

REFERENCES

Abraham, G., and R. J. J. Colonno. 1984. Many rhinovirus serotypes share the same cellular receptor. *Virology* **51**:340–345.

Andries, K., B. Dewindt, M. De Brabander, R. Stokbroekx, and P. A. J. Janssen. 1988. In vitro activity of R 61837, a new antirhinovirus compound. *Arch. Virol.* **101**:155–167.

Andries, K., B. Dewindt, J. Snoeks, and R. Willebrords. 1989. Lack of quantitative correlation between inhibition of replication of rhinoviruses by an antiviral drug and their stabilization. *Arch. Virol.* **106**:51–61.

Fox, J. P., M. K. Cooney, C. E. Hall, and H. M. Foy. 1985. Rhinoviruses in Seattle families, 1975–1979. *Am. J. Epidemiol.* **122**:830–846.

Lewi, P. J. 1976. Spectral mapping, a technique for classifying biological activity profiles of chemical compounds. *Arzneim. Forsch.* **26**:1295–1300.

Monto, A. S., E. R. Bryan, and J. Ohmit. 1987. Rhinovirus infections in Tecumseh, Michigan: frequency of illness and number of serotypes. *Infect. Dis.* **156**:43–49.

Palmenberg, A. C. 1989. Sequence alignments of picornaviral capsid proteins, p. 211–241. *In* B. Semler and E. Ehrenfeld (ed.), *Molecular Aspects of Picornavirus Infection and Detection*. American Society for Microbiology, Washington, D.C.

INDEX